Electromagnetic Fields

ELECTROMAGNETIC FIELDS

(Second Edition)

R. Meenakumari
R. Subasri
Kongu Engineering College
Tamilnadu

NEW AGE INTERNATIONAL (P) LIMITED, PUBLISHERS
New Delhi • Bangalore • Chennai • Cochin • Guwahati • Hyderabad
Jalandhar • Kolkata • Lucknow • Mumbai • Ranchi
Visit us at **www.newagepublishers.com**

Published by New Age International (P) Ltd., Publishers
First Edition: 2003
Second Edition: 2007
Reprint : 2009

Branches:

- 36, Malikarjuna Temple Street, Opp. ICWA, Basavanagudi, **Bangalore**. © (080) 26677815
- 26, Damodaran Street, T. Nagar, **Chennai**. © (044) 24353401
- Hemsen Complex, Mohd. Shah Road, Paltan Bazar, Near Starline Hotel, **Guwahati**. © (0361) 2543669
- No. 105, 1st Floor, Madhiray Kaveri Tower, 3-2-19, Azam Jahi Road, Nimboliadda, **Hyderabad**. © (040) 24652456
- RDB Chambers (Formerly Lotus Cinema) 106A, Ist Floor, S.N. Banerjee Road, **Kolkata**. © (033) 22275247
- 18, Madan Mohan Malviya Marg, **Lucknow**. © (0522) 2209578
- 142C, Victor House, Ground Floor, N.M. Joshi Marg, Lower Parel, **Mumbai**. © (022) 24927869
- 22, Golden House, Daryaganj, **New Delhi**. © (011) 23262370, 23262368

ISBN (10) : 81-224-2016-8
ISBN (13) : 978-81-224-2016-6

Rs. 225.00

C-09-01-3380

Printed in India at Print 'O' Pack, Delhi.
Typeset at Pagitek Graphics, Delhi.

PUBLISHING FOR ONE WORLD
NEW AGE INTERNATIONAL (P) LIMITED, PUBLISHERS
4835/24, Ansari Road, Daryaganj, New Delhi-110002
Visit us at **www.newagepublishers.com**

Dedicated
to
The Supreme Father
who plants and nourishes
"THE DEDICATION"

Preface to the Second Edition

Electromagnetic Fields aims at presenting the concept of Electromagnetic field more clearly and offers an easily understandable approach based on classroom experience of the authors.

The book is primarily meant for the undergraduate students of EEE, ECE, EIE, I&C and also for AMIE students.

The authors have taken into account the difficulties of students while teaching these subjects.

Vector analysis and co-ordinate systems which are so fundamental for learning electromagnetic field are covered right from the beginning.

Electromagnetics is dealt with separately in chapters covering topics such as capacitors, boundary value problems and image theory.

Another chapter is devoted to magnetostatics which deals with all fundamental laws and inductance. Finally, the time varying EM field, waves and their propagation are covered substantially. For the benefit of students, each chapter provides a summary of the chapter and tutorial problems with answers. A good deal of solved and worked out examples are also included in each chapter.

At the end of the book a number of one-word and detailed questions are given so as to equip the students thoroughly on the subject. The authors hope that the book will help the students in brushing up their subject and the students will like using the book for performing better in the area of electromagnetic field and its related elements. The authors welcome constructive suggestions and criticism for the betterment of the subject.

Authors

Preface to the First Edition

Electromagnetism or the electromagnetic field is the theory of science which unifies two kinds of forces, namely electric field and magnetic field. A non-zero electric field is produced by the presence of electrically charged particles, and gives rise to the electric force; this is the force that causes static electricity and drives the flow of electric charge (electric current) in electrical conductors. The magnetic field, on the other hand, can be produced by the motion of electric charges, or electric current, and gives rise to the magnetic force associated with magnets. When an appliance is plugged in and even when it is not turned on, an electric field will be created and it is measured in V/m. If the device is turned on, then a magnetic field will be created which is measured in Gauss or Tesla. In other words, a rest charge (q) creates electrostatic field, a moving charge (static current flow $i = \frac{dq}{dt}$) creates magnetostatic field and an accelerated charge (dynamic or time varying current flow $\left(\frac{d^2q}{dt^2} = \frac{di}{dt}\right)$ produces electromagnetic field.

Electromagnetic field is the study of interactions between electric charges at rest and in motion. The term 'electromagnetism' comes from the individual component electrical and magnetic forces involved. A changing magnetic field produces an electric field (this is the phenomenon of electromagnetic induction, which underlies the operation of electrical generators, induction motors, and transformers). Similarly, a changing electric field generates a magnetic field. James Clark Maxwell identified this interdependence of electric and magnetic fields and unified them as electromagnetic field.

Some of the applications of electromagnetic field are in microwaves, antennas, electric machines, satellite communication, radar meteorology, remote sensing and fiber optics. In medical fields, electromagnetic waves are used to relieve pathological conditions. In industry, electromagnetic heaters are used for melting, forging, annealing, surface hardening and soldering. Electromagnetic waves are used to change the taste of vegetables by reducing acidity in agricultural fields.

The devices working under the principle of electromagnetic field are transformers, electric relay, radio, T.V., telephone, mobile phones, electric motors and generators, transmission lines, computer, etc.

Electromagnetic waves cause harmful effects to humans. Sitting in front of the computer all day will create irritation, fatigue, dizziness, nausea and headaches. Electric workers may get affected by tumors, brain cancer and breast cancer. Exposure to electromagnetic field may be a cancer promoter rather than a cancer creator. Living near power lines can cause a risk of leukemia for children.

Genesis

The current book is an outgrowth of lecture notes collected over past 10 years for the course 'Field Theory' which was introduced by Bharathiar University. Initially, the lecture notes were compiled as a book for local circulation within the college, which attracted most of the students. It was Dr. A.M. Natarajan, Principal, Kongu Engineering College, Perundurai, Tamilnadu, who encouraged us to write this book for publication.

Scope

'Electromagnetic Fields' aims at presenting the concept of electromagnetism more clearly and offers an easily understandable approach based on classroom experience of the authors. This book offers comprehensive coverage of the subject Electromagnetism, with a clear exposition of the theory along with practical application. The presentation is very simple to facilitate independent learning by the students.

The book is primarily meant for the undergraduate engineering students of all circuit branches. The authors have taken into account the difficulties of students as well as teachers while teaching this subject.

For each topic, there are a large number of solved examples so as to aid the students in grasping the concepts. This book would also serve as a useful text for the students preparing for AMIE, M.Sc. (Physics) and for various competitive exams concerning the subject.

Organization of the book

This textbook starts from vector analysis and co-ordinate systems which are so fundamental for learning electromagnetic field.

The complete concept of electromagnetic field is dealt with by dividing into three major subdivisions—Electrostatics (static electric field), Magnetostatics (magnetostatic field) and Electromagnetic field (dynamic field). Electrostatics discusses the methods to determine the electric field intensity and potential due to various charge distributions using Coulomb's law and Gauss's law. In addition to this, various capacitors, boundary value problems and image theory are also dealt in detail. In Magnetostatics, the effect of various current distributions using Biot-Savart's law and Ampere circuital law are discussed. Various inductors and fundamentals of magnetic circuits are also covered.

The chapter Electromagnetic Field starts with Faraday's law of electromagnetic induction and subsequently covers the entire concept of time-varying (dynamic) fields along with Maxwell's equations of wave and wave propagation.

Finally, introduction to solving electromagnetic field problems using numerical methods is discussed. At the end of each chapter, a summary of the chapter and tutorial problems with answers are also included.

The revised edition includes:

- Expanded coverage of some topics in electrostatics and magnetostatics
- A new section on circuit theory and field theory
- A complete new set of solved problems in Chapter 7.

The authors welcome constructive suggestions for the betterment of the book.

Authors

Contents

1

Vector Analysis

1.1 VECTOR ANALYSIS

Introduction

In general, electromagnetic field problem involves three space variables, as a result of which the solutions tend to become complex. This can be overcome by the use of vector analysis. So, knowledge of vector analysis is an essential pre-requisite to the study of theory of electromagnetic fields.

1.1.1 Scalar and Vectors

A quantity is said to be a scalar if it has only magnitude. Quantities such as time, mass, distance, temperature, electric potential and population are scalars. A quantity is called vector if it has both magnitude and direction. Vector quantities include velocity, force, displacement and electric field intensity.

To distinguish between a scalar and a vector, it is customary to represent a vector by a letter with an arrow on the top, such as $\vec{A}$ and $\vec{B}$. A scalar is represented by a letter e.g., *A*, *B* and *U*.

Electromagnetic theory is essentially a study of some particular fields. We define a field as a function that specifies a particular quantity, everywhere in a region. Some examples of scalar fields are: temperature distribution in a building, sound intensity in a theatre and electric potential in a region. The gravitational force on a body in space and the velocity of raindrops in the atmosphere are examples of vector fields.

1.1.2 Unit Vector

A vector $\vec{A}$ has both magnitude and direction. The magnitude of $\vec{A}$ is a scalar written as A or $|A|$. A unit vector $\vec{a}_A$ along $\vec{A}$ is defined as a vector whose magnitude is unity and its direction is along $\vec{A}$. In general, any vector can be represented by its magnitude and its direction as follows.

$$\vec{A} = A\vec{a}_A, \qquad ...(1.1)$$

where $A \rightarrow$ magnitude of the vector $\vec{A}, \vec{a}_A \rightarrow$ direction of the vector $\vec{A}$.

This $\vec{a}_A$ is called a unit vector.

From eqn. (1.1), unit vector $\vec{a}_A = \dfrac{\vec{A}}{A} = \dfrac{\vec{A}}{|\vec{A}|}$. ...(1.2)

Note that $|\vec{a}_A| = 1$.

A vector $\vec{A}$ in cartesian (rectangular) co-ordinates may be represented as (A_x, A_y, A_z) or $A_x\vec{a}_x + A_y\vec{a}_y + A_z\vec{a}_z$, where A_x, A_y, A_z are called components of A in x, y, z directions respectively; a_x, a_y and a_z are unit vectors in the x, y and z directions respectively.

The magnitude of A is given by

$$A = |\vec{A}| = \sqrt{A_x^2 + A_y^2 + A_z^2}$$

and the unit vector along $\vec{A}$ is given by

$$\vec{a}_A = \frac{\vec{A}}{A} = \frac{\vec{A}}{|\vec{A}|}$$

$$\vec{a}_A = \frac{A_x\vec{a}_x + A_y\vec{a}_y + A_z\vec{a}_z}{\sqrt{A_x^2 + A_y^2 + A_z^2}}. \quad ...(1.3)$$

1.1.3 Vector Addition and Subtraction

Two vectors $\vec{A}$ and $\vec{B}$ can be added together to give another vector $\vec{C}$.

The vector addition is carried out component by component. Thus if $\vec{A} = (A_x, A_y, A_z)$ and $\vec{B} = (B_x, B_y, B_z)$,

$$C = (A_x + B_x)\vec{a}_x + (A_y + B_y)\vec{a}_y + (A_z + B_z)\vec{a}_z. \quad ...(1.4)$$

Vector subtraction is similarly carried out as

$$\vec{D} = \vec{A} - \vec{B} = \vec{A} + (-\vec{B})$$

$$\vec{D} = (A_x - B_x)\vec{a}_x + (A_y - B_y)\vec{a}_y + (A_z - B_z)\vec{a}_z. \quad ...(1.5)$$

Graphically, vector addition and subtraction are obtained by parallelogram rule as in Fig. 1.1 and Fig. 1.2.

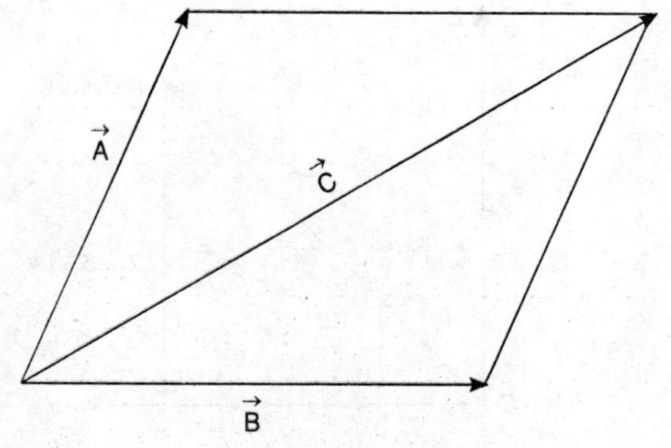

Fig. 1.1

$\vec{C} = \vec{A} + \vec{B}$

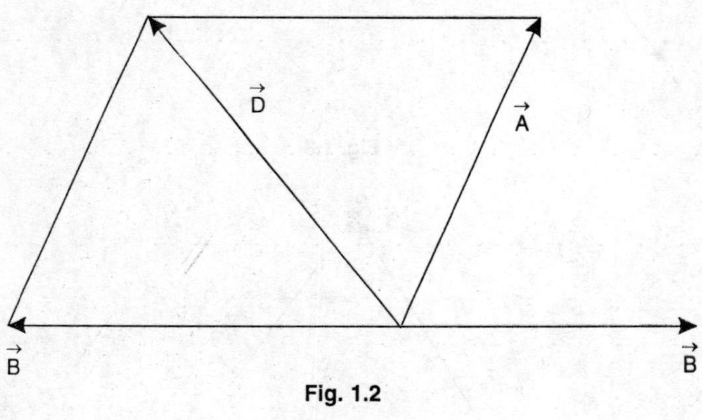

Fig. 1.2

$\vec{D} = \vec{A} - \vec{B}$

1.1.4 Position and Distance Vectors

A point P in cartesian coordinates may be represented by (x, y, z). The position vector $\vec{r}_p$ (radius vector) of point P is defined as the directed distance from the origin O to P, that is

$$\vec{r}_p = \vec{OP} = x\vec{a}_x + y\vec{a}_y + z\vec{a}_z \,. \qquad ...(1.6)$$

For example, the position vector for the point (3, 4, 5) is $3\vec{a}_x + 4\vec{a}_y + 5\vec{a}_z$ which is shown in Fig. 1.3.

If two points P and Q are given by (x_p, y_p, z_p) and (x_Q, y_Q, z_Q), the distance vector (or separation vector) between P and Q is the displacement from P to Q as in Fig. 1.4.

$$\vec{r}_{PQ} = \vec{r}_Q - \vec{r}_P$$

$$= (x_Q - x_P)\vec{a}_x + (y_Q - y_P)\vec{a}_y + (z_Q - z_P)\vec{a}_z \qquad ..(1.7)$$

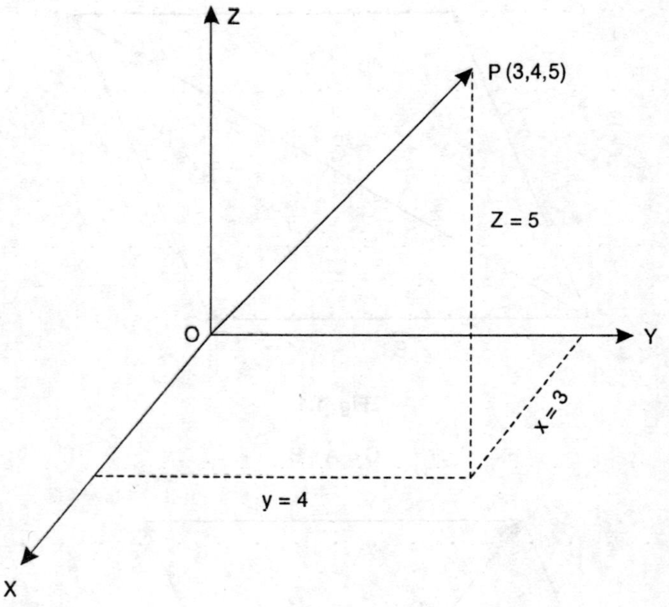

Fig. 1.3

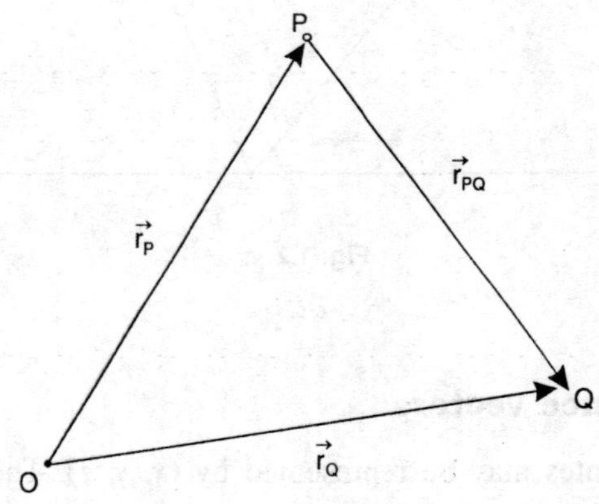

Fig. 1.4

Example 1.1

Given vectors $\vec{A} = 2\vec{a}_x + 3\vec{a}_y + 4\hat{a}_z$ and $\vec{B} = 5\vec{a}_x + 2\vec{a}_y - 6\hat{a}_z$, determine:

(*a*) $\left|\vec{A} + \vec{B}\right|$

(*b*) The component of $\vec{A}$ along $\vec{a}_y$

(*c*) A unit vector parallel to $3\vec{A} + \vec{B}$.

(a) $\vec{A}+\vec{B} = (2, 3, 4)+(5, 2, -6)$

$= (7, 5, -2)$

$|\vec{A}+\vec{B}| = \sqrt{7^2+5^2+(-2)^2}$

$= \sqrt{78}$

$= 8.83$

(b) The component of $\vec{A}$ along $\vec{a}_y$ is 3.

(c) A unit vector parallel to $3\vec{A}+\vec{B}$.

$3\vec{A}+\vec{B} = 3\ (2, 3, 4) + (5, 2, -6)$
$= (6, 9, 12) + (5, 2, -6)$
$= (11, 11, 6)$

Let $\vec{C} = 3\vec{A}+\vec{B}$

A unit vector parallel to $\vec{C}$ is

$$\vec{a}_c = \pm\frac{\vec{C}}{|\vec{C}|} = \pm\frac{(11, 11, 6)}{\sqrt{11^2+11^2+6^2}}$$

$$= \pm\left(0.659\,\vec{a}_x + 0.659\,\vec{a}_y + 0.359\,\vec{a}_z\right)$$

Note that $|\vec{a}_c| = 1$ as expected.

Example 1.2

Points P and Q are located at (1, –2, 4) and (–5, 4, 2). Determine

(a) The position vector of P.
(b) The distance vector from P to Q.
(c) The distance between P and Q.
(d) A vector parallel to PQ with magnitude of 10.

(a) $\vec{r}_P = \vec{a}_x - 2\vec{a}_y + 4\vec{a}_z$

(b) $\vec{r}_{PQ} = \vec{r}_Q - \vec{r}_P = (-5,\ 4,\ 2) - (1, -2, 4)$

$= (-6, 6, -2)$

$\vec{r}_{PQ} = -6\vec{a}_x + 6\vec{a}_y - 2\vec{a}_z$

(c) Distance between P and Q is the magnitude of vector $\vec{r}_{PQ}$.

$d = |\vec{r}_{PQ}| = \sqrt{6^2+6^2+4^2}$

$= 9.381.$

(*d*) Let the required vector be $\vec{A}$.

$\vec{A} = A\vec{a}_A$,

where A is the magnitude which is given as 10. Since $\vec{A}$ is parallel to *PQ*, it must have the same unit vector as $\vec{r}_{PQ}$ (or) $\vec{r}_{QP}$. Hence,

$$\vec{a}_A = \pm\frac{\vec{r}_{PQ}}{|r_{PQ}|} = \pm\frac{(-6, 6, -2)}{9.381}$$

$$= -0.639\,\vec{a}_x + 0.639\,\vec{a}_y - 0.213\,\vec{a}_z$$

$$\vec{A} = \pm 10\left(-0.639\,\vec{a}_x + 0.639\,\vec{a}_y - 0.213\,\vec{a}_z\right)$$

$$= \pm\left(-6.39\,\vec{a}_x + 6.39\,\vec{a}_y - 2.13\,\vec{a}_z\right).$$

1.1.5 Vector Multiplication

When two vectors $\vec{A}$ and $\vec{B}$ are multiplied, the result is either a scalar or a vector depending on how they are multiplied. There are two types of vector multiplication:

(*a*) Scalar (or dot) product: $\vec{A}\cdot\vec{B}$

(*b*) Vector (or cross) product: $\vec{A}\times\vec{B}$.

Multiplication of three vectors *A*, *B* and *C* can result in either

(*c*) Scalar triple product: $\vec{A}\cdot\left(\vec{B}\times\vec{C}\right)$

or

(*d*) Vector triple product: $\vec{A}\times\left(\vec{B}\times\vec{C}\right)$

(a) *Dot Product*

The dot product of two vectors $\vec{A}$ and $\vec{B}$, written as $\vec{A}\cdot\vec{B}$, is defined geometrically as the product of the magnitude of $\vec{B}$ and the projection of $\vec{A}$ onto $\vec{B}$ (or vice versa). Thus

$$\vec{A}\cdot\vec{B} = AB\cos\theta_{AB}, \qquad \text{...(1.8)}$$

where θ_{AB} is the smaller angle between *A* and *B*. The result $\vec{A}\cdot\vec{B}$ is called either the scalar product because it is scalar, or the dot product due to the dot sign.

If $\vec{A} = \left(A_x, A_y, A_z\right)$ and $\vec{B} = \left(B_x, B_y, B_z\right)$,

then

$$\vec{A}\cdot\vec{B} = A_x\,B_x + A_y\,B_y + A_z\,B_z, \qquad \text{...(1.9)}$$

which is obtained by multiplying $\vec{A}$ and $\vec{B}$, component by component.

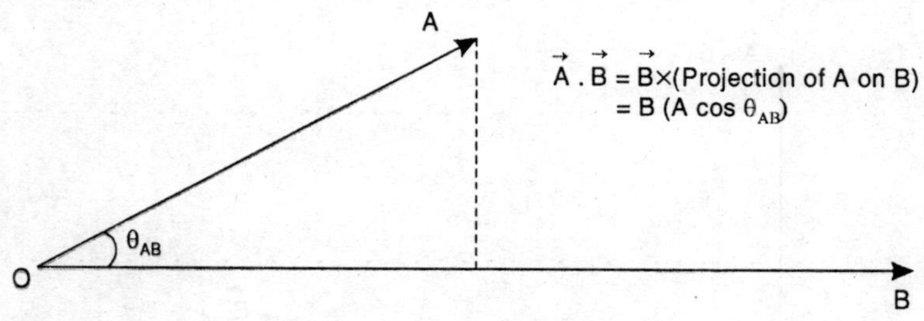

Fig. 1.5

The dot product obeys the following:

(*i*) Commutative law:

$$\vec{A} \cdot \vec{B} = \vec{B} \cdot \vec{A} \qquad \text{...(1.10)}$$

(*ii*) Distributive law:

$$\vec{A} \cdot (\vec{B} + \vec{C}) = \vec{A} \cdot \vec{B} + \vec{A} \cdot \vec{C} \qquad \text{...(1.11)}$$

(*iii*) $\vec{A} \cdot \vec{A} = (A)^2 = A^2$. ...(1.12)

Also note that

$$\vec{a}_x \cdot \vec{a}_y = \vec{a}_y \cdot \vec{a}_z = \vec{a}_z \cdot \vec{a}_x = 0$$

$$\vec{a}_x \cdot \vec{a}_x = \vec{a}_y \cdot \vec{a}_y = \vec{a}_z \cdot \vec{a}_z = 1 \qquad \text{...(1.13)}$$

(b) Cross Product

The cross product of two vectors $\vec{A}$ and $\vec{B}$ is defined as

$$\vec{A} \times \vec{B} = AB \sin\theta_{AB}\, \vec{a}_n, \qquad \text{...(1.14)}$$

where $\vec{a}_n$ is the unit vector normal to the plane containing *A* and *B*.

The direction of $\vec{a}_n$ is taken as the direction of right thumb when the fingers of the right hand rotate from $\vec{A}$ and $\vec{B}$. Alternatively, the direction of $\vec{a}_n$ is taken as that of advance of right-handed screw as $\vec{A}$ is turned into $\vec{B}$ as shown in Fig. 1.6.

The vector multiplication is called cross product due to the cross sign. It is also called as vector product because the result is a vector.

If $\vec{A} = (A_x, A_y, A_z)$ and $\vec{B} = (B_x, B_y, B_z)$, then

$$\vec{A} \times \vec{B} = \begin{vmatrix} \vec{a}_x & \vec{a}_y & \vec{a}_z \\ A_x & A_y & A_z \\ B_x & B_y & B_z \end{vmatrix}$$

$$= (A_y B_z - A_z B_y)\vec{a}_x + (A_z B_x - A_x B_z)\vec{a}_y + (A_x B_y - A_y B_x)\vec{a}_z. \qquad \text{...(1.15)}$$

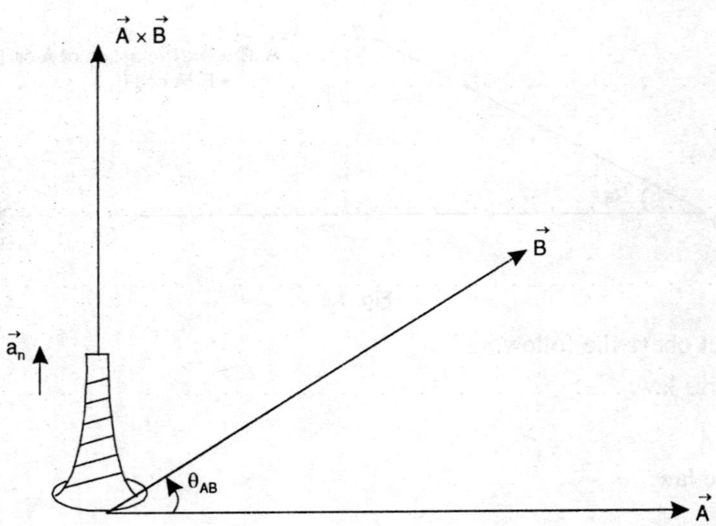

Fig. 1.6

Cross product has the following properties

(*i*) It is not commutative:

$$\vec{A} \times \vec{B} \neq \vec{B} \times \vec{A} \quad ...(1.16)$$

It is anti-commutative:

$$\vec{A} \times \vec{B} = -\vec{B} \times \vec{A} \quad ...(1.17)$$

(*ii*) It is not associative:

$$\vec{A} \times \left(\vec{B} \times \vec{C}\right) \neq \left(\vec{A} \times \vec{B}\right) \times \vec{C} \quad ...(1.18)$$

(*iii*) It is distributive:

$$\vec{A} \times \left(\vec{B} + \vec{C}\right) = \left(\vec{A} \times \vec{B}\right) + \left(\vec{A} \times \vec{C}\right) \quad ...(1.19)$$

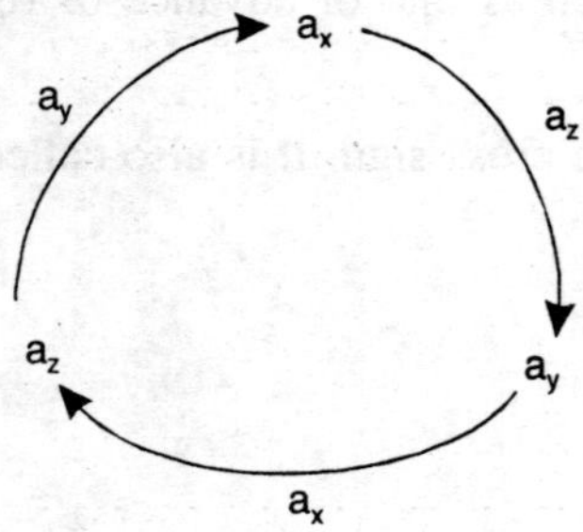

Fig. 1.7
(*a*) Moving clockwise leads to positive results

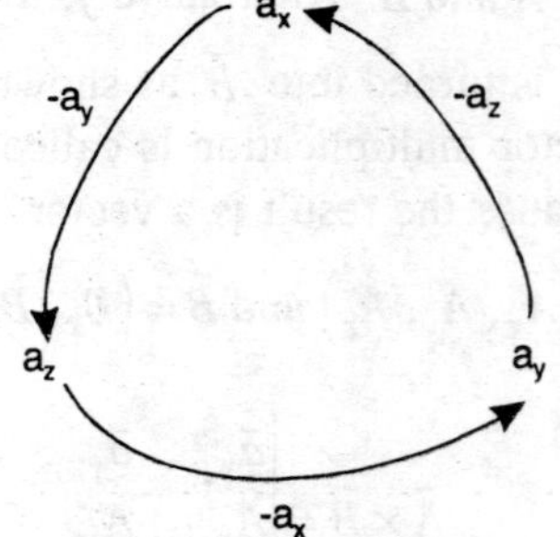

Fig. 1.8
(*b*) Moving counterclockwise results in negative results.

(iv) $\vec{A} \times \vec{A} = 0$ ...(1.20)

Also note that

$$\vec{a}_x \times \vec{a}_y = \vec{a}_z$$

$$\vec{a}_y \times \vec{a}_z = \vec{a}_x \qquad ...(1.21)$$

$$\vec{a}_z \times \vec{a}_x = \vec{a}_y$$

which are obtained in cyclic permutation and illustrated in Figs. 1.7 and 1.8.

(c) *Scalar triple product*

Given three vectors $\vec{A}$, $\vec{B}$ and $\vec{C}$, we define the scalar triple product as

$$\vec{A} \cdot (\vec{B} \times \vec{C}) = \vec{B} \cdot (\vec{C} \times \vec{A}) = \vec{C} \cdot (\vec{A} \times \vec{B}) \qquad ...(1.22)$$

obtained in cyclic permutation. If $\vec{A} = (A_x, A_y, A_z)$, $\vec{B} = (B_x, B_y, B_z)$ and $\vec{C} = (C_x, C_y, C_z)$, then $\vec{A} \cdot (\vec{B} \times \vec{C})$ is the volume of a parallelopiped having $\vec{A}$, $\vec{B}$ and $\vec{C}$ as edges and is easily obtained by finding the determinant of 3 × 3 matrix formed by $\vec{A}$, $\vec{B}$ and $\vec{C}$, that is,

$$\vec{A} \cdot (\vec{B} \times \vec{C}) = \begin{vmatrix} A_x & A_y & A_z \\ B_x & B_y & B_z \\ C_x & C_y & C_z \end{vmatrix} \cdot \qquad ...(1.23)$$

Since the result is scalar, this is known as scalar triple product.

(d) *Vector triple product*

We define vector triple product for vectors $\vec{A}$, $\vec{B}$ and $\vec{C}$ as

$$\vec{A} \times (\vec{B} \times \vec{C}) = \vec{B}(\vec{A} \cdot \vec{C}) - \vec{C}(\vec{A} \cdot \vec{B}) = \vec{B}(\vec{A} \cdot \vec{C}) - \vec{A}(\vec{B} \cdot \vec{C}) \qquad ...(1.24)$$

Note:

$$(\vec{A} \cdot \vec{B})\, \vec{C} \neq \vec{A}\, (\vec{B} \cdot \vec{C})$$

$$(\vec{A} \cdot \vec{B})\, \vec{C} = \vec{C}\, (\vec{A} \cdot \vec{B}) \qquad ...(1.25)$$

1.1.6 Components of a Vector

A direct application of vector product is its use in determining the projection (or component) of a vector in a given direction. The projection can be scalar or vector.

Given a vector $\vec{A}$, we define the scalar component A_B of $\vec{A}$ along vector $\vec{B}$ as in Fig. 1.9.

$$A_B = A\cos\theta_{AB} = |\vec{A}||\vec{a}_B|\cos\theta_{AB} \qquad ...(1.26)$$

$$A_B = \vec{A} \cdot \vec{a}_B. \qquad ...(1.27)$$

The vector component $\vec{A}_B$ of $\vec{A}$ along $\vec{B}$ is the scalar component multiplied by a unit vector along B, that is,

$$\vec{A}_B = A_B\, \vec{a}_B = \left(\vec{A} \cdot \vec{a}_B\right) \vec{a}_B. \qquad ...(1.28)$$

Notice from Fig. 1.9 and Fig. 1.10 that the vector $\vec{A}$ can be resolved into two orthogonal components; one component $\vec{A}_B$ parallel to $\vec{B}$, another $\left(\vec{A} - \vec{A}_B\right)$ perpendicular to $\vec{B}$.

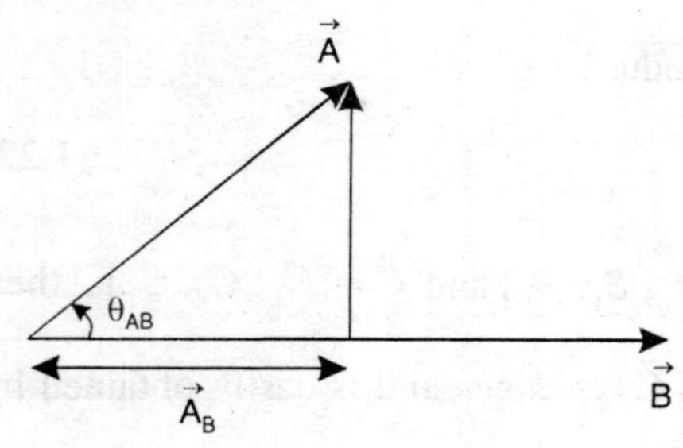

Fig. 1.9

(*a*) **Scalar component A_B**

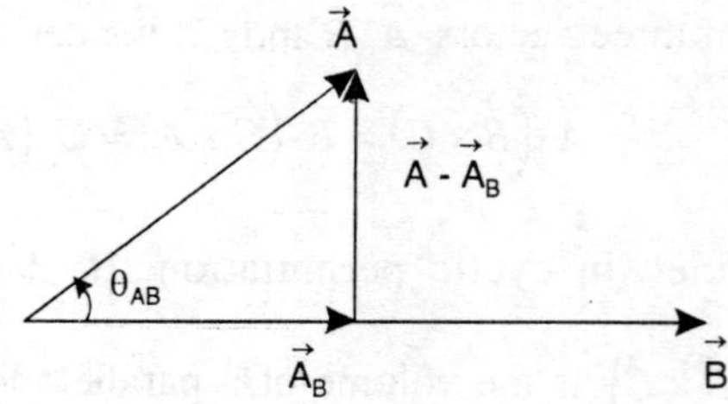

Fig. 1.10

(*b*) **Vector component**

Example 1.3

Given vectors $\vec{A} = 3\vec{a}_x + 4\vec{a}_y + \vec{a}_z$ and $\vec{B} = 2\vec{a}_x - 3\vec{a}_y + 5\vec{a}_z$, find the angle between $\vec{A}$ and $\vec{B}$.

The angle θ_{AB} can be found by either dot or cross product.

$$\vec{A} \cdot \vec{B} = (3, 4, 1) \cdot (2, -3, 5)$$

$$= 6 - 12 + 5$$

$$= -1.$$

$$\left|\vec{A}\right| = \sqrt{3^2 + 4^2 + 1} = \sqrt{26}$$

$$\left|\vec{B}\right| = \sqrt{2^2 + (-3)^2 + 5^2} = \sqrt{38}$$

$$\cos\theta_{AB} = \frac{\vec{A} \cdot \vec{B}}{\left|\vec{A}\right| \left|\vec{B}\right|} = \frac{-1}{\sqrt{26} \cdot \sqrt{38}} = -0.032$$

$$\theta_{AB} = 91.834.$$

Alternatively:

$$\vec{A} \times \vec{B} = \begin{vmatrix} \vec{a}_x & \vec{a}_y & \vec{a}_z \\ 3 & 4 & 1 \\ 2 & -3 & 5 \end{vmatrix}$$

$$= (20+3)\,\vec{a}_x - (15-2)\,\vec{a}_y + (-9-8)\,\vec{a}_z$$

$$= (23,\ -13,\ -17)$$

$$\left|\vec{A} \times \vec{B}\right| = \sqrt{(23)^2 + (-13)^2 + (-17)^2} = 31.41$$

$$\sin\theta_{AB} = \frac{\left|\vec{A} \times \vec{B}\right|}{\left|\vec{A}\right|\left|\vec{B}\right|} = \frac{31.41}{\sqrt{26} \cdot \sqrt{38}}$$

$$= 0.999$$

$$\theta_{AB} = 87.8°.$$

Example 1.4

Three fields are given by

$$\vec{P} = 2\vec{a}_x - \vec{a}_y + \vec{a}_z$$

$$\vec{Q} = 4\vec{a}_x - \vec{a}_z$$

$$\vec{R} = \vec{a}_x - \vec{a}_y$$

Determine:

(*a*) $\vec{Q} \cdot \vec{R} \times \vec{P}$

(*b*) A unit vector perpendicular to both $\vec{Q}$ and $\vec{R}$

(*c*) The component of $\vec{P}$ along $\vec{Q}$.

(*a*) The only way $\vec{Q} \cdot \vec{R} \times \vec{P}$ makes sense is

$$\vec{Q} \cdot \left(\vec{R} \times \vec{P}\right) = (4,\ 0,\ -1) \cdot \begin{vmatrix} \vec{a}_x & \vec{a}_y & \vec{a}_z \\ 1 & -1 & 0 \\ 2 & -1 & 1 \end{vmatrix}$$

$$= (4,\ 0,\ -1) \cdot (-1,\ -1,\ 1)$$

$$= -4 + 0 - 1 = -5.$$

(*b*) A unit vector perpendicular to both $\vec{Q}$ and $\vec{R}$ is given by

$$\vec{a} = \pm \frac{\vec{Q} \times \vec{R}}{\left|\vec{Q} \times \vec{R}\right|}$$

$$\vec{Q} \times \vec{R} = \begin{vmatrix} \vec{a}_x & \vec{a}_y & \vec{a}_z \\ 4 & 0 & -1 \\ 1 & -1 & 0 \end{vmatrix}$$

$$= -\vec{a}_x - \vec{a}_y - 4\vec{a}_z .$$

$$\left|\vec{Q} \times \vec{R}\right| = \sqrt{1^2 + 1^2 + 4^2} = \sqrt{18}$$

$$\vec{a} = \pm\left(\frac{-\vec{a}_x - \vec{a}_y - 4\vec{a}_z}{\sqrt{18}}\right)$$

$$\vec{a} = \pm\left(-0.235\vec{a}_x - 0.235\vec{a}_y - 0.942\vec{a}_z\right).$$

(*c*) The component of $\vec{P}$ along $\vec{Q}$ is

$$\vec{P}_Q = \left|\vec{P}\right| \cos\theta_{PQ}\, \vec{a}_Q$$

$$= \left(\vec{P} \cdot \vec{a}_Q\right)\vec{a}_Q$$

$$\vec{a}_Q = \frac{\vec{Q}}{\left|\vec{Q}\right|}.$$

$$\vec{P}_Q = \frac{\left(\vec{P} \cdot \vec{Q}\right)\vec{Q}}{|Q|^2}$$

$$= \frac{(2, -1, 1) \cdot (4, 0, -1)}{\sqrt{4^2 + (1)^2}}(4, 0, -1)$$

$$= \frac{(8 + 0 - 1)}{\sqrt{17}}(4, 0, -1) = \frac{7}{\sqrt{17}}(4, 0, -1)$$

$$= 6.79\vec{a}_x - 1.69\vec{a}_z.$$

Example 1.5

If $\vec{A} = \alpha\,\vec{a}_x + 2\vec{a}_y + 10\vec{a}_z$ and $\vec{B} = 4\alpha\,\vec{a}_x + 8\vec{a}_y - 2\alpha\,\vec{a}_z$, for what values of α are $\vec{A}$ and $\vec{B}$ perpendicular?

The condition for $\vec{A}$ and $\vec{B}$ to be perpendicular is $\vec{A} \cdot \vec{B} = 0$.

$$\vec{A} \cdot \vec{B} = (\alpha, 2, 10) \cdot (4\alpha, 8, -2\alpha) = 0$$

$$= 4\alpha^2 + 16 - 20\alpha = 0$$

$$\alpha^2 + 4 - 5\alpha = 0$$

$$\alpha^2 - 5\alpha + 4 = 0$$

$$(\alpha - 1)(\alpha - 4) = 0$$

$$\alpha = 1, 4.$$

Example 1.6

Find the scalar component of $6\vec{a}_x + 2\vec{a}_y - 3\vec{a}_z$ along $3\vec{a}_x - 4\vec{a}_y$.

Let $\quad \vec{A} = 6\vec{a}_x + 2\vec{a}_y - 3\vec{a}_z$

$$\vec{B} = 3\vec{a}_x - 4\vec{a}_y.$$

Component of A along B is

$$A_B = (\vec{A} \cdot \vec{a}_B)$$

$$\vec{a}_B = \frac{\vec{B}}{|\vec{B}|} = \frac{3\vec{a}_x - 4\vec{a}_y}{\sqrt{3^2 + (-4)^2}}$$

$$A_B = (\vec{A} \cdot \vec{a}_B) = (6, 2, -3) \cdot \left(\frac{3}{5}, \frac{-4}{5}, 0\right)$$

$$= \frac{18}{5} - \frac{8}{5} = 2.$$

1.2 COORDINATE SYSTEMS AND TRANSFORMATIONS

Introduction

In electromagnetics, all the quantities are functions of space and time. In order to describe the spatial variations of the quantities, all the points in space must be defined using an appropriate co-ordinate system.

Although there are different co-ordinate systems available, the three well-known co-ordinate systems are Cartesian (or rectangular) co-ordinates, circular cylindrical co-ordinates and spherical co-ordinates.

1.2.1 Cartesian Co-ordinates (X, Y, Z)

A point '*P*' is represented in cartesian or rectangular co-ordinates as (*X, Y, Z*).

From Fig. 1.11, it is known that any point in rectangular co-ordinate is the intersection of three plane (*i*) constant *X*-plane (*ii*) constant *Y*-plane and (*iii*) constant *Z*-plane which are mutually perpendicular.

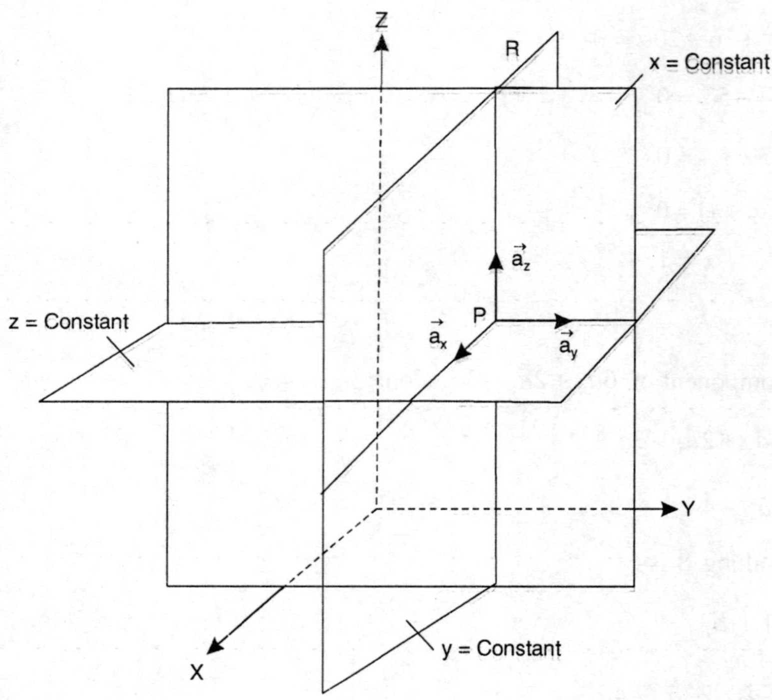

Fig. 1.11

The range of the co-ordinate variables x, y and z are

$$-\infty < x < \infty$$

$$-\infty < y < \infty$$

$$-\infty < z < \infty .$$

A vector $\vec{A}$ in cartesian co-ordinates can be written as (A_x, A_y, A_z) or

$$A_x \vec{a}_x + A_y \vec{a}_y + A_z \vec{a}_z ,$$

where $\vec{a}_x$, $\vec{a}_y$ and $\vec{a}_z$ are unit vectors along x, y and z directions respectively.

1.2.2 Circular Cylindrical Co-ordinates (ρ, φ, Z)

A point P in cylindrical co-ordinate is represented as (ρ, φ, Z).

From Fig. 1.12, it is known that any point 'P' in the cylindrical co-ordinates is an intersection of three planes viz. (*i*) constant 'ρ' plane (a cylinder) (*ii*) constant φ plane (plane perpendicular to XY plane and displaced by φ angle from X-axis) (*iii*) constant Z-plane (parallel to XY plane).

Each variable is defined as

(*i*) ρ is the radius of cylinder passing through P or the radial distance from Z-axis.

(*ii*) φ is called the azimuthal angle and is measured from the X-axis in the XY plane:

(*iii*) Z is the same as in cartesian system.

The ranges of variables are

$$0 \le \rho < \infty$$

$$0 \le \phi < 2\pi$$
$$-\infty < z < \infty \,.$$

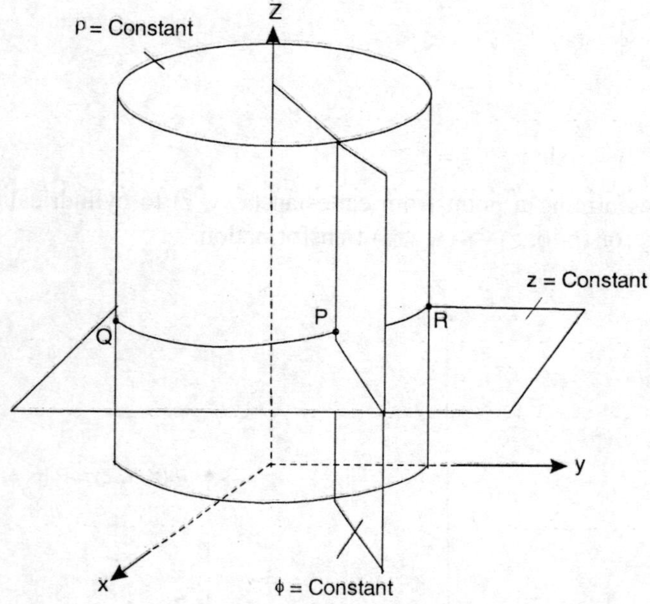

Fig. 1.12

A vector $\vec{A}$ in cylindrical co-ordinates can be written as (A_ρ, A_ϕ, A_z) or $A_\rho \vec{a}_\rho + A_\phi \vec{a}_\phi + A_z \vec{a}_z$, where $\vec{a}_\rho, \vec{a}_\phi$ and $\vec{a}_z$ are unit vectors in the ρ, ϕ, and z-directions as in Fig. 1.13.

$$\vec{a}_\rho \cdot \vec{a}_\rho = \vec{a}_\phi \cdot \vec{a}_\phi = \vec{a}_z \cdot \vec{a}_z = 1 \qquad \text{...(1.29)}$$
$$\vec{a}_\rho \cdot \vec{a}_\phi = \vec{a}_\phi \cdot \vec{a}_z = \vec{a}_z \cdot \vec{a}_\rho = 0$$

$$\vec{a}_\rho \times \vec{a}_\phi = \vec{a}_z$$
$$\vec{a}_\phi \times \vec{a}_z = \vec{a}_\rho \qquad \text{...(1.30)}$$
$$\vec{a}_z \times \vec{a}_\rho = \vec{a}_\phi$$

Note: $\vec{a}_\rho$, $\vec{a}_\phi$, $\vec{a}_z$ are mutually perpendicular and all point to the direction of increasing magnitude.

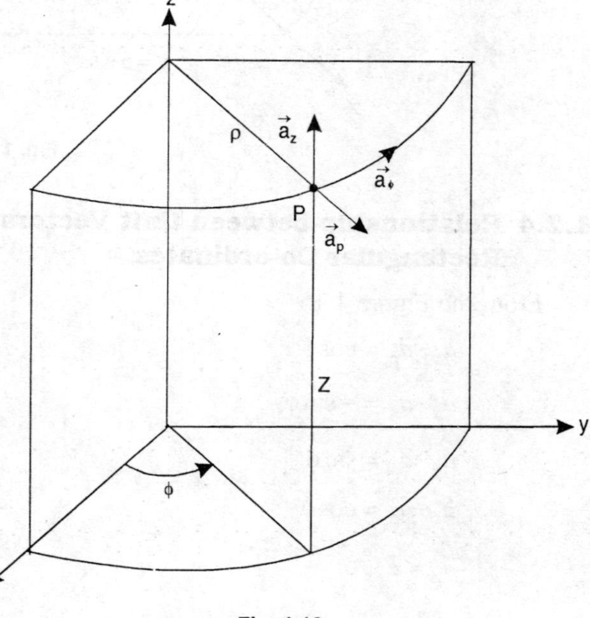

Fig. 1.13

1.2.3 Relationship between Cartesian and Cylindrical Co-ordinates

Relationship between the variables (*x, y, z*) of the cartesian co-ordinate system and those of the cylindrical system (ρ, ϕ, *z*) are easily obtained from Fig. 1.14 as

$$\rho = \sqrt{x^2 + y^2}, \quad \phi = \tan^{-1}\frac{y}{x}, \quad z = z \qquad ...(1.31)$$

(or)

$$x = \rho\cos\phi, \quad y = \rho\sin\phi, \quad z = z \qquad ...(1.32)$$

Eqn. (1.31) is for transforming a point from cartesian (*x, y, z*) to cylindrical (ρ, ϕ, *z*) co-ordinates, whereas Eqn. (1.32) is for (ρ, ϕ, *z*) → (*x, y, z*) transformation.

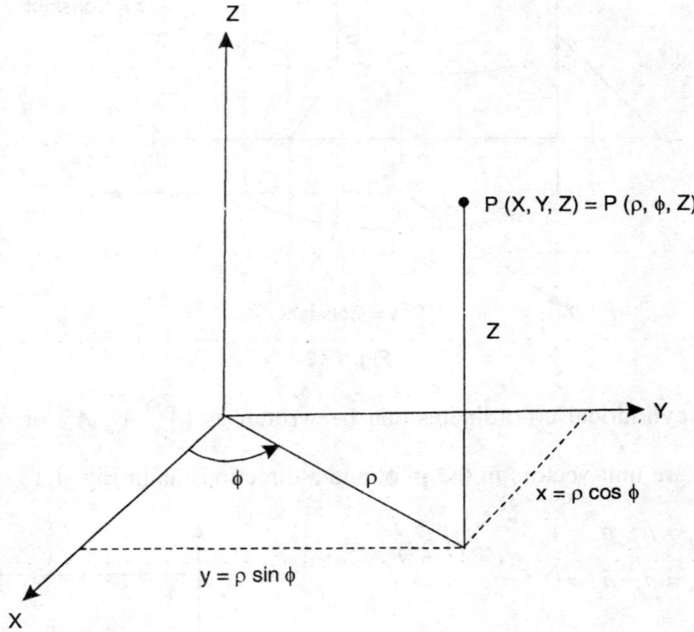

Fig. 1.14

1.2.4 Relationship between Unit Vectors of Cylindrical and Rectangular Co-ordinates

From the Figure 1.15

$$\vec{a}_x \cdot \vec{a}_\rho = \cos\phi$$

$$\vec{a}_x \cdot \vec{a}_\phi = -\sin\phi$$

$$\vec{a}_y \cdot \vec{a}_\rho = \sin\phi \qquad ...(1.33)$$

$$\vec{a}_y \cdot \vec{a}_\phi = \cos\phi$$

$$\vec{a}_z = \vec{a}_z$$

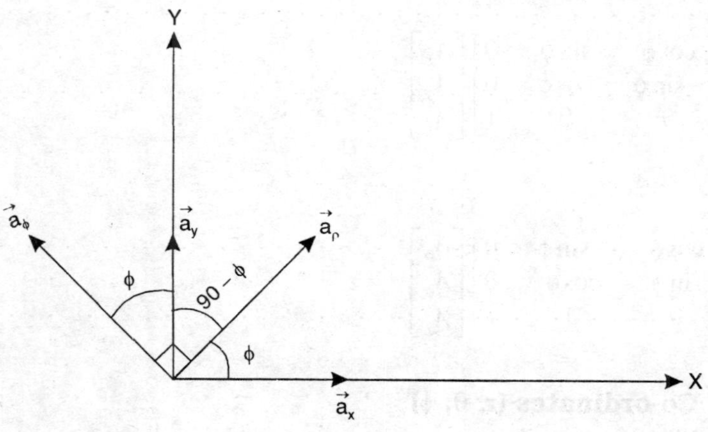

Fig. 1.15

Transformation of the given vector $\bar{A}$ (A_x, A_y, A_z) in cartesian into cylindrical.

$$\vec{A} = A_x\,\vec{a}_x + A_y\,\vec{a}_y + A_z\,\vec{a}_z$$

The component $\vec{A}$ along $\vec{a}_\rho$ is A_ρ

$$A_\rho = \left(\vec{A}\cdot\vec{a}_\rho\right)$$

$$= \left(A_x\vec{a}_x + A_y\vec{a}_y + A_z\vec{a}_z\right)\cdot\vec{a}_\rho$$

$$= A_x \cos\phi + A_y \sin\phi$$

$$\vec{A}_\rho = \left(A_x\cos\phi + A_y\sin\phi\right)\vec{a}_\rho .$$

The component of $\vec{A}$ along $\vec{a}_\phi$ is A_ϕ

$$A_\phi = \left(\vec{A}\cdot\vec{a}_\phi\right)$$

$$= \left(A_x\vec{a}_x + A_y\vec{a}_y + A_z\vec{a}_z\right)\cdot\vec{a}_\phi$$

$$= \left(-A_x\sin\phi + A_y\cos\phi\right)$$

$$\vec{A}_\phi = \left(-A_x\sin\phi + A_y\cos\phi\right)\vec{a}_\phi .$$

The component of $\vec{A}$ along $\vec{a}_z$ is A_z

$$A_z = \vec{A}\cdot\vec{a}_z$$

$$\vec{A}_z = A_z\,\vec{a}_z .$$

$$\begin{bmatrix} A_\rho \\ A_\phi \\ A_z \end{bmatrix} = \begin{bmatrix} \cos\phi & \sin\phi & 0 \\ -\sin\phi & \cos\phi & 0 \\ 0 & 0 & 1 \end{bmatrix} \begin{bmatrix} A_x \\ A_y \\ A_z \end{bmatrix} \qquad ...(1.34)$$

Similarly,

$$\begin{bmatrix} A_x \\ A_y \\ A_z \end{bmatrix} = \begin{bmatrix} \cos\phi & -\sin\phi & 0 \\ \sin\phi & \cos\phi & 0 \\ 0 & 0 & 1 \end{bmatrix} \begin{bmatrix} A_\rho \\ A_\phi \\ A_z \end{bmatrix} \qquad ...(1.35)$$

1.2.5 Spherical Co-ordinates (r, θ, φ)

Any point P can be represented as (r, θ, φ) *i.e.* intersection of three mutually perpendicular planes

(*i*) a sphere of radius r (centre at origin)

(*ii*) A cone having the Z-axis as its axis and vertex at origin.

(*iii*) a constant φ plane which is perpendicular to XY plane and displaced by φ angle from X-axis.

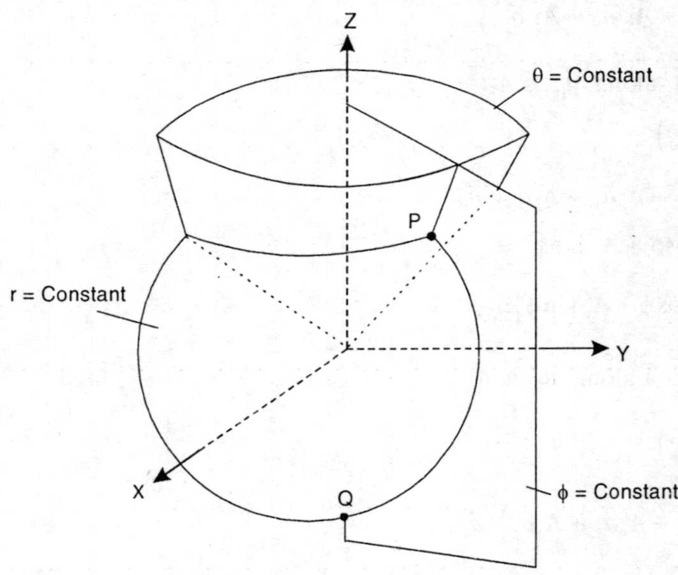

Fig. 1.16

From the Fig. 1.16, r is defined as the distance from the origin to the point P or radius of the sphere, centred at origin, and passing through P. θ (co-latitude) is the angle between the Z-axis and the position vector of P, and φ is the angle measured from X-axis (same angle in cylindrical coordinates).

The range of r, θ, φ are

$$0 \le r < \infty$$

$$0 \le \theta < \pi$$

$$0 \le \phi < 2\pi$$

The three unit vectors $\vec{a}_r$, $\vec{a}_\theta$ and $\vec{a}_\phi$ are mutually orthogonal and are shown in Fig. 1.17.

$$\vec{a}_r \times \vec{a}_\theta = \vec{a}_\phi$$

$$\vec{a}_\theta \times \vec{a}_\phi = \vec{a}_r \qquad \text{...(1.36)}$$

$$\vec{a}_\phi \times \vec{a}_r = \vec{a}_\theta$$

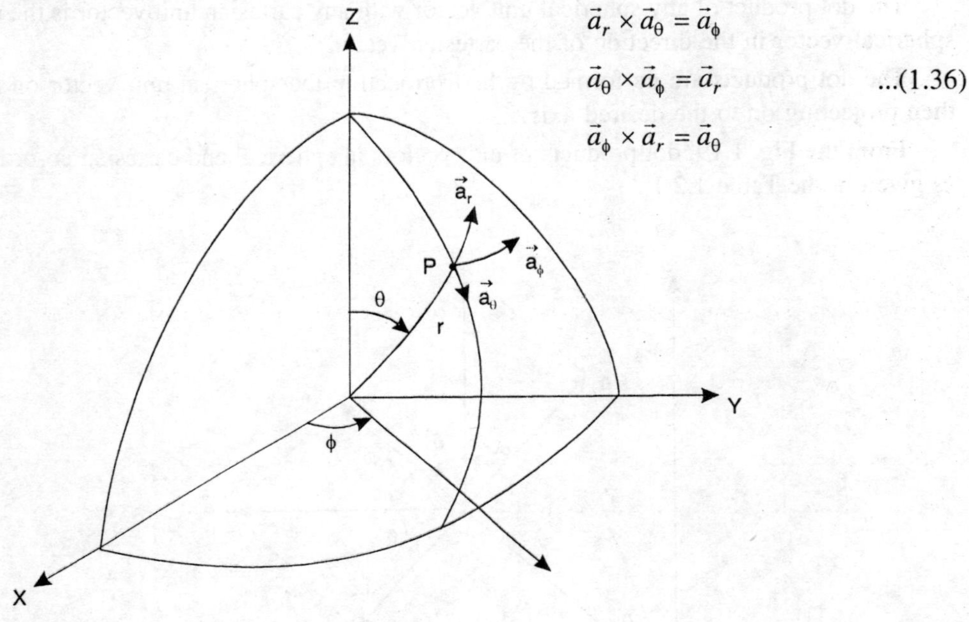

Fig. 1.17

1.2.6 Relationship between Cartesian, Cylindrical and Spherical Co-ordinates

From the Fig. 1.18,

$$r = \sqrt{x^2 + y^2 + z^2}$$

$$\theta = \tan^{-1} \frac{\sqrt{x^2 + y^2}}{z} \qquad \text{...(1.37)}$$

$$\phi = \tan^{-1} \frac{y}{z}.$$

And

$$x = r \sin\theta \cos\phi$$

$$y = r \sin\theta \sin\phi \qquad \text{...(1.38)}$$

$$z = r \cos\theta.$$

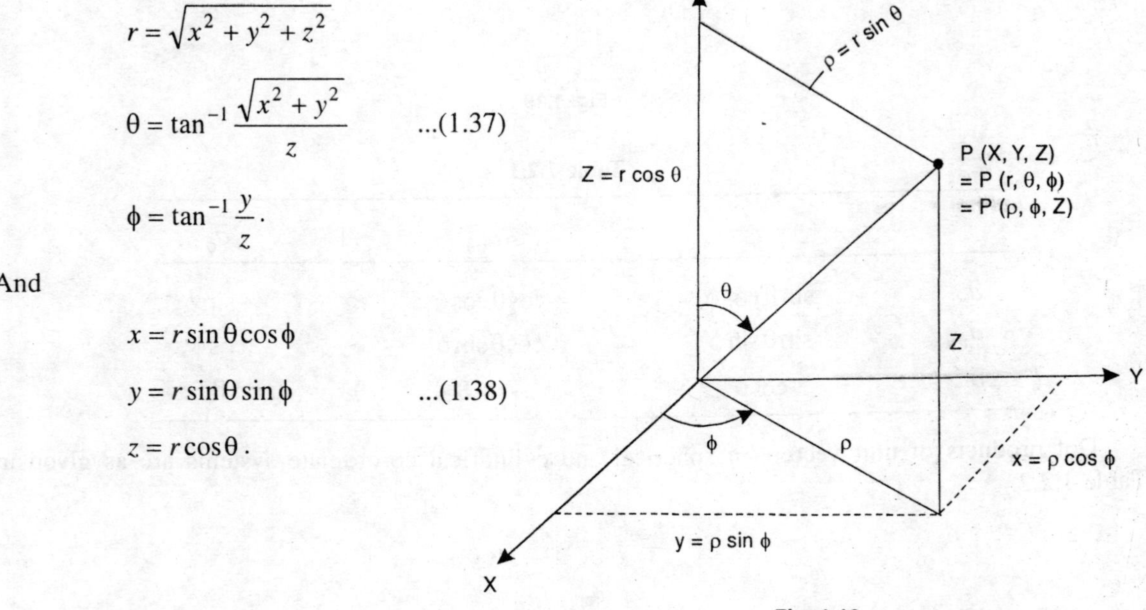

Fig. 1.18

1.2.7 Relationship between Unit Vectors of Three Co-ordinates

The transformation of vector requires the determination of the products of the unit vector in respective co-ordinates.

The dot product of any spherical unit vector with any cartesian unit vector is the component of the spherical vector in the direction of the cartesian vector.

The dot products are performed by first projecting the spherical unit vector on the *XY* plane and then projecting on to the desired axis.

From the Fig. 1.19, dot products of unit vectors in spherical and cartesian co-ordinate systems are as given in the Table 1.2.1.

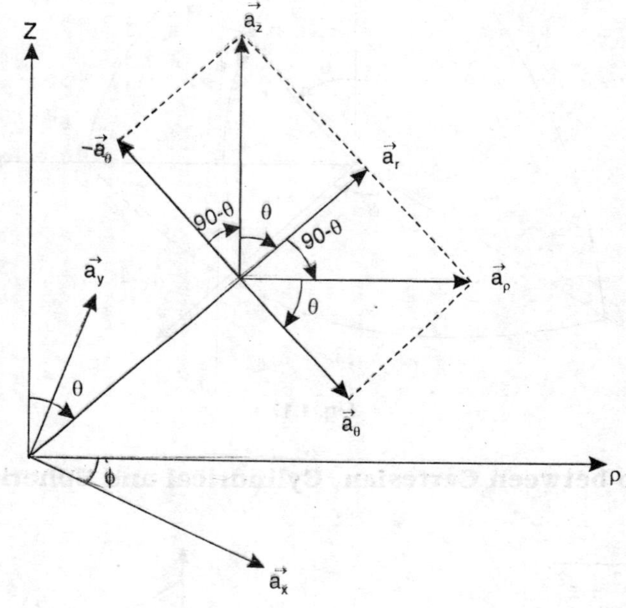

Fig. 1.19

Table 1.2.1

	$\vec{a}_r$	$\vec{a}_\theta$	$\vec{a}_\phi$
$\vec{a}_x \cdot$	$\sin\theta\cos\phi$	$\cos\theta\cos\phi$	$-\sin\phi$
$\vec{a}_y \cdot$	$\sin\theta\sin\phi$	$\cos\theta\sin\phi$	$\cos\phi$
$\vec{a}_z \cdot$	$\cos\theta$	$-\sin\theta$	0

Dot products of unit vectors in spherical and cylindrical co-ordinate systems are as given in Table 1.2.2.

Table 1.2.2

	$\vec{a}_r$	$\vec{a}_\theta$	$\vec{a}_\phi$
$\vec{a}_\rho \cdot$	$\sin\theta$	$\cos\theta\cos\phi$	0
$\vec{a}_\phi \cdot$	0	0	1
$\vec{a}_z \cdot$	$\cos\theta$	$-\sin\theta$	0

Using the Tables 1.2.1 and 1.2.2, the components of vector $\vec{A} = \left(A_x, A_y, A_z\right)$ and $\vec{A} = \left(A_r, A_\theta, A_\phi\right)$ are related as given in the matrix form:

$$\begin{bmatrix} A_r \\ A_\theta \\ A\phi \end{bmatrix} = \begin{bmatrix} \sin\theta\cos\phi & \sin\theta\sin\phi & \cos\theta \\ \cos\theta\cos\phi & \cos\theta\sin\phi & -\sin\theta \\ -\sin\phi & \cos\phi & 0 \end{bmatrix} \begin{bmatrix} A_x \\ A_y \\ A_z \end{bmatrix}. \quad ...(1.39)$$

The inverse transformation $(A_r, A_\theta, A_\rho) \rightarrow (A_x, A_y, A_z)$ is similarly obtained.

$$\begin{bmatrix} A_x \\ A_y \\ A_z \end{bmatrix} = \begin{bmatrix} \sin\theta\cos\phi & \cos\theta\cos\phi & -\sin\phi \\ \sin\theta\sin\phi & \cos\theta\sin\phi & \cos\phi \\ \cos\theta & -\sin\theta & 0 \end{bmatrix} \begin{bmatrix} A_r \\ A_\theta \\ A_\phi \end{bmatrix}. \quad ...(1.40)$$

Transformation from spherical (A_r, A_θ, A_ϕ) to form cylindrical (A_ρ, A_ϕ, A_z) is also obtained from the Table 1.2.2.

$$\begin{bmatrix} A_\rho \\ A_\phi \\ A_z \end{bmatrix} = \begin{bmatrix} \sin\theta & \cos\theta & 0 \\ 0 & 0 & 1 \\ \cos\theta & -\sin\theta & 0 \end{bmatrix} \begin{bmatrix} A_r \\ A_\theta \\ A_\phi \end{bmatrix} \quad ...(1.41)$$

Inverse transformation of cylindrical to spherical is similarly written as

$$\begin{bmatrix} A_r \\ A_\theta \\ A_\phi \end{bmatrix} = \begin{bmatrix} \sin\theta & 0 & \cos\theta \\ \cos\theta & 0 & -\sin\theta \\ 0 & 1 & 0 \end{bmatrix} \begin{bmatrix} A_\rho \\ A_\phi \\ A_z \end{bmatrix}. \quad ...(1.42)$$

Example 1.7

Given points A ($x = 2$, $y = 3$, $z = -1$) and B ($\rho = 4$, $\phi = -50°$, $z = 2$), find the distance from (a) A to the origin (b) B to the origin (c) A to B.

(i) A to the origin.

$$OA = \sqrt{x^2 + y^2 + z^2}$$

$$= \sqrt{(2)^2 + (3)^2 + (1)^2}$$

$$= 3.74.$$

(*ii*) *B* to origin.

Transform *B* (ρ, ϕ, *z*) cylindrical form into *B* (*x*, *y*, *z*) cartesian form.

$$x = \rho \cos\phi$$

$$y = \rho \sin\phi\,.$$

Given $B(\rho = 4,\ \phi = -50°,\ z = 2)$

$$x = 4\cos(-50°) = 2.57$$

$$y = 4\sin(-50°) = -3.06$$

$$z = 2$$

$$OB = \sqrt{(2.57)^2 + (-3.06)^2 + 2^2}$$

$$= 4.47$$

(*iii*) *A* to *B*

$$AB = \sqrt{(B_x - A_x)^2 + (B_y - A_y)^2 + (B_z - A_z)^2}$$

$$(B_x,\ B_y,\ B_z) = (2.57,\ -3.06,\ 2)$$

$$(A_x,\ A_y,\ A_z) = (2,\ 3,\ -1)$$

$$AB = 6.79.$$

Example 1.8

Transform each of the following vectors to cylindrical co-ordinates at the point specified:

(*a*) $5\vec{a}_x$ at $P(\rho = 4,\ \phi = 120°,\ z = 2)$

(*b*) $5\vec{a}_x$ at $Q(x = 3,\ y = 4,\ z = -1)$

(*c*) $4\vec{a}_x - 2\vec{a}_y - 4\vec{a}_z$ at $A\ (x = 2,\ y = 3,\ z = 5)$

Transformation from cartesian to cylindrical

$$\begin{bmatrix} A_\rho \\ A_\phi \\ A_z \end{bmatrix} = \begin{bmatrix} \cos\phi & \sin\phi & 0 \\ -\sin\phi & \cos\phi & 0 \\ 0 & 0 & 1 \end{bmatrix} \begin{bmatrix} A_x \\ A_y \\ A_z \end{bmatrix}.$$

Given *A* = (5, 0, 0)

$$\begin{bmatrix} A_\rho \\ A_\phi \\ A_z \end{bmatrix} = \begin{bmatrix} \cos\phi & \sin\phi & 0 \\ -\sin\phi & \cos\phi & 0 \\ 0 & 0 & 1 \end{bmatrix} \begin{bmatrix} 5 \\ 0 \\ 0 \end{bmatrix}$$

$$A_\rho = 5\cos\phi$$

$$A_\phi = -5\sin\phi$$

$$A_z = 0$$

$$\therefore\ A(\rho, \phi, z) = A(5\cos\phi, -5\sin\phi, 0).$$

$$\vec{A} = 5\cos\phi\ \vec{a}_\rho - 5\sin\phi\ \vec{a}_\phi$$

$$\vec{A} \text{ at } (\rho = 4, \phi = 120°, z = 2)$$

$$= -2.5\vec{a}_\rho - 4.33\vec{a}_\phi .$$

(*ii*) Transformation of $5\vec{a}_x$ to cylindrical is given in eqn. (5cos ϕ, –5sin ϕ, 0).
Given point P ($x = 3$, $y = 4$, $z = 1$).
From the eqn.

$$\rho = \sqrt{x^2 + y^2} = 5,\ \phi = \tan^{-1} y/x = 53.13.$$

Given point P in cylindrical co-ordinate is

$$P(\rho = 5, \phi = 53.13°, z = 1)$$

$$\vec{A} \text{ at } P = 5\cos(53.13)\ \vec{a}_\rho - 5\sin(53.13)\vec{a}_\phi$$

$$= 3\vec{a}_\rho - 4\vec{a}_\phi .$$

(*iii*) Let $\vec{A} = 4\vec{a}_x - 2\vec{a}_y - 4\vec{a}_z$

Using the transformation Eqn (1.34)

$$\begin{bmatrix} A_\rho \\ A_\phi \\ A_z \end{bmatrix} = \begin{bmatrix} \cos\phi & \sin\phi & 0 \\ -\sin\phi & \cos\phi & 0 \\ 0 & 0 & 1 \end{bmatrix} \begin{bmatrix} 4 \\ -2 \\ -4 \end{bmatrix}$$

$$\vec{A} = (4\cos\phi - 2\sin\phi)\ \vec{a}_\rho + (-4\sin\phi - 2\cos\phi)\vec{a}_\phi - 4\vec{a}_z .$$

Coordinates of A are given as ($x = 2$, $y = 3$, $z = 5$)

Cylindrical co-ordinates of A are $(\rho = 3.6, \phi = 56.3°, z = 5)$

$$\therefore\quad \vec{A} \text{ at } A = 0.555\ \vec{a}_\rho - 4.44\ \vec{a}_\phi - 4\ \vec{a}_z .$$

Example 1.9

Given the points A ($x = 2$, $y = 3$, $z = -1$) and B ($r = 4$, $\theta = 25°$, $\phi = 120°$), find the distance from A to B.

From eqn. (1.18)
$x = r\sin\theta\cos\phi$
$y = r\sin\theta\sin\phi$
$z = \cos\theta.$
The cartesian co-ordinates of B are

$$x = -0.845,\ y = 1.464,\ z = 3.63$$

$$AB = \sqrt{(B_x - A_x)^2 + (B_y - A_y)^2 + (B_z - A_z)^2}$$

$$= 5.64.$$

Example 1.10

Transform the vector $\vec{A} = 4\vec{a}_x - 2\vec{a}_y - 4\vec{a}_z$ at P ($x = +2$, $y = +3$, $z = 4$) to spherical co-ordinate.

Transformation is given by eqn. (1.39)

$$\begin{bmatrix} A_r \\ A_\theta \\ A_\phi \end{bmatrix} = \begin{bmatrix} \sin\theta\cos\phi & \sin\theta\sin\phi & \cos\theta \\ \cos\theta\cos\phi & \cos\theta\sin\phi & -\sin\theta \\ -\sin\phi & \cos\phi & 0 \end{bmatrix} \begin{bmatrix} A_x \\ A_y \\ A_z \end{bmatrix}$$

$$A_x = 4, \quad A_y = -2, \quad A_z = -4$$

$$\vec{A} = (4\sin\theta\cos\phi - 2\sin\theta\sin\phi - 4\cos\theta)\,\vec{a}_r$$

$$+(4\cos\theta\cos\phi - 2\cos\theta\sin\phi + 4\sin\theta)\vec{a}_\theta + (-4\sin\phi - 2\cos\phi)\vec{a}_\phi\,.$$

Coordinates of P is to be converted to spherical.

$$r = \sqrt{x^2 + y^2 + z^2} = 5.38$$

$$\theta = \tan^{-1}\frac{\sqrt{x^2 + y^2}}{z} = 42.03°$$

$$\phi = \tan^{-1}\frac{y}{x} = 56.3°$$

$\vec{A}$ at $(5.38, 42.03°, 56.3°)$

$$= 2.59\vec{a}_r + 3.11\vec{a}_\theta - 4.43\vec{a}_\phi\,.$$

1.3 DIFFERENTIAL LENGTH, AREA AND VOLUME

Differential elements in length, area and volume are useful in vector calculus. They are defined in cartesian, cylindrical and spherical coordinate systems.

1.3.1 Cartesian Co-ordinates

From the Fig. 1.20, if we move from P to Q, $\vec{dl} = d_y \, \vec{a}_y$ and if we move from Q to S, $\vec{dl} = d_y \, \vec{a}_y + d_z \, \vec{a}_z$. Similarly, to move from D to Q would mean that

$$d\vec{l} = dx \, a\vec{x} + dy \, a\vec{y} + dz \, a\vec{z} \cdot \quad ...(1.43)$$

The differential surface (or area) element ds may generally be defined as

$$ds = ds \, \vec{a}_n, \quad ...(1.44)$$

where ds is the area of the element and $\vec{a}_n$ is unit vector normal to the surface. For example, for the surface $ABCD$ in Fig. 1.20, $\vec{ds} = dy \, dz \, \vec{a}_x$, whereas for $PQRS$, $\vec{ds} = -dy \, dz \, \vec{a}_x$, because $\vec{a}_n = -\vec{a}_x$ is normal to $PQRS$.

Differential volume dv can be obtained from $d\vec{l}$ as the product of three components of $d\vec{l}$.

Note that $d\vec{l}$ and $d\vec{s}$ are vectors and dv is a scalar.

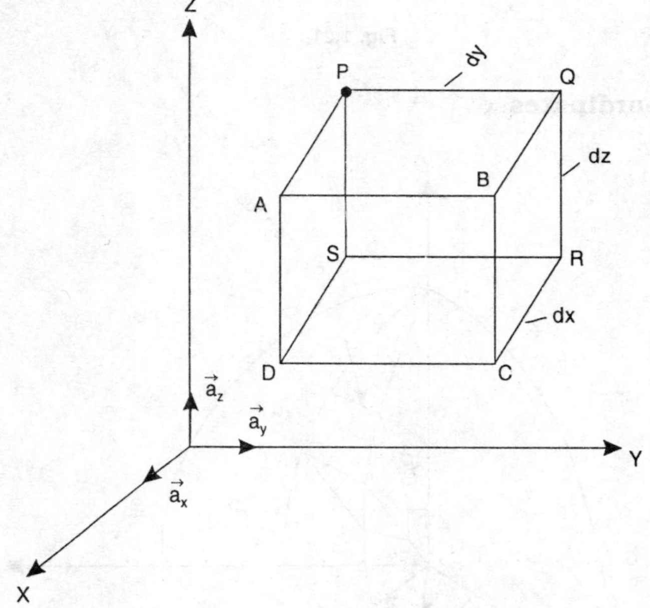

Fig. 1.20

$$d\vec{s} = dy \, dz \, \vec{a}_x$$

$$= dx \, dz \, \vec{a}_y \quad ...(1.45)$$

$$= dx \, dy \, \vec{a}_z$$

$$dv = dx \, dy \, dz. \quad ...(1.46)$$

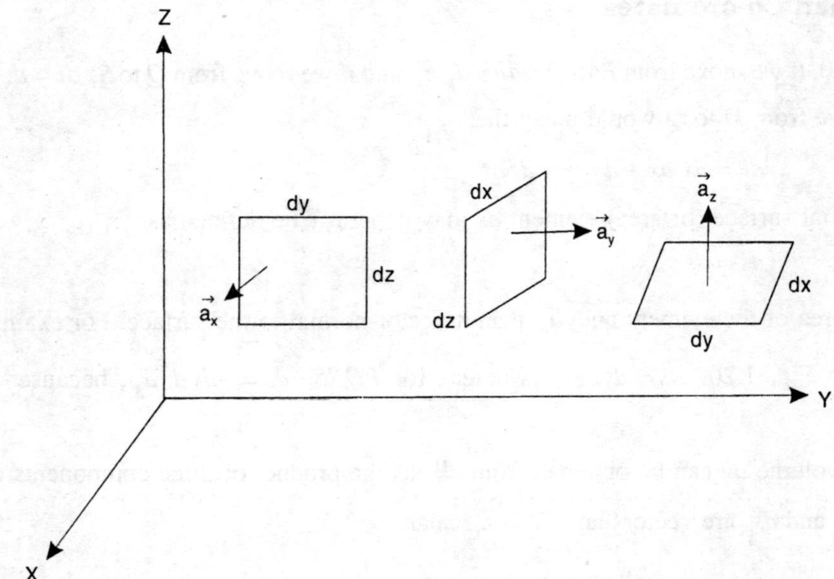

Fig. 1.21

1.3.2 Spherical Co-ordinates

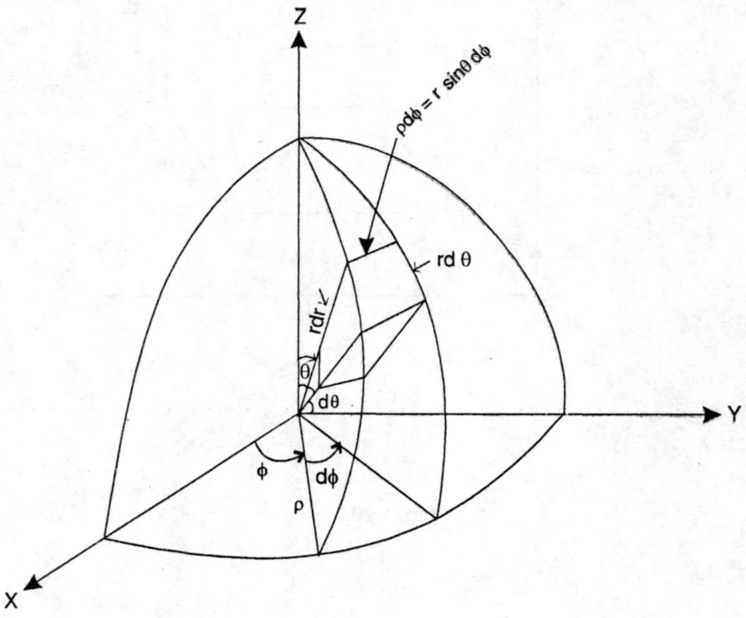

Fig. 1.22

(*i*) The differential displacement

$$d\vec{l} = dr\,\vec{a}_r + r\,d\theta\,\vec{a}_\theta + r\sin\theta\,d\phi\;\vec{a}_\phi \qquad ...(1.47)$$

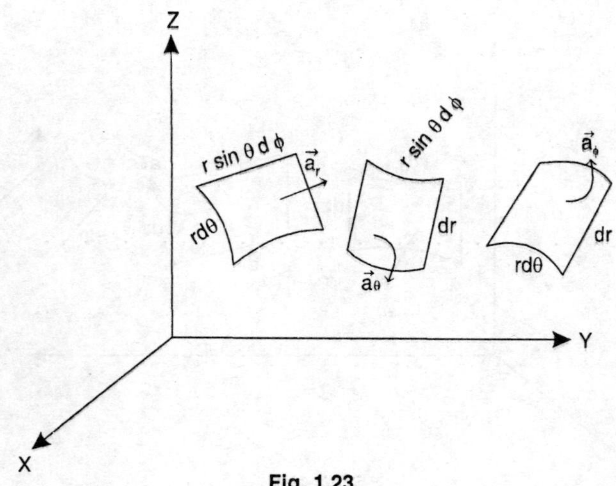

Fig. 1.23

(*ii*) The differential normal area

$$\vec{ds} = r^2 \sin\theta\, d\theta\, d\phi\, \vec{a}_r$$

$$= r \sin\theta\, dr\, d\phi\, \vec{a}_\theta \qquad ...(1.48)$$

$$= r\, dr\, d\theta\, \vec{a}_\phi .$$

(*iii*) The differential volume

$$dv = r^2 \sin\theta\, dr\, d\theta\, d\phi . \qquad ...(1.49)$$

1.3.3 Cylindrical Co-ordinates

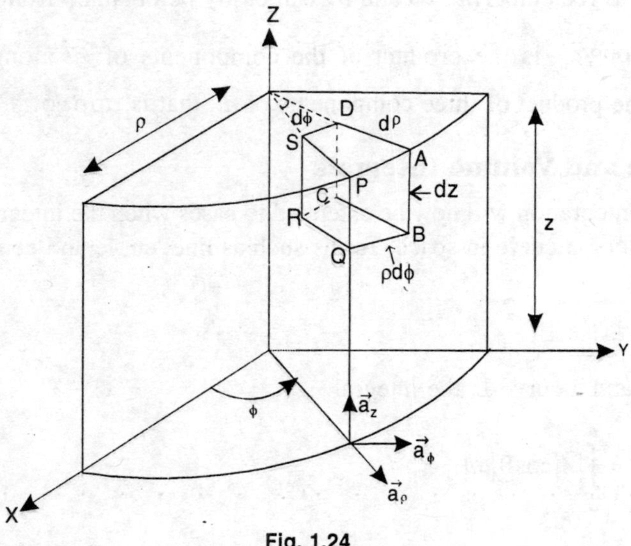

Fig. 1.24

(*i*) The differential displacement

$$\vec{dl} = d\rho\, \vec{a}_\rho + \rho d\phi\, \vec{a}_\phi + dz\, \vec{a}_z \qquad ...(1.50)$$

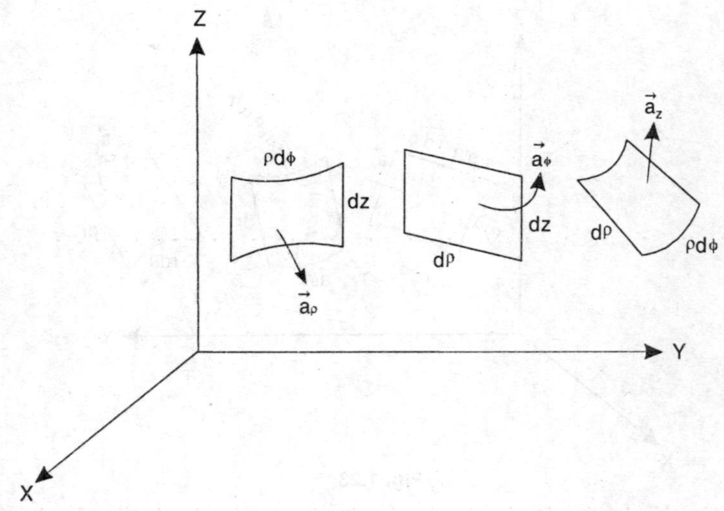

Fig. 1.25

(*ii*) The differential normal area

$$d\vec{S} = \rho d\phi\, dz\, \vec{a}_\rho$$

$$= d\rho\, dz\, \vec{a}_\phi \qquad ...(1.51)$$

$$= \rho\, d\phi\, d\rho\, \vec{a}_z\,.$$

(*iii*) The differential volume

$$dv = \rho\, d\rho\, d\phi\, dz\,. \qquad ...(1.52)$$

It is only necessary to remember $d\vec{l}$; $d\vec{s}$ and dv can easily be obtained from $d\vec{l}$.

For example, $d\vec{s}$ along $\vec{a}_z$ is the product of the components of $d\vec{l}$ along $\vec{a}_\rho$ and $\vec{a}_\phi$, that is, $d\rho\, \rho\, d\phi\, \vec{a}_z$. Also, dv is the product of three components of dl, that is, $d\rho\, \rho\, d\phi$.

1.3.4 Line, Surface and Volume Integrals

The familiar concept of integration will now be extended to cases when the integrand involves a vector. A line means the path along a curve in space. Terms such as line, curve and contour can also be used interchangeably.

(a) *Line Integral*

Given a vector field $\vec{A}$ and a curve L, the integral

$$\int_L \vec{A} \cdot d\vec{l} = \int_a^b |A| \cos\theta\, |dl| \qquad ...(1.53)$$

is defined as the line integral of $\vec{A}$ around L. It is the tangential component of A along the curve L. If the path of integration is a closed curve such as **abca** as in Fig. 1.26, then the integral becomes a closed contour integral $\oint_L \vec{A} \cdot d\vec{l}$, which is called circulation of $\vec{A}$ around L.

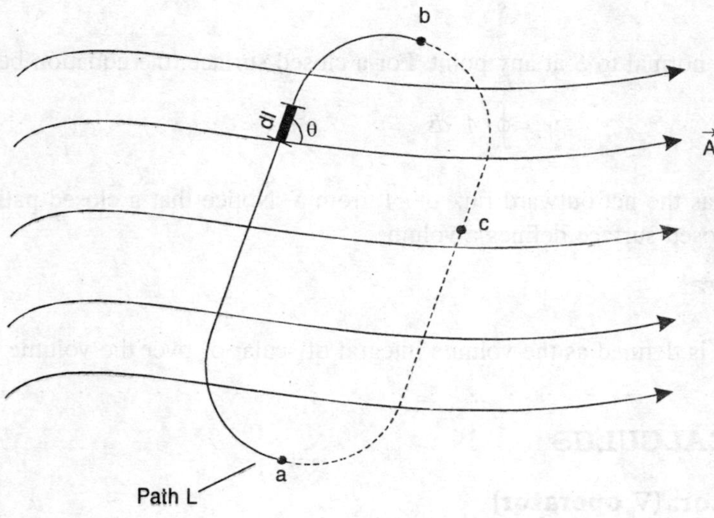

Fig. 1.26

(b) *Surface Integral:*

Consider Fig. 1.27. A vector field $\vec{A}$ continuous in a region containing the smooth surface S, is given. The surface integral or flux of A through S is defined as

$$\psi = \int_S |A| \cos\theta \, ds = \int_S \vec{A} \cdot \vec{a}_n \, ds \quad ...(1.54)$$

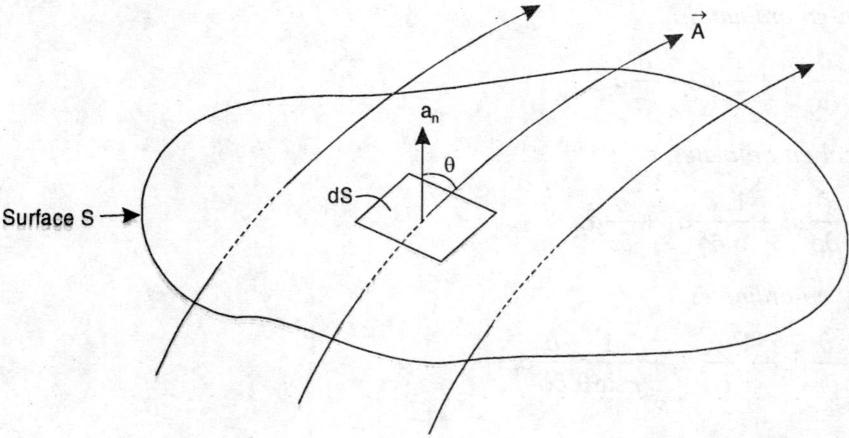

Fig. 1.27

or simply

$$\psi = \int_S \vec{A} \cdot \vec{ds},$$

where $\vec{a}_n$ is the unit normal to S at any point. For a closed surface, the equation becomes

$$\psi = \oint_s A \cdot ds$$

which is referred to as the net outward flux of $\vec{A}$ from S. Notice that a closed path defines an open surface whereas a closed surface defines a volume.

(c) *Volume Integral*

The integral $\int_V \rho_v \, dv$ is defined as the volume integral of scalar ρ_v over the volume v.

1.4 VECTOR CALCULUS

1.4.1 Del Operator: (∇ operator)

The del operator, written ∇ is the vector differential operator. This vector differential operator otherwise known as the gradient operator is not a vector in itself, but when it operates on a scalar function, it produces a vector.

The operator is useful in defining

1. The gradient of a scalar V, written as ∇V.
2. The divergence of a vector $\vec{A}$, $\nabla \cdot \vec{A}$.
3. The curl of a vector $\vec{A}$, $\nabla \times \vec{A}$.
4. The Laplacian of a scalar V, $\nabla^2 V$.

∇, the del operator is defined in various co-ordinates as below.

In cartesian co-ordinates,

$$\nabla = \frac{\partial}{\partial x}\vec{a}_x + \frac{\partial}{\partial y}\vec{a}_y + \frac{\partial}{\partial z}\vec{a}_z . \qquad ...(1.55)$$

In cylindrical co-ordinates,

$$\nabla = \frac{\partial}{\partial \rho}\vec{a}_\rho + \frac{1}{\rho}\frac{\partial}{\partial \phi}\vec{a}_\phi + \frac{\partial}{\partial z}\vec{a}_z . \qquad ...(1.56)$$

In spherical co-ordinates.

$$\nabla = \frac{\partial}{\partial r}\vec{a}_r + \frac{1}{r}\frac{\partial}{\partial \theta}\vec{a}_\theta + \frac{1}{r\sin\theta}\frac{\partial}{\partial \phi}\vec{a}_\phi . \qquad ...(1.57)$$

1.4.2 Gradient of a Scalar

The gradient of a scalar field V is a vector that represents both the magnitude and the direction of the maximum space rate of increase of V.

$$\text{grad } V = \nabla V = \frac{\partial V}{\partial x}\vec{a}_x + \frac{\partial V}{\partial y}\vec{a}_y + \frac{\partial V}{\partial z}\vec{a}_z. \quad ...(1.58)$$

For cylindrical co-ordinates

$$\nabla V = \frac{\partial V}{\partial \rho}\vec{a}_\rho + \frac{1}{\rho}\frac{\partial V}{\partial \phi}\vec{a}_\phi + \frac{\partial V}{\partial z}\vec{a}_z. \quad ...(1.59)$$

For spherical co-ordinates

$$\nabla V = \frac{\partial V}{\partial r}\vec{a}_r + \frac{1}{r}\frac{\partial V}{\partial \theta}\vec{a}_\theta + \frac{1}{r\sin\theta}\frac{\partial V}{\partial \phi}\vec{a}_\phi. \quad ...(1.60)$$

Properties of Gradient of a Scalar Field

1. ∇V points in the direction of the maximum rate of change in V.
2. The projection (or component) of ∇V in the direction of a unit vector $\vec{a}$ is $\nabla V \cdot \vec{a}$, and is called the directional derivative of V along $\vec{a}$. This is the rate of change of V in the direction of $\vec{a}$.
3. If $\vec{A} = \nabla V$, V is said to be the scalar potential of $\vec{A}$.

1.4.3 Divergence of a Vector

Divergence of a vector field is defined as the net outflow of flux per unit volume over a closed incremental surface.

In other words, the divergence of A at a given point P is the outward flux per unit volume as the volume shrinks about P. Hence,

$$\text{div} A = \nabla \cdot A = \underset{\Delta v \to 0}{Lt} \frac{\oint_s A \cdot ds}{\Delta v}, \quad ...(1.61)$$

where Δv is the volume enclosed by the closed surface S in which P is located.

Physically, the divergence of a vector quantity represents the rate of change of the field strength in the direction of the field.

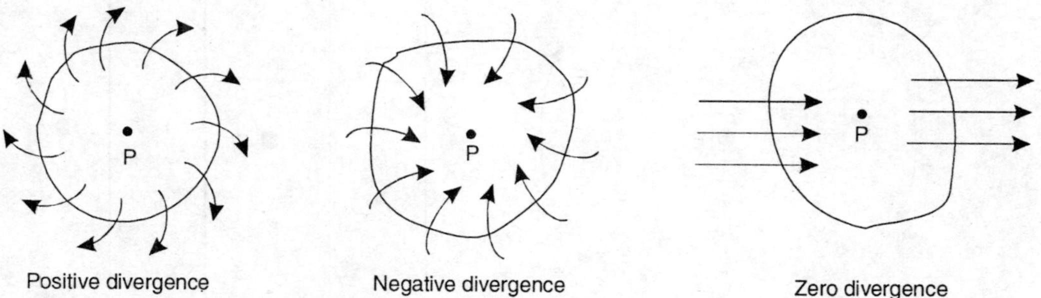

Positive divergence Negative divergence Zero divergence

Fig. 1.28

If the divergence of a vector field is positive at a point 'P', then something is diverging from a small volume surrounding that point and that point is acting as a source. If it is negative, then something is converging into a small volume surrounding that point and that point is acting as a sink. If the divergence is zero at a point '*P*', then the rate at which something entering a small volume surrounding that point is equal to the rate at which it is leaving that volume.

The divergence of a vector $\vec{A}$ can be represented as $\nabla \cdot \vec{A}$ (or) div$\vec{A}$.

In cartesian system, $\nabla \cdot \vec{A}$ is represented as

$$\nabla \cdot \vec{A} = \frac{\partial A_x}{\partial y} + \frac{\partial A_y}{\partial y} + \frac{\partial A_z}{\partial z}. \quad \text{...(1.62)}$$

Properties of Divergence of a Vector Field

1. It produces a scalar field.
2. The divergence of a scalar *V*, div *V* makes no sense.
3. $\nabla \cdot (\vec{A} + \vec{B}) = \nabla \cdot \vec{A} + \nabla \cdot \vec{B}$ ...(1.63)
4. The vector field whose divergence is zero is called solenoidal field.

1.4.4 Curl of a Vector

Curl of a vector field $\vec{A}$ is defined as an axial (or rotational) vector whose magnitude is the maximum circulation of $\vec{A}$ per unit area as the area tends to zero and the direction is the normal direction of the area.

$$\text{Curl } \vec{A} = \nabla \times \vec{A} = \left(\underset{\Delta S \to 0}{Lt} \frac{\oint_L \vec{A} \cdot dl}{\Delta S} \right) \vec{a}_{n \text{ max.}}, \quad \text{...(1.64)}$$

where the area ΔS is bounded by the curve *L* and $\vec{a}_n$ is the unit vector normal to the surface ΔS and is determined using the right-hand rule.

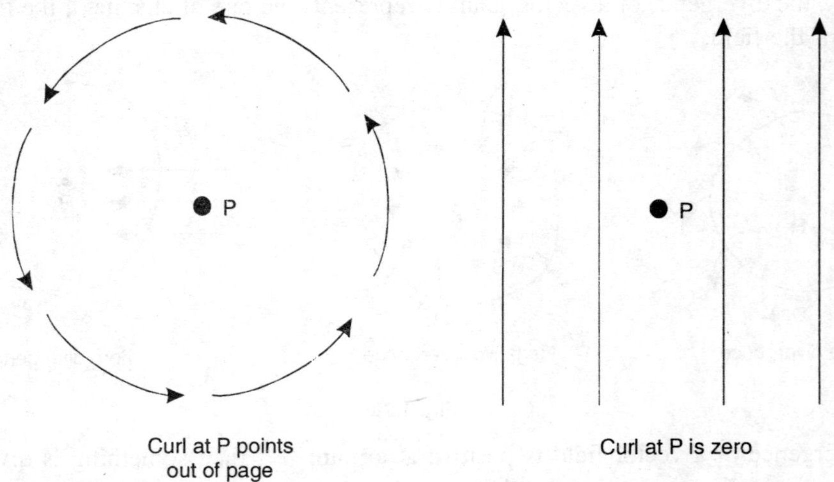

Fig. 1.29

Physically, the curl of a vector field represents the rate of change of field strength in a direction at right angles to the field and is a measure of rotation of something in a small volume surrounding a particular point. For streamline motions, the curl is zero, while it is maximum near eddies and whirl-pools.

The vector fields whose curl is zero are called irrotational. Irrotational fields are also called as conservative fields.

In cartesian system, the curl of vector field can be found using

$$\nabla \times \vec{A} = \begin{vmatrix} \vec{a}_x & \vec{a}_y & \vec{a}_z \\ \partial/\partial x & \partial/\partial y & \partial/\partial z \\ A_x & A_y & A_z \end{vmatrix}. \qquad ...(1.65)$$

Properties of Curl

1. The curl of a vector field is another vector field.
2. The curl of a scalar field V, $\nabla \times V$ makes no sense.
3. $\nabla \times \left(\vec{A} + \vec{B}\right) = \nabla \times \vec{A} + \nabla \times \vec{B}$ ...(1.66)
4. The divergence of the curl of a vector field vanishes, that is,

   $\nabla \cdot \left(\nabla \times \vec{A}\right) = 0$ ...(1.67)
5. The curl of the gradient of a scalar field vanishes, that is,

   $\nabla \times \nabla V = 0$ ...(1.68)
6. If $\nabla \times \vec{A} = 0$, $\vec{A}$ is irrotational.

1.4.5 Laplacian of a Scalar

The Laplacian of a scalar field V can be written as $\nabla^2 V$. It is defined as the divergence of the gradient of V.

In cartesian coordinates,

Laplacian $V = \nabla \cdot \nabla V = \nabla^2 V$

$$\nabla^2 V = \frac{\partial^2 V}{\partial x^2} + \frac{\partial^2 V}{\partial y^2} + \frac{\partial^2 V}{\partial z^2}. \qquad ...(1.69)$$

Properties

1. Laplacian of a scalar field is another scalar field.
2. If Laplacian of a scalar field is zero in a given region, then the scalar field is said to be harmonic in that region.

Example 1.11

Prove that div curl $\vec{F} = 0$ where $\vec{F} = F_x \vec{a}_x + F_y \vec{a}_y + F_z \vec{a}_z$

$\text{div curl}\, \vec{F} = \nabla \cdot \left(\nabla \times \vec{F}\right)$

$$\text{Curl}\,\vec{F} = \nabla \times \vec{F} = \begin{vmatrix} \vec{a}_x & \vec{a}_y & \vec{a}_z \\ \partial/\partial x & \partial/\partial y & \partial/\partial z \\ F_x & F_y & F_z \end{vmatrix}$$

$$= \left(\frac{\partial F_z}{\partial y} - \frac{\partial F_y}{\partial z}\right)\vec{a}_x - \vec{a}_y\left(\frac{\partial F_z}{\partial x} - \frac{\partial F_x}{\partial z}\right) + \left(\frac{\partial F_y}{\partial x} - \frac{\partial F_x}{\partial y}\right)\vec{a}_z$$

$$\text{div}\left(\text{curl}\,\vec{F}\right) = \nabla\cdot\left(\nabla\times\vec{F}\right) = \left(\frac{\partial}{\partial x}\vec{a}_x + \frac{\partial}{\partial y}\vec{a}_y + \frac{\partial}{\partial z}\vec{a}_z\right)\cdot$$

$$\left\{\left(\frac{\partial F_z}{\partial y} - \frac{\partial F_y}{\partial z}\right)\vec{a}_x - \left(\frac{\partial F_z}{\partial x} - \frac{\partial F_x}{\partial z}\right)\vec{a}_y + \left(\frac{\partial F_y}{\partial x} - \frac{\partial F_x}{\partial y}\right)\vec{a}_z\right\}$$

$$= \frac{\partial^2 F_z}{\partial x\,\partial y} - \frac{\partial F_y}{\partial x\,\partial z} - \frac{\partial^2 F_z}{\partial x\,\partial y} + \frac{\partial^2 F_x}{\partial y\,\partial z} + \frac{\partial^2 F_y}{\partial x\,\partial z} - \frac{\partial^2 F_x}{\partial y\,\partial z}$$

$$= 0$$

Thus proved.

Example 1.12

Prove that curl grad $\phi = 0$, where ϕ is a scalar field.

Curl grad $\phi = \nabla \times \nabla\phi$

$$\nabla\phi = \frac{\partial\phi}{\partial x}\vec{a}_x + \frac{\partial\phi}{\partial y}\vec{a}_y + \frac{\partial\phi}{\partial z}\vec{a}_z$$

$$\nabla\times\nabla\phi = \begin{vmatrix} \vec{a}_x & \vec{a}_y & \vec{a}_z \\ \partial/\partial x & \partial/\partial y & \partial/\partial z \\ \partial\phi/\partial x & \partial\phi/\partial y & \partial\phi/\partial z \end{vmatrix}$$

$$= \left(\frac{\partial^2\phi}{\partial y\,\partial z} - \frac{\partial^2\phi}{\partial y\,\partial z}\right)\vec{a}_x - \vec{a}_y\left(\frac{\partial^2\phi}{\partial x\,\partial z} - \frac{\partial^2\phi}{\partial x\,\partial z}\right) + \left(\frac{\partial^2\phi}{\partial x\,\partial y} - \frac{\partial^2\phi}{\partial x\,\partial y}\right)\vec{a}_z$$

$$= 0.$$

Thus proved.

Example 1.13

If a scalar potential is given by the expression $\phi = xyz$, determine the potential gradient and also prove that the vector is irrotational.

Scalar potential $\phi = xyz$.

Potential gradient $= \nabla\phi$

$$\nabla\phi = \frac{\partial\phi}{\partial x}\vec{a}_x + \frac{\partial\phi}{\partial y}\vec{a}_y + \frac{\partial\phi}{\partial y}\vec{a}_z$$

$$\frac{\partial\phi}{\partial x} = yz; \ \frac{\partial\phi}{\partial y} = xz; \ \frac{\partial\phi}{\partial z} = xy$$

$$\nabla\phi = yz\,\vec{a}_x + xz\,\vec{a}_y + xy\,\vec{a}_z\,.$$

Then, $\vec{F} = \nabla\phi\,.$

If $\nabla\times\vec{F}$ is zero, then $\vec{F}$ is said to be irrotational.

$$\nabla\times\vec{F} = \begin{vmatrix} \vec{a}_x & \vec{a}_y & \vec{a}_z \\ \partial/\partial x & \partial/\partial y & \partial/\partial z \\ yz & xz & xy \end{vmatrix}$$

$$= \left(\frac{\partial}{\partial y}(xy) - \frac{\partial}{\partial z}(xz)\right)\vec{a}_x - \left(\frac{\partial}{\partial x}(xy) - \frac{\partial}{\partial z}(yz)\right)\vec{a}_y + \left(\frac{\partial}{\partial x}(xz) - \frac{\partial}{\partial y}(yz)\right)\vec{a}_z$$

$$= 0.$$

Thus, $\vec{F}$ is an irrotational field.

1.4.6 Divergence Theorem

Divergence theorem states that the integral of the normal component of a vector function over a closed surface is equal to the integral of the divergence of that vector throughout the volume 'v' enclosed by the surface 'S'.

$$\oint_S \vec{D}\cdot\vec{ds} = \int_v \nabla\cdot\vec{D}\,dv\,. \qquad ...(1.70)$$

Proof:

Consider a small volume $\Delta x\,\Delta y\,\Delta z = \Delta v$ placed in an electric field with flux density $\vec{D}$, having components D_x, D_y and D_z in the three co-ordinate directions as shown in Fig. 1.30.

The total flux density is $\vec{D}$ related to its components by $\vec{D} = D_x\,\vec{a}_x + D_y\,\vec{a}_y + D_z\,\vec{a}_z$. The normal outward component of $\vec{D}$ at the backface is $-D_x$, since the field is directed outwards.

If the field changes between the back and front faces, the normal component at the front face can be represented using Taylor's theorem (neglecting higher-order terms) as

$$D_x + \frac{\partial D_x}{\partial x}\Delta x. \qquad ...(1.71)$$

Similarly, the normal component of $\vec{D}$ at the left-side face is $-D_y$ and at the right-side face is

$$D_y + \frac{\partial D_y}{\partial y}\Delta y\,. \qquad ...(1.72)$$

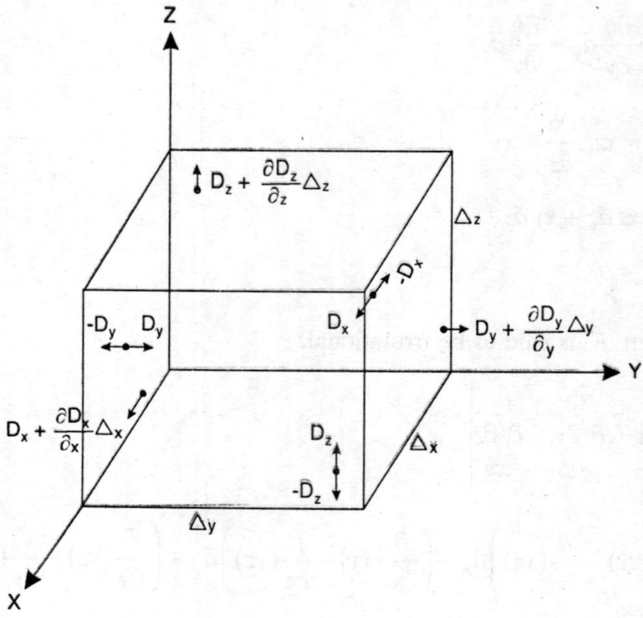

Fig. 1.30

In the same manner, at the bottom face it is $-D_z$ and at the top face it is

$$D_z + \frac{\partial D_z}{\partial z}\Delta z. \quad \text{...(1.73)}$$

The outward flux of $\vec{D}$ over the back face is $-D_x\,\Delta y\,\Delta z$, and over the front face is $\left(D_x + \frac{\partial D_x}{\partial x}\Delta x\right)\Delta y\,\Delta z$.

Adding up the outward flux of $\vec{D}$ over the entire volume, the total flux becomes

$$\Delta\psi = \left(-D_x + D_x + \frac{\partial D_x}{\partial x}\Delta x\right)\Delta y\,\Delta z + \left(-D_y + D_y + \frac{\partial D_y}{\partial y}\Delta y\right)\Delta x\,\Delta z$$

$$+\left(-D_z + D_z + \frac{\partial D_z}{\partial z}\Delta z\right)\Delta x\,\Delta y. \quad \text{...(1.74)}$$

On simplifying,

$$\Delta\psi = \left(\frac{\partial D_x}{\partial x} + \frac{\partial D_y}{\partial y} + \frac{\partial D_z}{\partial z}\right)\Delta x\,\Delta y\,\Delta z. \quad \text{...(1.75)}$$

This total electric flux over the surface of the volume is also equal to the normal component of $\vec{D}$ over the surface of the volume.

That is, $\Delta\psi = \oint D_n \cdot ds$

$$= \left(\frac{\partial D_x}{\partial x} + \frac{\partial D_y}{\partial y} + \frac{\partial D_z}{\partial z}\right)\Delta x\,\Delta y\,\Delta z\,. \quad ...(1.76)$$

Now, RHS of the Divergence Theorem can be expanded in cartesian co-ordinates as

$$\nabla \cdot \vec{D} = \left(\vec{a}_x \frac{\partial}{\partial x} + \vec{a}_y \frac{\partial}{\partial y} + \vec{a}_z \frac{\partial}{\partial z}\right) \cdot \left(D_x\,\vec{a}_x + D_y\,\vec{a}_y + D_z\,\vec{a}_z\right)$$

$$= \frac{\partial D_x}{\partial x} + \frac{\partial D_y}{\partial y} + \frac{\partial D_z}{\partial z}\,. \quad ...(1.77)$$

So,

$$\oint_V \nabla \cdot \vec{D}\,dv = \left(\frac{\partial D_x}{\partial x} + \frac{\partial D_y}{\partial y} + \frac{\partial D_z}{\partial z}\right)\Delta x\,\Delta y\,\Delta z. \quad ...(1.78)$$

Hence, from eqns. (1.76) and (1.78), $\oint_S \vec{D} \cdot ds = \oint_V \nabla \cdot \vec{D}\,dv.$

Example 1.14

Determine the divergence of these vector fields.

$$\vec{P} = x^2 yz\,\vec{a}_x + xz\,\vec{a}_z$$

$$\vec{Q} = \rho \sin\phi\,\vec{a}_\rho + \rho^2 z\,\vec{a}_\phi + z\cos\phi\,\vec{a}_z$$

$$\vec{T} = \frac{1}{r^2}\cos\theta\,\vec{a}_r + r\sin\theta\cos\phi\,\vec{a}_\theta + \cos\theta\,\vec{a}_\phi$$

(*i*) $\nabla \cdot \vec{P} = \frac{\partial}{\partial x}(P_x) + \frac{\partial}{\partial y}(P_y) + \frac{\partial}{\partial z}(P_z)$

$= 2xyz + 0 + x$

$= 2xyz + x.$

(*ii*) $\nabla \cdot \vec{Q} = \frac{1}{\rho}\frac{\partial}{\partial \rho}(\rho Q_\rho) + \frac{1}{\rho}\frac{\partial}{\partial \phi}(Q_\phi) + \frac{\partial}{\partial z}(Q_z)$

$$= \frac{1}{\rho}\frac{\partial}{\partial \rho}(\rho^2 \sin\phi) + \frac{1}{\rho}\frac{\partial}{\partial \phi}(\rho^2 z) + \frac{\partial}{\partial z}(z\cos\phi)$$

$$= 2\sin\phi + 0 + \cos\phi$$

$$= 2\sin\phi + \cos\phi$$

(iii) $$\nabla\cdot\vec{T}=\frac{1}{r^2}\frac{\partial}{\partial r}\left(r^2T_r\right)+\frac{1}{r\sin\theta}\frac{\partial}{\partial\theta}(T_\theta\sin\theta)+\frac{1}{r\sin\theta}\frac{\partial}{\partial\phi}\left(T_\phi\right)$$

$$=\frac{1}{r^2}\frac{\partial}{\partial r}(\cos\theta)+\frac{1}{r\sin\theta}\frac{\partial}{\partial\theta}\left(r\sin^2\theta\cos\phi\right)+\frac{1}{r\sin\theta}\frac{\partial}{\partial\phi}(\cos\theta)$$

$$=0+2\cos\theta\cos\phi+0$$

$$=2\cos\theta\cos\phi.$$

Example 1.15

Determine curl of the vector fields in the previous example.

(i) $$\nabla\times\vec{P}=\begin{vmatrix}\vec{a}_x & \vec{a}_y & \vec{a}_z\\ \partial/\partial x & \partial/\partial y & \partial/\partial z\\ A_x & A_y & A_z\end{vmatrix}$$

$$=\begin{vmatrix}\vec{a}_x & \vec{a}_y & \vec{a}_z\\ \partial/\partial x & \partial/\partial y & \partial/\partial z\\ x^2yz & 0 & xz\end{vmatrix}$$

$$=(0)\vec{a}_x-\vec{a}_y\left(z-x^2y\right)+\vec{a}_z\left(x^2z\right)$$

$$=\left(x^2y-z\right)\vec{a}_y+x^2z\vec{a}_z.$$

(ii) $$\nabla\times\vec{Q}=\frac{1}{\rho}\begin{vmatrix}\vec{a}_\rho & \rho\vec{a}_\phi & \vec{a}_z\\ \partial/\partial\rho & \partial/\partial\phi & \partial/\partial z\\ A_\rho & \rho A_\phi & A_z\end{vmatrix}$$

$$\nabla\times\vec{Q}=\frac{1}{\rho}\begin{vmatrix}\vec{a}_\rho & \rho\vec{a}_\phi & \vec{a}_z\\ \partial/\partial\rho & \partial/\partial\phi & \partial/\partial z\\ \rho\sin\phi & \rho^3z & z\cos\phi\end{vmatrix}$$

$$=\frac{1}{\rho}\left\{-\left(z\sin\phi+\rho^3\right)\vec{a}_\rho+\left(3\rho^2z-\rho\cos\phi\right)\vec{a}_z\right\}$$

$$=-\frac{1}{\rho}(z\sin\phi)\vec{a}_\rho+(3\rho z-\cos\phi)\vec{a}_z.$$

$$\nabla\times T=\frac{1}{r^2\sin\theta}\begin{vmatrix}\vec{a}_r & r\vec{a}_\theta & r\sin\theta\vec{a}_\phi\\ \partial/\partial r & \partial/\partial\theta & \partial/\partial\phi\\ A_r & rA_\theta & r\sin\theta A_\phi\end{vmatrix}$$

$$= \frac{1}{r^2 \sin\theta} \begin{vmatrix} \vec{a}_r & r\vec{a}_\theta & r\sin\theta\,\vec{a}_\phi \\ \partial/\partial r & \partial/\partial\theta & \partial/\partial\phi \\ \frac{1}{r^2}\cos\theta & r^2\sin\theta\cos\phi & r\sin\theta\cos\theta \end{vmatrix}$$

$$= \frac{1}{r^2 \sin\theta}\left[\left\{r\cos 2\theta + r^2 \sin\theta\sin\phi\right\}\vec{a}_r - (\sin\theta\cos\theta)\, r\,\vec{a}_\theta \right.$$

$$\left. + \left(2r\sin\theta\cos\phi + \frac{1}{r^2}\sin\theta\right) r\sin\theta\,\vec{a}_\phi\right]$$

$$= \left(\frac{\cos 2\theta}{r\sin\theta} + \sin\phi\right)\vec{a}_r - \frac{\cos\theta}{r}\vec{a}_\theta + \left(2\cos\phi\sin\theta + \frac{1}{r^3}\sin\theta\right)\vec{a}_\phi .$$

Example 1.16

Given that $\vec{D} = \left(\frac{10\rho^3}{4}\right)\vec{a}_\rho$ in cylindrical co-ordinates, evaluate both sides of divergence theorem for the volume enclosed by $\rho = 2$, $z = 0$ and $z = 10$.

Divergence theorem

$$\oint_S \vec{D}\cdot\vec{ds} = \int_v \nabla\cdot\vec{D}\,dv.$$

As $\vec{D}$ has no Z component, $\vec{D}\cdot\vec{ds}$ is zero for the top and bottom circular surfaces.

So, $\oint_S \vec{D}\cdot\vec{ds} = \oint_S D\vec{a}_\rho \cdot \rho\, d\phi\, dz\,(\vec{a}_\rho).$

On the inner cylindrical surface, $\vec{ds}$ is in the direction $(-\vec{a}_\rho)$. Here, consider only the outer surface which is in $\vec{a}_\rho$ direction.

$$\oint_S \vec{D}\cdot\vec{ds} = \int_{z=0}^{10}\int_{\phi=0}^{2\pi} \frac{10\rho^3}{4}\vec{a}_\rho \cdot \rho\, d\phi\, dz\ \vec{a}_\rho.$$

Put $\rho = 2$

$$= \int_{z=0}^{10}\int_{\phi=0}^{2\pi} \frac{10\times 2^4}{4}\, d\phi\, dz$$

$$= 40 \times 2\pi \times 10$$

$$= 800\pi.$$

$$\int_v \nabla \cdot \vec{D} dv = \int_{z=0}^{10} \int_{\phi=0}^{2\pi} \int_{\rho=0}^{2} \nabla \cdot \vec{D} \rho d\rho d\phi dz.$$

$$\nabla \cdot \vec{D} = \frac{1}{\rho} \frac{\partial}{\partial \rho} (\rho D_\rho)$$

$$= \frac{1}{\rho} \frac{\partial}{\partial \rho} \left(\frac{10\rho^4}{4} \right)$$

$$= \frac{1}{\rho} (10\rho^3) = 10\rho^2.$$

$$\int_v \nabla \cdot \vec{D} dv = \int_{z=0}^{10} \int_{\phi=0}^{2\pi} \int_{\rho=0}^{2} 10\rho^3 \, d\rho d\phi dz$$

$$= \int_{z=0}^{10} \int_{\phi=0}^{2\pi} \left(\frac{10\rho^4}{4} \right)_0^2 d\phi dz$$

$$= 40 \times 2\pi \times 10$$

$$= 800\pi.$$

Thus, Divergence Theorem is verified.

Example 1.17

Given that $\vec{D} = \left(\frac{5r^2}{4} \right) \vec{a}_r$ in spherical co-ordinates. Verify divergence for the volume enclosed between $r = 1$ and $r = 2$.

$$\oint_S \vec{D} \cdot \vec{ds} = \int_v \nabla \cdot \vec{D} dv.$$

As $\vec{D}$ has only radial component, $\vec{D} \cdot \vec{ds}$ has a non-zero value only on the surfaces of $r = 1$ and $r = 2$. Further on the inner surface, $\vec{ds}$ is in the direction of $-\vec{a}_r$.

$$\oint_S \vec{D} \cdot \vec{ds} = \int_{\phi=0}^{2\pi} \int_{\theta=0}^{\pi} D\vec{a}_r \cdot r^2 \sin\theta \, d\theta \, d\phi \, \vec{a}_r$$

$$= \int_{\phi=0}^{2\pi} \int_{\theta=0}^{\pi} \frac{5(2)^2}{4} \vec{a}_r \cdot (2)^2 \sin\theta \, d\theta \, d\phi \, (\vec{a}_r) - \int_{\phi=0}^{2\pi} \int_{\theta=0}^{\pi} \frac{5(1)^2}{4} \vec{a}_r \cdot (1)^2 \sin\theta \, d\theta \, d\phi \, (\vec{a}_r)$$

$$= \int_{\phi=0}^{2\pi} \int_{\theta=0}^{\pi} (20) \sin\theta \, d\theta \, d\phi - \int_{\phi=0}^{2\pi} \int_{\theta=0}^{\pi} \frac{5}{4} \sin\theta \, d\theta \, d\phi$$

$$= -\frac{5}{4}(-\cos\theta)_0^{\pi} \times 2\pi + 20(-\cos\theta)_0^{\pi} \times 2\pi$$

$$= -5\pi + 80\pi$$

$$= 75\pi.$$

$$\int_v \nabla \cdot \vec{D} \, dv = \int_{\phi=0}^{2\pi} \int_{\theta=0}^{\pi} \int_{r=1}^{2} \nabla \cdot \vec{D} r^2 \sin\theta \, dr \, d\theta \, d\phi$$

$$\nabla \cdot \vec{D} = \frac{1}{r^2} \frac{\partial}{\partial r}\left(r^2 D_r\right)$$

$$= \frac{1}{r^2} \frac{\partial}{\partial r}\left(\frac{5r^4}{4}\right)$$

$$= \frac{1}{r^2} 5r^3$$

$$\nabla \cdot \vec{D} = 5r.$$

$$\int_v \left(\nabla \cdot \vec{D}\right) dv = \int_{\phi=0}^{2\pi} \int_{\theta=0}^{\pi} \int_{r=1}^{2} 5r \cdot r^2 \sin\theta \, dr \, d\theta \, d\phi$$

$$= \int_{\phi=0}^{2\pi} \int_{\theta=0}^{\pi} \int_{r=1}^{2} 5r^3 \sin\theta \, dr \, d\theta \, d\phi$$

$$= \int_{\phi=0}^{2\pi} \int_{\theta=0}^{\pi} \left[\frac{5r^4}{4}\right]_1^2 \sin\theta \, d\theta \, d\phi$$

$$= \left(20 - \frac{5}{4}\right) \int_{\phi=0}^{2\pi} \int_{\theta=0}^{\pi} \sin\theta \, d\theta \, d\phi$$

$$= \frac{75}{4} \times 2 \times 2\pi$$

$= 75\pi$. Hence, verified.

1.4.7 Stokes Theorem

The line integral of the tangential component of a vector $\vec{A}$ around a closed path L is equal to the surface integral of the normal component of curl $\vec{A}$ over the surface enclosed by the path L.

$$\oint_L \vec{A} \cdot d\vec{l} = \iint_S \left(\text{curl}\,\vec{A}\right) \cdot \vec{n}\, ds, \qquad ...(1.79)$$

where L is the closed path which bounds the surface S. The positive direction of $\overrightarrow{ds} = \vec{n}\, ds$ is related to the positive sense of describing the contour L in accordance with the right hand rule.

Proof:

Consider a arbitrary surface S illustrated in Fig. 1.31. Let us subdivide the surface S into a set of small approximately rectangular area.

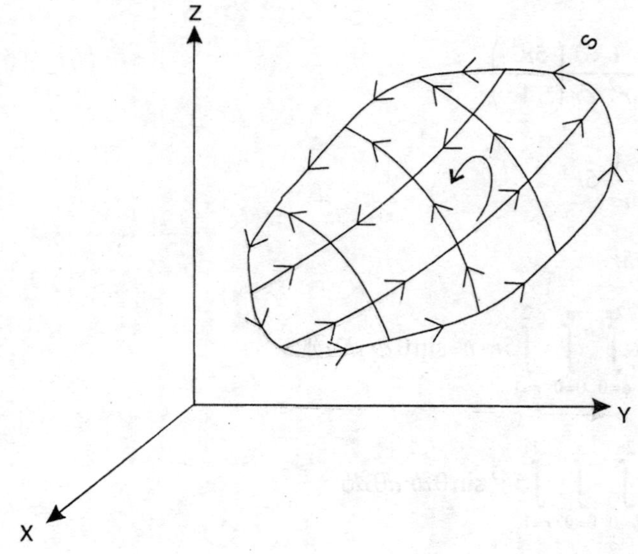

Fig. 1.31

Let us consider a separate rectangular area as in Fig. 1.32.

If $\vec{A} = A_x\,\vec{a}_x + A_y\,\vec{a}_y + A_z\,\vec{a}_z$ at the corner of ΔS closest to the origin (point 1), then

$$\oint_L \vec{A} \cdot d\vec{l} = \int_1^2 + \int_2^3 + \int_3^4 + \int_4^1$$

$$= A_y \Delta y + \left(A_z + \frac{\partial A_z}{\partial y} \Delta y \right) \Delta z + \left(A_y + \frac{\partial A_y}{\partial z} \Delta z \right) (-\Delta y) + A_z (-\Delta z) \quad ...(1.80)$$

$$= \left(\frac{\partial A_z}{\partial y} - \frac{\partial A_y}{\partial z} \right) \Delta y \, \Delta z \, . \quad ...(1.81)$$

The Y and Z components can be determined in a similar fashion as

$$\left(\frac{\partial A_x}{\partial z} - \frac{\partial A_z}{\partial x} \right) \Delta z \, \Delta x \quad ...(1.82)$$

and
$$\left(\frac{\partial A_y}{\partial x} - \frac{\partial A_x}{\partial y} \right) \cdot \Delta x \, \Delta y \, . \quad ...(1.83)$$

$$\therefore \quad \oint_L \vec{A} \cdot d\vec{l} = \left(\frac{\partial A_z}{\partial y} - \frac{\partial A_y}{\partial z} \right) \Delta y \, \Delta z + \left(\frac{\partial A_x}{\partial z} - \frac{\partial A_z}{\partial x} \right) \Delta z \, \Delta x + \left(\frac{\partial A_y}{\partial x} - \frac{\partial A_x}{\partial y} \right) \Delta x \, \Delta y \, . \quad ...(1.84)$$

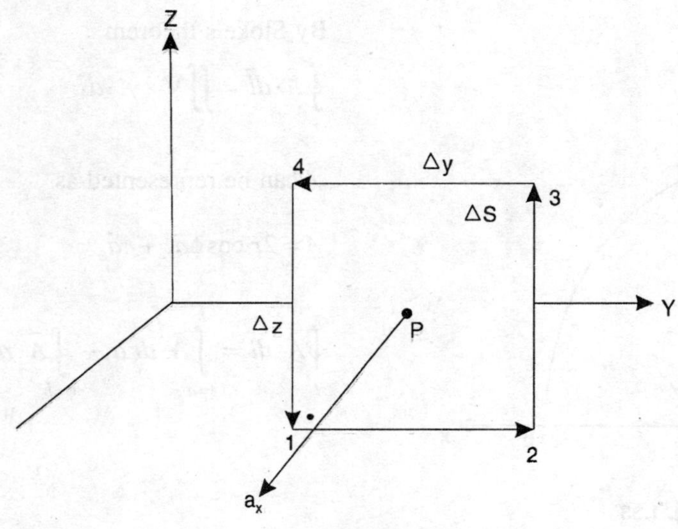

Fig. 1.32

Curl $\vec{A}$ in cartesian co-ordinate can be given by

$$\nabla \times \vec{A} = \begin{vmatrix} \vec{a}_x & \vec{a}_y & \vec{a}_z \\ \partial/\partial x & \partial/\partial y & \partial/\partial z \\ A_x & A_y & A_z \end{vmatrix} \quad ...(1.85)$$

$$= \left(\frac{\partial A_z}{\partial y} - \frac{\partial A_y}{\partial z} \right) \vec{a}_x + \left(\frac{\partial A_x}{\partial z} - \frac{\partial A_z}{\partial x} \right) \vec{a}_y + \left(\frac{\partial A_y}{\partial x} - \frac{\partial A_x}{\partial y} \right) \vec{a}_z. \quad ...(1.86)$$

The differential surface area in cartesian co-ordinate is described as

$$\vec{ds} = \Delta x \Delta y \vec{a}_z + \Delta y \Delta z \vec{a}_x + \Delta z \Delta x \vec{a}_y. \quad ...(1.87)$$

$$\therefore \quad \iint (\nabla \times \vec{A}) \cdot \vec{n}\, ds = \left(\frac{\partial A_z}{\partial y} - \frac{\partial A_y}{\partial z} \right) \Delta y \Delta z + \left(\frac{\partial A_z}{\partial x} - \frac{\partial A_x}{\partial z} \right) \Delta z \Delta x + \left(\frac{\partial A_y}{\partial x} - \frac{\partial A_x}{\partial y} \right) \Delta x \Delta y. \quad ...(1.88)$$

Comparing eqn. (1.84) and eqn. (1.88)

$$\oint_L \vec{A} \cdot d\vec{l} = \iint (\nabla \cdot \vec{A}) \cdot \vec{n}\, ds. \quad ...(1.89)$$

Example 1.18

Given $\vec{A} = 2r \cos\phi\ \vec{I}_r + r\vec{J}_\phi$ in cylindrical co-ordinates. For the contour shown in Fig. 1.33, verify Stoke's theorem. **(Anna University, Oct. 2002)**

Solution:

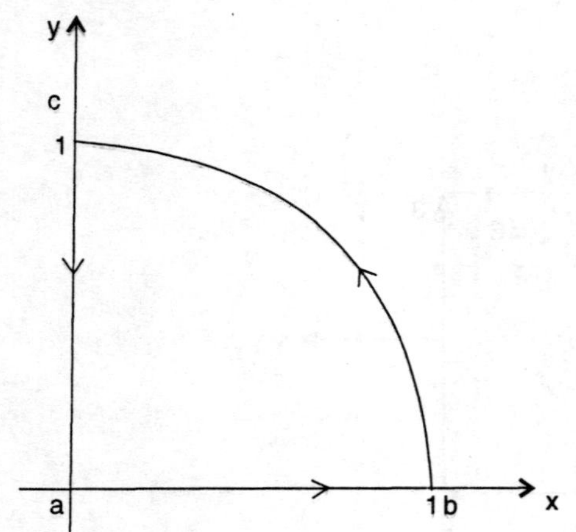

Fig. 1.33

By Stoke's theorem

$$\oint_r \vec{A} \cdot d\vec{l} = \iint_S \nabla \times \vec{A} \cdot d\vec{s}$$

$\vec{A}$ can be represented as

$$\vec{A} = 2r \cos\phi \vec{a}_r + r\vec{a}_\phi$$

$$\oint_l \vec{A} \cdot d\vec{l} = \underbrace{\int_{l=a}^{b} \vec{A} \cdot dr\, \vec{a}_r}_{\text{I}} + \underbrace{\int_{\phi=b}^{c} \vec{A} \cdot r d\phi\, \vec{a}_\phi}_{\text{II}} + \underbrace{\int_{l=c}^{a} \vec{A} \cdot dr\, \vec{a}_r}_{\text{III}}$$

(I) $$\int_{r=0}^{1} (2r\cos\phi\, \vec{a}_r + r\vec{a}_\phi) \cdot dr\, \vec{a}_r = \int_{r=0}^{1} 2r\cos\phi\, dr = r^2 \cos\phi \Big)_0^1 = \cos\phi \quad [\vec{a}_\phi \cdot \vec{a}_r = 0]$$

(II) $$\int_{\phi=0}^{\pi/2} (2r\cos\phi\, \vec{a}_r + r\vec{a}_\phi) \cdot r d\phi\, \vec{a}_\phi = \int_{\phi=0}^{\pi/2} r^2 d\phi = r^2 \cdot \frac{\pi}{2} = \frac{\pi}{2} \quad [\because r = 1]$$

(III) $\int_{r=1}^{0}\left(2r\cos\phi\,\vec{a}_r + r\vec{a}_\phi\right)\cdot dr\,\vec{a}_r = \int_{r=1}^{0} 2r\cos\phi\,dr = -\cos\phi$

$$\oint \vec{A}\cdot d\vec{l} = \frac{\pi}{2} \qquad \text{...R.H.S}$$

$$\nabla\times\vec{A} = \frac{1}{\rho}\begin{vmatrix} \vec{a}_\rho & \rho\vec{a}_\phi & \vec{a}_z \\ \frac{\partial}{\partial\rho} & \frac{\partial}{\partial\phi} & \frac{\partial}{\partial z} \\ A_\rho & \rho A_\phi & A_z \end{vmatrix} = \frac{1}{r}\begin{vmatrix} \vec{a}_r & r\vec{a}_\phi & \vec{a}_z \\ \frac{\partial}{\partial r} & \frac{\partial}{\partial\phi} & \frac{\partial}{\partial z} \\ 2r\cos\phi & r^2 & 0 \end{vmatrix}.$$

The radius of cylinder 'ρ' is given as 'r'.

$$= \frac{1}{r}\left[\vec{a}_r(0-0) - r\vec{a}_\phi(0-0) + \vec{a}_z(2r + 2r\sin\phi)\right]$$

$$= 2(1+\sin\phi)\vec{a}_z$$

$$\iint_S \left(\nabla\times\vec{A}\right)\cdot\vec{ds} = \int_{\phi=0}^{\pi/2}\int_{r=0}^{1} 2(1+\sin\phi)\vec{a}_z\cdot r\,dr\,d\phi\,\vec{a}_z\,.$$

$$= \int_{\phi=0}^{\pi/2} 2(1+\sin\phi)\left(\frac{r^2}{2}\right)_0^1 d\phi = \int_0^{\pi/2}\frac{2(1+\sin\phi)}{2}d\phi$$

$$= \left[\phi - \cos\phi\right]_0^{\pi/2} = \frac{\pi}{2} = \text{L.H.S}$$

Hence, verified.

SUMMARY

1. A vector $\vec{A}$ is uniquely specified by its magnitude and a unit vector along it. $\vec{A} = \left|\vec{A}\right|\vec{a}_A$.
2. Scalar product of two vectors $\vec{A}\cdot\vec{B} = \left|\vec{A}\right|\left|\vec{B}\right|\cos\theta_{AB}$.

   Vector product of two vectors $\vec{A}\times\vec{B} = \left|\vec{A}\right|\left|\vec{B}\right|\sin\theta_{AB}\vec{a}_n$.
3. The scalar projection of vector $\vec{A}$ onto $\vec{B}$ is $A_B = \vec{A}\cdot\vec{a}_B$., whereas vector projection of $\vec{A}$ onto $\vec{B}$ is $\vec{A}_B = A_B\,\vec{a}_B$.

4. A point P is represented as $P(x, y, z)$, $P(\rho, \phi, z)$ and $P(r, \theta, \phi)$ in cartesian, cylindrical and spherical co-ordinates, respectively.
5. The differential displacement in the cartesian, cylindrical and spherical systems are

$$d\vec{r} = dx\,\vec{a}_x + dy\,\vec{a}_y + dz\,\vec{a}_z$$

$$d\vec{r} = d\rho\,\vec{a}_\rho + \rho\,d\phi\,\vec{a}_\phi + dz\,\vec{a}_z$$

$$d\vec{r} = dr\,\vec{a}_r + r\,d\theta\,\vec{a}_\phi + r\sin\theta\,d\phi\vec{a}_\phi.$$

6. The differential normal areas in the three systems are:

$$d\vec{s} = dy\,dz\;\vec{a}_x$$

$$= dx\,dz\;\vec{a}_y$$

$$= dy\,dz\;\vec{a}_x$$

$$d\vec{s} = \rho\;d\phi\;dz\;\vec{a}_\rho$$

$$= d\rho\;dz\;\vec{a}_\phi$$

$$= \rho\,d\rho\;d\phi\;\vec{a}_z$$

$$d\vec{s} = r^2\;\sin\theta\,d\theta\;d\phi\,\vec{a}_r$$

$$= r\sin\theta\,dr\,d\phi\,\vec{a}_\phi.$$

$$= r dr\,d\theta\,\vec{a}_\phi.$$

$d\vec{s}$ can be in positive or negative direction depending on the surface under consideration.

7. The differential volumes in the three systems are:

$$dv = dx\,dy\,dz$$

$$dv = \rho\,d\rho\,d\phi\,dz$$

$$dv = r^2\,\sin\theta dr\,d\theta\,d\phi.$$

8. The sine integral along a path L of a vector $\vec{A}$ is given by $\int_L \vec{A}\cdot d\vec{l}$. If the path is closed, then

$$\int_L \vec{A}\cdot d\vec{l}.$$

9. The surface integral of a vector $\vec{A}$ across a surface S is given by $\int_s \vec{A}\cdot d\vec{s}$. If the surface is closed,

then $\oint_s \vec{A}\cdot d\vec{s}$.

10. The volume integral of a scalar ρ_v over a volume v is defined $\int_v \rho_v dv$.

11. Vector differentiation is performed using the vector differential operator ∇ (del).

    Gradient of a scalar (V) ∇V results in a vector.

    Divergence of a vector ($\vec{A}$) $\nabla \cdot \vec{A}$ results in a scalar.

    Curl of a vector $\vec{A}$ is $\nabla \times \vec{A}$ results in a vector.

    Laplacian of a scalar is $\nabla^2 V$ results in a scalar.

12. The divergence theorem is $\oint_s \vec{A} \cdot d\vec{s} = \int_v \nabla \cdot \vec{A}\, dv$

13. The Stoke's theorem is $\oint_L \vec{A} \cdot d\vec{l} = \int_s \vec{A} \cdot d\vec{s}$.

14. If $\nabla^2 V = 0$, then the scalar field (V) is known as harmonic.

15. If $\nabla \times \vec{A} = 0$, then the vector field is known as irrotational or conservative.

16. If $\nabla \cdot \vec{A} = 0$, then the vector field $\vec{A}$ is known as solenoidal.

17. Vector identities $\nabla \cdot \nabla \times \vec{A} = 0$ and $\nabla \times \nabla V = 0$.

18. Gradient of a scalar (V) is represented in three co-ordinate systems as

$$\nabla V = \frac{\partial V}{\partial x}\vec{a}_x + \frac{\partial V}{\partial y}\vec{a}_y + \frac{\partial V}{\partial z}\vec{a}_z$$

$$\nabla V = \frac{\partial V}{\partial \rho}\vec{a}_\rho + \frac{1}{\rho}\frac{\partial V}{\partial \phi}\vec{a}_\phi + \frac{\partial V}{\partial z}\vec{a}_z.$$

$$\nabla V = \frac{\partial V}{\partial r}\vec{a}_r + \frac{1}{r}\frac{\partial V}{\partial \theta}\vec{a}_\theta + \frac{1}{r\sin\theta}\frac{\partial V}{\partial \phi}\vec{a}_\phi.$$

19. The divergence of a vector $\vec{A}$ in three co-ordinate systems are:

$$\nabla \cdot \vec{A} = \frac{\partial A_x}{\partial x} + \frac{\partial A_y}{\partial y} + \frac{\partial A_z}{\partial z}.$$

$$\nabla \cdot \vec{A} = \frac{1}{\rho}\frac{\partial}{\partial \rho}\left(\rho A_\rho\right) + \frac{1}{\rho}\frac{\partial A_\phi}{\partial \phi} + \frac{\partial A_z}{\partial z}.$$

$$\nabla \cdot \vec{A} = \frac{1}{r^2}\frac{\partial}{\partial r}\left(r^2 Ar\right) + \frac{1}{r\sin\theta}\frac{\partial}{\partial \theta}\left(A_\theta \sin\theta\right) + \frac{1}{r\sin\theta}\frac{\partial A_\phi}{\partial \phi}.$$

20. The curl of a vector $\vec{A}$ in three coordinate systems are

$$\nabla \times \vec{A} = \begin{vmatrix} \vec{a}_r & r\vec{a}_\theta & r\sin\theta\,\vec{a}_\phi \\ \dfrac{\partial}{\partial r} & \dfrac{\partial}{\partial \theta} & \dfrac{\partial}{\partial \phi} \\ Ar & rA\theta & r\sin\theta\,A_\phi \end{vmatrix}, \nabla \times \vec{A} = \begin{vmatrix} \vec{a}_x & \vec{a}_y & \vec{a}_z \\ \dfrac{\partial}{\partial x} & \dfrac{\partial}{\partial y} & \dfrac{\partial}{\partial z} \\ A_x & A_y & A_z \end{vmatrix} \text{ and } \nabla \times \vec{A} = \begin{vmatrix} \vec{a}_\rho & \rho\vec{a}_\phi & \vec{a}_z \\ \dfrac{\partial}{\partial \rho} & \dfrac{\partial}{\partial \phi} & \dfrac{\partial}{\partial z} \\ A_\rho & \rho A_\phi & A_z \end{vmatrix}.$$

21. The relationship between the variables (x, y, z) of the cartesian system and those of cylindrical system (ρ, ϕ, z) are:

$$\rho = \sqrt{x^2 + y^2}, \quad \phi = \tan^{-1} y/x, \; z = z$$

$$x = \rho\cos\phi, \; y = \rho\sin\phi, \; z = z$$

22. The relationship between the variables (x, y, z) of the cartesian system and those of spherical system (r, θ, ϕ) are

$$r = \sqrt{x^2 + y^2 + z^2}, \; \theta = \tan^{-1}\frac{\sqrt{x^2 + y^2}}{z}, \; \phi = \tan^{-1} y/x.$$

$$x = r\sin\theta\cos\phi, \; y = r\sin\theta\sin\phi, \; z = r\cos\theta.$$

EXERCISE

1. Vectors $\vec{A} = 5\vec{a}_r + 7\vec{a}_\theta + 3\vec{a}_\phi$ and $\vec{B} = \vec{a}_r + 2\vec{a}_\theta + 3\vec{a}_\phi$ are situated at a point $(r, \theta, \phi) = (2, 60°, 45°)$

   Find *(a)* $\vec{A} + \vec{B}$ *(b)* $\vec{A} \cdot \vec{B}$ (c) $\vec{A} \times \vec{B}$.
2. Find the distance between two points *A* at (2, 6, 4) and *B* at (7, 55°, 3).
3. A vector $\vec{A} = 4\vec{a}_r + \vec{a}_\theta + 6\vec{a}_\theta$ is located at the point (5, 45°, 45°). Express $\vec{A}$ in terms of rectangular components.
4. Find the distance between two points *A* (3, 60°, 30°) and *B* (3, 30°, 120°).
5. Convert the vector into cartesian co-ordinates

$$\vec{C} = z\sin\phi\;\vec{a}_\rho - \rho\cos\phi\vec{a}_\phi + 2\rho z\vec{a}_z.$$

6. Transform the vector to cylindrical and spherical co-ordinates.

$$\vec{A} = \left(y^2 - x^2\right)\vec{a}_x + xyz\;\vec{a}_y + \left(x^2 - z^2\right)\vec{a}_z.$$

7. Given $\vec{A} = 10\vec{a}_y + 3\vec{a}_z$ and $\vec{B} = 5\vec{a}_x + 4\vec{a}_y$. Find the projection of *A* on *B*.

8. Find the angle between $\vec{A} = 10\vec{a}_y + 3\vec{a}_z$ and $\vec{B} = -4\vec{a}_y + 0.5\vec{a}_z$, using both dot product and the cross product.

9. Find the gradient of the following scalar field

   (*a*) $A = x^2 y + xyz$

   (*b*) $B = \rho z \sin\phi + z^2 \cos\phi + \rho^2$

   (*c*) $C = \cos\theta \sin\phi + r^2 \phi$.

10. Determine the divergence of the following vector fields at specified points.

    (*a*) $\vec{A} = yz\vec{a}_x + 4xy\vec{a}_y + y\vec{a}_z$ at (1, – 2, 3)

    (*b*) $\vec{B} = \rho z \sin\phi\, \vec{a}_\rho + 3\rho z^2 \cos\phi\, \vec{a}_\phi$ at (5, π/2, 1)

    (*c*) $\vec{C} = 2r\cos\theta\cos\phi\, \vec{a}_r + r^2\, \vec{a}_\phi$ at (1, π/6, π/3).

11. Determine the curl of the vector field at specified points

    (*a*) $\vec{A} = \vec{a}_x + y\vec{a}_y + (4y - z)\vec{a}_z$ at (1, – 2, 11)

    (*b*) $\vec{B} = -b\rho z \cos\phi\, \vec{a}_\rho + \rho \sin\phi\, \vec{a}_\phi$ at (0, π, 0)

    (*c*) $\vec{C} = r^2 \cot\theta\, \vec{a}_r + 2\sin\theta\cos\phi\, \vec{a}_\phi$ at (2, π/6, π/4).

12. Determine the flux of $\vec{D} = \rho^2 \cos^2\phi\, \vec{a}_\rho + z\sin\phi\, \vec{a}_\phi$ over the closed surface of the cylinder $0 \le z \le 1$, $\rho = 4$. Also verify the divergence theorem.

13. Verify the divergence theorem for the following case

$$\vec{A} = xy^2\vec{a}_x + y^3\vec{a}_y + y^2 z\vec{a}_z,$$

    and the surface is a cuboid defined by $0 < x < 1$, $0 < y < 1$ and $0 < z < 1$.

14. Given that $\vec{F} = x^2 y\vec{a}_x - y\vec{a}_y$

    (*a*) Find

    $\oint_L \vec{F} \cdot d\vec{l}$ where L is shown in Fig. E-14.

(*b*) Verify Stoke's theorem.

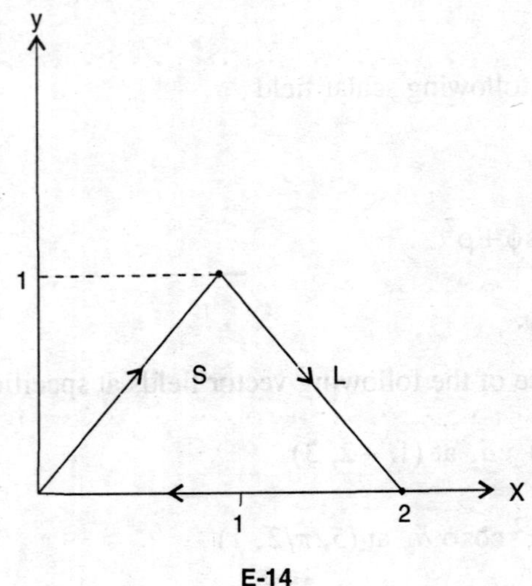

E-14

15. Find the directional derivative of $V = r z^2 \cos 2\phi$ along the direction $A = 2\,\vec{a}_r - \vec{a}_z$ and evaluate it at $(1, \pi/2, 2)$. Find $(dV/dr)\big|(1, \pi/2, 2)$. $\left[4/\sqrt{5}\right]$

2

Electrostatics

2.1 ELECTROSTATICS

Introduction

The electromagnetic field is the combined study of electric and magnetic effects. So, this study of two fields are carried out separately by considering static (or) time-independent fields in free space. The present chapter is devoted to static electric field in free space.

2.1.1 Electric Field Concept

Just as the gravitational field is associated with the physical property known as 'mass', a force field is associated with the physical property known as 'charge' merely by virtue of its existence. This force field is known as the electric field.

If a charge of 'q' coulomb is placed in the electric field of a larger charge Q, then it will experience a force $\vec{F}$, which is given by

$$\vec{F} = q\vec{E}, \qquad \text{...(2.1)}$$

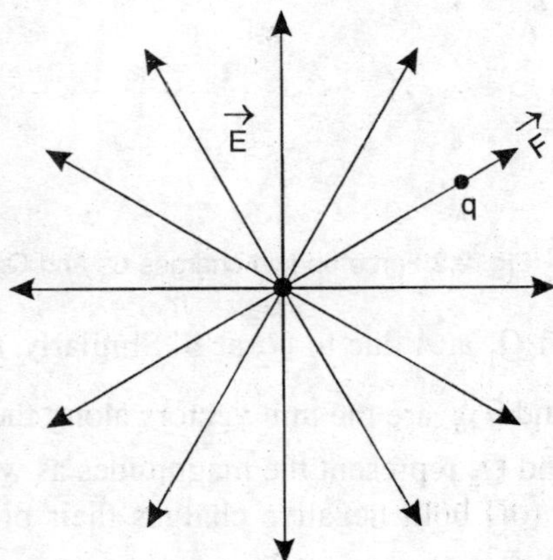

Fig. 2.1

where $\vec{E}$ is the intensity of electric field which is analogous to the gravitational field '*g*'.

2.1.2 Coulomb's Law

In the previous section, it was mentioned that a charge placed in the electric field of a larger charge experiences a force. Actually, the larger charge also experiences a force just as two masses attract each other. This fact was proved experimentally by Coulomb. Coulomb's law relates the force between two charged bodies which are very small in size compared to their separation. Ideally, the charged bodies must be so small that they can be considered as 'point charges'.

Coulomb's law states that:

1. Like charges repel where as unlike charges attract.
2. The magnitude of the force is proportional to the product of the magnitudes of the charges.
3. The magnitude of the force is inversely proportional to the square of the distance between the charges.
4. The direction of the force is along the line joining the charges.
5. The force depends upon the medium in which the charges are placed.

If two point charges Q_1 and Q_2 are considered to be situated at *A* and *B*, separated by a distance *R* m as shown in Fig 2.2, then according to Coulomb's law

$$\vec{F}_{BA} = k\frac{Q_1 Q_2}{R^2}\vec{a}_{BA} \quad ...(2.2)$$

$$\vec{F}_{AB} = k\frac{Q_1 Q_2}{R^2}\vec{a}_{AB}\ , \quad ...(2.3)$$

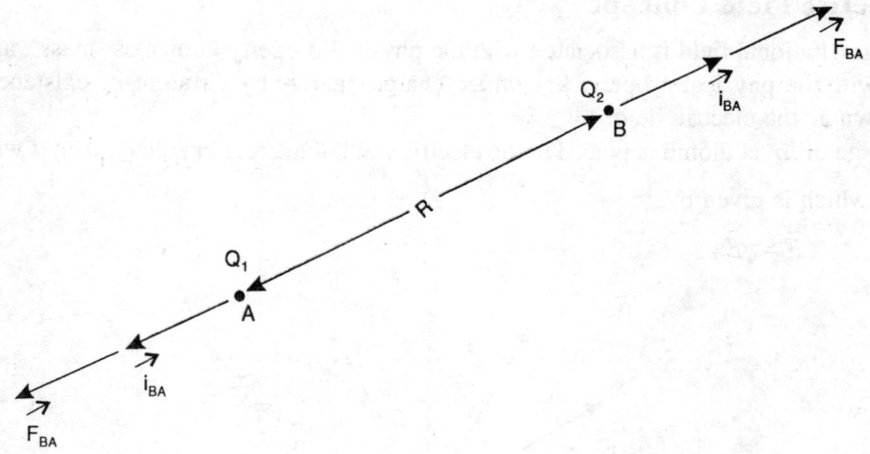

Fig. 2.2 Force on two charges Q_1 and Q_2

where $\vec{F}_{BA}$ is the force acting on Q_1 at *A* due to Q_2 at *B*. Similarly, $\vec{F}_{AB}$ is the force acting on Q_2 at *B* due to the charge Q_1 at *A*. $\vec{a}_{BA}$ and $\vec{a}_{AB}$ are the unit vectors along the line joining *A* and *B* and *k* is the constant of proportionality. Q_1 and Q_2 represent the magnitudes as well as signs of the charges. If both Q_1 and Q_2 are positive charges (or) both negative charges their product will be positive and hence positive forces act along $\vec{a}_{BA}$ and $\vec{a}_{AB}$. If one of the two charges is negative, then the product Q_1Q_2 will be negative, hence negative forces act along $\vec{a}_{BA}$ and $\vec{a}_{AB}$ or positive forces act along the directions opposite to $\vec{a}_{BA}$ and $\vec{a}_{AB}$, respectively.

The constant of proportionality k is equal to $1/4\pi\varepsilon_0$ of for free space in MKS units. The quantity ε_0 is known as the permittivity of free space and its value is 8.854×10^{-12} farad/m or approximately equal to $10^{-9}/36\pi$ farad/m. Substituting for k in eqn. (2.2) and eqn. (2.3),

$$\vec{F}_{BA} = \frac{Q_1 Q_2}{4\pi\varepsilon_0 R^2}\vec{a}_{BA} \text{ N} \quad \text{...(2.4)}$$

$$\vec{F}_{AB} = \frac{Q_1 Q_2}{4\pi\varepsilon_0 R^2}\vec{a}_{AB} \text{ N} \quad \text{...(2.5)}$$

If the location of point charges Q_1 and Q_2 are represented by their position vectors $\vec{r}_1$ and $\vec{r}_2$ respectively, as shown in Fig 5.3, then the Coulomb's law equations (eqn. 2.4 and eqn. 2.5) are modified as

$$\vec{F}_{BA} = \frac{Q_1 Q_2}{4\pi\varepsilon_0 R^2}\vec{a}_{RBA} \quad \text{...(2.6)}$$

$$\vec{F}_{AB} = \frac{Q_1 Q_2}{4\pi\varepsilon_0 R^2}\vec{a}_{RAB}, \quad \text{...(2.7)}$$

where $\vec{a}_{RBA}$ and $\vec{a}_{RAB}$ are the unit vectors which are given as follows:

$$\vec{a}_{RAB} = \frac{\vec{R}_{AB}}{\left|\vec{R}_{AB}\right|} \quad \text{...(2.8)}$$

$$\vec{a}_{RBA} = \frac{\vec{R}_{BA}}{\left|\vec{R}_{BA}\right|} \quad \text{...(2.9)}$$

$$\vec{a}_{RAB} = \vec{r}_B - \vec{r}_A \quad \text{...(2.10)}$$

$$\vec{a}_{RBA} = \vec{r}_A - \vec{r}_B \quad \text{...(2.11)}$$

$$\left|\vec{R}_{AB}\right| = \left|\vec{R}_{BA}\right| = R. \quad \text{...(2.12)}$$

Frorn eqn. (2.8) to eqn. (2.12),

$$\vec{F}_{BA} = \frac{Q_1 Q_2}{4\pi\varepsilon_0} \cdot \frac{\vec{r}_A - \vec{r}_B}{\left|\vec{r}_A - \vec{r}_B\right|^3} \quad \text{...(2.13)}$$

$$\vec{F}_{AB} = \frac{Q_1 Q_2}{4\pi\varepsilon_0} \cdot \frac{\vec{r}_B - \vec{r}_A}{\left|\vec{r}_B - \vec{r}_A\right|^3} \quad \text{...(2.14)}$$

$$\vec{F}_{BA} = \frac{Q_1 Q_2}{4\pi\varepsilon_o R^3} \cdot \vec{R}_{BA} \qquad ...(2.15)$$

$$\vec{F}_{AB} = \frac{Q_1 Q_2}{4\pi\varepsilon_o R^3} \cdot \vec{R}_{AB} \,. \qquad ...(2.16)$$

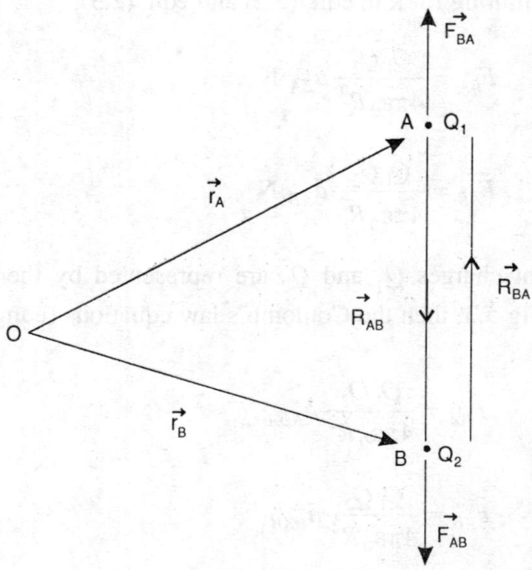

Fig. 2.3 Force on charges Q_1 and Q_2

If there are several point charges Q_1, Q_2, Q_3, ..., Q_n located at different points as shown in Fig. 2.4, then by using superposition principle, the force $\vec{F}$ experienced by a test charge situated at a point P is the vector sum of forces experienced by the test charge due to individual charges

$$\vec{F} = \frac{Q_1 q}{4\pi\varepsilon_o R_1^2} \vec{a}_{R_1} + \frac{Q_2 q}{4\pi\varepsilon_o R_2^2} \vec{a}_{R_2} + \frac{Q_3 q}{4\pi\varepsilon_o R_3^2} \vec{a}_{R_3} + \ldots + \frac{Q_3 q}{4\pi\varepsilon_o R_n^2} \vec{a}_{R_n} \qquad ...(2.17)$$

$$\vec{F} = \sum_{j=1}^{n} \frac{Q_j q}{4\pi\varepsilon_o R_j^2} \vec{a}_{R_j} \,. \qquad ...(2.18)$$

Similarly, if there are N charges Q_1, Q_2, Q_3,... Q_N located at points with position vectors $\vec{r}_1, \vec{r}_2, \ldots \vec{r}_n$ respectively, then the resultant force $\vec{F}$ on a test charge q at a point with the position vector $\vec{r}$ is the vector sum of the forces exerted on the test charge 'q' by all the other charges individually.

$$\vec{F} = \frac{qQ_1}{4\pi\varepsilon_0 \cdot |\vec{r} - \vec{r}_1|^3} \cdot (\vec{r} - \vec{r}_1) + \frac{qQ_2}{4\pi\varepsilon_0 (\vec{r} - \vec{r}_2)^3} \cdot (\vec{r} - \vec{r}_2) \qquad ...(2.19)$$

$$+ \frac{qQ_3}{4\pi\varepsilon_0 \cdot |\vec{r} - \vec{r}_1|^3} \cdot (\vec{r} - \vec{r}_3) + \ldots\ldots + \frac{qQ_N}{4\pi\varepsilon_0 (\vec{r} - \vec{r}_N)^3} \cdot (\vec{r} - \vec{r}_N)$$

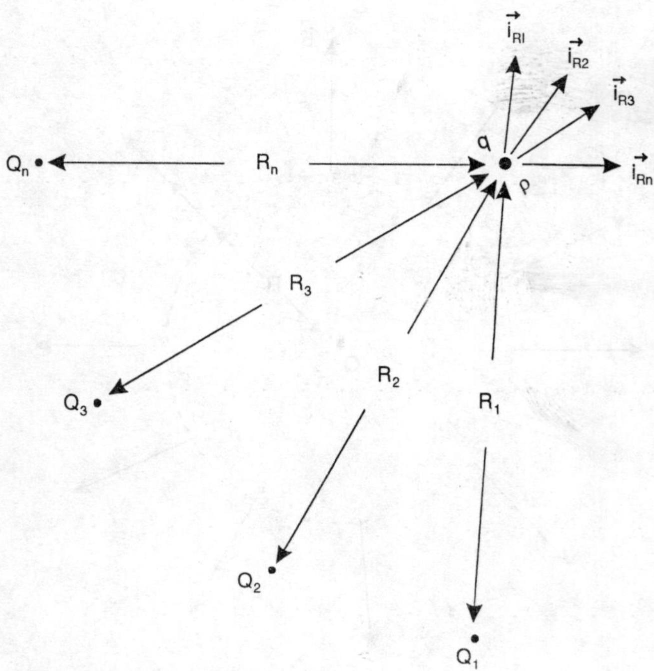

Fig. 2.4 Force and electric field intensity at *P* due to assembly of charges.

$$\vec{F} = \frac{q}{4\pi\varepsilon_0} \sum_{k=1}^{N} \frac{Q_k(\vec{r} - \vec{r}_k)}{|\vec{r} - \vec{r}_k|^3}. \qquad ...(2.20)$$

2.1.3 Electric Field Intensity

Consider a point charge Q_1 at the origin. If another point charge Q_2 is brought into the vicinity of Q_1, then Q_2 is acted upon by a force according to Coulomb's law. It may be said that Q_1 has a field around it where forces act.

If Q_2 is a positive test charge, the resulting force per unit charge is defined as the electric field intensity $\vec{E}$. Thus,

$$\vec{E} = \frac{\vec{F}}{Q_2} = \frac{Q_1}{4\pi\varepsilon_0 r^2}\vec{a}_r \text{ newtons/coulomb } (\text{N/C})(or)\,\text{volt/m}(\text{V/m}) \qquad ...(2.21)$$

where r is the distance of Q_2 from Q_1.

In general, electric field intensity at any point due to a point charge of Q coulomb is given by

$$\vec{E} = \frac{Q}{4\pi\varepsilon_0 R^2}\vec{a}_r \text{ N/C}, \qquad ...(2.22)$$

where R is the distance from the point charge to the point at which the field intensity is to be computed and $\vec{a}_r$ is the unit vector along the direction joining of the line the two points under consideration and directed away from the point charge. The electric field intensity of a point charge is thus directed everywhere radially away from the point charge, and on any spherical surface centered at the point charge its magnitude is constant as it is illustrated in Fig. 2.5.

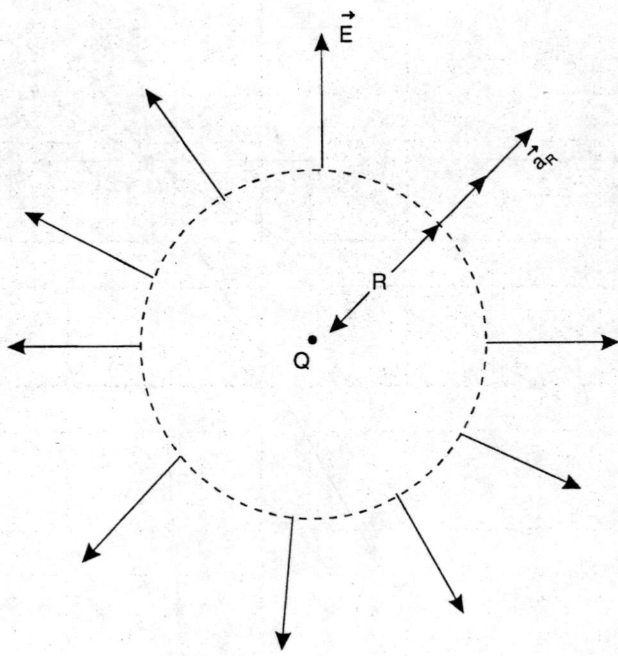

Fig. 2.5 Electric field of a point charge *Q*.

Also, for N point charges Q_1, Q_2, Q_3....., Q_N, the electric field intensity on a test charge is given by the superposition principle.

So, combining the eqns. from (2.17) to (2.20), the expression for $\vec{E}$ is written as $\vec{E} = \frac{\vec{F}}{q}$ N/C.

$$\vec{E} = \frac{Q_1}{4\pi\varepsilon_0 R_1^2}\vec{a}_{R_1} + \frac{Q_2}{4\pi\varepsilon_0 R_2^2}\vec{a}_{R_2} + \ldots\ldots + \frac{Q_n}{4\pi\varepsilon_0 R_n^2}\vec{a}_{R_n} \qquad \text{...(2.23)}$$

$$\vec{E} = \sum_{j=1}^{n} \frac{Q_j}{4\pi\varepsilon_0 R_j^2}\vec{a}_{R_j} \text{ N/C.} \qquad \text{...(2.24)}$$

Similarly, in terms of position vectors

$$\vec{E} = \frac{Q_1}{4\pi\varepsilon_0 |\vec{r} - \vec{r}_1|^3}\cdot(\vec{r} - \vec{r}_1) + \frac{Q_2}{4\pi\varepsilon_0 |\vec{r} - \vec{r}_2|^3}\cdot(\vec{r} - \vec{r}_2)$$

$$+ \frac{Q_3}{4\pi\varepsilon_0 |\vec{r} - \vec{r}_3|^3}\cdot(\vec{r} - \vec{r}_3) + \ldots\ldots + \frac{Q_N}{4\pi\varepsilon_0 |\vec{r} - \vec{r}_N|^3}\cdot(\vec{r} - \vec{r}_N) \qquad \text{...(2.25)}$$

$$\vec{E} = \sum_{k=1}^{N} \frac{Q_k}{4\pi\varepsilon_0 |\vec{r} - \vec{r}_k|^3}\cdot(\vec{r} - \vec{r}_k) \quad \text{N/C.} \qquad \text{...(2.26)}$$

Example 2.1

Find the force on a charge Q_1 given by 20 μC due to charge Q_2 given by 300 μC, where Q_1 is at (0, 1, 2) m and Q_2 at (2, 0, 0) m

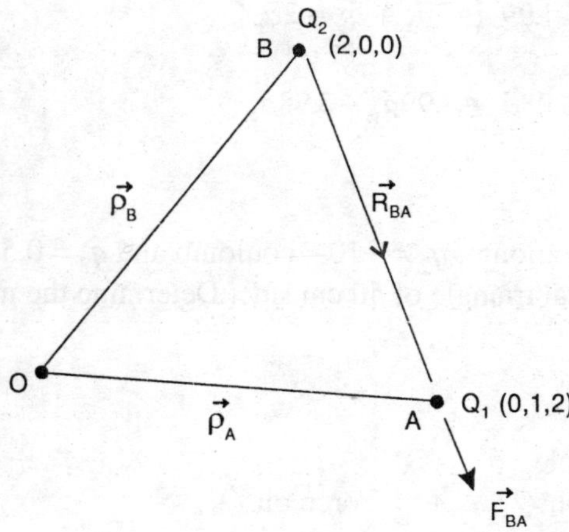

Fig. 2.6

Solution:

Since position vectors of Q_1 and Q_2 are given, force is found out using the eqn. (2.15), and eqns. (2.11) and (2.12).

$$\vec{F}_{BA} = \frac{Q_1 Q_2}{4\pi\varepsilon_0 R^3} \cdot \vec{R}_{BA}$$

$$\vec{R}_{BA} = \vec{r}_A - \vec{r}_B$$

$$R = \left|\vec{R}_{BA}\right|$$

$$Q_1 = 20 \times 10^{-6} \text{ C}$$

$$Q_2 = -300 \times 10^{-6} \text{ C}$$

$$\vec{r}_A = (0, 1, 2), \;\; \vec{r}_B = (2, 0, 0)$$

$$\vec{R}_{BA} = (0, 1, 2) - (2, 0, 0)$$

$$= (-2, 1, 2)$$

$$R = \sqrt{(-2)^2 + 1^2 + 2^2}$$

$$= \sqrt{9} = 3 \text{ m}$$

$$\vec{F}_{BA} = \frac{20 \times 10^{-6} \times (-300) \times 10^{-6}}{4\pi \times 8.87 \times 10^{-12} \times 3^3} \cdot (-2, 1, 2)$$

$$= -1.99 \cdot \left(-2\vec{a}_x + \vec{a}_y + 2\vec{a}_z\right)$$

$$\vec{F}_{BA} = 3.98\vec{a}_x - 1.99\vec{a}_y - 3.98\vec{a}_z .$$

Example 2.2

Three point charges $q_1 = 10^{-6}$ coulomb, $q_2 = -10^{-6}$ coulomb and $q_3 = 0.5 \times 10^{-6}$ coulomb are located in air at the corners of an equilateral triangle of 50 cm side. Determine the magnitude and direction of the force on q_3.

Solution:

Using superposition principle,

Force on Q_3 = Force on Q_3 due to Q_1 + Force on Q_3 due to Q_2

$$\vec{F} = \vec{F}_{13} + \vec{F}_{23}$$

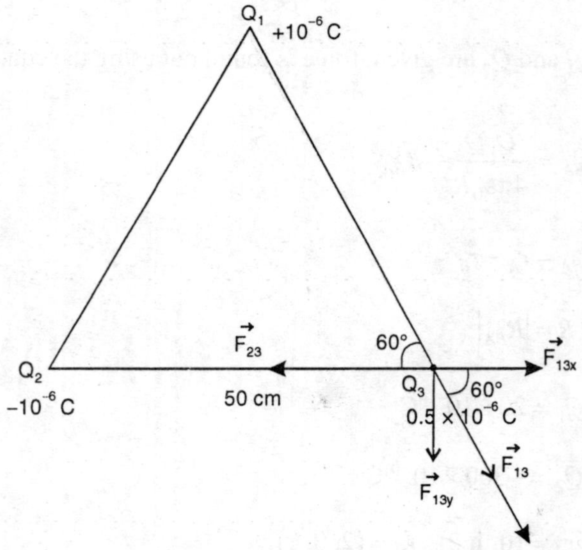

Fig. 2.7

$$\vec{F}_{13} = \frac{Q_1 Q_3}{4\pi\varepsilon_0 R_{13}^2} \vec{a}_{R_{13}}$$

$$= \frac{10^{-6} \times 0.5 \times 10^{-6}}{4\pi \times 8.87 \times 10^{-12} \times (0.5)^2} \vec{a}_{R_{13}}$$

$$= 0.0179 \ \vec{a}_{R_{13}} .$$

$\vec{F}_{13}$ has both x and y components.

From the Fig. 2.7,

$$x \text{ component of } \vec{F}_{13} = \vec{F}_{13x} = \vec{F}_{13}\cos 60° \ \cos 60°$$

$$\vec{F}_{13x} = 0.0179\cos 60°$$

$$= 0.0089\,\vec{a}_x\,\text{N}$$

$$y \text{ component of } \vec{F}_{13} = \vec{F}_{13y} = \vec{F}_{13}\sin 60°$$

$$= 0.0155\,\vec{a}_y\,\text{N}$$

$$\vec{F}_{13} = \vec{F}_{13x} + \vec{F}_{13y}$$

$$\vec{F}_{23} = \frac{Q_2 Q_3}{4\pi\varepsilon_0 R_{23}^2}\cdot\vec{a}_{23}$$

$$R_{23} = 0.5 \text{ m.}$$

Unit vector $\vec{a}_{R23}$ is along x axis $(\vec{a}_x)$.

$$\vec{F}_{23} = \frac{-10^{-6}\times 0.5\times 10^{-6}}{4\pi\times 8.87\times 10^{-12}\times (0.5)^2}\cdot(\vec{a}_x)$$

$$= -0.0178\vec{a}_x\,\text{N}$$

$$= 0.0178(-\vec{a}_x)\text{N}$$

$$\vec{F} = \vec{F}_{13} + \vec{F}_{23}$$

$$= 0.0089\vec{a}_x + 0.0155\vec{a}_y - 0.0178\vec{a}_x$$

$$\vec{F} = -0.0089\vec{a}_x + 0.0155\vec{a}_y$$

$$\left|\vec{F}\right| = 0.0178\,\text{N}$$

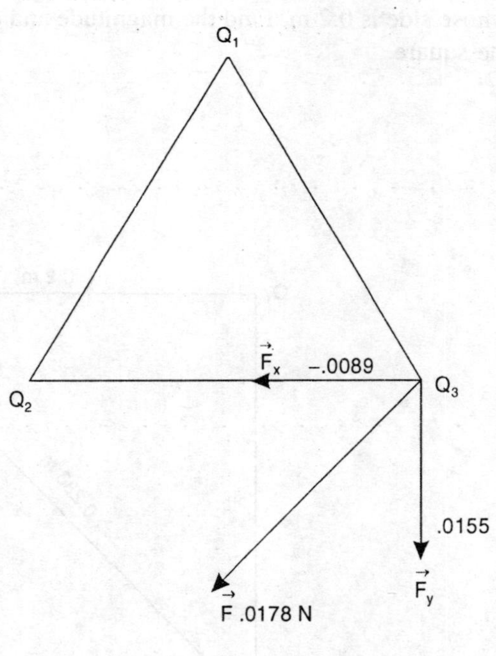

Fig. 2.8

Magnitude of the force on Q_3 is 0.0178 N and its direction is parallel to the line joining Q_1 to Q_2.

Example 2.3

The force on a point charge situated 10 cm away from another point charge of the same magnitude in a dielectric medium of relative permittivity 81 is 0.1 newton. Determine the magnitude of the charge.

$$\vec{F} = \frac{Q_1 Q_2}{4\pi\varepsilon_0\varepsilon_r R^2}\vec{a}_R$$

So far, we have considered the point charges in free space, for which $\varepsilon_r = 1$. In this problem, the given charges are in the dielectric medium whose relative permittivity is 81.

$$\vec{F} = 0.1\,\text{N}$$

$$Q_1 = Q_2$$

$$R = 10\,\text{cm} = 0.1\,\text{m}.$$

Using the equation

$$Q_1\,Q_2 = 0.1 \times 4 \times \pi \times 8.87 \times 10^{-12} \times 81 \times (.1)^2$$

$$Q^2 = 9 \times 10^{-12}$$

$$Q = 3 \times 10^{-6}\ \text{C}$$

Magnitude of the charges are 3 μC.

Example 2.4

Point charges of 3×10^3 micro-micro coulomb are placed at each of the three corners of a square whose side is 0.2 m. Find the magnitude and direction of the electric field at the vacant corner point of the square.

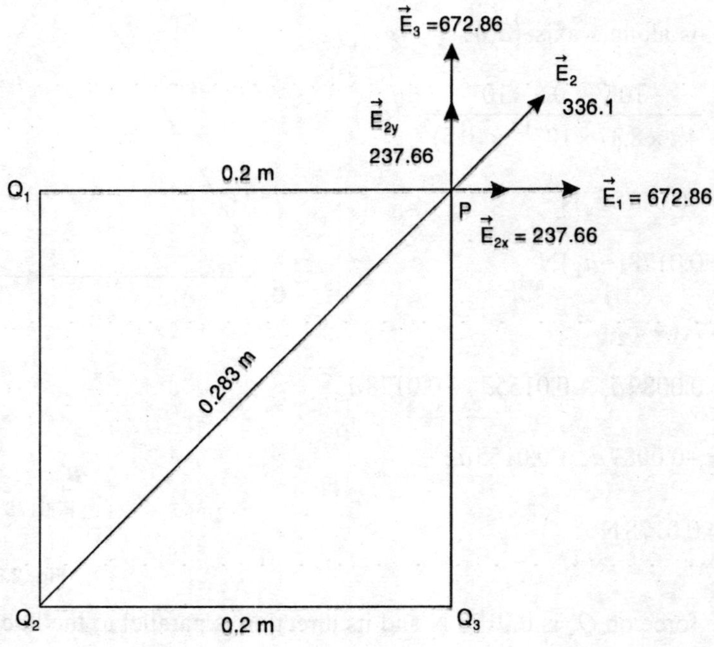

Fig. 2.9

$Q_1 = Q_2 = Q_3 = 3 \times 10^3$ micro-micro C

Electric field at $P, \vec{E}$ = Electric field at P due to $[Q_1 + Q_2 + Q_3]$

$$= \vec{E}_1 + \vec{E}_2 + \vec{E}_3$$

From the eqn. (2.22)

$$\vec{E}_1 = \frac{Q_1}{4\pi\varepsilon_0 R_1^2}\vec{a}_{R_1}$$

where $\vec{a}_{R_1}$ is the unit vector which is along X-axis.

$$\therefore \quad \vec{E}_1 = \frac{3\times10^3\times10^{-6}\times10^{-6}}{4\pi\times8.87\times10^{-12}\times(0.2)^2}\vec{a}_z$$

$$= 672.86\,\vec{a}_x \text{ V/m}$$

$$\vec{E}_2 = \frac{Q_2}{4\pi\varepsilon_0 R_2^2}\vec{a}_{R_2}$$

$$R_2 = \sqrt{(0.2)^2+(0.2)^2} = 0.283\,\text{m}.$$

The unit vector $\vec{a}_{R_2}$ has both x component and y component.

From the Fig. (2.9),

x component of $E_2 = E_2 \cos 45° \; \vec{a}_x$

y component of $E_2 = E_2 \sin 45° \; \vec{a}_y$

$$\vec{E}_2 = \frac{3\times10^3\times10^{-12}}{4\pi\times\left(8.87\times10^{-12}\right)\times(0.283)^2}\vec{a}_{R_2}$$

$$= 336.1 \text{ V/m}.$$

$$\vec{E}_{2x} = 336.1\times\cos 45°$$

$$= 237.66 \text{ V/m } \vec{a}_x$$

$$\vec{E}_{2y} = 336.1\times\sin 45°$$

$$= 237.66 \text{ V/m } \vec{a}_y$$

$$\vec{E}_3 = \frac{Q_3}{4\pi\times\varepsilon_0\times R_3^2}\vec{a}_{R_3}$$

The unit vector is along Y-axis.

$$\therefore \quad \vec{E}_3 = \frac{3\times10^3\times10^{-12}}{4\pi\times8.87\times10^{-12}\times(0.2)^2}\vec{a}_y$$

$$= 672.86\,\vec{a}_y \text{ V/m}.$$

$$\vec{E} = \vec{E}_1 + \vec{E}_{2x} + \vec{E}_{2y} + \vec{E}_3$$

$$= 672.86\,\vec{a}_x + 237.66\,\vec{a}_x + 237.66\,\vec{a}_y + 672.86\,\vec{a}_y$$

$$\vec{E} = 910.52\,\vec{a}_x + 910.52\,\vec{a}_y$$

$$|\vec{E}| = 1287.67 \text{ V/m}.$$

The direction of electric field intensity at P is along the line joining Q_2 to P.

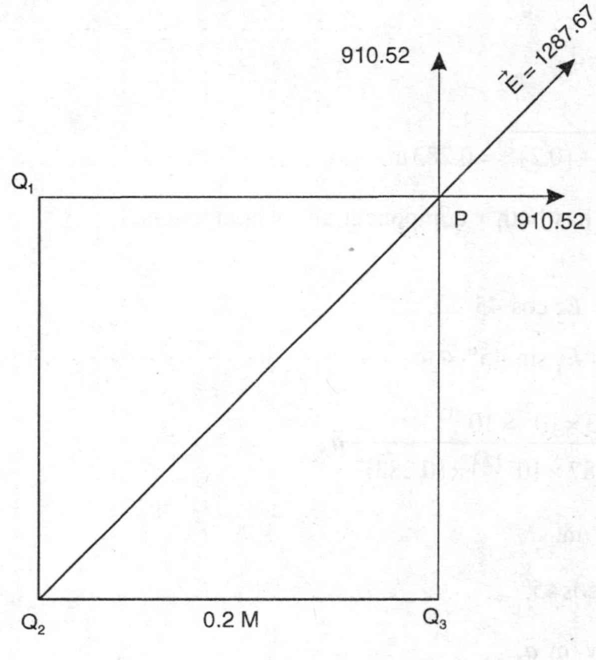

Fig. 2.10

2.1.4 Electric Field Due to Continuous Charge Distributions

So far, we have only considered Forces and Fields due to point charges. It is also possible to have continuous charge distribution along a line or a surface or in a volume.

It is customary to denote the line charge density, surface charge density and volume charge density by ρ_L in C/m, ρ_s in C/m^2 and ρ_v in C/m^3.

The electric field due to each of the charge distributions ρ_L, ρ_s and ρ_v may be regarded as the summation of field contributed by the numerous point charges making up charge distributions.

Thus the electric field intensity for these continuous charge distribution systems can be found by modifying the eqn. (2.22).

Electric field intensity can be found by replacing dQ with differential charge element $\rho_L\, dl$, $\rho_s\, ds$ or $\rho_v\, dv$ and integrating.

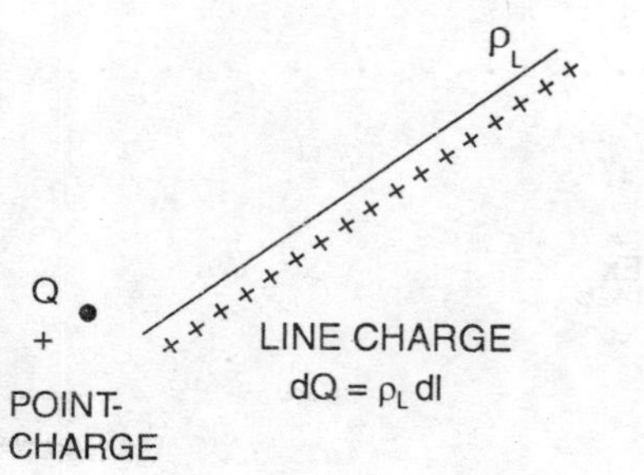

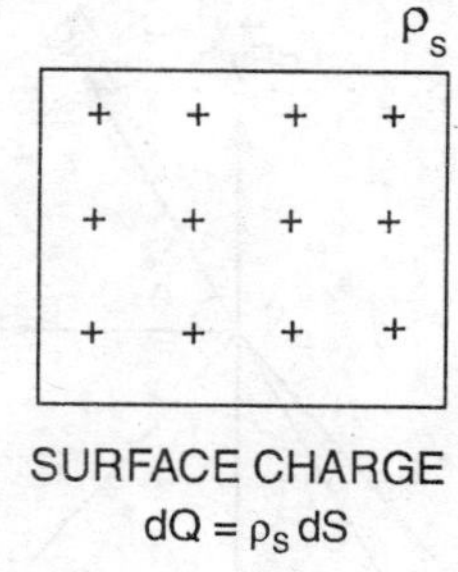

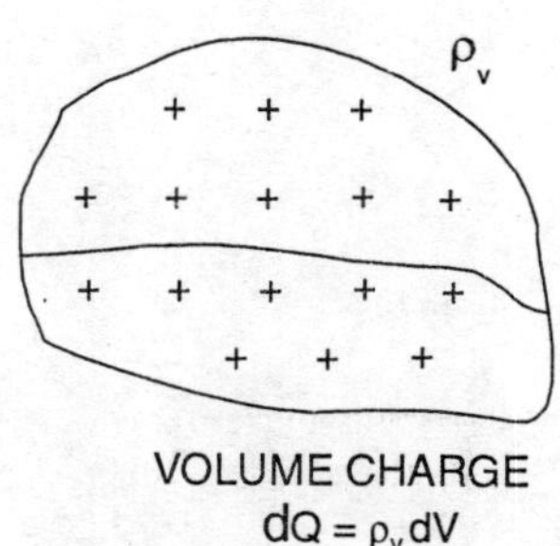

Fig. 2.11

$$\vec{E} = \int \frac{dQ}{4\pi\varepsilon_0 R^2}\vec{a}_R \text{ V/m.}$$

For line charge distributions

$$\vec{E} = \int_L \frac{\rho_L dl}{4\pi\varepsilon_0 R^2}\vec{a}_R \text{ V/m.} \quad ...(2.27)$$

For surface charge distribution

$$\vec{E} = \int_S \frac{\rho_s ds}{4\pi\varepsilon_0 R^2}\vec{a}_R \text{ V/m.} \quad ...(2.28)$$

For volume charge distribution

$$\vec{E} = \int_v \frac{\rho_v dv}{4\pi\varepsilon_0 R^2}\vec{a}_R \text{ V/m.} \quad ...(2.29)$$

Example 2.5

Find the electric field intensity of a straight uniformly-charged wire of length 'L'm and having linear charge density of $+\lambda$ coulombs/m at any point 'P' at a distance of 'h'm. The angles made by the two ends of the charged wire with the point 'P' are given as α_1 and α_2 with respect to X-axis.

Let the point 'P' at a distance of L_1m from one end of the charged wire along X-axis and at a height of 'h'm along Y-axis.

Consider a small elemental length dx at a distance x m from the left end of the wire. The corresponding charge element $dq = \lambda dx$. The wire can be divided into a large number of such infinitesimally small elements, each of which will render its contribution towards production of field at P.

Let dE be the field due to charge element λdx. It has a component dE_x along X-axis and dE_y along Y-axis.

From the Fig. 2.12,

$$dE_x = dE \cos\theta$$

$$dE_y = dE \sin\theta$$

$$dE = \frac{\lambda dx}{4\pi\varepsilon_0 r^2}\vec{a}_r \text{ V/m} \qquad [\because \text{ In Fig. 2.12, } R = r]$$

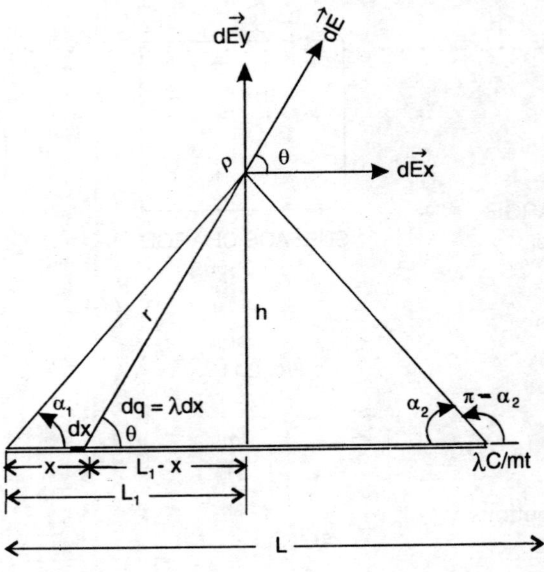

Fig. 2.12

$$d\vec{E}_x = \frac{\lambda dx}{4\pi\varepsilon_0 r^2}\cos\theta$$

$$\vec{E}_x = \int_{x=0}^{L}\frac{\lambda dx}{4\pi\varepsilon_0 r^2}\cos\theta\,\vec{a}_x.$$

Integrating using substitution, from the Fig. 2.12

$L_1 - x = h\cot\theta$

$-dx = -h\,\mathrm{cosec}^2\theta\,d\theta.$

Also,

$\sin\theta = h/r.$

$r = h\,\mathrm{cosec}\,\theta.$

So,

$$\vec{E}_x = \int\frac{\lambda h\,\mathrm{cosec}^2\theta\cdot d\theta}{4\pi\,\varepsilon_0\,(h\,\mathrm{cosec}\,\theta)^2}\cdot\cos\theta\,\vec{a}_x.$$

Substituting the limits for θ as from α_1 to $\pi-\alpha_2$,

$$\vec{E}_x = \int_{\alpha_1}^{\pi-\alpha_2}\frac{\lambda}{4\pi\varepsilon_0 h}\cos\theta\,d\theta$$

$$= \frac{\lambda}{4\pi\varepsilon_0 h}[\sin\theta]_{\alpha_1}^{\pi-\alpha_2}$$

$$= \frac{\lambda}{4\pi\varepsilon_0 h}\left[\sin(\pi-\alpha_2)-\sin\alpha_1\right]$$

$$E_x = \frac{\lambda}{4\pi\varepsilon_0 h}\left[\sin\alpha_2 - \sin\alpha_1\right] \text{ V/m}. \qquad ...(2.30)$$

Similarly,

$$d\vec{E}_y = \frac{\lambda dx}{4\pi\varepsilon_0 r^2}\sin\theta$$

$$\vec{E}_y = \int_{x=o}^{l} \frac{\lambda dx}{4\pi\varepsilon_0 r^2}\sin\theta.$$

Again substituting for θ and dx

$$\vec{E}_y = \int_{\alpha_1}^{\pi-\alpha_2} \frac{\lambda h \operatorname{cosec}^2\theta\, d\theta}{4\pi\varepsilon_0 \left(h^2\operatorname{cosec}^2\theta\right)}\sin\theta \quad \vec{a}_y$$

$$= \int_{\alpha_1}^{\pi-\alpha_2} \frac{\lambda}{4\pi\varepsilon_0 h}\sin\theta\, d\theta \quad \vec{a}_y$$

$$= \frac{\lambda}{4\pi\varepsilon_0 h}[1-\cos\theta]_{\alpha_1}^{\pi-\alpha_2} \quad \vec{a}_y$$

$$\vec{E}_y = \frac{\lambda}{4\pi\varepsilon_0 h}\left[\cos\alpha_1 + \cos\alpha_2\right] \vec{a}_y \text{ V/m}. \qquad ...(2.31)$$

Case I

If the point P is along the perpendicular bisector of wire, then $\alpha_1 = \alpha_2$. Therefore the x component of field get vanished. The y component of field always add to each other and so $\vec{E}_y$ becomes

$$\vec{E}_y = \vec{E} = \frac{\lambda}{4\pi\varepsilon_0 h}(\cos\alpha + \cos\alpha)$$

$$\vec{E} = \frac{\lambda}{2\pi\varepsilon_0 h}\cos\alpha\, \vec{a}_y \text{ V/m}. \qquad ...(2.32)$$

Case II

In case of infinitely long conductor, the field is given by substituting $\alpha = 0$

$$\vec{E} = \frac{\lambda}{2\pi\varepsilon_0 h}\vec{a}_y \text{ V/m}. \qquad ...(2.33)$$

Example 2.6

A circular disc of radius '*a*'m is charged uniformly with a charge density of σ C/m²*t*. Find the electric field at a point '*h*'m from the disc along its axis.

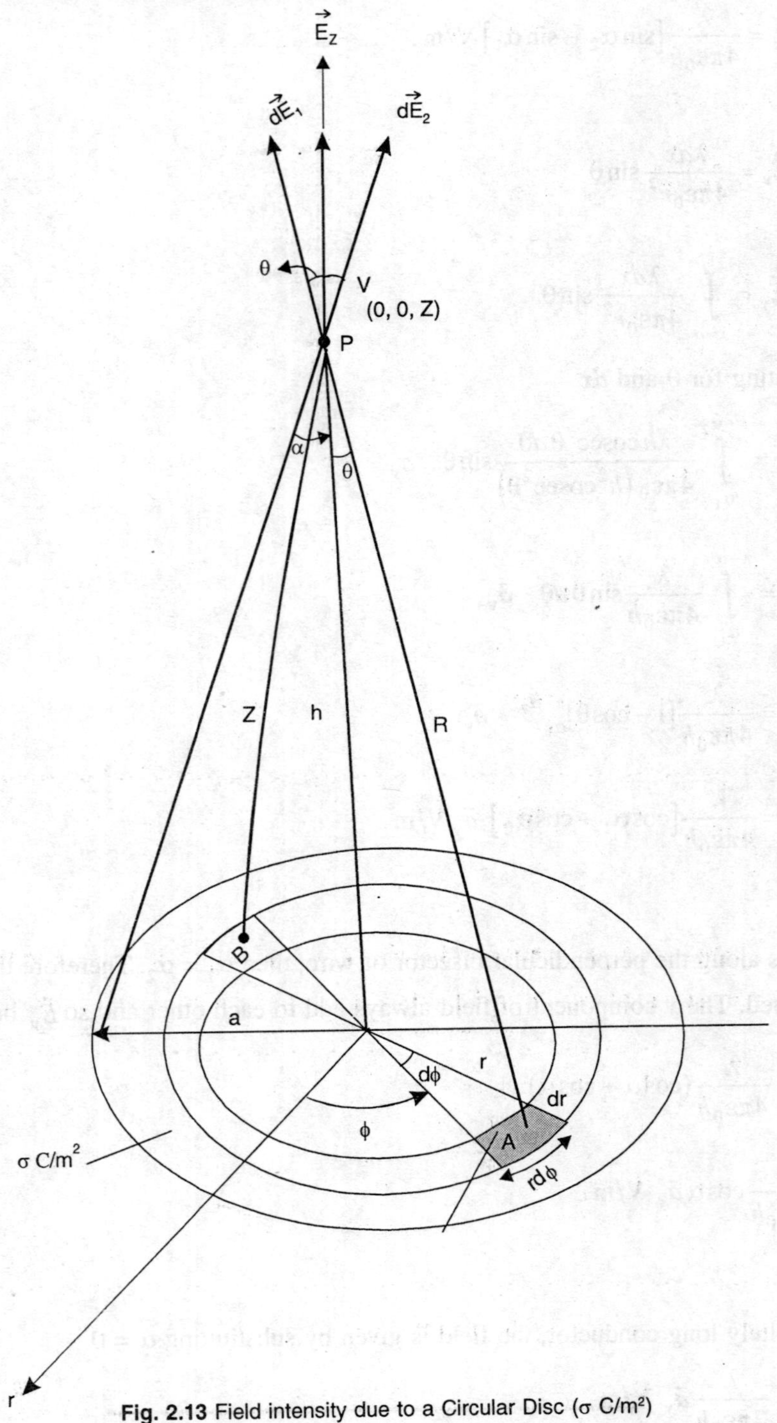

Fig. 2.13 Field intensity due to a Circular Disc (σ C/m²)

The point '*P*' is at '*h*' m along the axis of the disc. Let 'α' be the angle made by any point on the circumference of disc with the point '*P*'.

Consider the disc surface area consisting of a number of annular area elements of varying radius '*r*' m and radial thickness '*dr*' m.

From the Fig. 2.13, it can be seen that for every small elemental charge in the annular area of disc, there will be an equal and opposite charge in the diametrically opposite end. So, the horizontal component of the field gets neutralised. Consequently, the field intensity at *P* is given by summation of vertical component alone.

Consider an elemental area $ds = 2\pi\, rdr$ and its contribution in producing the field dE at *P*. It is vertical component is $dE \cos\theta$ as in Fig. 2.13.

We know from eqn. (2.28)

$$d\vec{E} = \frac{\sigma ds}{4\pi\varepsilon_0 R^2}\vec{a}_R$$

$$d\vec{E}_y = \frac{\sigma ds}{4\pi\varepsilon_0 R^2}\cos\theta.$$

Total field at *P* is

$$\vec{E}_y = \int_{r=0}^{r=a} \frac{\sigma}{4\pi\varepsilon_0 R^2} 2\pi r dr \cos\theta$$

$$\vec{E}_y = \frac{\sigma}{4\pi\varepsilon_0}\cdot 2\pi \int_{r=0}^{r=a} \frac{r}{R^2} dr \cos\theta$$

$$= \frac{\sigma}{2\varepsilon_0}\int_{r=0}^{a} \frac{\cos\theta}{R^2} r\, dr.$$

From Fig 2.13

$r = h\tan\theta$ and $R = h\cdot\sec\theta$

$dr = h\sec^2\theta\, d\theta$

$$\text{Also, } \vec{E}_y = \frac{\sigma}{2\varepsilon_0}\int_{\theta=0}^{\theta=\alpha} \frac{(h\tan\theta)\cdot(h\sec^2\theta)}{(h\sec\theta)^2}\cos\theta\, d\theta\, \vec{a}_y$$

$$= \frac{\sigma}{2\varepsilon_0}\int_{\theta=0}^{\alpha} \frac{h^2\sec^2\theta\tan\theta\cos\theta}{h^2\sec^2\theta} d\theta\, \vec{a}_y$$

$$= \frac{\sigma}{2\varepsilon_0}\int_{\theta=0}^{\alpha} \sin\theta\, d\theta\, \vec{a}_y$$

$$= \frac{\sigma}{2\varepsilon_0}[-\cos\theta]_0^{\alpha}\, \vec{a}_y$$

$$= \frac{\sigma}{2\varepsilon_0}\left[-\cos\alpha - (-\cos 0)\right] \vec{a}_y$$

$$\vec{E}_0 = \frac{\sigma}{2\varepsilon_0}[1 - \cos\alpha]\, \vec{a}_y. \quad \text{...(2.34)}$$

The result may be extended to an infinite plane sheet of charge.

Using the above result and substituting $\alpha = 90°$ for an infinite plane sheet electric field will become

$$\vec{E} = \frac{\sigma}{2\varepsilon_0}\vec{a}_y. \quad \text{...(2.35)}$$

Example 2.7

An infinitely long line charge of uniform density ρ_L C/m is situated along the Z-axis as shown. Obtain the electric field intensity at any point due to this line charge.

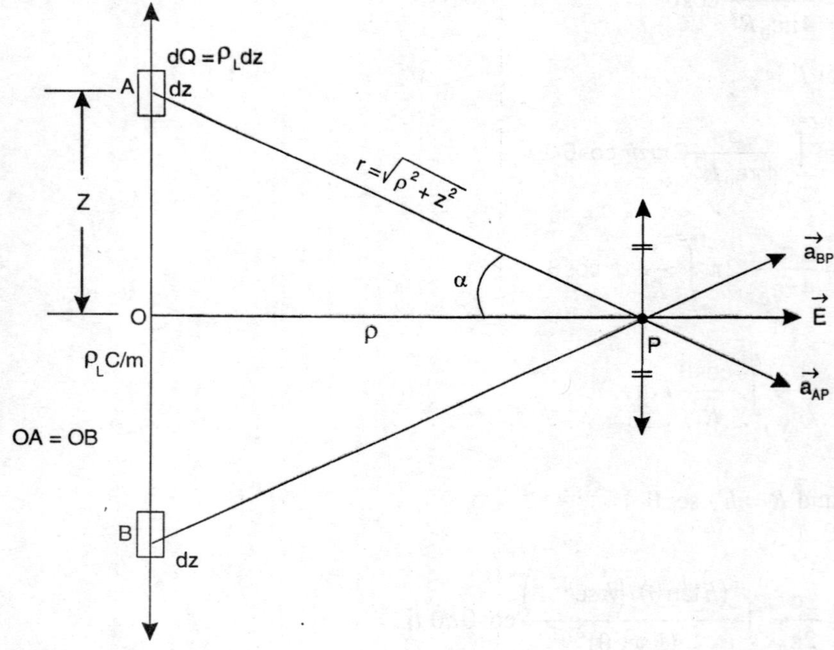

Fig. 2.14

First divide the line into a number of infinitesimal segments each of length dz, such that the charge $dq = \rho_L\, dz$ coulomb in each segment can be considered as a point charge. From the eqn. (2.22) the electric field intensity due to each point charge is directed radially away from that point charge and varies inversely as the square of the distance from that charge.

Consider a point '*P*' at a distance 'ρ' from the Z-axis, with the projection of the point '*P*' on to the Z-axis being 'O'.

The electric field intensity vectors at '*P*' due to infinitesimal segment above 'O' and below 'O' have equal magnitudes and makes equal angles with the line OP. The components of these two vectors, perpendicular to OP (parallel to Z-axis) will cancel each other. So, the resultant electric field intensity at 'P' due to any other two segments which are equidistant from 'O' with one above it, is normal to the axis of line charge.

Thus the electric field intensity of an infinite line charge of uniform density at any point is directed radially away from the line charge and is dependent only on the distance of point 'P' from the Z-axis.

To find magnitude of field intensity, we know

$$d\vec{E} = \frac{dq}{4\pi\varepsilon_0 r^2}\vec{a}_R .$$

From the Fig. 2.14,

$$d\vec{E} = \frac{\rho_L dz}{4\pi\varepsilon_0 r^2}\vec{a}_{AP} \quad \text{...(2.36)}$$

$$r = \sqrt{\rho^2 + z^2} . \quad \text{...(2.37)}$$

The component of E along OP is

$$d\vec{E}_{OP} = \frac{\rho_L dz}{4\pi\varepsilon_0 \left(\rho^2 + z^2\right)}\cos\alpha\, \vec{a}_{OP}$$

$$\cos\alpha = \frac{\rho}{\sqrt{\rho^2 + z^2}}$$

$$d\vec{E}_{OP} = \frac{\rho_L dz}{4\pi\varepsilon_0 \left(\rho^2 + z^2\right)} \cdot \frac{\rho}{\sqrt{\rho^2 + z^2}}\vec{a}_\rho .$$

The unit vector $\vec{a}_{OP}$ is replaced by $\vec{a}_\rho$ since it is radially away from the line charge.

The magnitude of resultant $d\vec{E}$ at 'P' due to two segments at A and B is

$$d\vec{E} = \frac{2\rho_L \rho dz}{4\pi\varepsilon_0 \left(\rho^2 + z^2\right)^{3/2}}\vec{a}_\rho$$

$$\vec{E} = \int_{z=0}^{z=\infty} dE\, \vec{a}_\rho$$

$$= \frac{2\rho_L}{4\pi\varepsilon_0}\int_0^\infty \frac{\rho dz}{\left(\rho^2 + z^2\right)^{3/2}}\vec{a}_\rho .$$

Integrating by substitution, from the Fig. 2.14

$$z = \rho\tan\alpha$$

$$dz = \rho\sec^2\alpha\, d\alpha$$

$$\rho^2 + z^2 = \rho^2 \sec^2\alpha .$$

Changing the limits, when $z = 0$

$\tan\alpha = 0, \quad \therefore \alpha = 0°$

When $z = \infty$

$\tan \alpha = \infty, \therefore \alpha = 90°$

$$\therefore \quad \vec{E} = \frac{2\rho_L}{4\pi\varepsilon_0} \int_{0°}^{90°} \frac{\rho \cdot \rho \cdot \sec^2 \alpha \, d\alpha}{\left(\rho^2 \sec^2 \alpha\right)^{3/2}} \vec{a}_\rho$$

$$= \frac{\rho_L}{2\pi\varepsilon_0} \int_{0°}^{90°} \cos\alpha \, d\alpha \, \vec{a}_\rho$$

$$= \frac{\rho_L}{2\pi\varepsilon_0 \rho} [\sin\alpha]_{0°}^{90°} \, \vec{a}_\rho \qquad \text{...(2.38)}$$

$$\vec{E} = \frac{\rho_L}{2\pi\varepsilon_0 \rho} \vec{a}_\rho .$$

Example 2.8

A sheet of charge having uniform charge density ρ_s C/m^2 extends over the entire *XY* plane. Find the electric field intensity due to this infinite sheet charge at any point '*P*'.

Let us consider a point '*P*' at a distance '*Z*' from *XY* plane with the projection of '*P*' on *XY* plane being '*O*'.

The field intensity at '*P*' due to two point charges situated at the diametrically opposite point *A* and *B* as shown in Fig. 2.15 have equal magnitudes, but their directions are such that the resultant field intensity is directed along *OP* and away from the sheet charge.

In fact, for any point charge on the ring of radius *r* m, there is a diametrically opposite point charge which results in a resultant electric field intensity entirely along *OP*.

Thus the electric field intensity at '*P*' due to the charge on the entire ring of radius '*r*' and with '*dr*' is directed normally away from the sheet charge.

So, we divide the area of *XY* plane into several rings each of width '*dr*' and divide each ring into angular increments of $d\phi$, thus creating infinitesimal area $rdrd\phi$ having point charges $dq = \rho_S \, r \, dr \, d\phi$ coulomb.

If we move '*P*' sideways while keeping *z* constant, the situation remains unchanged so that '*E*' is independent of '*P*' in plane parallel to sheet of charge.

To find magnitude of electric field intensity:

$$d\vec{E} = \frac{dq}{4\pi\varepsilon_0 R^2} \vec{a}_R$$

From the Fig. 2.15,

$$d\vec{E} = \frac{\rho_s r \, dr \, d\phi}{4\pi\varepsilon_0 \left(r^2 + z^2\right)} \vec{a}_{BP} . \qquad \text{...}(i)$$

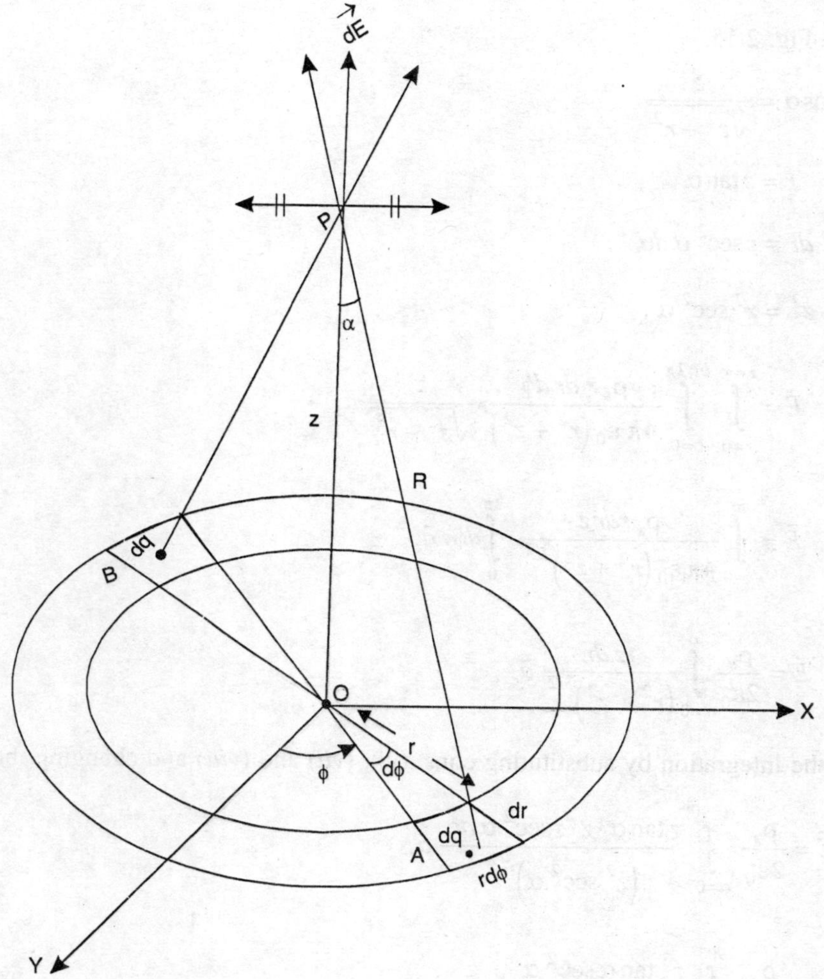

Fig. 2.15

The vertical component of E along (or) parallel to Z-axis is given by

$$d\vec{E}_z = \frac{\rho_s r\, dr\, d\phi}{4\pi\varepsilon_0\left(r^2 + z^2\right)} \cdot \cos\alpha \cdot \vec{a}_{OP}\,. \qquad \text{...}(ii)$$

Since the unit vector is parallel to Z-axis, this can be replaced by $\vec{a}_z$. So, eqn. (*ii*) can be written as

$$d\vec{E} = \frac{\rho_s\, r dr\, d\phi}{4\pi\varepsilon_0\left(r^2 + z^2\right)} \cos\alpha\, \vec{a}_z \qquad \text{...}(iii)$$

$$\vec{E} = \int_{r=0}^{r=\infty} \int_{\phi=0}^{\phi=2\pi} d\vec{E}\;\; \vec{a}_z\,. \qquad \text{...}(iv)$$

From the Fig. 2.15,

$$\cos\alpha = \frac{z}{\sqrt{z^2 + r^2}} \quad ...(v)$$

$$r = z\tan\alpha \quad ...(vi)$$

$$dr = z\sec^2\alpha\, d\alpha \quad ...(vii)$$

$$r^2 + z^2 = z^2\sec^2\alpha\,. \quad ...(viii)$$

$$\vec{E} = \int_{r=0}^{r=\infty}\int_{\phi=0}^{\phi=2\pi} \frac{\rho_s\, r\, dr\, d\phi}{4\pi\,\varepsilon_0\left(r^2 + z^2\right)} \cdot \frac{z}{\sqrt{z^2 + r^2}}\,\vec{a}_z \quad ...(ix)$$

$$\vec{E} = \int_{r=0}^{\infty} \frac{\rho_s\, r dr\, z}{4\pi\,\varepsilon_0\left(r^2 + z^2\right)^{3/2}} \cdot \int_0^{2\pi} d\phi\ \vec{a}_z \quad ...(x)$$

$$\vec{E} = \frac{\rho_s}{2\varepsilon_0}\int_{r=0}^{\infty} \frac{rz\, dr}{\left(r^2 + z^2\right)^{3/2}}\,\vec{a}_z\,. \quad ...(xi)$$

Solving the integration by substituting eqns. (*vi*), (*vii*) and (*viii*) and changing the limits

$$\vec{E} = \frac{\rho_s}{2\varepsilon_0}\int_{\alpha=0°}^{\alpha=90°} \frac{z\tan\alpha \cdot z \cdot z\sec^2\alpha\, d\alpha}{\left(z^2\sec^2\alpha\right)^{3/2}}\,\vec{a}_z \quad ...(xii)$$

$$= \frac{\rho_s}{2\varepsilon_0}\int_{\alpha=0°}^{90°} \frac{z^3\tan\alpha\sec^2\alpha}{z^3\sec^3\alpha}\, d\alpha\,\vec{a}_z$$

$$\vec{E} = \frac{\rho_s}{2\varepsilon_0}\int_{0°}^{90°} \sin\alpha\, d\alpha\,\vec{a}_z$$

$$\vec{E} = \frac{\rho_s}{2\varepsilon_0}\vec{a}_z\,. \quad ...(2.39)$$

If the sheet of charge occupies the $Z = Z_0$ plane, it follows that

$$\vec{E} = \begin{cases} \dfrac{\rho_s}{2\varepsilon_0}\vec{a}_z & \text{for} \quad z > z_0 \\ -\dfrac{\rho_s}{2\varepsilon_0}\vec{a}_z & \text{for} \quad z > z_0 \end{cases} \quad ...(2.40)$$

Example 2.9

A charged wire with linear charge density + λ C/m length is in the form of a semicircle of radius '*a*' m. Deduce an expression for the electric field at the centre of the circle.

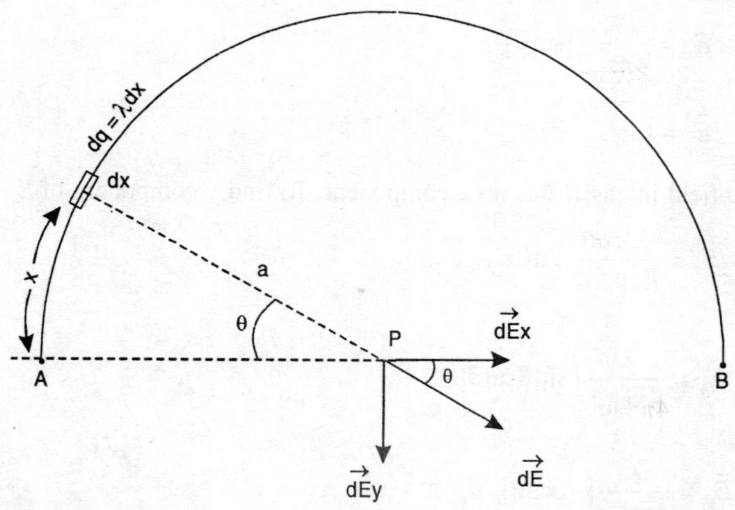

Fig. 2.16

In the Fig. 2.16, let *dx* be a portion of wire at a distance *x* from *A* and so the charge $dq = \lambda dx$.

We can divide the wire into a large number of such infinitesimally small elements, each of which contribute towards the electric field at *P*.

$$d\vec{E} = \frac{dq}{4\pi\varepsilon_0 R^2}\vec{a}_R.$$

Substitute $dq = \lambda dx$, R = *a*

$$d\vec{E} = \frac{\lambda dx}{4\pi\varepsilon_0 a^2}\vec{a}_{xp}. \quad ...(i)$$

This electric field $d\vec{E}$ has both *x* and *y* components.

$$d\vec{E}_x = d\vec{E}\cos\theta \quad ...(ii)$$

$$d\vec{E}_y = d\vec{E}\sin\theta. \quad ...(iii)$$

From Fig. 2.16

$$x = a\theta \text{ and}$$

$$dx = a\, d\theta. \quad ...(iv)$$

To find *x*-component of $\vec{E}$

$$d\vec{E}_x = \frac{\lambda dx}{4\pi\varepsilon_0 a^2}\cos\theta\, \vec{a}_x$$

$$\vec{E}_x = \frac{\lambda}{4\pi\varepsilon_0 a}\int_{\theta=0}^{\pi}\cos\theta\, d\theta \quad \vec{a}_x$$

$$\vec{E}_x = \frac{\lambda}{4\pi\varepsilon_0 a}[\sin\theta]_0^{\pi}$$

$$\vec{E}_x = 0.$$

The electric field intensity has no x-component. To find y-component of $\vec{E}$,

$$d\vec{E}_y = \frac{\lambda\, a\, d\theta}{4\pi\varepsilon_0 a^2}\sin\theta\, \vec{a}_y$$

$$\vec{E}_y = \frac{\lambda}{4\pi\varepsilon_0 a}\int_0^{\pi}\sin\theta\, d\theta\, \vec{a}_y$$

$$\vec{E}_y = \frac{\lambda}{4\pi\varepsilon_0 a}[-\cos\theta]_0^{\pi}\, \vec{a}_y$$

$$\vec{E}_y = \frac{\lambda}{2\pi\varepsilon_0 a}\vec{a}_y.$$

The electric field intensity at the centre of semicircular wire is along Y-axis and is away from the wire.

$$\vec{E} = \frac{\lambda}{2\pi\varepsilon_0 a}\vec{a}_y. \qquad ...(2.41)$$

2.1.5 Electric Flux

We have already seen that an electric field is set up around a point charge. The field intensity given by eqn. (2.22) is reproduced below:

$$\vec{E} = \frac{Q}{4\pi\varepsilon_0 R^2}\vec{a}_R.$$

Multiplying by ε_0 gives

$$\varepsilon_0\vec{E} = \frac{Q}{4\pi R^2}\vec{a}_R,$$

where $4\pi R^2$ is the area of spherical surface with radius R, and Q is the charge in coulomb. So, $\varepsilon_0\vec{E}$ may be considered as surface charge density.

The electric field lines or the lines of force as in Fig. 2.17 can be designated as "electric flux" which may be taken as equal to the charge itself. Thus, a charge of Q coulomb can be said to produce an electric flux of $\psi = Q$ coulomb. The quantity $\varepsilon_0\vec{E}$ can be designated by the symbol $\vec{D}$ which has the dimension of charge-density and is generally designated as "electric flux-density". Thus,

$$\vec{D} = \frac{Q}{4\pi R^2}\vec{a}_R \qquad ...(2.42)$$

$$\vec{D} = \varepsilon_0 \vec{E} \quad \text{(in free space).} \qquad ...(2.43)$$

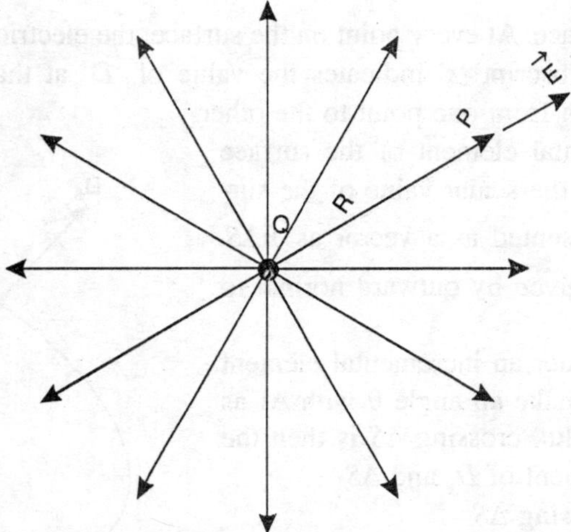

Fig. 2.17 Electric field lines or flux lines originating in an isolated point charge

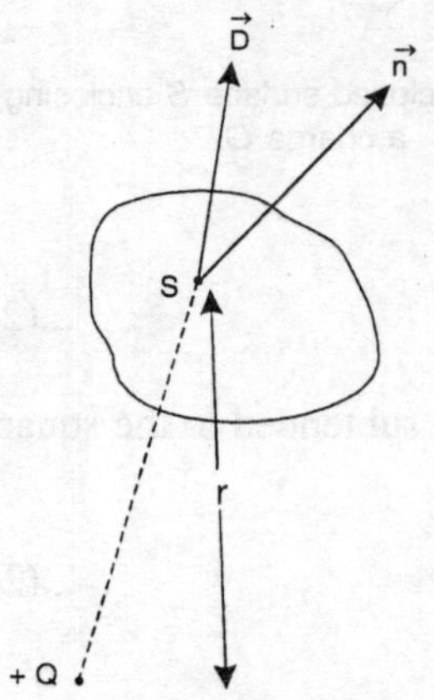

Fig. 2.18 Electric flux over A surface S due to A point charge

Eqn (2.43) indicates that the electric flux density is a vector which has the same direction as the electric field intensity.

Consider a small surface area S at an average distance r from an isolated charge $+Q$ as in Fig. 2.18. The electric flux over the surface S is given by the integral of the normal component of $\vec{D}$ over that surface. Area S can be represented as vector $\vec{n}_s$, where $\vec{n}$ is a unit vector given by outwards normal to the surface, and S is the scalar value of the surface area

Thus, $\qquad \psi = \vec{D}\cdot\vec{n}\,S. \qquad ...(2.44)$

In general, for the case of a large surface area, ψ can be obtained by integration of $\vec{D}$ over the whole surface. Thus

$$\psi = \iint_S \vec{D}\cdot\vec{n}ds \qquad ...(2.45)$$

The integral of normal component of a vector over a surface gives the 'flux' of the vector over that surface. Unit for electric flux is 'coulomb' and for $\vec{D}$ is coulomb/m^2.

2.1.6 Gauss's Law

Although Coulomb's law is applicable in finding the electric field due to any charge distribution, it is easier to use Gauss's law when charge distribution is symmetrical.

It states that: 'The electric flux passing through closed surface is equal to the total charge enclosed by that surface'.

Proof:

Assume any closed surface in an isotropic medium which consists of charge.

If the total charge inside the surface is Q coulomb, then Q coulomb of electric flux will pass through the enclosing surface. At every point on the surface, the electric flux density $\vec{D}$ will have same value of D_s where the subscript '*s*' indicates the value of '*D*' at that surface and D_s will vary in magnitude and in direction from one point to the other.

Consider an incremental element of the surface with an area of ΔS. ΔS is the scalar value of the surface area. It can be represented as a vector as $\vec{n}\Delta S$ where $\vec{n}$ is a unit vector given by outward normal to the surface.

At any point P, consider an incremental element of surface ΔS and let D_s make an angle θ with Δs as shown in Fig. 2.19. The flux crossing ΔS is then the product of normal component of D_s and ΔS.

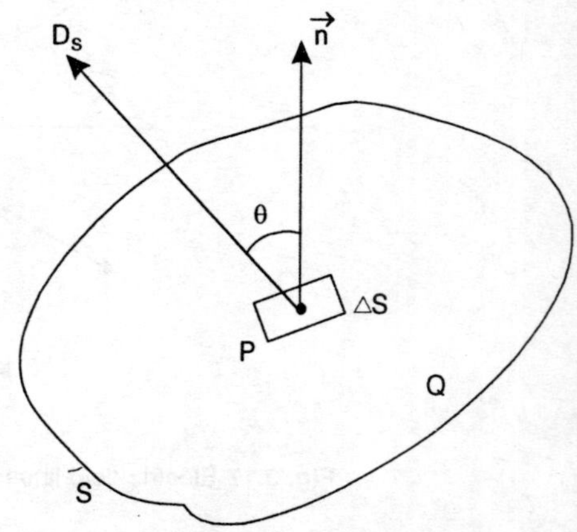

Fig. 2.19 A closed surface *S* enclosing a charge *Q*

$$\Delta\psi = \text{flux crossing } \Delta S$$

$$\Delta\psi = D_s \cdot \vec{n}\,\Delta S \qquad \text{...(2.46)}$$

$$\Delta\psi = D_s\,\Delta S\cos\theta. \qquad \text{...(2.47)}$$

From eqn. (2.42), $\vec{D} = \dfrac{Q}{4\pi R^2}\,\vec{a}_R$

$$\Delta\psi = \frac{Q}{4\pi R^2}\Delta S\cos\theta. \qquad \text{...(2.48)}$$

In solid geometry, the solid angle is defined as the ratio of surface area subtended to the square of the radius.

$$d\Omega = \frac{\Delta S\cos\theta}{R^2}. \qquad \text{...(2.49)}$$

From eqn. (2.46) and eqn. (2.48)

$$\Delta\psi = D_S \cdot \vec{n}\Delta S = \frac{Q}{4\pi R^2}\Delta S\cos\theta \qquad \text{...(2.50)}$$

$$D_S \cdot \vec{n}\Delta S = \frac{Q}{4\pi}d\Omega. \qquad \text{...(2.51)}$$

In solid geometry, the total angle subtended by any closed surface is 4π steradians.

$$\int \Delta\psi = \oint_S D_S \cdot \vec{n}\,\Delta S = \frac{Q}{4\pi}\cdot\oint d\Omega \qquad \text{...(2.52)}$$

$$\psi = \oint_S D_S \cdot \vec{n}\,\Delta S = \frac{Q}{4\pi} \times 4\pi$$

$$\psi = \oint_S D_S \cdot \hat{n}\Delta S = Q. \qquad \text{...(2.53)}$$

A small circle on the integral sign indicates that the integration is to be performed over a closed surface. Such surface is often called Gaussian surface.

Mathematical formulation of Gauss's law is

$$\psi = \oint \vec{D}_S \cdot \Delta\vec{S} = \text{Charge enclosed} = Q.$$

$\Delta\vec{S}$ can be replaced by $d\vec{s}$ for convenience.

$$\therefore \quad \psi = \oint_S \vec{D}_S \cdot d\vec{s} = \text{charge enclosed} = Q.$$

Gauss's law may be written in terms of the charge distribution as

$$\oint_S \vec{D}_S \cdot d\vec{s} = \int_{vol} \rho_v dv. \qquad \text{...(2.54)}$$

2.1.7 Applications of Gauss's Law

Gauss's law is used to find out the electric field intensity (or) electric flux density only for symmetrical charge distributions.

Consider the Gauss's law equation.

$$Q = \oint_S \vec{D}_S \cdot d\vec{s}. \qquad \text{...(2.55)}$$

It can be also written in terms of electric field intensity $\vec{E}$ as

$$Q = \oint_s \varepsilon_0 \vec{E}_s \cdot d\vec{s}, \qquad \text{since } \vec{D} = \varepsilon_0 \vec{E}.$$

$$\therefore \quad \oint_s \vec{E}_s \cdot d\vec{s} = \frac{Q}{\varepsilon_0}. \qquad \text{...(2.56)}$$

From the above two equations, it is known that the unknown quantity, either $\vec{D}_s$ or $\vec{E}_s$, is inside the integral.

2.1.8 Conditions of Gauss's Law

The solution is easy if a closed surface which satisfies the following two conditions is correctly chosen. The conditions are:

1. D_s is everywhere either normal or tangential to the closed surface so that the integral becomes either zero or $D_s ds$.

2. On that portion of the closed surface for which $\vec{D}_s \cdot d\vec{s}$ is not zero, $\vec{D}_s$ should be constant.

The above conditions is used to replace the dot product with the product of scalars $\vec{D}_s$ and $d\vec{s}$ and then to bring $\vec{D}_s$ outside the integral sign. The remaining integral is then $\int_s ds$ over that portion of closed surface which $\vec{D}_s$ crosses normally.

By remembering that the electric field intensity due to a positive point charge is directed radially outward from the point charge, it is easy to choose a closed surface which is known as Gaussian surface.

As already mentioned, Gauss's law is applied for only symmetrical distribution of charges. A charge distribution is said to be symmetrical, if the resultant direction of electric field intensity is along only one direction. From Example 2.7, the infinite line charge has the electric field intensity along cylindrical radial $(\vec{a}_\rho)$ axis. But the finite length of line charge has both x and y components as in example 2.5. So, infinite line charge is said to be having symmetrical distribution, but not finite line charge. Similarly, the infinite surface charge and spherical volume charge are said to be of symmetrical distribution.

As the direction of electrical field intensity due to infinite line charge is along cylindrical radial axis, it is said to be having cylindrical symmetry and the Gaussian surface to be selected is of closed circular cylinder. Similarly, as the electric field intensity due to infinite surface charge is along $\vec{a}_z$ axis, and due to spherical volume charge is along spherical radial $(\vec{a}_r)$, they are said to be having rectangular (cartesian) symmetry and spherical symmetry respectively. Hence a rectangular box is a Gaussian surface for infinite surface and a sphere is a Gaussian surface for spherical volume charge distribution.

Let us now apply this Gauss's law for finding out electric field intensity due to a point charge.

Assume the point charge of Q coulomb is at the origin of spherical co-ordinate system. A closed Gaussian surface which will meet the above-mentioned two requirements should be selected. The suitable surface is obviously a spherical surface, centered at origin of any radius r. For this surface, $\vec{D}_s$ is everywhere normal and has the same value at all points on the surface.

Thus $$Q = \oint_s \vec{D}_s \cdot d\vec{s}$$

$$= \oint_{sph} D_s\, \vec{a}_r \cdot ds\, \vec{a}_r \qquad ...(2.57)$$

From eqn. (1.48), the differential area of spherical surface $ds\,\vec{a}_r$ is given by

$$d\vec{s} = r^2 \sin\theta\, d\theta\, d\phi\, \vec{a}_r$$

$$Q = \vec{D}_s \oint_{sph} ds = \vec{D}_s \int_{\phi=0}^{\phi=2\pi} \int_{\theta=0}^{\theta=\pi} r^2 \sin\theta\, d\theta\, d\phi \qquad ...(2.58)$$

$$Q = 4\pi r^2 D_s.$$

Hence, $\vec{D}_s = \dfrac{Q}{4\pi r^2}\vec{a}_r.$

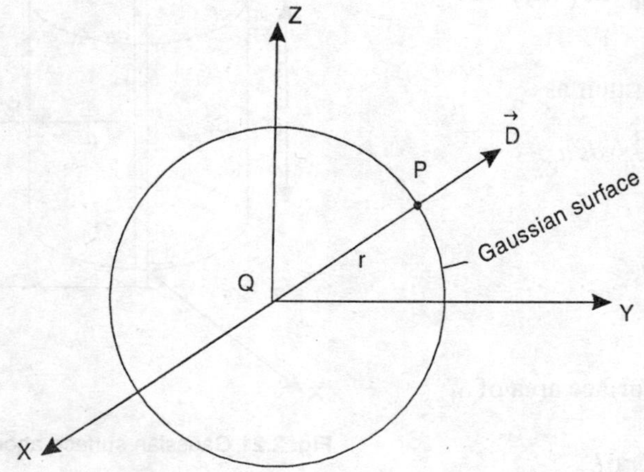

Fig. 2.20 Gaussian surface about a point charge.

Since r may have any value and since $\vec{D}_s$ is directed radially outward, it is along $\vec{a}_r$.

$$\vec{D} = \frac{Q}{4\pi r^2}\,\vec{a}_r,\ \vec{E} = \frac{Q}{4\pi\varepsilon_0 r^2}\,\vec{a}_r. \qquad ...(2.59)$$

Example 2.10

An infinitely long line charge of uniform density ρ_L C/m is situated along the Z-axis. Obtain the electric field intensity due to this charge using Gauss's law.

According to the qualitative arguments in Example 2.7, it is known that $\vec{E}$, due to the infinite line charge of uniform density, is directed radially away from the line charge and its magnitude is dependent only on its distance from the line charge (ρ). Thus,

$$\vec{E} = E_\rho(\rho)\vec{a}_\rho. \qquad ...(2.60)$$

To satisfy the symmetry condition, a closed cylindrical surface is chosen as a Gaussian surface. The cylindrical surface is the only surface to which $\vec{E}_\rho$ is everywhere normal and constant. It may be closed by plane surfaces normal to the Z-axis. A closed right circular cylinder of radius ρ, extending from $Z = 0$ to $Z = L$ is shown in Fig. 2.21.

$$\underset{\substack{\text{surface of}\\ \text{cylinder, } S}}{\oint \vec{E}\cdot d\vec{s}} = \underset{\substack{\text{curved}\\ \text{surface } S_1}}{\oint \vec{E}\cdot d\vec{s}} + \underset{\substack{\text{Plane (circular)}\\ \text{surfaces } S_2,\ S_3}}{\oint \vec{E}\cdot d\vec{s}} \qquad ...(2.61)$$

The second integral on the right side is zero, since $\vec{E}$ is normal to the surfaces. That is,

$$\int_{S_2} \vec{E}\cdot d\vec{s} = \int E_\rho \vec{a}_\rho \cdot ds\,\vec{a}_z = 0$$

$$\int_{S_3} \vec{E}\cdot d\vec{s} = \int E_\rho \vec{a}_\rho \cdot ds(-\vec{a}_z) = 0.$$

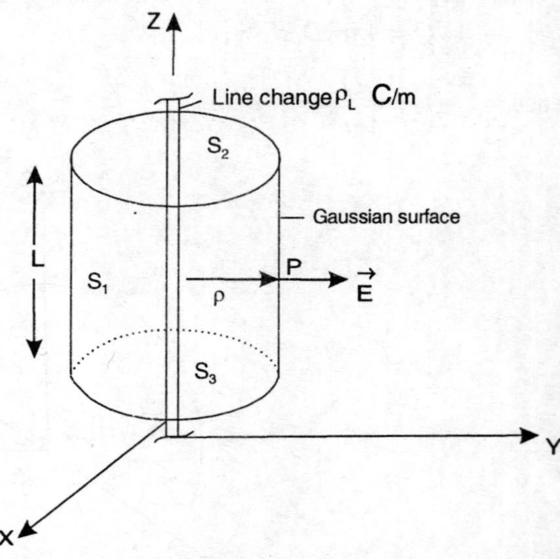

Fig. 2.21 Gaussian surface about an infinite line charge.

First integral can be written as

$$\int_{S_1} \vec{E}\cdot d\vec{s} = \int_{S_1} E_\rho\, \vec{a}_\rho \cdot ds\,\vec{a}_\rho$$

$$= \vec{E}_\rho \int_{S_1} ds$$

$$= \vec{E}_\rho \times \text{surface area of } S_1$$

$$= \vec{E}_\rho \times 2\pi\rho L$$

$$\oint \vec{E}\cdot d\vec{s} = \vec{E}_\rho\,(2\pi\rho L). \qquad \text{...(2.62.1)}$$

From Gauss's law

$$\oint \vec{E}\cdot d\vec{s} = \frac{\text{Charge enclosed by } S \text{ (Gaussian surface)}}{\varepsilon_0}$$

$$\oint_S E\cdot ds = \frac{\rho_L : L}{\varepsilon_0} \qquad \text{...(2.62.2)}$$

Comparing eqns. (2.62.1) and (2.62.2),

$$\vec{E} = \frac{\rho_L}{2\pi\varepsilon_0\rho}\vec{a}_\rho\,. \qquad \text{...(2.63)}$$

Example 2.11

A sheet charge of uniform density ρ_s C/m^2 extends over the entire *XY* plane. Obtain the electric field intensity due to this infinite sheet charge using Gauss's law.

From the qualitative arguments in Example 2.8, $\vec{E}$ due to the infinite sheet charge of uniform density is directed normally away from the sheet charge that is along *Z*-axis and it is uniform in planes parallel to the sheet charge.

Thus, $\quad \vec{E} = E_z\,\vec{a}_z$

A rectangular box that is cut symmetrically by the sheet of charge and has two of its faces parallel to the sheet as shown in Fig. 2.22.

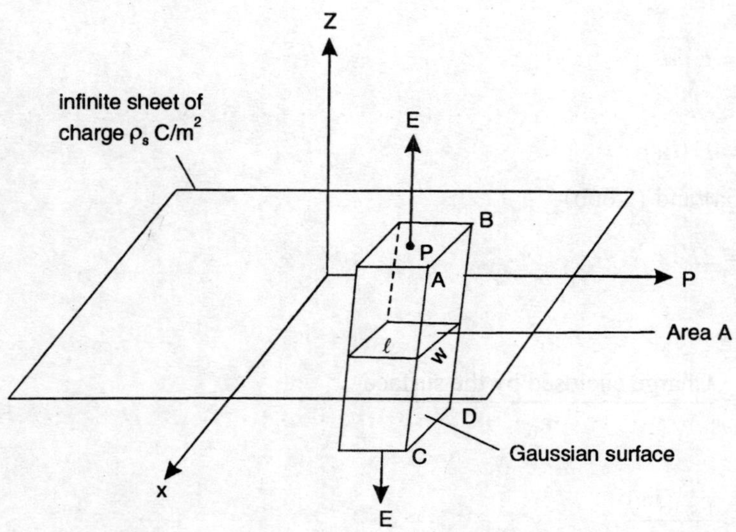

Fig. 2.22 Gaussian surface about an infinite line sheet of charge.

Applying Gauss's law.

$$\oint_S \vec{E}\cdot d\vec{s} = \oint_{\substack{\text{top}\\\text{Surface}}} \vec{E}\cdot d\vec{s} + \oint_{\substack{\text{bottom}\\\text{surface}}} \vec{E}\cdot d\vec{s} + \oint_{\substack{\text{side}\\\text{surfaces}}} \vec{E}\cdot d\vec{s} \qquad ...(2.64)$$

Last integral will become zero, since $\vec{E}$ is normal to all the side surfaces. For example, take one of the side surfaces *ABCD.*

The surface area is along $\vec{a}_y$.

$$\therefore \quad \int_S \vec{E}\cdot d\vec{s} = \int_S E(\vec{a}_z)\cdot d\vec{s}\,\vec{a}_y = 0 \qquad ...(2.65)$$

From the Fig. 2.22,

$$\int_{\substack{\text{top}\\\text{surface}}} \vec{E}\cdot d\vec{s} = \int_S E\vec{a}_z \cdot ds(\vec{a}_z)$$

$$= E\int_S ds$$

$$= E\ (lw) \qquad ...(2.66a)$$

$$\int_{\substack{\text{bottom}\\\text{surface}}} \vec{E}\cdot d\vec{s} = \int_S E(-\vec{a}_z)\cdot ds(-\vec{a}_z)$$

$$= E\int_S ds$$

$$= E\,(lw)\,. \qquad ...(2.66b)$$

From eqns. (2.66a) and (2.66b)

$$\therefore \quad \oint_S \vec{E}_z \cdot d\vec{s} = 2E\,lw\,. \qquad ...(2.67)$$

From Gauss's law

$$\oint_S \vec{E}_z \cdot d\vec{s} = \frac{\text{Charge enclosed by the surface}}{\varepsilon_0}$$

$$\oint_S \vec{E}_z \cdot d\vec{s} = \frac{\rho_s \cdot (lw)}{\varepsilon_0}\,. \qquad ...(2.68)$$

Comparing eqns. (2.67) and (2.68),

$$E_z = \frac{\rho_s}{2\varepsilon_0}$$

$$\vec{E} = \frac{\rho_s}{2\varepsilon_0}\vec{a}_z\,. \qquad ...(2.69)$$

Example 2.12

A volume charge is distributed throughout a sphere of radius '*a*' with uniform density ρ_v C/m^3. Obtain the electric field intensity due to this volume charge using Gauss's law.

Electric field intensity due to spherical volume charge of uniform density is directed radially away from the centre of the charge and is a function of the distance from the centre of sphere. Thus

$$\vec{E} = E_r(r)\vec{a}_r\,,$$

choosing the Gaussian surface S as the surface of sphere of radius $r > a$ and $r < a$. For $r > a$, consider the point is outside the sphere, and for $r < a$ the point is inside the sphere.

Case 1: $r < a$ (inside).

Consider the point(P) is inside the given spherical volume of radius '*a*'.

Consider the point is at a distance of '*r*' m from the centre of sphere.

Construct the Gaussian sphere of radius '*r*' m as in Fig. 2.23.

$$\oint \vec{E} \cdot d\vec{s} = \int_{Sph} E_r\,\vec{a}_r \cdot ds\,\vec{a}_r$$

$$= \vec{E}_r \int_{Sph} d\vec{s}$$

$$= \vec{E}_r \text{ (surface area of Gaussian sphere)}$$

$$= E_r \left(4\pi r^2\right). \qquad \text{...(2.70)}$$

From Gauss's law

$$\oint_S \vec{E} \cdot d\vec{s} = \frac{\text{Charge enclosed by Gaussian sphere } S}{\varepsilon_0}$$

$$= \frac{\rho_v \times \text{volume of Gaussian sphere of radius } r}{\varepsilon_0}$$

$$= \frac{\rho_v \times \frac{4}{3}\pi r^3}{\varepsilon_0}. \qquad \text{...(2.71)}$$

Comparing eqns. (2.70) and (2.71),

$$\vec{E}_r = \frac{\rho_v r}{3\varepsilon_0}$$

$$\vec{E}_r = \frac{\rho_v r}{3\varepsilon_0}\vec{a}_r. \qquad \text{...(2.72)}$$

Case (*ii*) $r > a$ (outside the sphere).
Consider the point 'P' is outside the given spherical volume of radius 'a'.

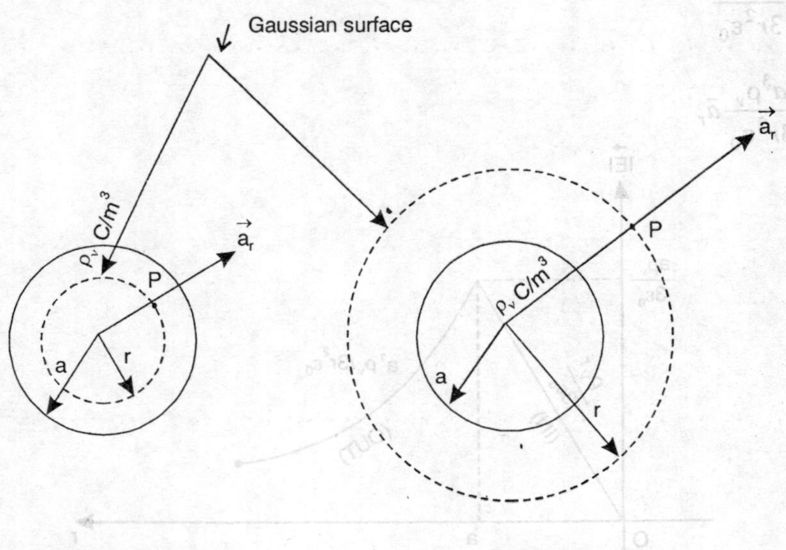

Fig. 2.23 **Fig. 2.24**

Let the point be at a distance of 'r' m from the centre of the sphere and construct a spherical Gaussian surface at this point with the radius of 'r' m as in Fig 2.24. Then,

$$\oint_S \vec{E}\cdot d\vec{s} = \int_{sph} \vec{E}_r\,\vec{a}_r \cdot ds\,\vec{a}_r$$

$$= \vec{E}_r \int_{sph} d\vec{s}$$

$$= \vec{E}_r \text{ (surface area of Gaussian sphere of radius '}r\text{' m)}$$

$$= \vec{E}_r \cdot 4\pi r^2. \quad \text{...(2.73)}$$

From Gauss's law

$$\oint_S \vec{E}\cdot d\vec{s} = \frac{\text{Charge enclosed by the Gaussian surface } S \text{ of radius } r}{\varepsilon_0}$$

Since $r > a$, charge enclosed by the Gaussian surface of radius 'r' is same as the charge enclosed by the given spherical volume of radius 'a'.

$$\oint \vec{E}\cdot d\vec{s} = \frac{\rho_v \times \text{volume of given sphere of radius '}a\text{'}}{\varepsilon_0}$$

$$= \frac{\rho_v \times \frac{4}{3}\pi a^3}{\varepsilon_0}. \quad \text{...(2.74)}$$

Comparing eqns. (2.73) and (2.74),

$$\vec{E}_r = \frac{a^3 \rho_v}{3r^2 \varepsilon_0}$$

$$\vec{E} = \frac{a^3 \rho_v}{3r^2 \varepsilon_0}\vec{a}_r \quad \text{...(2.75)}$$

Fig. 2.25 Variation of $|\vec{E}|$ against r

$$\vec{E} = \begin{cases} \dfrac{r\rho_v}{3\varepsilon_0}\vec{a}_r & 0 < r \le a \\ \dfrac{a^3\rho_v}{3r^2\varepsilon_0}\vec{a}_r & r \ge a \end{cases} \quad ...(2.76)$$

Gaussian surface form uniformly-charged sphere when $r \le a$ and $r \ge a$.

SOLVED PROBLEMS

Problem 1

Two small identical conducting spheres have charges of 2×10^{-9} coulomb and -0.5×10^{-9} coulomb respectively. If they are brought into contact and then separated by 4 cm, what is the force between them?

Solution:

When two identical conducting spheres are brought into contact and then separated, the charge in each sphere is given by

$$Q_1 = Q_2 = Q.$$

$$= \frac{2 \times 10^{-9} + (-0.5 \times 10^{-9})}{2}$$

$$= 0.75 \times 10^{-9}\ \text{C}$$

The force $\vec{F}$ when they are separated by 4 cm apart will be

$$\vec{F} = \frac{Q_1 Q_2}{4\pi\varepsilon_0 r^2}\vec{a}_r$$

$$= \frac{(0.75 \times 10^{-9})^2}{4\pi \times 8.854 \times 10^{-12} \times (4 \times 10^{-2})^2}\vec{a}_r$$

$$\vec{F} = 0.3164 \times 10^{-5}\ \text{N}$$

Problem 2

Two particles, each of mass '*m*' and carrying a charge '*q*', are suspended by a string of length *l* from a common point. Show that angle θ which each string makes with the vertical is obtained from

$$\frac{\tan^3\theta}{1 + \tan^2\theta} = \frac{q^2}{16\pi\varepsilon_0 l^2\ \text{mg}}.$$

Solution:

Two charged particles are suspended from a common point as shown in Fig. P-2.

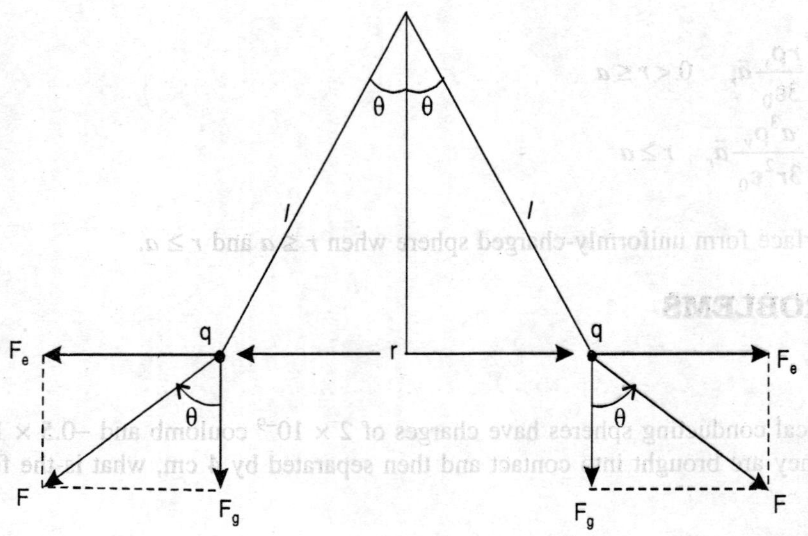

Fig. P-2

Since the two particles are similarly charged, there will be a force of repulsion between them. Also, each particle will experience gravitational force F_g due to its mass 'm'.

The resultant force $\vec{F} = \vec{F}_e + \vec{F}_g$

$$\left|\vec{F}_g\right| = \text{mg}, \quad \left|\vec{F}_e\right| = \frac{q^2}{4\pi\varepsilon_0 r^2}$$

where
g = acceleration due to gravity
r = distance between the particle.
l = length of each string.
θ = angle made by each string with the vertical.
r = $l\sin\theta + l\sin\theta = 2l\sin\theta$

$$\therefore \quad \vec{F}_e = \frac{q^2}{4\pi\varepsilon_0\,(2l\rho\sin\theta)^2}$$

$$\left|\vec{F}\right| = \sqrt{\left|\vec{F}_e\right|^2 + \left|\vec{F}_g\right|^2}$$

$$\tan\theta = \frac{F_e}{F_g}.$$

$$\therefore \quad \frac{q^2}{4\pi\varepsilon_0\left(4l^2\sin^2\theta\right)} = F_g\tan\theta$$

$$\frac{q^2}{16\pi\varepsilon_0 l^2\sin^2\theta} = \text{mg}\tan\theta$$

$$\frac{q^2}{16\pi\varepsilon_0 l^2 mg} = \sin^2\theta \cdot \tan\theta$$

$$= \sin^2\theta \cdot \tan\theta \cdot \frac{\tan^2\theta}{\tan^2\theta}$$

$$= \frac{\tan^3\theta \cdot \sin^2\theta}{\frac{\sin^2\theta}{\cos^2\theta}}$$

$$= \frac{\tan^3\theta}{\sec^2\theta}$$

$$= \frac{\tan^3\theta}{1+\tan^2\theta}$$

$$\therefore \quad \frac{q^2}{16\pi\varepsilon_0 l^2 mg} = \frac{\tan^3\theta}{1+\tan^2\theta}. \text{ Thus proved.}$$

Problem 3

Find the value of point charge placed at the centre of an equilateral triangle which holds the three point charges '$+q$' at the corners of triangle in equilibrium.

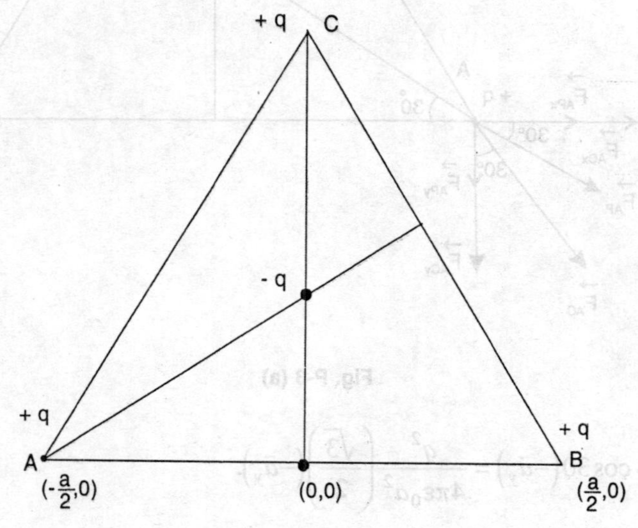

Fig. P-3

Solution:

Let us consider a point charge '$-q$' is placed at the centroid. The force acting on any charge due to all the other charges will be zero for equilibrium. Let us assume the point at A.

Force on A due to B,

$$\vec{F}_{AB} = \frac{q^2}{4\pi\varepsilon_0 a^2}\left(-\vec{a}_x\right).$$

Force on A due to C,

$$\vec{F}_{AC} = \frac{q^2}{4\pi\varepsilon_0 a^2}\vec{a}_{AC}.$$

Resolving $\vec{F}_{AC}$ into two components,

$$\vec{F}_{ACx} = \vec{F}_{AC}\sin 30\left(-\vec{a}_x\right) = \frac{q^2}{4\pi\varepsilon_0 a^2}\left(\frac{1}{2}\right)\left(-\vec{a}_x\right)$$

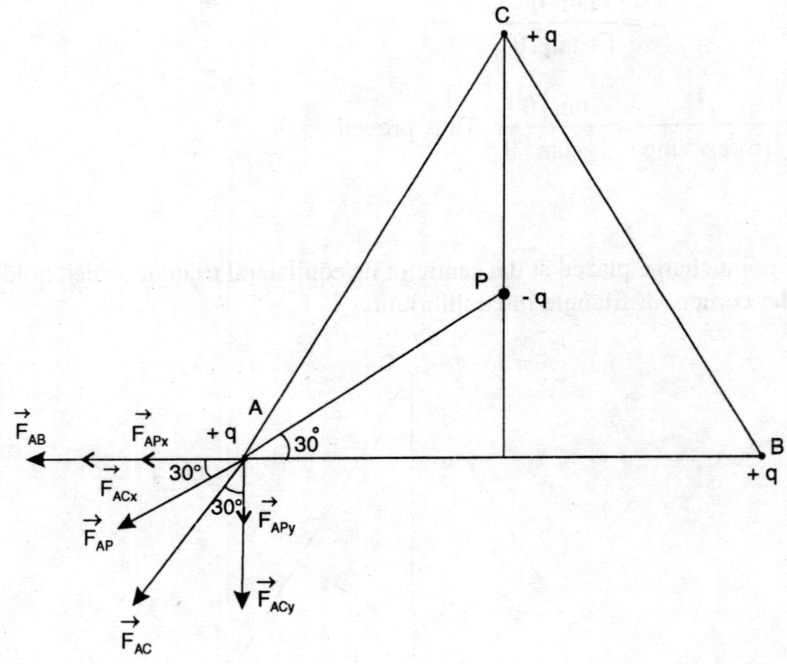

Fig. P-3 (a)

$$\vec{F}_{ACy} = \vec{F}_{AC}\cos 30\left(-\vec{a}_y\right) = \frac{q^2}{4\pi\varepsilon_0 a^2}\left(\frac{\sqrt{3}}{2}\right)\left(-\vec{a}_y\right).$$

Similarly, force on A due to $-q$ at P,

$$\vec{F}_{AP} = \frac{-qQ}{4\pi\varepsilon_0 r_{AP}^2}\vec{a}_{AP}.$$

From Figure P-3,

$$\cos 30° = \frac{a/2}{AP}$$

$$AP = \frac{a}{2} \cdot \frac{2}{\sqrt{3}} = \frac{a}{\sqrt{3}}$$

$$\therefore \quad \vec{F}_{AP} = \frac{-qQ}{4\pi\varepsilon_0 \left(\frac{a^2}{3}\right)} \vec{a}_{AP} \, .$$

Resolving $\vec{F}_{AP}$ into two components,

$$\vec{F}_{APx} = \vec{F}_{AP} \quad \cos 30 \cdot (-\vec{a}_x)$$

$$= \frac{3qQ}{4\pi\varepsilon_0 a^2} \quad \vec{a}_x \cdot \frac{\sqrt{3}}{2}$$

$$\vec{F}_{APy} = \vec{F}_{AP} \sin 30 \cdot (-\vec{a}_y)$$

$$= \frac{3qQ}{4\pi\varepsilon_0 a^2} \cdot \left(\frac{1}{2}\right) \vec{a}_y \, .$$

For equilibrium,

$$\vec{F}_A = \vec{F}_{AB} + \vec{F}_{AC} + \vec{F}_{AP} = 0$$

$$\left[\frac{-q^2}{4\pi\varepsilon_0 a^2} - \frac{q^2}{4\pi\varepsilon_0 a^2}\left(\frac{1}{2}\right) + \frac{3\sqrt{3}\,Qq}{8\pi\varepsilon_0 a^2}\right]\vec{a}_x + \left[\frac{-q^2}{4\pi\varepsilon_0 a^2} \cdot \frac{\sqrt{3}}{2} + \frac{3qQ}{8\pi\varepsilon_0 a^2}\right]\vec{a}_y = 0$$

$$= \frac{q}{4\pi\varepsilon_0 a^2}\left\{\left(-q + \left(-\frac{q}{2}\right) + \frac{3\sqrt{3}\,Q}{2}\right)a_x + \left(\frac{-q\sqrt{3}}{2} + \frac{3Q}{2}\right)\vec{a}_y\right\}$$

$= 0.$

For the above function to be zero,

$$-q + \left(\frac{-q}{2}\right) + \frac{3\sqrt{3}\,Q}{2} = 0 \qquad \text{...(1)}$$

$$\frac{-q\sqrt{3}}{2} + \frac{3Q}{2} = 0. \qquad \text{...(2)}$$

Condition (1) gives

$$\frac{3\sqrt{3}\,Q}{2} = \frac{3q}{2}$$

(or) $q = \sqrt{3}\,Q$.

Condition (2) gives

$$\frac{-q\sqrt{3}}{2} + \frac{3Q}{2} = 0$$

(or) $q = \sqrt{3}\,Q$.

So, when the negative charge of value $\sqrt{3}\,Q$ is placed at the centroid, three charges placed at the corners of the triangle will be in equilibrium.

Problem 4

A circular disc of 10 cm radius is charged uniformly with a total charge of 10^{-6} C. Find the electric intensity at a point 30 cm away from the disc along the axis.

Solution:

From Example 2.6,

Electric field intensity $\vec{E} = \dfrac{\sigma}{2\varepsilon_0}(1 - \cos\alpha)\vec{a}_n$

$$\sigma = \frac{Q}{\pi a^2}$$

$$\cos\alpha = \frac{h}{\sqrt{h^2 + a^2}}.$$

Substituting $Q = 10^{-6}$, $h = 0.3$ m, $a = 0.1$ m.

$$\vec{E} = \frac{Q}{\pi a^2 \times 2\varepsilon_0}(1 - \cos\alpha)\vec{a}_n$$

$$= \frac{10^{-6}}{\pi \times (0.1)^2 \times 2 \times 8.854 \times 10^{-12}}\left[1 - \frac{0.3}{\sqrt{(0.3)^2 + (0.1)^2}}\right]$$

$$\vec{E} = 922.85\ \text{kV/m}.$$

Problem 5

Two charges are in the positions shown in the figure. At what point on the *X*-axis is the electric field intensity equal to zero?

Solution:

Case 1: If point *A* is between *OP*, the electric field intensity due to Q_1 and Q_2 are along the same direction and so it never becomes zero.

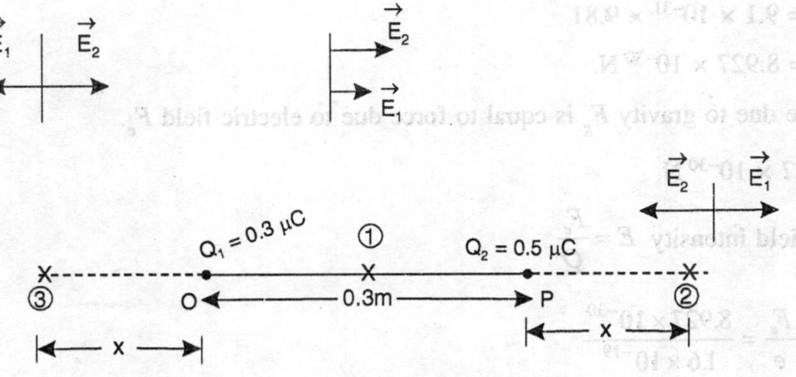

Fig. P-5

Case 2: Point *A* at any position beyond *P*,

$$\vec{E} \text{ due to } Q_1,\ \vec{E}_1 = \frac{Q_1}{4\pi\varepsilon_0 r_1^2}\vec{a}_x$$

$$\vec{E} \text{ due to } Q_2,\ \vec{E}_2 = \frac{Q_2}{4\pi\varepsilon_0 r_2^2}(-\vec{a}_x), \text{ where } r_1 = r \text{ and } r_2 = r + x.$$

Since $Q_1 < Q_2$ and $r_1 > r_2$, the field due to $Q_1\left(\vec{E}_1\right)$ is smaller in magnitude than due to $Q_2\left(\vec{E}_2\right)$ at all such points. Consequently, there cannot be zero field point beyond *P*.

Case 3: Point *A* at any position to the left of O. Here $Q_1 < Q_2$, but $r_1 < r_2$. So, there will be a point *A* at distance *x* to the left of O, where the two fields get neutralised.

Equating the magnitudes of the fields $\left|\vec{E}_1\right|$ and $\left|\vec{E}_2\right|$,

$$\frac{Q_1}{4\pi\varepsilon_0 x^2} = \frac{Q_2}{4\pi\varepsilon_0 (x+0.3)^2}$$

$$\frac{0.3}{4\pi\varepsilon_0 x^2} = \frac{0.5}{4\pi\varepsilon_0 (x+0.3)^2}.$$

Solving for *x*,

$$x = 1.034 \text{ m}.$$

Problem 6

Calculate the electric field intensity that will just be sufficient to balance the gravitational force of an election ($e = 1.6 \times 10^{-19}$ C and $m = 9.1 \times 10^{-31}$ kg for an election).

Solution:

$$m = 9.1 \times 10^{-31} \text{ kg}$$

$$g = 9.81 \text{ N} - \text{m}.$$

$$F_g = m \times g.$$

$= 9.1 \times 10^{-31} \times 9.81$

$= 8.927 \times 10^{-30}$ N.

This force due to gravity F_g is equal to force due to electric field F_e.

$F_e = 8.927 \times 10^{-30}$ N.

Electric field intensity $E = \dfrac{F_e}{Q}$

$$= \frac{F_e}{e} = \frac{8.927 \times 10^{-30}}{1.6 \times 10^{-19}}$$

$= 5.5 \times 10^{-11}$ V/m.

Problem 7

On the line described by $x = 2$ m, $y = -4$ m, there is a uniform charge distribution of density $\rho_L = 20$ nC/m. Determine the electric field at (–2, –1, 4) m.

Solution:

Since the line is parallel to $\vec{a}_z$, the field has no Z component.

$$\vec{E} = \frac{\rho_L}{2\pi\varepsilon_0\rho}\vec{a}_\rho,$$

where P is the distance of the point from the line charge

$$\vec{R} = (-2, -1) - (2, -4)$$

$$= (-4, 3)$$

$$= -4\vec{a}_x + 3\vec{a}_y$$

$$\vec{a}_\rho = \frac{\vec{R}}{|\vec{R}|} = \frac{-4\vec{a}_x + 3\vec{a}_y}{5}.$$

$$\therefore \quad \vec{E} = \frac{20 \times 10^{-9}}{2\pi \times 8.54 \times 10^{-12} \times 5} \times \left(\frac{-4\vec{a}_x + 3\vec{a}_y}{5}\right)$$

$$\vec{E} = -57.69\vec{a}_x + 43.2\vec{a}_y \text{ V/m}.$$

Problem 8

Two uniform line charges of density $\rho_l = 4$ nC/m lie in the $x = 0$ plane at $y = 4$ m. Find $\vec{E}$ at (4, 0, 10) m.

Solution:

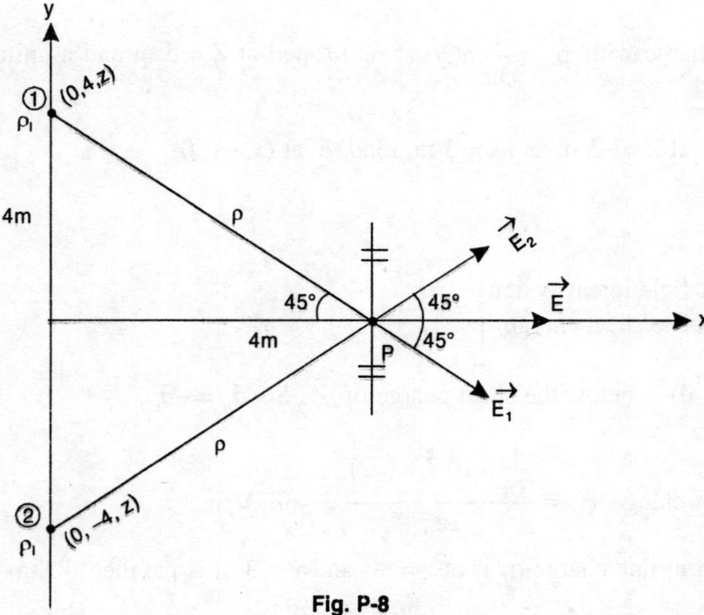

Fig. P-8

The line charges are both parallel to $\vec{a}_z$. The electric field intensities are radial and parallel to the *XY* plane.

The electric field intensity due to the line charge along $y = 4$ is

$$\vec{E}_1 = \frac{\rho_l}{2\pi\varepsilon_0 \rho}\vec{a}_{1P}$$

$$= \frac{4\times 10^{-9}}{2\pi \times 8.854\times 10^{-12} \times 4\sqrt{2}}\vec{a}_{1P}.$$

Similarly, electric field intensity due to the line charge along $y = -4$ is

$$\vec{E}_2 = \frac{\rho_l}{2\pi\varepsilon_0 \rho}\vec{a}_{2P}$$

$$= \frac{4\times 10^{-9}}{2\pi \times 8.854\times 10^{-12} \times \left(4\sqrt{2}\right)}\vec{a}_{2P}.$$

The vertical component of two intensities are equal and opposite. Hence, they cancel each other. The horizontal components are added together.

$$\vec{E} = E_1 \cos 45 + E_2 \cos 45$$

$$= \frac{2\times 4\times 10^{-09}}{2\pi \times 8.854\times 10^{-12} \times 4\sqrt{2}} \times \cos 45$$

$$\vec{E} = 18\vec{a}_x \text{ V/m}.$$

Problem 9

A uniform sheet charge with $\rho_s = \frac{1}{3\pi} \text{nC/m}^2$ is located at $Z = 5$ m and a uniform line charge with $\rho_L = \left(\frac{-25}{9}\right) \text{nC/m}$ at $Z = -3$ m and $y = 3$ m. Find $\vec{E}$ at $(x, -1, 0)$.

Solution:

$$\left.\begin{array}{l}\text{Electric field intensity at a}\\\text{point due to sheet charge}\end{array}\right\} = \frac{\rho_s}{2\varepsilon_0}\vec{a}_n$$

Point $P\,(x, -1, 0)$ is below the sheet charge of $\vec{Z}$. So $\vec{a}_n = -\vec{a}_z$.

$$\vec{E} \text{ due to sheet charge } \vec{E}_S = \frac{\frac{1}{3\pi} \times 10^{-9} \times (\vec{a}_z)}{2\varepsilon_0} = -6\vec{a}_z \text{ V/m}.$$

Since the uniform line charge ρ_L is at $z = -3$ and $y = 3$, it is parallel to X-axis.

Electric field intensity due to line charge $= \vec{E}_L = \frac{\rho_L}{2\pi\varepsilon_0\rho}\vec{a}_\rho$, where $\vec{a}_\rho$ is the unit vector of line joining the line charge at $(3, -3)$ to $(-1, 0)$ $\vec{a}_\rho = \frac{4\vec{a}_y - 3\vec{a}_z}{5}$

$$\therefore \quad \vec{E}_L = \frac{\frac{-25}{9} \times 10^{-9}}{2\pi\varepsilon_0 \times 5} \cdot \frac{4\vec{a}_y - 3\vec{a}_z}{5}$$

$$\vec{E}_L = 8\vec{a}_y = 6\vec{a}_z.$$

$$\therefore \quad \text{Total electric field} = \vec{E} = \vec{E}_L + \vec{E}_S$$

$$= 8\vec{a}_y - 12\vec{a}_z \text{ V/m}.$$

Problem 10

A charge of uniform density $\rho_s = 0.3 \text{ nC/m}^2$ covers the plane $2x - 3y + z = 6$. Find $\vec{E}$ on the side of the plane containing the origin.

Solution:

Electric field intensity due to uniform sheet $= \frac{\rho_s}{2\varepsilon_0}\vec{a}_n$

$$= \frac{0.3 \times 10^{-9}}{2 \times 8.854 \times 10^{-12}} \vec{a}_n$$

$$= 17.0 \vec{a}_n \text{ V/m}.$$

The unit normal vector for a plane $Ax + By + Cz = D$ are

$$\vec{a}_n = \frac{\pm A\vec{a}_x + B\vec{a}_y + C\vec{a}_z}{\sqrt{A^2 + B^2 + C^2}}.$$

∴ Unit normal vector for the plane $2x - 3y + z = 6$ will be

$$\vec{a}_n = \pm \frac{2 \cdot \vec{a}_x + (-3) \cdot \vec{a}_y + 1 \cdot \vec{a}_z}{\sqrt{14}}.$$

Since the point (0, 0, 0) is below this plane, the direction of unit normal vector is negative.

$$\therefore \quad \vec{E} = \frac{17 \times \left(-2\vec{a}_x + 3 \cdot \vec{a}_y - \vec{a}_z\right)}{\sqrt{14}}$$

$$\vec{E} = -9\vec{a}_x + 13.62\vec{a}_y - 4.54\vec{a}_x \text{ V/m}.$$

Problem 11

A uniform line charge with $\rho_L = 5$ μC/m lies along the *X*-axis. Find $\vec{D}$ at (3, 2, 1) m.

Solution:

Since the line charge lies along *X*-axis, *y* and *z* component will be zero.

The co-ordinates of line joining line charge along *X*-axis and the point (3, 2, 1) is

$(2, 1) - (0, 0) = (2, 1) = 2\vec{a}_y + \vec{a}_z$.

$$\vec{a}_\rho = \frac{2\vec{a}_y + \vec{a}_z}{\sqrt{5}}.$$

$$\therefore \quad \vec{E} = \frac{\rho_L}{2\pi\varepsilon_0\rho} \vec{a}_\rho$$

$$= \frac{5 \times 10^{-6}}{2\pi \times 8.54 \times 10^{-12} \times \sqrt{5}} \times \frac{2\vec{a}_y + \vec{a}_z}{\sqrt{5}}$$

$$\vec{E} = 17{,}975.5\left(2\vec{a}_y + \vec{a}_z\right).$$

Problem 12

Determine the electric field intensity $\vec{E}$ at a point $2a$ along the axis perpendicular to the plane of a circular wire charged uniformly at λC/m which has a radius '*a*'.

Solution:

Consider a differential charge (dq) of length dl ($dq = \lambda dl$) at a point A on the circular ring.

The electric field intensity at P due to the charge at A

$$d\vec{E}_A = \frac{\lambda dl}{4\pi\varepsilon_0 r^2}\vec{a}_{AP}$$

$$d\vec{E}_A = \frac{\lambda dl}{4\pi\varepsilon_0\left(a^2+(2a)^2\right)}\vec{a}_{AP}.$$

The electric field intensity at P due to the charge at B which is diametrically opposite to A is

$$d\vec{E}_B = \frac{\lambda dl}{4\pi\varepsilon_0\left(a^2+(2a)^2\right)}\vec{a}_{BP}.$$

From the Fig. P-12, the resultant electric field intensity at 'P' has only Z component.

$$d\vec{E} = d\vec{E}_A + d\vec{E}_B$$

$$d\vec{E}_z = \frac{2\lambda dl}{4\pi\varepsilon_0\left\{a^2+4a^2\right\}}\cos\theta\,\vec{a}_z$$

$$d\vec{E}_z = \frac{\lambda\, dl}{2\pi\varepsilon_0 5a^2}\cdot\frac{2a}{\sqrt{a^2+4a^2}}\vec{a}_z$$

$$E_z = \oint\frac{\lambda\, dl}{5\sqrt{5}\,a^2\varepsilon_0\pi}\vec{a}_z$$

$$\vec{E}_z = \frac{\lambda}{5\sqrt{5}\,\varepsilon_0\pi a^2}\oint dl\;\vec{a}_z$$

$$\vec{E}_z = \frac{\lambda}{5\sqrt{5}\,\varepsilon_0\pi a^2}\times\pi a\,\vec{a}_z.$$

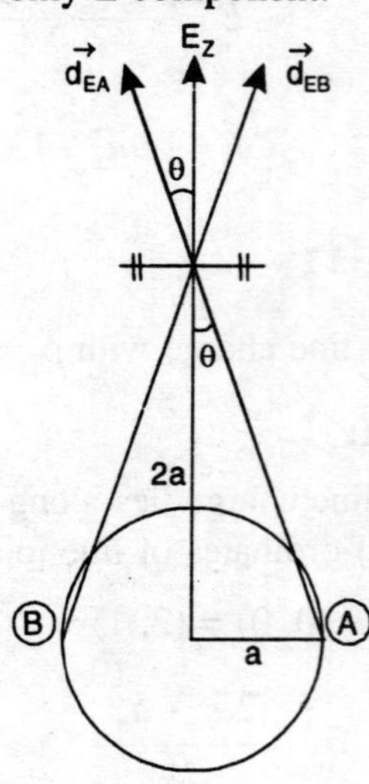

Fig. P-12

Since two diametrically opposite charges are considered $\oint dl = \pi a$,

$$\vec{E}_z = \frac{\lambda}{5\sqrt{5}\,\varepsilon_0 a}\vec{a}_z\ \text{V/m}.$$

Problem 13

A positive charge Q_v coulomb/m^3 occupies the volume of sphere. At a point in the interior at a distance of r m from the center, a small probe charge of $+q$ is inserted. What is the force acting on the probe charge?

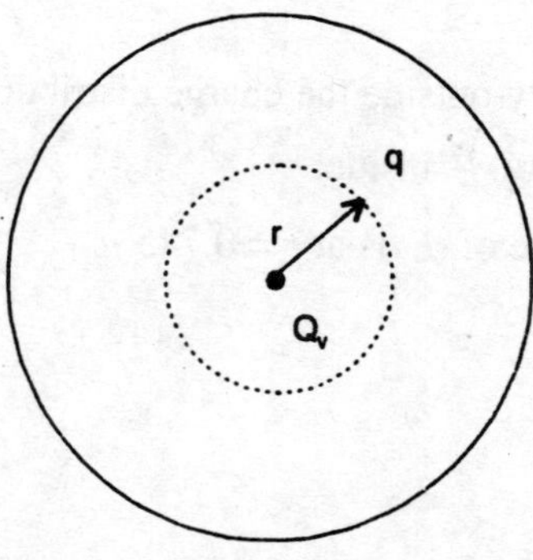

Fig. P-13

Solution:

Force $\vec{F}$ on the probe charge 'q' at a point P at a distance r from the centre as in Fig. P-13 is $\vec{F} = q\vec{E}$.

Applying Gauss's law to find $\vec{E}$ at the point P, consider the Gaussian spherical surface at point P with radius as 'r'.

$$\oint \vec{E} \cdot d\vec{s} = Q/\varepsilon_0 .$$

Charge enclosed by this Gaussian surface is a portion of the charge enclosed by the given sphere. Since the given sphere consists of charge density Q_v C/m^3, charge enclosed by the Gaussian sphere is $Q_v \times 4/3\pi r^3$.

$$\oint \vec{E} \cdot d\vec{s} = \frac{Q_v \times 4/3\pi r^3}{\varepsilon_0}$$

$$\vec{E} \times 4\pi r^2 = \frac{Q_v 4\pi r^3}{3\varepsilon_0}$$

$$\vec{E} = \frac{Q_v r}{3\varepsilon_0} \vec{a}_r .$$

So,
$$\vec{F} = \frac{Q_v q r}{3\varepsilon_0} \vec{a}_r .$$

Problem 14

A spherical volume charge density distribution is given by

$$\rho = \rho_0 \left(1 - \frac{r^2}{a^2}\right) ; (r \le a)$$

$$= 0 (r > a)$$

(*i*) Calculate the total charge Q.

(*ii*) Find the electric field intensity outside the charge distribution.

(*iii*) Find the electric field intensity $\vec{E}$ inside.

(*iv*) Show that the maximum value of $\vec{E}$ is at $r = 0.745\ a$.

Solution:

(*i*) Calculation of total charge.

$$Q = \iiint_{vol.} \rho \cdot dV.$$

In spherical coordinate

$$Q = \int_{\phi=0}^{2\pi} \int_{\theta=0}^{\pi} \int_{r=0}^{a} \rho \cdot r^2 \sin\theta \cdot dr\, d\theta\, d\phi$$

$$= \iiint \rho_0 \left(1 - \frac{r^2}{a^2}\right) \cdot r^2 \sin\theta\, dr\, d\theta\, d\phi$$

$$= \iiint \rho_0 r^2 \sin\theta\, dr\, d\theta\, d\phi - \iiint \rho_0 \frac{r^4}{a^2} \sin\theta \cdot dr\, d\theta\, d\phi$$

$$= \iint \left(\frac{\rho_0 r^3}{3}\right)_0^a \sin\theta \cdot d\theta \cdot d\phi - \iint \left(\frac{\rho_0}{a^2} \cdot \frac{r^5}{5}\right)_0^a \sin\theta \cdot d\theta \cdot d\phi$$

$$= \int_0^{2\pi} \int_0^{\pi} \frac{\rho_0 a^3}{3} \sin\theta \cdot d\theta \cdot d\phi - \int_0^{2\pi} \int_0^{\pi} \frac{\rho_0 a^3}{5} \sin\theta \cdot d\theta \cdot d\phi$$

$$= \int_0^{2\pi} \frac{-\rho_0 a^3}{3} (\cos\theta)_0^{\pi} \cdot d\phi - \int_0^{2\pi} -\frac{\rho_0 a^3}{5} (\cos\theta)_0^{\pi} \cdot d\phi$$

$$= \int_0^{2\pi} 2\,\frac{\rho_0 a^3}{3} d\phi + \int_0^{2\pi} \frac{\rho_0 a^3}{5} (-2)\, d\phi$$

$$= \frac{2\rho_0 a^3}{3} (\phi)_0^{2\pi} - \frac{2\rho_0 a^3}{5} (\phi)_0^{2\pi}$$

$$= \frac{2\rho_0 a^3}{3} \times 2\pi - \frac{2\rho_0 a^3}{5} (2\pi)$$

$$= 4\rho_0 a^3 \pi \left(\frac{1}{3} - \frac{1}{5}\right)$$

$$= \frac{8\pi\rho_0 a^3}{15} \text{ coulomb.}$$

(*ii*) To find electric field intensity outside. Consider a Gaussian sphere of radius $r > a$.

Using Gauss's law,

$$\oint \vec{E} \cdot ds = \frac{Q}{\varepsilon_0}$$

$$\oint E \cdot \vec{a}_r \cdot ds\, \vec{a}_r = \frac{8\pi\rho_0 a^3}{15\varepsilon_0}$$

$$E \int ds = \frac{8\pi\rho_0 a^3}{15\varepsilon_0}.$$

For a spherical surface $\int ds = 4\pi r^2$

$$\therefore \quad E = \frac{8\pi\rho_0 a^3}{15\varepsilon_0} \times \frac{1}{4\pi r^2}$$

$$\vec{E} = \frac{2\rho_0 a^3}{15\varepsilon_0 r^2} \vec{a}_r .$$

(*iii*) To find the electric field intensity inside the sphere.

Consider a Gaussian sphere of radius $r < a$ and applying Gauss law.

$$\oint \vec{E} \cdot ds = \frac{Q}{\varepsilon_0}$$

$$= E_{in} \times 4\pi r^2 = \iiint \frac{\rho\, dv}{\varepsilon_0}$$

$$E_{in} \times 4\pi\varepsilon_0 r^2 = \iiint \frac{\rho_0}{\varepsilon_0}\left(1 - \frac{r^2}{a^2}\right) r^2 \cdot \sin\theta \cdot d\theta \cdot d\phi \cdot dr$$

$$= \int_0^{2\pi}\int_0^{\pi}\int_0^{r} \rho_0 r^2 dr \cdot \sin\theta\, d\theta\, d\phi - \int_0^{2\pi}\int_0^{\pi}\int_0^{r} \frac{\rho_0 r^4}{a^2} dr \cdot \sin\theta\, d\theta\, d\phi$$

$$= \int_0^{2\pi}\int_0^{\pi} \frac{\rho_0 r^3}{3} \cdot \sin\theta\, d\theta\, d\phi - \int_0^{2\pi}\int_0^{\pi} \frac{\rho_0\, r^5}{5a^2} \cdot \sin\theta\, d\theta\, d\phi.$$

Cancelling r^2 on both sides

$$= \int_0^{2\pi} \frac{-\rho_0 r}{3} (\cos\theta)_0^{\pi}\, d\phi + \int_0^{2\pi} \frac{\rho_0 r^3}{5a^2} (\cos\theta)_0^{\pi}\, d\phi$$

$$= \int_0^{2\pi} \frac{2\rho_0 r}{3} d\phi - \int_0^{2\pi} \frac{2\rho_0 r^3}{5a^2} d\phi$$

$$= \frac{2\rho_0 r}{3} (\phi)_0^{2\pi} - \frac{\rho_0 r^3}{5a^2} (\phi)_0^{2\pi}$$

$$= \frac{4\pi\rho_0 r}{3} - \frac{4\pi r^3 \rho_0}{5a^2}.$$

Cancelling 4π on both sides,

$$E_{in} = \frac{1}{\varepsilon_0} \rho_0 r \left[\frac{1}{3} - \frac{r^2}{5a^2}\right]$$

$$\vec{E}_{in} = \frac{\rho_0 r}{\varepsilon_0} \left[\frac{1}{3} - \frac{r^2}{5a^2}\right] \vec{a}_r \text{ V/m}.$$

(*iv*) To find the maximum value of $\vec{E}$

$$\frac{dE_{in}}{dr} = 0$$

$$\frac{\rho_0}{3\varepsilon_0} - \frac{\rho_0 \cdot 3r^2}{5a^2} = 0$$

$$\frac{\rho_0}{\varepsilon_0} \left(\frac{1}{3} - \frac{3}{5} \frac{r^2}{a^2}\right) = 0$$

$$\frac{1}{3} - \frac{3}{5} \frac{r^2}{a^2} = 0$$

$$\frac{3}{5} \frac{r^2}{a^2} = \frac{1}{5}$$

$$\frac{r^2}{a^2} = \frac{5}{9}.$$

$\therefore \quad r = 0.745\ a.$

$\vec{E}$ is maximum at $r = 0.745$.

Problem 15

The components of $\vec{E}$ in a certain region are: $E_x = x$, $E_y = y$, $E_z = z$. Show that the charge contained in a cube of side a situated with one of its corners at the origin and all its faces parallel to the co-ordinate planes is $3\varepsilon_0\, a^3$.

Solution:

As per Gauss's law, the charge inside the cube $Q = \oint\!\!\oint_S \varepsilon_0 \vec{E} \cdot d\vec{s}$.

As per divergence theorem,

$$\oint\!\!\oint_S \vec{E} \cdot d\vec{s} = \iiint_V \nabla \cdot \vec{E}\ dv$$

$$\therefore \qquad \oint\!\!\oint_S \vec{E} \cdot d\vec{s} = \frac{Q}{\varepsilon_0} = \iiint_V \nabla \cdot \vec{E}\, dv.$$

Given $\vec{E} = x\vec{a}_x + y\vec{a}_y + z\vec{a}_z$,

$$\nabla \cdot \vec{E} = \left(\frac{\partial}{\partial x}\vec{a}_x + \frac{\partial}{\partial y}\vec{a}_y + \frac{\partial}{\partial z}\vec{a}_z\right)(x\vec{a}_x + y\vec{a}_y + z\vec{a}_z)$$

$$= 1 + 1 + 1$$

$$= 3.$$

$$\therefore \qquad \iiint_V \nabla \cdot \vec{E}\, dv = \iiint 3\, dv$$

$$= 3 \times a^3 \qquad [\because \text{ volume of cube of side 'a'} = a^3]$$

$$\iiint \nabla \cdot \vec{E}\, dv = \frac{Q}{\varepsilon_0}$$

$$3a^3 = \frac{Q}{\varepsilon_0}$$

$$Q = 3\varepsilon_0 a^3 \text{ coulomb.}$$

2.1.9 Electric Potential

As far as charges are concerned, electric field is a force field and hence work is associated with the movement of charges in an electric field. If a test charge is moved along the direction of the field, work is done by the field and hence it accelerates the test charge. If the charge is moved against the direction of the field, an external agent has to supply the energy to overcome the force exerted on the charge by the field.

Let us consider the displacement of a test charge 'Q' by an infinitesimal distance 'dl' from A to B at an angle α with the electric field $\vec{E}$ at the point 'A' as shown in Fig. 2.26.

The force exerted on the test charge by the field has magnitude $Q\vec{E}$ and is directed along $\vec{E}$. Its component along the line from A to B is $Q\vec{E}\cos\alpha$. If the charge is moved from A to B, the amount of work dW done by the field is the product of the force and the displacement, that is,

$$dW = Q\left|\vec{E}\right|\cos\alpha\left|d\vec{l}\right| \quad ...(2.77)$$

$$= Q\vec{E}\cdot d\vec{l},$$

where $d\vec{l}$ is the vector from A to B.

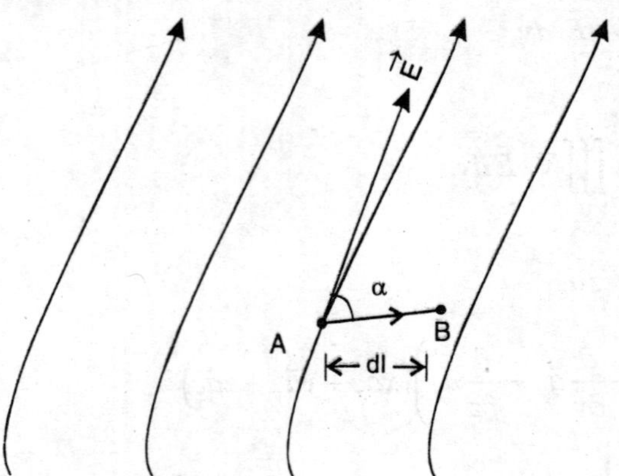

Fig. 2.26 Movement of charge in an electric field

Note that dW is positive if $0 < \alpha < 90°$, so that work is done by the field; dW is negative if $90° < \alpha < 180°$, so that negative work is done by the field, that is, work is done by an external agent against the field.

Thus the total work done (or) the potential energy required in moving Q from A to B against the field is

$$W = -Q\int_A^B \vec{E}\cdot d\vec{l}. \quad ...(2.78)$$

When W is divided by Q, it gives the potential energy per unit charge. This quantity is denoted by V_{AB} and is known as potential difference between A and B.

$$V_{AB} = \frac{W}{Q} = -\int_A^B \vec{E}\cdot d\vec{l}. \quad ...(2.79)$$

V_{AB} is measured in joule per coulomb, commonly referred to as volt.

2.1.10 The Potential Field of Point Charges

Consider two point charges A and B in the electric field of a point charge Q situated at distances r_A and r_B respectively, from the point charge as shown in Fig. 2.27.

Using the eqn. (2.79), the potential difference V_{AB} can be computed for any specified path from A to B.

$\vec{E} = \dfrac{Q}{4\pi\varepsilon_0 r^2}\vec{a}_r$ and $d\vec{l}$ in spherical co-ordinates is from eqn. (1.47)

$$d\vec{l} = dr\,\vec{a}_r + r d\theta \vec{a}_\theta + r\sin\theta\, d\phi\, \vec{a}_\phi.$$

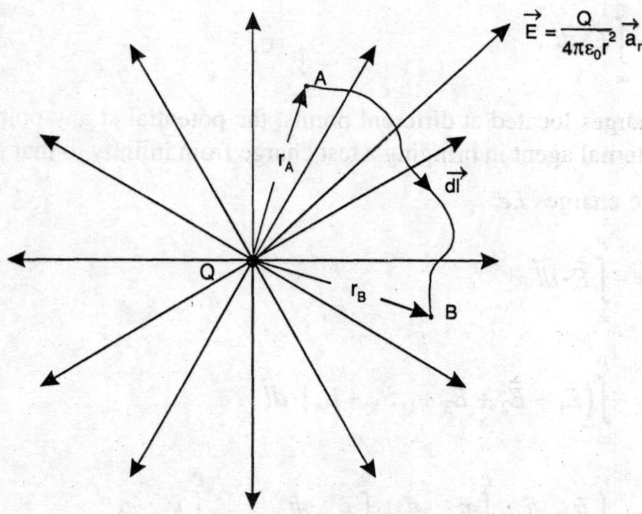

Fig. 2.27 Potential difference between two points in the electric field of point charge.

$$V_{AB} = -\int_A^B \vec{E}\cdot d\vec{l} = -\int_A^B \frac{Q}{4\pi\varepsilon_0 r^2}\vec{a}_r \cdot \left(dr\,\vec{a}_r + r d\theta\,\vec{a}_\theta + r\sin\theta\, d\phi\,\vec{a}_\phi\right)$$

$$= -\int_A^B \frac{Q}{4\pi\varepsilon_0 r^2} dr$$

$$V_{AB} = \frac{Q}{4\pi\varepsilon_0 r_B} - \frac{Q}{4\pi\varepsilon_0 r_A} \qquad ...(2.80)$$

$$V_{AB} = V_B - V_A, \qquad ...(2.81)$$

where $V_A = \dfrac{Q}{4\pi\varepsilon_0 r_A}$

$$V_B = \frac{Q}{4\pi\varepsilon_0 r_B}.$$

From eqn. (2.80), it is known that, for a given charge Q, the potential difference between two points is dependent only upon their distances from the point charge and not on the path from A to B chosen for its evaluation.

The potential at any point is the potential difference between that point and an arbitrary reference point at which the potential is zero. In case of a point charge, a convenient reference point is $r_0 = \infty$.

$$V_{(r)} = \frac{Q}{4\pi\varepsilon_0 r}. \qquad ...(2.82)$$

The potential at a distance 'r' from the point charge is thus the work done per unit charge by the external agent in bringing a test charge from infinity to that point, that is,

$$V_{(r)} = -\int_{\infty}^{r} \vec{E} \cdot d\vec{l} \; . \qquad ...(2.83)$$

For several point charges located at different points, the potential at any point P is the work done per unit charge by an external agent in bringing a test charge from infinity to that point in the combined electric field $\vec{E}$ of all the charges *i.e.*

$$V(P) = -\int_{\infty}^{P} \vec{E} \cdot d\vec{l}$$

$$= -\int_{\infty}^{P} \left(\vec{E}_1 + \vec{E}_2 + \vec{E}_3 + ...+ \vec{E}_n\right) \cdot d\vec{l}$$

$$= -\int_{\infty}^{P} \vec{E}_1 \cdot d\vec{l} - \int_{\infty}^{P} \vec{E}_2 \cdot d\vec{l} - \int_{\infty}^{P} \vec{E}_3 \cdot d\vec{l} ......- \int_{\infty}^{P} \vec{E}_n \cdot dl \, , \qquad ...(2.84)$$

where $\vec{E}_1$, $\vec{E}_2$ and $\vec{E}_3$ ......$\vec{E}_n$ are the electric field intensities due to the individual point charges Q_1, Q_2, Q_3, ..., Q_n, respectively.

Eqn. (2.84) can be written as

$$V(P) = \frac{Q_1}{4\pi\varepsilon_0 R_1} + \frac{Q_2}{4\pi\varepsilon_0 R_2} + ...+ \frac{Q_n}{4\pi\varepsilon_0 R_n}$$

$$= \sum_{j=1}^{n} \frac{Q_j}{4\pi\varepsilon_0 R_j} \; . \qquad ...(2.85)$$

In the vector notation,

$$V(\vec{r}) = \sum_{j=1}^{n} \frac{Q_j}{4\pi\varepsilon_0 \left|\vec{r} - \vec{r}_j\right|} \; . \qquad ...(2.86)$$

For a continuous distribution of charges, Q_j can be replaced with charge element $\rho_L\, dl$, $\rho_s ds$ (or) $\rho_v dV$ and summation becomes an integration.

Thus, the potential at 'r' for a line charge distribution of ρ_L C/m is

$$V = \frac{1}{4\pi\varepsilon_0} \int_L \frac{\rho_L\, dl}{r} \text{ volt.} \qquad ...(2.87)$$

For surface charge distribution of ρ_s C/m^2 is

$$V = \frac{1}{4\pi\varepsilon_0} \int_S \frac{\rho_s\, ds}{r} \text{ volt.} \qquad ...(2.88)$$

For volume charge distribution of ρ_v C/m^3 is

$$V = \frac{1}{4\pi\varepsilon_0}\int_V \frac{\rho_v\, dv}{r} \text{ volt.} \qquad \text{...(2.89)}$$

2.1.11 Potential Gradient

The potential difference between points *A* and *B* is independent of the path taken. Hence,

$$V_{AB} = -V_{BA}.$$

From the Fig. 2.28,

$$V_{AB} + V_{BA} = \oint_l \vec{E}\cdot d\vec{r}$$

$$= 0$$

$$\oint_l \vec{E}\cdot d\vec{l} = 0. \qquad \text{...(2.90)}$$

This shows that the line integral of $\vec{E}$ along a closed path must be zero.

That is, no work is done in moving a charge along a closed path in an electrostatic field.

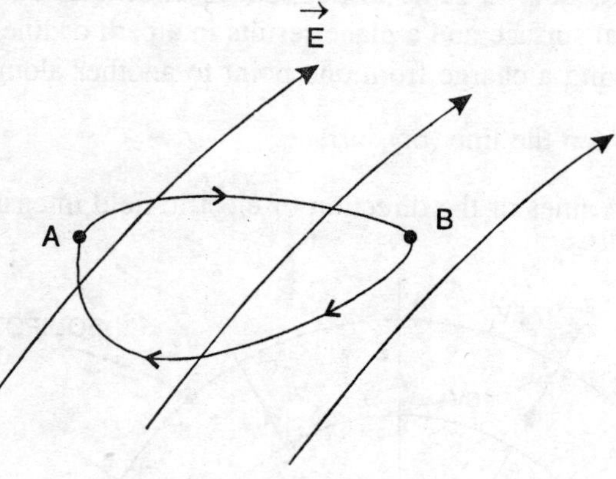

Fig. 2.28

Applying Stoke's theorem,

$$\oint_l \vec{E}\cdot d\vec{l} = \oint_S \left(\nabla\times\vec{E}\right)\cdot d\vec{s} = 0$$

$$\nabla\times\vec{E} = 0. \qquad \text{...(2.91)}$$

Any vector field that satisfies either of two equations, eqn.(2.90) or (2.91), is said to be conservative field (or) irrotational field.

Thus an electrostatic field is a **conservative field.**

$$V = -\oint_l \vec{E}\cdot d\vec{l}.$$

Differentiate on both sides,

$$dV = -\vec{E}\cdot d\vec{l}$$

$$= -\left(E_x\vec{a}_x + E_y\vec{a}_y + E_z\vec{a}_z\right)\cdot\left(dx\,\vec{a}_x + dy\,\vec{a}_y + dz\,\vec{a}_z\right)$$

$$dV = -E_x dx - E_y dy - E_z dz\,. \qquad ...(2.92)$$

But dV can be written as

$$dV = \frac{\partial V}{\partial x}dx + \frac{\partial V}{\partial y}dy + \frac{\partial V}{\partial z}dz\,. \qquad ...(2.93)$$

Comparing eqns. (2.92) and (2.93)

$$E_x = -\frac{\partial V}{\partial x},\quad E_y = -\frac{\partial V}{\partial y},\quad E_z = -\frac{\partial V}{\partial z}.$$

In general,

$$\vec{E} = -\nabla V\,. \qquad ...(2.94)$$

That is, electric field intensity is the gradient of V (or) potential gradient. The negative sign shows that the direction of $\vec{E}$ is opposite to the direction in which V increases.

2.1.12 Equipotential Surface

Any surface on which the potential is same throughout is known as an equipotential surface. The intersection of an equipotential surface and a plane results in a path or line known as an equipotential line. No work is done in moving a charge from one point to another along an equipotential line (or) surface and hence $\int_l \vec{E}\cdot d\vec{l} = 0$ on the line (or) surface.

So, lines of force (or) flux lines or the direction of electric field intensity $\vec{E}$ are always normal to the equipotential surfaces.

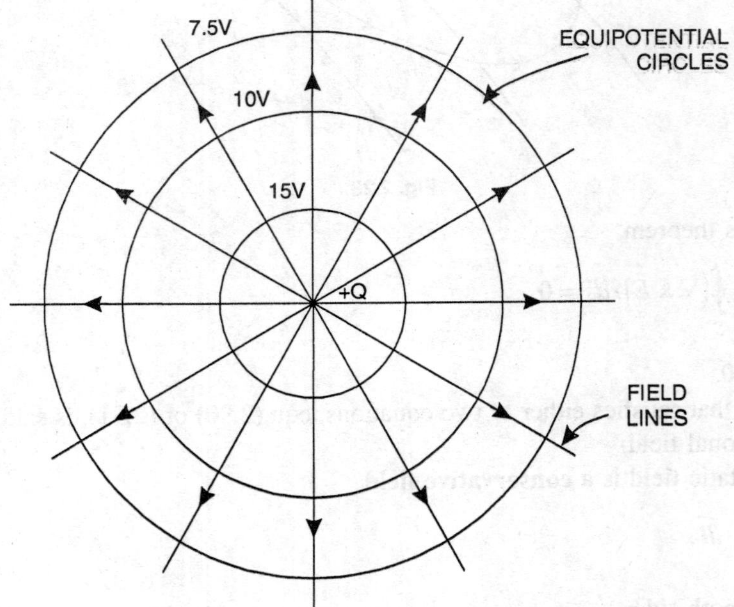

Fig. 2.29 Equipotential lines of a point charge

Example 2.13

Determine the potential difference between two points '*a*' and '*b*' which are at a distance of 0.5 m and 0.1 m from the positive charge of 20×10^{-10} coulomb.

Potential difference between '*a*' and '*b*' is given as $V_{ab} = -\int_a^b \frac{Q}{4\pi\varepsilon_0 r^2} dr$.

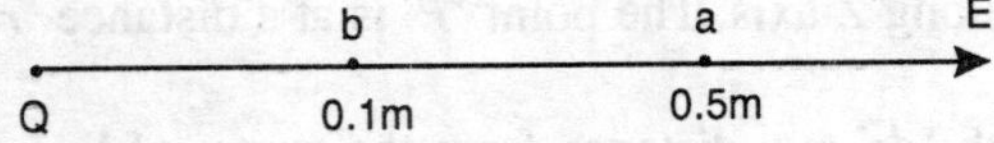

Fig. 2.30

where *a* is the initial point and *b* is the final point. Here, work is done against the field.

$$V_{ab} = \frac{Q}{4\pi\varepsilon_0}\left[\frac{1}{b} - \frac{1}{a}\right]$$

$$= \frac{20 \times 10^{-10}}{4\pi \times 8.87 \times 10^{-12}}\left[\frac{1}{0.1} - \frac{1}{0.5}\right]$$

$$V_{ab} = 143.5 \text{ V}$$

Example 2.14

Calculate the potential at a point *p* ($x = 0$, $y = 0$) due to point charges Q_1 and Q_2. $Q_1 = 10^{-12}$ C at ($x = 0.5$, $y = 0$) and $Q_2 = -10^{-11}$ C at ($x = -0.5$, $y = 0$)

(0,0)

Q_2 (-0.5,0) P Q_1 (0.5,0)

Fig. 2.31

Potential at P = sum of potential at P due to Q_1 and Q_2

$$V = \frac{Q_1}{4\pi\varepsilon_0 r_1} + \frac{Q_2}{4\pi\varepsilon_0 r_2}$$

$$r_1 = r_2 = 0.5 \text{ mt}$$

$$Q_1 = 10^{-12} \text{ C}$$

$$Q_2 = -10^{-11} \text{ C}$$

$$V = \frac{10^{-12}}{4\pi\varepsilon_0 (0.5)} - \frac{10^{-11}}{4\pi\varepsilon_0 (0.5)}$$

$$= -0.162 \text{ V}$$

Example 2.15

A line of length l m carries charge λ C/m. Show that potential in the median plane can be written as

$$V = \frac{\lambda}{4\pi\varepsilon_0} \ln \frac{1+\sin\alpha}{1-\sin\alpha}, \text{ where } \tan\alpha = \frac{l}{2r}.$$

Solution

Let the line charge of λC/m be along Z-axis. The point 'P' is at a distance 'r' m from the centre of the line charge.

Consider a differential length 'dz' at a distance from the centre of line charge.

The potential at P due to line charge is given by eqn. (2.87) as

$$V = \frac{1}{4\pi\varepsilon_0} \int_l \frac{\rho_L \, dl}{r},$$

where 'r' is the distance of point from elemental length dl.

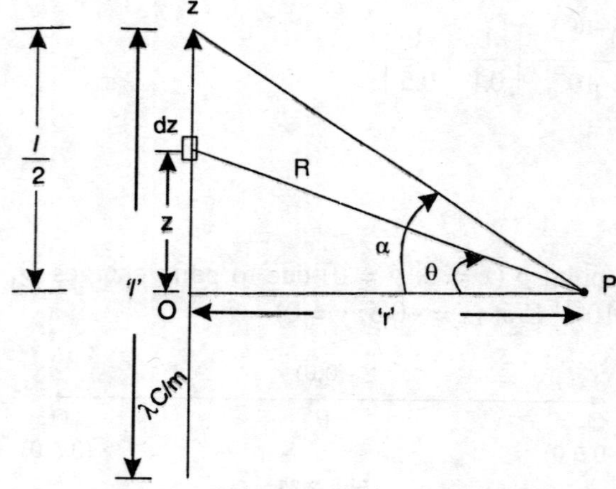

Fig. 2.32

The potential at P due to small elemental length 'dz' of line charge λC/m is given by

$$dV = \frac{1}{4\pi\varepsilon_0} \frac{\lambda dz}{R}.$$

From the Fig. 2.32,

$$V = \frac{1}{4\pi\varepsilon_0} \int_{-l/2}^{l/2} \frac{\lambda dz}{\sqrt{r^2 + z^2}}$$

$$= \frac{\lambda}{4\pi\varepsilon_0} \int_{-l/2}^{l/2} \frac{dz}{\sqrt{r^2 + z^2}}$$

Integrating by substituting as

$$z = r \tan\theta$$

$$dz = r \sec^2\theta \, d\theta$$

$$r^2 + z^2 = r^2 \sec^2\theta.$$

Also, θ varies from $-\alpha$ to α.

Now,

$$V = \frac{\lambda}{4\pi\varepsilon_0} \int_{-\alpha}^{\alpha} \frac{r \sec^2\theta \, d\theta}{\sqrt{r^2 \sec^2\theta}}$$

$$= \frac{\lambda}{4\pi\varepsilon_0} \int_{-\alpha}^{\alpha} \sec\theta \, d\theta$$

$$= \frac{\lambda}{4\pi\varepsilon_0} \ln[\sec\theta + \tan\theta]_{-\alpha}^{\alpha}$$

$$= \frac{\lambda}{4\pi\varepsilon_0} \ln\left[\frac{1+\sin\theta}{\cos\theta}\right]_{-\alpha}^{\alpha}$$

$$= \frac{\lambda}{4\pi\varepsilon_0} \ln\left[\frac{1+\sin\alpha}{1-\sin\alpha}\right]. \qquad ...(2.95)$$

Example 2.16

A thin square loop carries a uniform charge of ρ_L. Show that the potential at the center of the loop is

$$V = \frac{2\rho_L}{\pi\varepsilon_0} \ln\left(1+\sqrt{2}\right).$$

From eqn. (2.95), the potential at P due to line charge of λ is

$$V = \frac{\lambda}{4\pi\varepsilon_0} \ln\left[\frac{1+\sin\alpha}{1-\sin\alpha}\right].$$

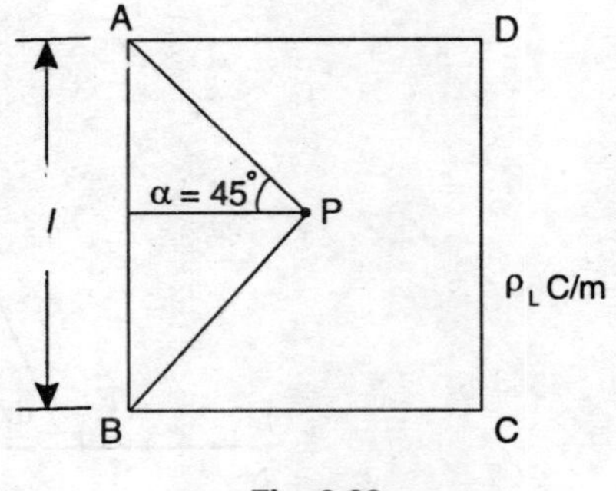

Fig. 2.33

From the Fig 2.33, the potential at P due to a square loop is given by

4 × potential at P due to one side.

Potential at P due to AB of length 'l' having charge density ρ_L C/m is

$$V_{AB} = \frac{\rho_L}{4\pi\varepsilon_0} \ln\left[\frac{1+\sin 45°}{1-\sin 45°}\right],$$

where $\alpha = 45°$ from the Fig 2.33.

Total potential at P

$$V = 4 \times \frac{\rho_L}{4\pi\varepsilon_0} \ln\left[\frac{1+\sqrt{2}}{\sqrt{2}-1}\right]$$

$$= \frac{\rho_L}{\pi\varepsilon_0} \ln\left(1+\sqrt{2}\right)^2$$

$$V = \frac{2\rho_L}{\pi\varepsilon_0} \ln\left(1+\sqrt{2}\right) \text{ volt.}$$

Example 2.17

A line charge is '$2a$' m long and carries uniformly distributed charge of λ_L C/m. Find the potential at a point distant 'h' from the line on the plane which bisects the line. Prove that it is

$$V = \frac{\lambda_L}{4\pi\varepsilon_0} \ln\left[\frac{\sqrt{a^2+h^2}+a}{\sqrt{a^2+h^2}-a}\right].$$

Consider a differential length 'dx' at a distance of 'x' m from one end of the line charge. The distance between the point 'P' and the differential length dx is 'r' m.
From eqn. (2.87),

$$V = \int_l \frac{1}{4\pi\varepsilon_0} \frac{\rho_L dl}{r}$$

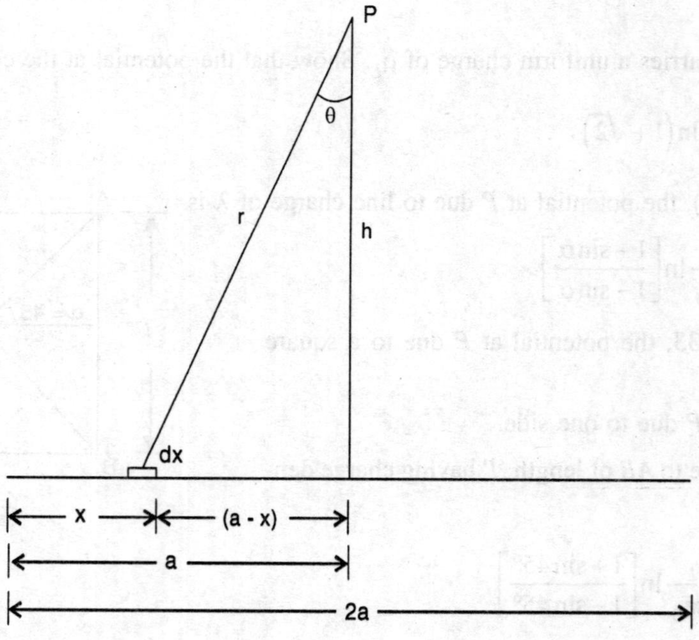

Fig. 2.34

$$V = \frac{\rho_L}{4\pi\varepsilon_0}\int_0^{2a}\frac{dx}{r}.$$

From Fig. 2.34, substitute

$$a - x = h\tan\theta$$

$$dx = -h\sec^2\theta\, d\theta$$

$$r = \sqrt{(a-x)^2 + h^2}$$

and integrating, V becomes

$$V = \frac{\rho_L}{4\pi\varepsilon_0}\int\frac{h\sec^2\theta\, d\theta}{\sqrt{h^2\sec^2\theta}}$$

$$= \frac{-\rho_L}{4\pi\varepsilon_0}\int\sec\theta\, d\theta$$

$$= \frac{-\rho_L}{4\pi\varepsilon_0}\ln[\sec\theta + \tan\theta]$$

$$= \frac{-\rho_L}{4\pi\varepsilon_0}\ln\left[\sqrt{1+\tan^2\theta} + \tan\theta\right]$$

$$= \frac{-\rho_L}{4\pi\varepsilon_0}\ln\left[\sqrt{1+\frac{(a-x)^2}{h^2}} + \frac{a-x}{h}\right]_{x=0}^{x=2a}$$

$$= \frac{-\rho_L}{4\pi\varepsilon_0}\ln\left[\sqrt{\frac{h^2+(a-x)^2}{h^2}} + \frac{a-x}{h}\right]_{x=0}^{x=2a}$$

$$= \frac{\rho_L}{4\pi\varepsilon_0}\ln\left[\frac{\sqrt{h^2+a^2}+a}{\sqrt{h^2-a^2}-a}\right].$$

Example 2.18

Find the potential at any point along the axis of uniformly charged disc of σ C/m^2. The disc has radius of 'a' m.

Solution:

Let the point 'P' be along the axis of circular disc at a distance of 'h' m.

Potential at P due to the circular disc can be evaluated by considering the circular disc as composed of a large number of elemental ring of thickness 'dr'.

Consider one annular ring of radius '*r*' m and thickness '*dr*', the surface area is equal to $(2\pi r)\ dr$ and this surface element carries a charge $dq = \sigma 2\pi r dr$.

Potential dV at P due to dq is given by

$$dV = \frac{dq}{4\pi\varepsilon_0 R} = \frac{\sigma 2\pi r\, dr}{4\pi\varepsilon_0 \sqrt{h^2 + r^2}}$$

$$V = \int_{r=0}^{r=a} \frac{\sigma\, 2\pi\, r\, dr}{4\pi\varepsilon_0 \sqrt{h^2 + r^2}}.$$

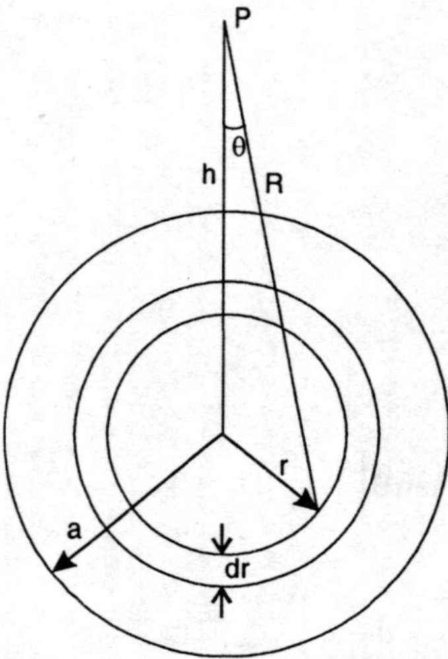

Fig. 2.35

From the Fig. 2.35, substituting

$$r = h\tan\theta$$

$dr = h\, sec^2\,\theta\ d\theta$ and integrating

$$V = \int \frac{2\pi\sigma\, h\tan\theta}{4\pi\varepsilon_0\ h\sec\theta} \cdot h\sec^2\theta\, d\theta$$

$$= \frac{\sigma h}{2\varepsilon_0}\int_{r=0}^{r=a} \tan\theta \cdot \sec\theta\, d\theta$$

$$= \frac{\sigma h}{2\varepsilon_0}[\sec\theta]_{r=0}^{r=a}$$

$$= \frac{\sigma h}{2\varepsilon_0}\left[\sqrt{1+\tan^2\theta}\right]_{r=0}^{r=a}$$

$$= \frac{\sigma h}{2\varepsilon_0}\left[\sqrt{1+\frac{r^2}{h^2}}\right]_{r=0}^{r=a}$$

$$V = \frac{\sigma h}{2\varepsilon_0}\left[\frac{\sqrt{a^2+h^2}}{h}-1\right].$$

Example 2.19

Find the potential at a point (*a*) outside (*b*) inside a uniformly charged sphere of radius '*a*' having total charge of Q_t coulomb.

(*i*) Outside : $r > a$

The field produced at a point outside the sphere of radius 'a' is given by

$$V = -\int_{\infty}^{r} \vec{E}\cdot d\vec{l}. \qquad ...(2.96)$$

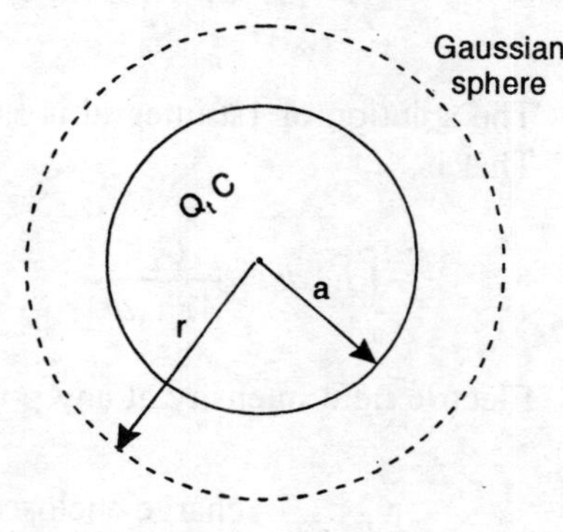

Fig. 2.36

To find electric field $\vec{E}$ at any point '*P*' outside the sphere is found by using Gauss's law. Construct a Gaussian sphere at the point '*P*' of radius '*r*' m.

From Gauss's law

$$\oint \vec{E}\cdot d\vec{s} = \frac{\text{charge enclosed by Gaussian surface}}{\varepsilon_0}$$

$\oint \vec{E}\cdot d\vec{s}$ is evaluated as in Example 2.12

$$\oint \vec{E}\cdot d\vec{s} = \vec{E}\times 4\pi r^2.$$

Charge enclosed by Gaussian surface = Q_t.

$$\therefore \qquad \vec{E}\times 4\pi r^2 = \frac{Q_t}{\varepsilon_0}$$

$$\vec{E} = \frac{Q_t}{4\pi\varepsilon_0 r^2}\vec{a}_r. \qquad ...(2.97)$$

Substituting eqn. (2.97) in (2.96), we get

$$V = -\int_{\infty}^{r} \frac{Q_t}{4\pi\varepsilon_0 r^2}\vec{a}_r \cdot dr\,\vec{a}_r$$

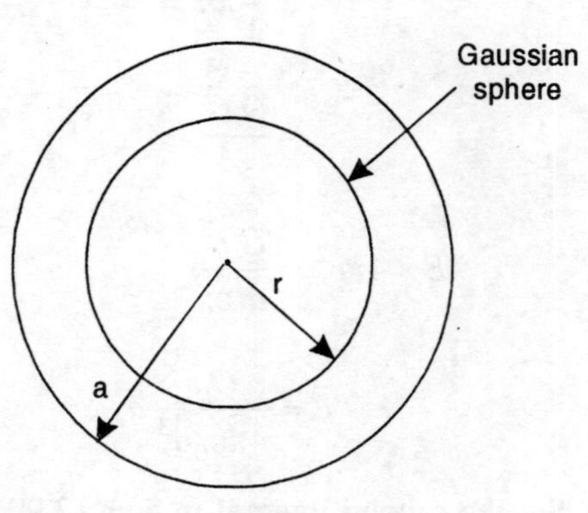

Fig. 2.37

$$V = \frac{Q_t}{4\pi\varepsilon_0 r} \text{ volt} \qquad ...(2.98)$$

Inside: $r < a$.

The field produced at a point inside the sphere of radius 'a' is given by

$$V = -\int_{\infty}^{r} \vec{E} \cdot d\vec{l}$$

$$= -\int_{\infty}^{a} \vec{E} \cdot d\vec{l} - \int_{a}^{r} \vec{E} \cdot d\vec{l}\,. \quad \text{...(2.99)}$$

The solution of 1st integral is same as in eqn. (2.98). Here 'r' is replaced by 'a'. That is,

$$-\int_{\infty}^{a} \vec{E} \cdot d\vec{r} = \frac{Q_t}{4\pi\varepsilon_0 a}. \quad \text{...(2.100)}$$

Electric field intensity at any point inside the spherical volume charge is found out using Gauss's law.

$$\oint \vec{E} \cdot d\vec{s} = \frac{\text{charge enclosed by the Gaussian sphere}}{\varepsilon_0}.$$

The solution for integral $\oint \vec{E} \cdot d\vec{s}$ is same as in eqn. (2.70)

$$\oint \vec{E} \cdot d\vec{s} = E \times 4\pi r^2\,.$$

Now, charge enclosed by the Gaussian sphere of radius 'r' = (Total charge density in C/m^3) × (Volume of the Gaussian sphere of radius r)

$$= \frac{Q_t}{\frac{4}{3}\pi a^3} \times \frac{4}{3}\pi r^3$$

$$= \frac{Q_t r^3}{a^3}.$$

So,

$$E \times 4\pi r^2 = \frac{Q_t r^3}{a^3 \varepsilon_0}$$

$$\vec{E} = \frac{Q_t r}{4\pi\varepsilon_0 a^3}\vec{a}_r.$$

So, the second integral in eqn. (2.99) becomes

$$-\int_{a}^{r} \vec{E} \cdot d\vec{r} = -\int_{a}^{r} \frac{Q_t r}{4\pi\varepsilon_0 a^3}\vec{a}_r \cdot dr\,\vec{a}_r$$

$$= -\left[\frac{Q_t r^2}{\pi\varepsilon_0 a^3}\right]_a^r$$

$$= -\frac{Qt}{8\pi\varepsilon_0 a^3}\left[r^2 - a^2\right]. \quad \text{...(2.101)}$$

Substituting eqns. (2.100) and (2.101) in (2.99)

$$\therefore \quad V = \frac{Q_t}{4\pi\varepsilon_0 a} + \frac{Q_t}{8\pi\varepsilon_0 a^3}\left[a^2 - r^2\right]. \quad \text{...(2.102)}$$

Example 2.20

Determine the variation of field from point to point due to

(*a*) Single spherical shell of charge with radius R_1

(*b*) Two concentric spherical shells of charge of radii R_1 (inner) and R_2 (outer).

Let us assume that a total charge Q coulomb to be uniformly distributed over an imaginary shell of radius R_1 in a medium of free space ($\varepsilon = \varepsilon_0$).

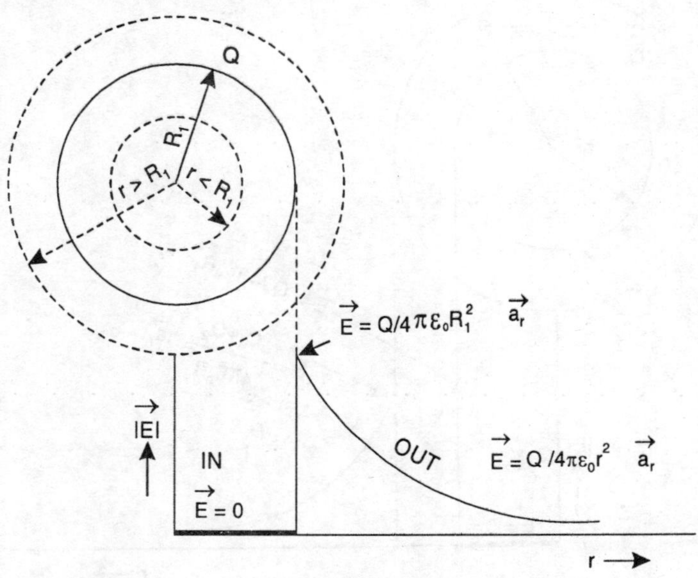

Fig. 2.38

(*a*) Single spherical shell of charge with radius R_1

Case 1: $r < R_1$ (inside)

At any radius $r < R_1$ (inside the shell of charge), no charge is enclosed by this surface. So the flux density and electric field intensity $\vec{E}$ are zero.

According to Gauss's law, $\oint_S \vec{E} \cdot d\vec{s} = \dfrac{Q}{\varepsilon_0} = 0$ [because of $Q = 0$]

$$\vec{E} = 0 \text{ for } (r < R_1).$$

Case 2: $r \geq R_1$ (outside)

At any radius $r \geq R_1$, construct a Gaussian sphere. The charge enclosed by the Gaussian sphere is equal to the charge over the spherical shell Q coulomb.

Applying Gauss's law,

$$\oint_S \vec{E} \cdot d\vec{s} = \frac{Q}{\varepsilon_0}$$

$$\vec{E} \times 4\pi r^2 = \frac{Q}{\varepsilon_0}$$

$$\vec{E} = \frac{Q}{4\pi\varepsilon_0 r^2} \vec{a}_r \quad (r \geq R_1).$$

The solution is identical with the point charge.

So, the field produced just outside a spherical shell of charge Q uniformly distributed over the surface is same as if the charge were concentrated at the centre of the sphere as a point charge.

(*b*) Two concentric spherical shells of charge of radii R_1 (inner) and R_2 (outer)

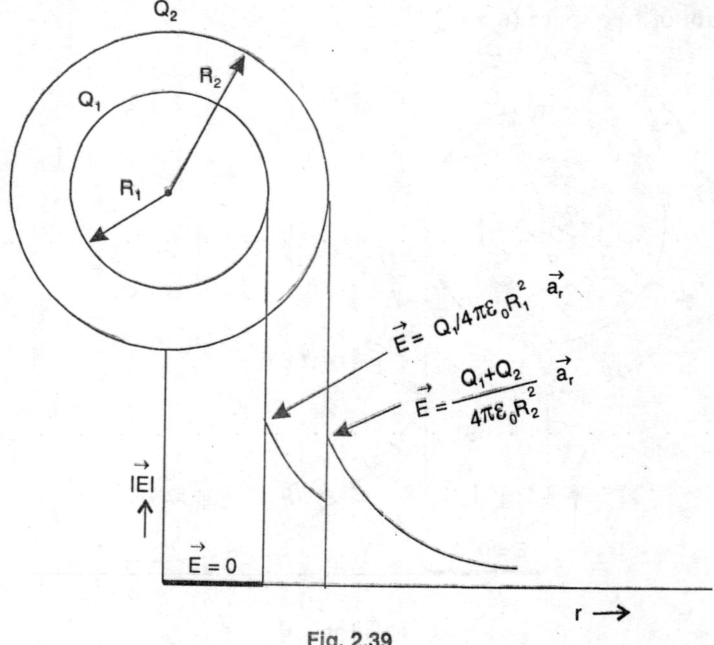

Fig. 2.39

Consider the two spherical shells of charges Q_1 and Q_2 coulomb at radii R_1 and R_2 respectively. The charges are uniformly distributed.

Case 1: $r < R_1$

As no charge is enclosed within Gaussian sphere of radius $r < R_1$

$$\vec{E} = 0 \quad \text{for} \quad (r \leq R_1).$$

Case 2: $R_1 < r < R_2$

In the region between the spherical shells, the charge enclosed is Q_1 coulomb.

So, according to Gauss's law,

$$\oint_s \vec{E}\cdot d\vec{s} = Q_1/\varepsilon_0$$

$$\vec{E} = \frac{Q}{4\pi\varepsilon_0 r^2} \quad (R_1 \le r \le R_2).$$

Case 3: $r > R_2$

In the region just outside the shells of charges Q_1 and Q_2, the charge enclosed is $(Q_1 + Q_2)$ coulomb.

By Gauss's law $\oint_s \vec{E}\cdot d\vec{s} = Q_1 + Q_2/\varepsilon_0$.

$$\vec{E} = \frac{Q_1 + Q_2}{4\pi\varepsilon_0 r^2} \quad (r \ge R_2).$$

2.1.13 Alternative Form of Gauss's law

From Gauss's law

$$\oint_S \vec{D}\cdot d\vec{s} = Q.$$

Assume the Gaussian surface contains the charge which is distributed uniformly over a volume.

Then, $$Q = \int_V \rho_v dv$$

$$\therefore \quad \oint_S \vec{D}\cdot d\vec{s} = \int_v \rho_v dv.$$

Using Divergence Theorem left-hand side can be written as

$$\oint_S \vec{D}\cdot d\vec{s} = \int_v \nabla\cdot\vec{D}\, dv$$

$$\therefore \quad \int_v \nabla\cdot\vec{D}\, dv = \int_v \rho_v\, dv$$

$$\nabla\cdot\vec{D} = \rho_v. \qquad \text{...(2.103)}$$

This equation is known as differential form of Gauss's law.

2.1.14 Poisson's Equation and Laplace's Equation

Poisson's and Laplace's equations are easily derived from Gauss's law for a linear material medium.

From eqn. (2.103),

$$\nabla\cdot\vec{D} = \rho_v$$

$$\nabla\cdot\vec{D} = \nabla\cdot\varepsilon\vec{E} = \rho_v$$

and from eqn. (2.94) $\vec{E} = -\nabla V$.
Then,

$$\nabla \cdot \varepsilon \vec{E} = \nabla \cdot (-\varepsilon \nabla V)$$

$\nabla \cdot (-\varepsilon \nabla V) = \rho_v$ for an inhomogeneous medium. For homogeneous medium, the equation becomes

$$\nabla^2 V = \frac{-\rho_v}{\varepsilon}. \qquad ...(2.104)$$

This is known as Poisson's equation. A special case of this equation occurs when $\rho_v = 0$ (i.e. for free charge region). Eqn. (5.104) becomes

$$\nabla^2 V = 0. \qquad ...(2.105)$$

This equation is known as Laplace's equation.

SOLVED PROBLEMS

Problem 16

Find the potential at the point P at the centre of the square 1 m side. The square has a point charge $Q_1 = +10^{-12}$C at the upper left corner, a point charge $Q_2 = -10^{-11}$C at the lower left corner and a line distribution of charge of uniform density 1 = 10^{-11} C/m along the right edge.

Solution:

The potential at P due to Q_1 is $V_{P_1} = \dfrac{Q_1}{4\pi\varepsilon_0 r_1}$

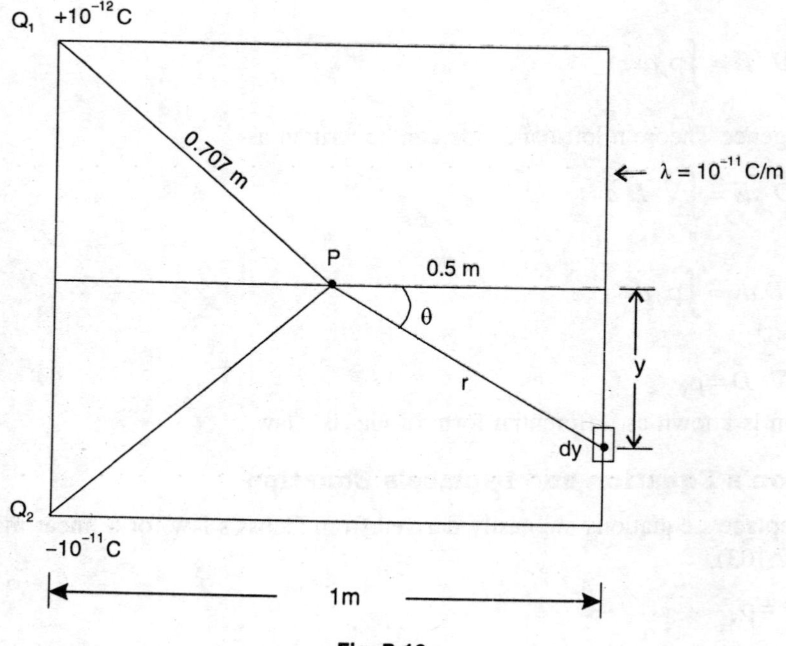

Fig. P-16

$$= \frac{10^{-12}}{4\pi\varepsilon_0 (0.707)} = 0.0127\,\text{V}.$$

The potential at P due to Q_2 is

$$V_{P_2} = \frac{Q_2}{4\pi\varepsilon_0 r_2}$$

$$= \frac{-10^{-11}}{4\pi\varepsilon_0 (0.707)} = -0.127\,\text{V}.$$

The potential at P by the line distribution of charge is

$$V_L = \int_l \frac{\lambda\, dl}{4\pi\varepsilon_0 r}.$$

Here, the line charge is along Y axis, $dl = dy$.
Consider an elemental line charge of dy at a distance y as in Fig. P-17.

$$r = \sqrt{(0.5)^2 + y^2}$$

$$V_L = \frac{\lambda_0}{4\pi\varepsilon_0} \int_{y=-0.5}^{y=0.5} \frac{dy}{\sqrt{(0.5)^2 + y^2}}.$$

From the Fig. P-17,

$$\tan\theta = y/0.5$$

$$y = 0.5 \tan\theta$$

$$dy = 0.5 \sec^2\theta\, d\theta$$

$$V_L = \frac{\lambda}{4\pi\varepsilon_0} \int \frac{0.5\sec^2\theta d\theta}{0.5\sec\theta}$$

$$= \frac{\lambda}{4\pi\varepsilon_0} \ln(\sec\theta + \tan\theta)$$

$$= \frac{\lambda}{4\pi\varepsilon_0} \ln\left[\frac{\sqrt{0.25 + y^2}}{0.5} + \frac{y}{0.5}\right]_{y=-0.5}^{y=+0.5}$$

$$= \frac{\lambda}{4\pi\varepsilon_0} \left\{ \ln\left[\frac{\sqrt{0.5}}{0.5} + 1\right] - \ln\left[\frac{\sqrt{0.5}}{0.5} - 1\right] \right\}$$

$$= \frac{10 \times 10^{-12}}{4\pi \times 8.85 \times 10^{-12}} \times 1.76$$

$V_L = 0.158$ V.

The total potential at P is

$$V = V_{P1} + V_{P2} + V_L$$

$$= 0.0127 - 0.127 + 0.158$$

$$V = 0.043$$

$$= 43 \text{ mV}.$$

Problem 17

A total charge of $\left(\frac{40}{3}\right)$nC is uniformly distributed in the form of a circular disk of radius 2 m. Find the potential due to this charge at a point on the axis, 2 m from the disk. Compare this potential that results if all the charges are present at the centre of the disk.

Solution:

From the Example 2.17, the potential at a point h from the uniformly charged disc of σ C/m^2 along its axis of radius 'a' mt is

$$V = \frac{\sigma h}{2\varepsilon_0}\left[\frac{\sqrt{a^2 + h^2}}{h} - 1\right] \text{ volt.}$$

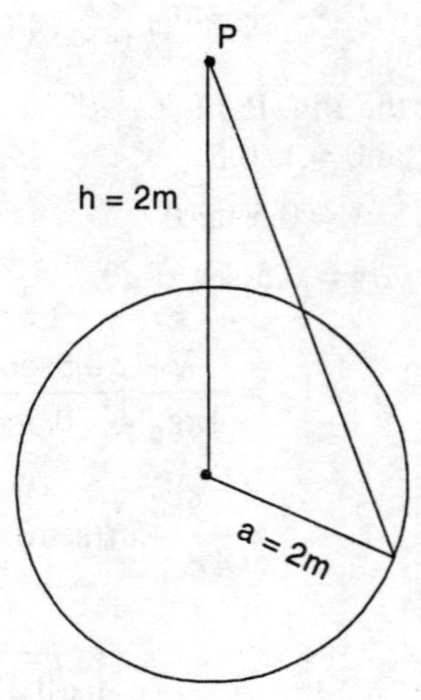

Fig. P-17

Given the total charge uniformly distributed on the disc surface

$$Q = \left(\frac{40}{3}\right) \times 10^{-9} \text{ C.}$$

$$\text{Charge density } \sigma = \frac{Q}{\text{Area}} = \frac{Q}{\pi a^2}$$

$$\sigma = \frac{\frac{40}{3} \times 10^{-9}}{\pi \times 4} = 1.06 \times 10^{-9} \text{ C/m}^2$$

$$V = \frac{1.06 \times 10^{-9} \times 2}{2 \times 8.85 \times 10^{-12}}\left[\frac{\sqrt{8}}{2} - 1\right]$$

$$V = 49.6 \text{ volt.}$$

If all the charges are concentrated at the centre of the disk, the potential at a point of distance 'r' m due to a point charge (Q) is

$$V = \frac{\frac{40}{3} \times 10^{-9}}{4\pi \times 8.85 \times 10^{-12} \times 2} = 60 \text{ V}$$

Problem 18

Find the potential at $r = 40$ cm and $r = 10$ cm from a charge $Q = 2 \times 10^{-4}$ μC and also the potential difference between these two points.

At $r = 40$ cm $= 0.4$ m

$$V = \frac{Q}{4\pi\varepsilon_0 r} = \frac{2 \times 10^{-4} \times 10^{-6}}{4\pi \times 8.85 \times 10^{-12} \times 0.4}$$

$$= 4.49 \text{ V.}$$

At $r = 10$ cm ($r = 0.1$ m)

$$V = \frac{2 \times 10^{-4} \times 10^{-6}}{4\pi \times 8.85 \times 10^{-12} \times 0.1}$$

$$= 17.98 \text{ V.}$$

Potential difference $= 17.98 - 4.49$

$= 13.49$ V.

Problem 19

A positive point charge of 100 μμC is located in air at $x = 0$, $y = 0.1$ m and another such charge at $x = 0$, $y = -0.1$ m. What are the magnitude and direction of $\vec{E}$ and what is the absolute potential V at $x = 0.2$ m, $y = 0$?

Solution:

From the diagram,

$$\vec{E}_{1P} = \frac{Q_1}{4\pi\varepsilon_0 r_{1P}^2}\vec{a}_{1P}$$

$$= \frac{100 \times 10^{-12}}{4\pi \times 8.85 \times 10^{-12} \times (0.223)^2}\vec{a}_{1P}$$

$$= 17.98 \text{ V/m } \vec{a}_{1P}.$$

$$\vec{E}_{1P_x} = \vec{E}_{1P}\cos\theta$$

$$= 17.98 \times \cos 26.56°$$

$$= 16.08 \text{ V/m } \vec{a}_x.$$

$$\vec{E}_{1P_y} = \vec{E}_{1P}\sin\theta$$

$$= 8 \text{ V/m } (-\vec{a}_y).$$

Similarly,

$$\vec{E}_{2P} = \frac{Q_2}{4\pi\varepsilon_0 r^2}\vec{a}_{2P}$$

$$= 17.98\vec{a}_{2P}$$

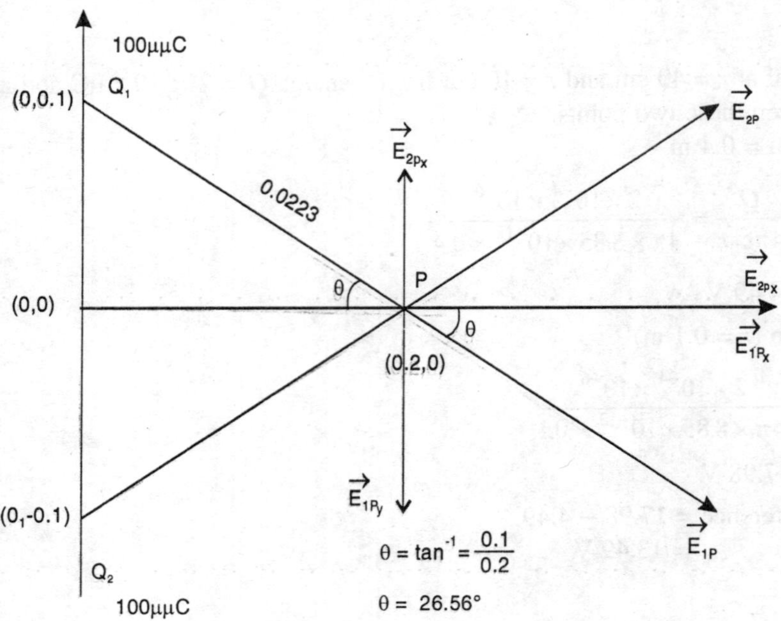

Fig. P-19

$$\vec{E}_{2P_x} = 16.08 \text{ V/m} \quad \vec{a}_x$$

$$\vec{E}_{2P_y} = 8 \text{ V/m} \quad \vec{a}_y$$

$$\vec{E} = \left(E_{1P_x} + E_{2P_x}\right)\vec{a}_x + \left(\vec{E}_{1P_y} + E_{2P_y}\right)\vec{a}_y$$

$$= (16.08 + 16.08)\vec{a}_x + (-8 + 8)\vec{a}_y$$

$$\vec{E} = 32.16 \text{ V/m}.$$

To find potential:

$$V = V_1 + V_2$$

$$= \frac{Q_1}{4\pi\varepsilon_0 r_1} + \frac{Q_2}{4\pi\varepsilon_0 r_2}$$

$$= \left(\frac{100 \times 10^{-12}}{4\pi \times 8.85 \times 10^{-12} \times 0.223}\right) \times 2$$

$$= 8.06 \text{ V}.$$

Problem 20

A spherical volume R = 1m has a uniform charge density of ρ = 1 C/m^3. Calculate the potential V at r = 50 cm.

Solution:

Potential at a point inside a sphere of radius 'R' is

$$V = \frac{Q_t}{4\pi\varepsilon_0 R} + \frac{Q_t}{8\pi\varepsilon_0 R^3}\left(R^2 - r^2\right),$$

where Q_t = total charge in coulomb in the given sphere

$Q_t = \rho \times$ volume of sphere

$= 1 \times 4/3\ \pi\ (1)^3$

$= 4/3\ \pi$

$R = 1$ m

$r = 0.5$ m

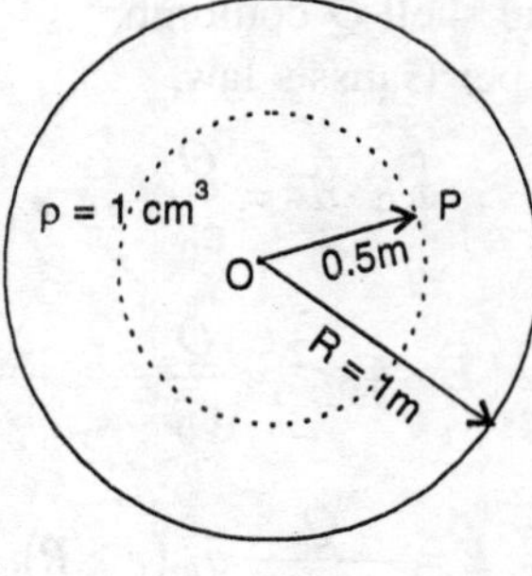

Fig. P-20

$$V = \frac{Q_t}{4\pi\varepsilon_0 R}\left[1 + \frac{1}{2R^2}\left(R^2 - r^2\right)\right]$$

$$= \frac{\frac{4}{3}\pi}{4\pi\varepsilon_0 \times 1}\left[1 + \frac{1}{2}(1 - 0.25)\right]$$

$$= \frac{1}{3\varepsilon_0}\left[1 + \frac{3}{8}\right]$$

$$= \frac{11}{24\varepsilon_0} = 5.17 \times 10^{10} \text{ V.}$$

Problem 21

Determine the variation of field and potential from point to point due to a single spherical shell of charge (Q) with radius R_1.

Solution:

Let us assume that a total charge Q coulomb to be uniformly distributed over an imaginary shell of radius R_1 in a medium of free space.

Case 1: $r < R_1$ (inside)

At any radius $r < R_1$, no charge is enclosed by this surface.

According to Gauss's law

$$\oint \vec{E} \cdot d\vec{s} = \frac{Q}{\varepsilon_0} \qquad [\because Q = 0]$$

$$= 0$$

$$\vec{E} = 0$$

Fig. P-21 (a)

So, the flux density $\vec{D}$ and electric field intensity $\vec{E}$ are zero.

Case 2: $r > R_1$ (outside)

Construct a Gaussian surface at any radius $r \geq R_1$.

The charge enclosed by the Gaussian sphere is equal to the charge over the spherical shell Q coulomb.

As per Gauss's law,

$$\oint \vec{E} \cdot d\vec{s} = \frac{Q}{\varepsilon_0}$$

$$\vec{E} \times 4\pi r^2 = \frac{Q}{\varepsilon_0}$$

$$\vec{E} = \frac{Q}{4\pi\varepsilon_0 r^2}\vec{a}_r \; (r \geq R).$$

The solution is identical with point charge. So, the field produced just outside a spherical shell of charge Q coulomb uniformly distributed over the surface is same as if the charge were concentrated at the centre of the sphere as a point charge.

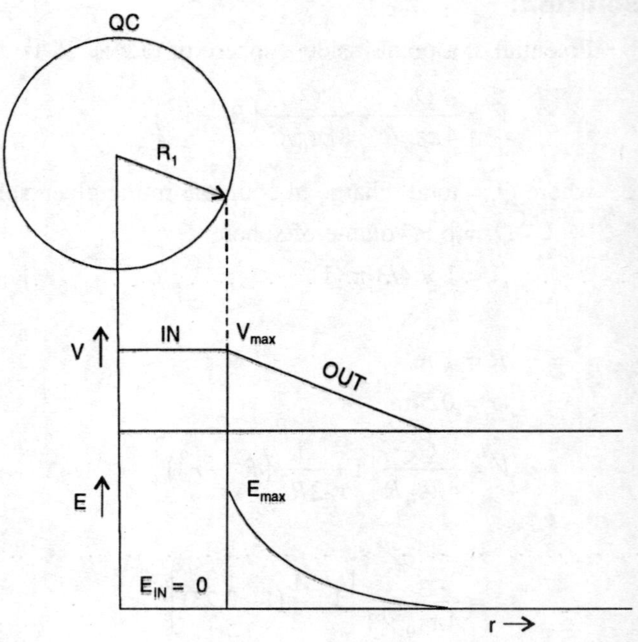

Fig. P-21 (b)

The absolute potential at a point outside the shell is

$$V = -\int_{\infty}^{r} \vec{E} \cdot d\vec{l} \qquad r \geq R_1$$

Substituting $\vec{E}$ from the above equation

$$V = -\frac{Q}{4\pi\varepsilon_0}\int_{\infty}^{r}\frac{dr}{r^2} = \frac{Q}{4\pi\varepsilon_0 r}.$$

At the surface of the shell, $r = R_1$

$$V = \frac{Q}{4\pi\varepsilon_0 R_1} \text{V}.$$

Since $\vec{E}$ inside the shell is zero, it requires no work to move a test charge inside and therefore the potential is constant.

We can say it is the other way that, $\vec{E} = -\nabla V$, as if $\vec{E} = 0, -\nabla V = 0, V =$ constant.

The potential inside the shell is equal to the value at the surface of the shell.

$$V = \frac{Q}{4\pi\varepsilon_0 R_1} \qquad \text{inside} \qquad r \leq R_1$$

$$V = \frac{Q}{4\pi\varepsilon_0 r} \qquad \text{outside} \qquad r \geq R_1.$$

Problem 22

A positive point charge Q is at the center of a spherical conducting shell of an inner radius R_i and an outer radius R_0. Determine $\vec{E}$ and V as functions of the radial distance R.

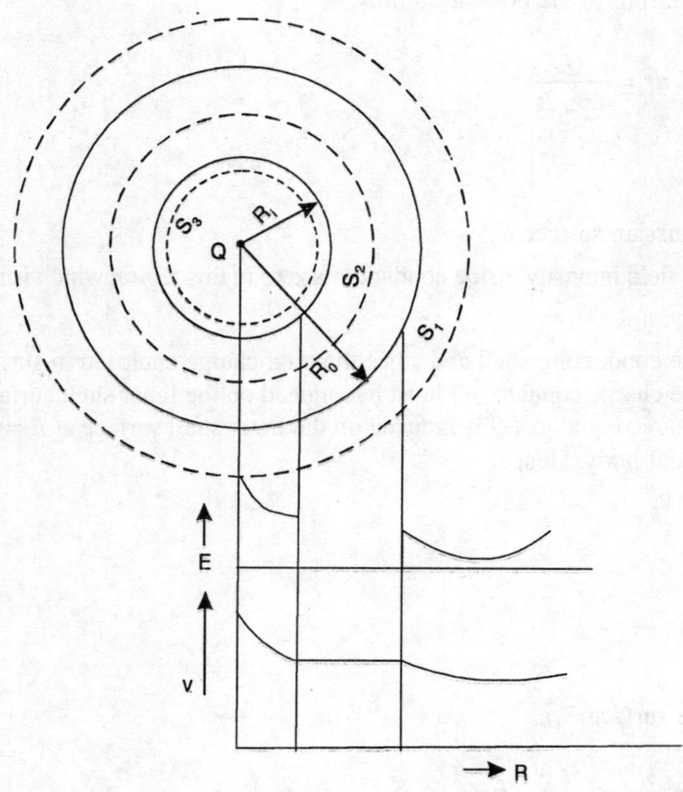

Fig. P-22

Since there is spherical symmetry, it is simplest to use Gauss's law to determine $\vec{E}$ and then find V by integration.

There are three cases (*i*) $R > R_0$ (*ii*) $R_i < R < R_0$ (*iii*) $R < R_i$.

Since spherical Gaussian surfaces are to be constructed, in all these regions $\vec{E} = E_r \vec{a}_r$.

Case (i)

$R > R_0$ (Gaussian surface S_1):

By Gauss's law, $\oint_{S_1} \vec{E} \cdot d\vec{s} = \dfrac{Q}{\varepsilon_0}$.

The charge enclosed by the Gaussian sphere is Q.

$$E\left(4\pi R^2\right) = \frac{Q}{\varepsilon_0}$$

$$\vec{E} = \frac{Q}{4\pi\varepsilon_0 R^2}\,\vec{a}_r.$$

This field intensity is same as that of a point charge Q without the presence of the shell. The potential referring to the point at infinity is

$$V_1 = -\int_{\infty}^{R} E\,dr = \frac{Q}{4\pi\varepsilon_0 R}.$$

Case (ii)

$R_i < R < R_0$ (Gaussian surface S_2)

Since the electric field intensity inside conductor is zero in this region which is inside the conducting shell, $\vec{E} = 0$.

Since $\rho = 0$ in the conducting shell and since the total charge enclosed in surface S_2 must be zero, an amount of negative charge equal to $-Q$ must be induced on the inner shell surface at $R = R_i$ (also an amount of positive charge equal to $+Q$ is induced on the outer shell surface at $R = R_0$). The conducting shell is an equipotential body. Hence,

$$V_2 = V_1 \text{ at } R = R_0$$

$$= \frac{Q}{4\pi\varepsilon_0 R_0}.$$

Case (iii)

$R < R_i$ (Gaussian surface S_3):

By applying Gauss's law $\oint_{S_1} \vec{E}\cdot d\vec{s} = \frac{Q}{\varepsilon_0}$.

Since the charge enclosed by this surface S_1 is also Q, $\vec{E} = \frac{Q}{4\pi\varepsilon_0 R^2}$.

The potential in this region is

$$V_3 = -\int_{\infty}^{R_i} E_{r_3}\,dR - \int_{R_i}^{R} E_{r_1}\,dR$$

$$= \left[\frac{Q}{4\pi\varepsilon_0 R_0} + \frac{Q}{4\pi\varepsilon_0 R}\right]_{R_i}^{R}$$

$$= \frac{Q}{4\pi\varepsilon_0 R_0} + \frac{Q}{4\pi\varepsilon_0 R} - \frac{Q}{4\pi\varepsilon_0 R_i}$$

$$V_3 = \frac{Q}{4\pi\varepsilon_0}\left[\frac{1}{R_0} + \frac{1}{R} - \frac{1}{R_i}\right].$$

Problem 23

A total charge of 10^{-8} C is distributed uniformly along a ring of radius of 5 m. Calculate the potential on the axis of the ring at a point 5 m from the centre of the ring.

Solution:

Consider the ring has a distributed charge of λ C/m along its circumference.

Take a differential length (dl) at any point on the ring.

We know that the potential V due to line charge distribution is

$$V = \int_l \frac{\lambda dl}{4\pi\varepsilon_0 r}$$

$$r = \sqrt{a^2 + h^2}$$

$$V = \frac{\lambda}{4\pi\varepsilon_0 \sqrt{a^2 + h^2}} \oint dl$$

$$= \frac{\lambda}{4\pi\varepsilon_0 \sqrt{a^2 + h^2}}(2\pi a)$$

$$V = \frac{\lambda a}{2\varepsilon_0 \sqrt{a^2 + h^2}}.$$

The linear charge density $\lambda = \frac{Q}{l} = \frac{Q}{2\pi a}$.

$$\therefore \quad V = \frac{(Q/2\pi a)\, a}{2\varepsilon_0 \sqrt{r^2 + h^2}}$$

Now, substitute $Q = 10^{-8}$ C_1, $a = 5$ m, $h = 5$ m.

$$V = \frac{10^{-8}}{4\pi \times 8.85 \times 10^{-12} \times \sqrt{5^2 + 5^2}} = 12.72 \text{ volt.}$$

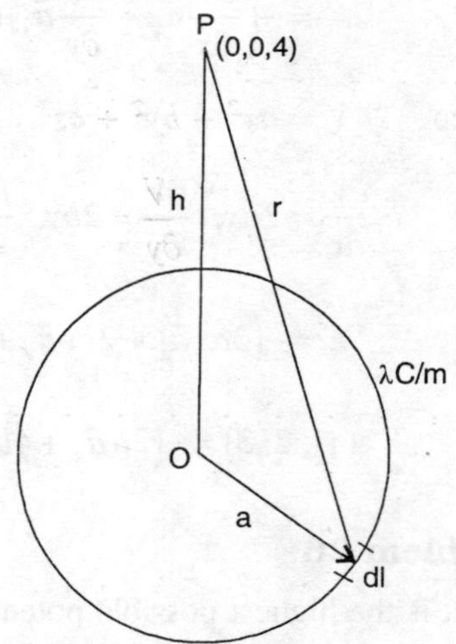

Fig. P-23

Problem 24

An electric field is given by $\vec{E} = 10^2\left(\vec{i} + 3\vec{j} + 2\vec{k}\right)$ V/m. Show that no work done in moving a test charge from the origin to $r = 0.06\vec{i} - 0.04\vec{j} + 0.03\vec{k}$.

Solution:

Work done/charge $= \vec{E} \cdot d\vec{l}$ newton.

'$d\vec{l}$' is the displacement from origin $\Big\} = \left(0.06\vec{i} - 0.04\vec{j} + 0.03\vec{k}\right) - \left(0\vec{i} + 0\vec{j} + 0\vec{k}\right)$

$$\text{work done/charge} = \left(100\vec{i} + 300\vec{j} + 200\vec{k}\right) \cdot \left(0.06\vec{i} - 0.04\vec{j} + 0.03\vec{k}\right)$$
$$= 6 - 12 + 6$$
$$= 0 \text{ newton.}$$

Problem 25

The electric potential near the origin of a system of coordinates is $V = ax^2 + by^2 + cz^2$. Find the field at (1, 2, 3).

Solution:

We know

$$\vec{E} = -\nabla V$$

$$= -\left(\frac{\partial V}{\partial x}\vec{a}_x + \frac{\partial V}{\partial y}\vec{a}_y + \frac{\partial V}{\partial z}\vec{a}_z\right).$$

Given $V = ax^2 + by^2 + cz^2$.

$$\frac{\partial V}{\partial x} = 2ax, \quad \frac{\partial V}{\partial y} = 2by, \quad \frac{\partial V}{\partial z} = 2cz.$$

$$\vec{E} = -\left(2ax\,\vec{a}_x + 2by\,\vec{a}_y + 2cz\,\vec{a}_z\right)$$

$$\vec{E} \text{ at } (1, 2, 3) = -\left(2a\,\vec{a}_x + 4b\,\vec{a}_y + 6c\,\vec{a}_z\right) \text{V/m}.$$

Problem 26

What is the highest possible potential of an isolated spherical conductor of radius 0.2 m? Assume the strength of air (max. value) is 30 kV/cm.

Solution:

We know the electric field intensity due to a spherical conductor of charge Q is

$$\vec{E} = \frac{Q}{4\pi\varepsilon_0 r^2}\vec{a}_r.$$

Similarly, the potential is

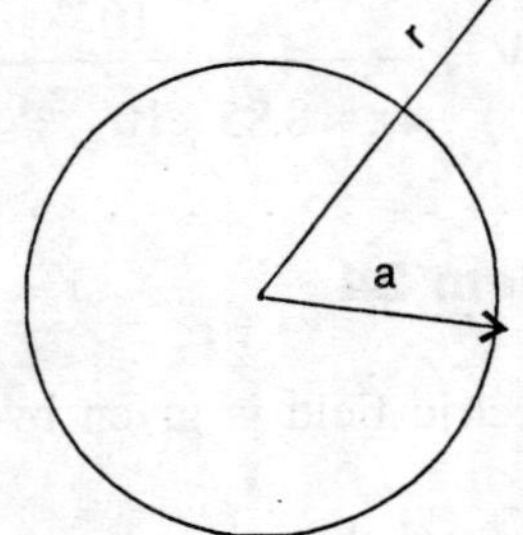

Fig. P-26

$$V = \frac{Q}{4\pi\varepsilon_0 r},\quad \text{where } r \text{ is radius of any point outside the sphere.}$$

$$|E| = \frac{V}{r}$$

$$|E|_{max} = \frac{V}{a}$$ As r increases, electric field intensity or electric strength decreases.

The minimum value of r is ($r = a$), at the surface of the conductor.

Given $a = 0.2$ m

$E_{max} = 30$ kV/cm.

$\therefore \quad V = 30 \times 0.2 \times 100 = 600$ kV.

Problem 27

Find the greatest charge per meter length that can be placed on a straight long wire of diameter 2 cm in air, given that air becomes ionized at $\vec{E} = 3 \times 10^6$ V/m (30 kV/cm).

Solution:

We know electric field intensity at any point due to a long straight wire is

$$\vec{E} = \frac{\lambda}{2\pi\varepsilon_0 \rho}\vec{a}_\rho,$$

where 'ρ' is the distance of a point outside the wire.

As the distance is inversely proportional to $\vec{E}$, the maximum stress will occur at $\rho = a$C at the surface of the wire).

$$|E_{max}| = \frac{\lambda}{2\pi\varepsilon_0 a}$$

$$3 \times 10^6 = \frac{\lambda}{2\pi \times 8.85 \times 10^{-12} \times 0.01}$$

$$\lambda = 1.6\,\mu\text{C/m}.$$

Problem 28

A hollow sphere is charged to 12 μC of electricity. Find the potential (*a*) at its surface (*b*) inside the sphere (*c*) at a distance of 0.3 m from the surface. The radius of sphere is 0.1 m.

Solution:

(*a*) Potential at the surface:

We know potential at any point is $V = \dfrac{Q}{4\pi\varepsilon_0 r}$.

If $r = a$

$$V = \frac{Q}{4\pi\varepsilon_0 a} = \frac{12 \times 10^{-6}}{4\pi \times 8.85 \times 10^{-12} \times 0.1}$$

$$V = 1.08 \times 10^6 \text{ V}$$

(*b*) Potential inside: For a hollow sphere, charge inside the sphere is zero. So, electric field intensity $\vec{E}$ is also zero.

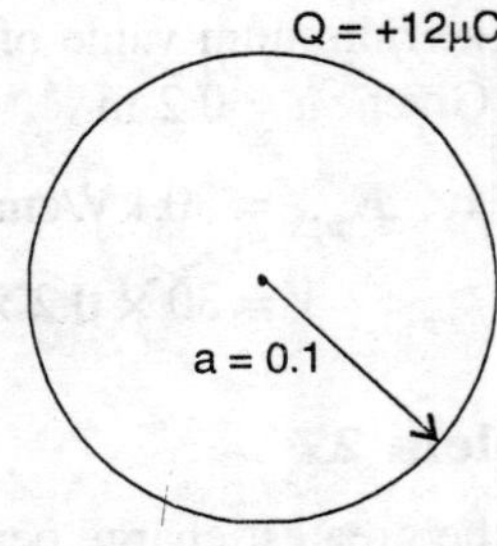

Fig. P-28

Since $\vec{E} = -\nabla V$,

if $\vec{E} = 0$

V = constant.

So, potential inside is same as the potential at the surface.
Potential at a point outside: $r > a$.
Given $r = 0.3$ from the surface,

$$V = \frac{Q}{4\pi\varepsilon_0 r}$$

$r = 0.4$ m (0.1 + 0.3)

$$V = \frac{12 \times 10^{-6}}{4\pi \times 8.85 \times 10^{-12} \times 0.4} = 0.27 \times 10^6 \text{ V}.$$

Problem 29

Given field

$\vec{E} = \left(\frac{-6y}{x^2}\right)\vec{I}_x + \left(\frac{6}{x}\right)\vec{I}_y + 5\vec{I}_z$ V/m , find the potential difference V_{AB}, given A (−7, 2, 1) and B (4, 1, 2). **(Anna. Univ. - Oct 2002)**

Solution:

$$\vec{E} = -\left(\frac{\partial V}{\partial x}\right)\vec{a}_x - \frac{\partial V}{\partial y}\vec{a}_y - \frac{\partial V}{\partial z}\vec{a}_z$$

$$\frac{dV}{dx} = \frac{6y}{x^2} \qquad V = \int \frac{6y}{x^2}dx = \frac{-6y}{x} + C_1$$

$$\frac{dV}{dx} = \frac{-6}{x} \qquad V = \int \frac{-6}{x}dx = \frac{-6y}{x} + C_2$$

$$\frac{dV}{dx} = -5 \qquad V = -\int 5\,dz = -5z + C_3$$

$$V = \frac{-6y}{x} - 5z + C$$

$$V = \frac{-6y}{x} - 5z \qquad \text{[Let } C = 0\text{]}$$

$$V_A = V(-7, 2, 1) = \frac{-6 \times 2}{-7} - 5$$

$$= \frac{12}{7} - 5 = -3.29$$

$$V_B = V(4, 1, 2) = \frac{-6}{4} - 10 = -11.5 \text{ V}$$

$$V_{AB} = V_B - V_A$$

$$= -11.5 + 3.29 = -8.21 \text{ V.}$$

Problem 30

The gradient of a scalar potential function ϕ is given by $\nabla\phi = 3\left[\left(x^2 - yz\right)\vec{a}_x + \left(y^2 - zx\right)\vec{a}_y + \left(z^2 - xy\right)\vec{a}_z\right]$. If the point (1, 1, 1) is at zero potential, determine the potential function ϕ.

Solution:

$\vec{E} = -\nabla V$. Here, potential function is given as ϕ.

$\vec{E} = \nabla\phi$.

$$\vec{E} = \frac{\partial\phi}{\partial x}\vec{a}_x + \frac{\partial\phi}{\partial y}\vec{a}_y + \frac{\partial\phi}{\partial z}\vec{a}_z$$

$$E_x = \frac{d\phi}{dx}, \quad E_y = \frac{d\phi}{dy}, \quad E_z = \frac{d\phi}{dz}.$$

$$E_x = \frac{d\phi}{dx} = 3\left(x^2 - yz\right),$$

$$\phi = \int 3\left(x^2 - yz\right)dx = 3\left(x^3/3 - xyz\right) + C_1$$

$$E_y = \frac{d\phi}{dy} = 3\left(y^2 - zx\right)$$

$$\phi = \int 3\left(y^2 - zx\right)dy = 3\left(\frac{y^3}{3} - xyz\right) + C_2$$

$$E_z = \frac{d\phi}{dz} = 3\left(z^2 - xy\right)$$

$$\phi = \int 3\left(z^2 - xy\right)dz = 3\left(\frac{z^3}{3} - xyz\right) + C_3$$

$$\phi = x^3 + y^3 + z^3 - 3xyz + C$$

$$\phi(1, 1, 1) = 0 = 3 - 3 + C$$

$$C = 0$$

$$\phi = x^3 + y^3 + z^3 - 3xyz.$$

2.2 ELECTRIC FIELDS IN MATERIAL SPACE

Introduction

So far, we have considered electrostatic fields in free space or space that has no materials in it. Just as electric fields can exist in free space, they can exist in material media also. So, a brief discussion of properties of materials, both conductors and dielectric materials is essential for understanding the behaviour of electric field in material media.

2.2.1 Conduction Current

When an electric field is applied to a conductor, conduction current occurs due to the drift motion of electrons. As the electrons move, they encounter some damping forces called resistance. The average drift velocity of the elections is directly proportional to the applied field. Thus, conduction current density is given by

$$J = \sigma E \qquad \text{...(2.106)}$$

where σ is the conductivity of the materials in mhos per meter or siemens per meter and J is known as conduction current density (in A/m^2).

Eqn (2.106) is referred to as **Ohm's law in point form.**

Based on this law, the resistance of a conducting material can be derived. The basic definition for resistance is the ratio of potential difference 'V' between two ends of conductor to the current through the conductor.

The total current through a prescribed surface 's' is given by

$$I = \int_S \vec{J} \cdot d\vec{s}. \qquad \text{...(2.107)}$$

The resistance of a conductor of non-uniform cross-section is

$$R = \frac{V}{I} = \frac{\int \vec{E} \cdot d\vec{l}}{\int \vec{J} \cdot d\vec{s}}$$

$$R = \frac{\int \vec{E} \cdot d\vec{l}}{\int \sigma \vec{E} \cdot d\vec{s}}. \quad ...(2.108)$$

However, the resistance of conductor of uniform cross-section is also obtained from the same definition.

Consider a length '*l*' of a conductor of uniform cross-section *s* applied with the potential difference of '*V*' volts.

The direction of electric field produced is same as the direction of flow of current *I* and its magnitude is given by

$$|\vec{E}| = \frac{V}{l}. \quad ...(2.109)$$

Since the conductor has a uniform cross-section

$$\vec{J} = \frac{I}{S}. \quad ...(2.110)$$

Substituting eqn. (2.110) in eqn. (2.106)

$$\sigma E = \frac{I}{S}$$

$$\frac{\sigma V}{l} = \frac{I}{S}.$$

Hence, $$R = \frac{V}{I} = \frac{l}{\sigma s}$$

$$R = \frac{\rho_c l}{s}, \quad ...(2.111)$$

where $\rho_c = \frac{1}{\sigma}$ is the resistivity of the material.

2.2.2 Convection Current

Convection current, as distinct from conduction current, does not involve conductors and consequently does not satisfy Ohm's law. It occurs when current flows through an insulating medium such as liquid, vacuum etc. For example, a beam of electrons in a vaccum tube is a convection current.

The current through a given area is defined as the electric charge passing through the area per unit time. That is,

$$I = \frac{dQ}{dt}. \quad ...(2.112)$$

Thus in a current of one ampere, charge is being transferred at the rate of one coulomb per second.

If there is a flow of charge of density ρ_v in a filament at a velocity u, then current through the filament is given by eqn. (2.112) as

$$\Delta I = \frac{\Delta Q}{\Delta t} \qquad ...(2.113)$$

$$\Delta Q = \rho_v \Delta v,$$

where,

Δv is a differential volume of filament in m^3

$\Delta Q = \rho_v\,(\Delta s\ \Delta l)$.

$\therefore$ Eqn. (2.113) can be written as

$$\Delta I = \rho_v\ \Delta s \frac{\Delta l}{\Delta t}$$

$$\Delta I = \rho_v\ \Delta s\ u. \qquad ...(2.114)$$

As current density at a point is the current through the area at that point, the eqn. (2.114) can be modified as

$$\frac{\Delta I}{\Delta S} = J_d = \rho_v u.$$

Hence, in general,

$$J_d = \rho_v u. \qquad ...(2.115)$$

'J_d' is convection current density or displacement current density is A/m^2.

2.2.3 Dielectrics in an Electric Field (OR) Polarisation in Di-electrics

Consider an atom of dielectric consists of a negative charge $-Q$ (nucleus). In the absence of electric field, the nucleus is in a position of stable equilibrium, since the centre of gravity of its positive charge coincides with that of negative charge. If the atom is placed in an external electric field, the field will push the positively-charged nucleus slightly in the direction of field and the negatively-charged electrons in the opposite direction. As soon as the centre of the electron cloud moves away from the nucleus, a strong electrostatic force of attraction is generated between the nucleus and the electron cloud which tends to restore the displaced charges to their original position. Until the internal restoring force balances the external field, the centres of positive and negative charges no longer coincide and the atom behaves like a dipole. The process of forming dipoles is known as **polarisation**.

In general, insulating materials can be classified into two broad groups as polar and non-polar. Molecules which do not behave like di-poles unless they are subjected to an external electric field are called non-polar. Examples of such dielectrics are hydrogen, oxygen and nitrogen. Other types of molecules such as water, sulphur dioxide and hydrochloric acid have permanent dipoles which are randomly oriented are said to be polar.

In polar molecules, owing to the thermal agitation, the dipoles are randomly oriented and as a consequence the material is left unpolarised. With the application of the field, the dipoles tend to orient themselves in the direction of the external field. This type of polarisation is said to be **oriental**

polarisation. Some materials become permanently polarised in the direction of field. It retain to the polarisation even after the removal of the field, they are called as electrets.

The amount of dipoles induced is proportional to the applied electric field. If the separation of the charges is '*d*' m and the magnitude of charges are *Q* coulomb, then the dipole moment is the product of magnitude of charge and the separation '*d*'.

$$p = Qd \text{ coulomb-mt,} \quad (2.116)$$

where *p* denotes dipole moment.

The polarisation density (or) polarisation vector $(\vec{P})$ is defined as the dipole moment per unit volume.

$$\vec{P} = \frac{p}{v} = \frac{Qd}{v} \text{ coulomb/m}^2 \quad ...(2.117)$$

Thus the application of electric field to the dielectric material creates the process of polarisation which will increase the flux density greater than that of free space.

So, the net effect of dielectric on the electric field is to increase $\vec{D}$ inside it by the amount $\vec{P}$. That is, eqn. (2.43) is modified as

$$\vec{D} = \varepsilon_0 \vec{E} + \vec{P}. \quad ...(2.118)$$

It should be noted that definition of $\vec{D}$ in eqn. (2.43) for free space is a special case of that in eqn. (2.118) because $\vec{P}$ is zero in free-space. As polarisation $\vec{P}$ varies directly as the applied electric field $\vec{E}$,

$$\vec{P} \propto \vec{E}$$

$$\vec{P} = \chi_e \varepsilon_0 \vec{E}, \quad ...(2.119)$$

where χ_e is the electric susceptibility of the material. It is a measure of sensitivity of a given dielectric to electric fields.

Substituting eqn. (2.119) in eqn. (2.118)

$$\vec{D} = \varepsilon_0 (1 + \chi_e) \vec{E}$$

$$\vec{D} = \varepsilon_0 \, \varepsilon_r \, \vec{E}$$

$$\vec{D} = \varepsilon \vec{E} \quad ...(2.120)$$

where $\varepsilon = \varepsilon_0 \, \varepsilon_r$

$$\varepsilon_r = 1 + \chi_e. \quad ...(2.121)$$

ε is the permittivity of the dielectric, ε_0 is the permittivity of the free space and ε_r is the dielectric constant or relative permittivity. It should be noted that χ_e and ε_r are dimensionless, but ε and ε_0 are in farad/m.

When the applied electric field in a dielectric is sufficiently large, it begins to pull electrons completely out of molecules and the dielectric becomes conducting. When a dielectric becomes conducting, it is said that dielectric breakdown has occurred. The dielectric strength of dielectric materials is defined as the maximum electric field which a dielectric can tolerate or withstand breakdown.

A dielectric material is said to be linear, if $\vec{D}$ varies linearly with $\vec{E}$, otherwise it is non-linear.

Dielectric materials for which the permittivity ε does not vary in the region being considered and is therefore the same at all points are said to be homogeneous. If the permittivity ε depends on the space co-ordinates, then the materials are inhomogeneous. The atmosphere is a typical example of an inhomogeneous medium, since its permittivity varies with altitude.

Materials for which $\vec{D}$ and $\vec{E}$ are in the same direction are said to be isotropic and they have same properties in all directions. For anisotropic materials, $\vec{D}$, $\vec{E}$ and $\vec{P}$ are not parallel. Crystalline materials are anisotropic.

2.2.4 Law of Conservation of Charge

The law of conservation of charge states that current due to the flow of charge out of a closed surface '*s*' bounding a volume '*v*' must be equal to the time rate of decrease of charge enclosed by the surface.

If the current flowing out of the surface is $[I_c]_s$ and the charge enclosed by the surface *s* is *Q* coulomb, then

$$[I_c]_s = -\frac{dQ}{dt}. \qquad ...(2.122)$$

From Gauss's law

$$\oint_S \vec{E}\cdot d\vec{s} = \frac{Q}{\varepsilon_0}$$

$$Q = \oint_s \varepsilon_0 \vec{E}\cdot d\vec{s}.$$

Eqn. (2.122) becomes

$$[I_c]_s = -\frac{d}{dt}\oint_s \varepsilon_0 \vec{E}\cdot d\vec{s}$$

$$[I_c]_s + [I_d]_s = 0,$$

where $[I_d]_s = \frac{d}{dt}\oint_s \varepsilon_0 \vec{E}\cdot d\vec{s}$ is the displacement current. ...(2.123)

Thus the law of conservation of charge states that the sum of the current due to the flow of charges and the displacement current across any closed surface must be equal to zero.

2.2.5 Equation of Continuity

According to the law of conservation of charges

$$[I_c]_s = \frac{-dQ}{dt}.$$

In terms of current density *J* and charge density ρ, $[I_c]_s$ and *Q* are given by

$$[I_c]_s = \oint_S J\cdot ds$$

$$Q = \int_v \rho\, dv,$$

where v is the volume bounded by s.

$$\oint_S \vec{J} \cdot d\vec{s} = -\frac{d}{dt}\int_v \rho\, dv. \qquad ...(2.124)$$

Applying divergence theorem to the left-hand side and interchanging the differentiation and integrating on the right side, eqn. (2.124) becomes

$$\int_v \nabla \cdot \vec{J}\, dv = -\int_v \frac{\partial \rho}{\partial t} dv$$

$$\int_v \left(\nabla \cdot \vec{J} + \frac{\partial \rho}{\partial t} \right) dv = 0$$

$$\nabla \cdot \vec{J} + \frac{\partial \rho}{\partial t} = 0$$

$$\nabla \cdot \vec{J} = -\frac{\partial \rho}{\partial t}. \qquad ...(2.125)$$

Eqn. (2.125) is called continuity equation (or) the law of conservation of charge in differential form. Eqn. (2.124) is called law of conservation charge in integral form.

For static fields, $\frac{\partial \rho}{\partial t} = 0$

$\therefore \qquad \nabla \cdot J = 0. \qquad ...(2.126)$

2.2.6 Conductors in an Electric Field

We know that free electrons in a conductor will drift under the influence of an electric field.

Consider an arbitrary shaped conductor of uniform conductivity σ placed in a static electric field.

The free electrons in the conductor move opposite to the direction of lines of electric field. Consider the conductor to be bounded by free space, in which case the electrons are held at the boundary from moving further by the atomic forces within the conductor and by the insulating property of free space. Thus a negative surface charges form on the part of boundary through which the electric field lines enter the conductor originally. Since the conductor as a whole is neutral, an equal amount of positive

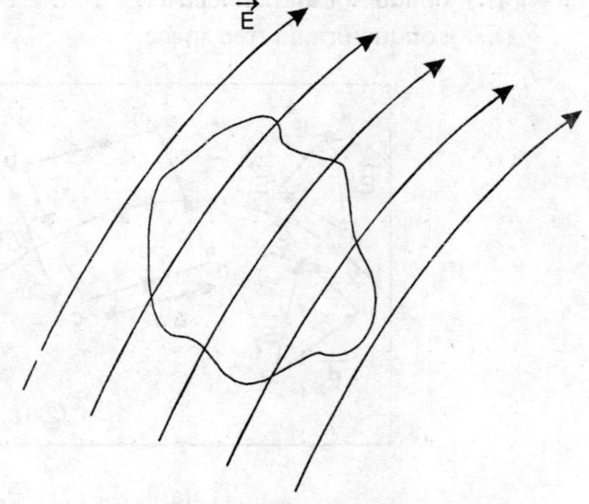

Fig. 2.40 Conductor in electric field

surface charges appear on the part of the boundary through which the electric field lines leave the conductor originally.

The surface charge distributions formed in this manner produces a secondary electric field which opposes the applied field inside the conductor and after a particular time the surface charge distribution reduces the field inside the conductor to zero. But this secondary charge distribution produces a secondary field outside the conductor which modifies the applied field.

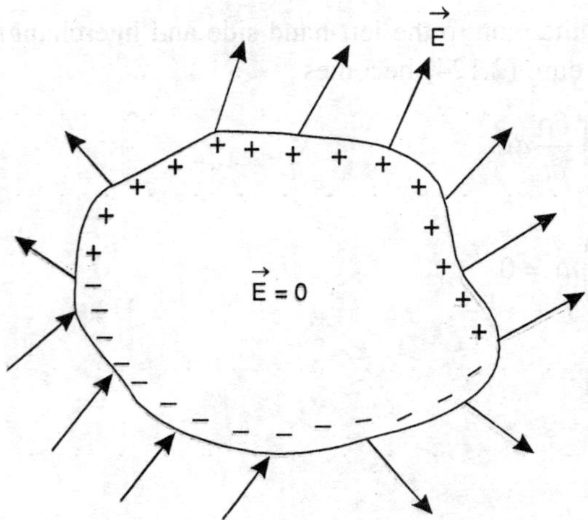

Fig. 2.41 Production of secondary field

2.2.7 Boundary Conditions

So far, existence of electric field in a homogeneous medium is considered. If the field exists in a region consisting of two different media, the conditions that the field must satisfy at the interface separating the media are called boundary conditions.

These conditions are helpful in determining the field on one side of the boundary if the field on the other side is known. Depending up on the types of material, the media are made of

(*i*) dielectric ε_{r1} and dielectric ε_{r2}

(*ii*) conductor and dielectric.

(*iii*) conductor and free space.

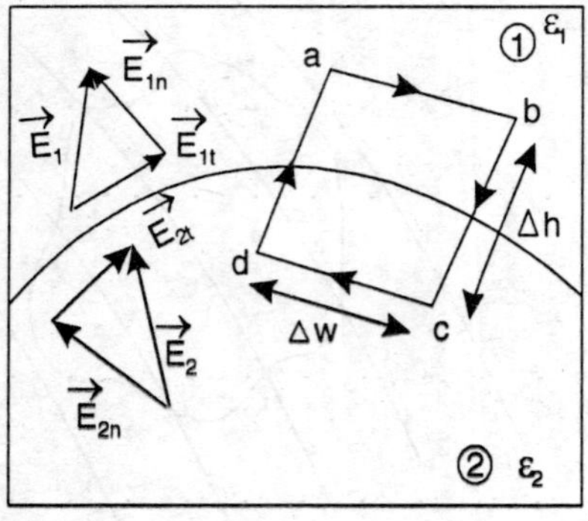

(a)

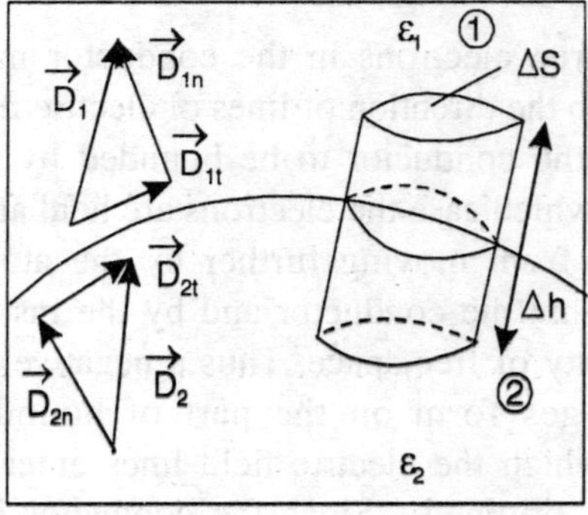

(b)

Fig. 2.42 Di-electric, Di-electric boundary

To determine the boundary conditions the following two equations of electrostatic field are used.

(*i*) Conservative property

$$\oint_l \vec{E}\cdot d\vec{l} = 0. \quad ...(2.127)$$

(*ii*) Gauss's law

$$\oint_s \vec{D}\cdot d\vec{s} = Q_{\text{enc.}}. \quad ...(2.128)$$

Also, the electric field and electric flux density are decomposed into two orthogonal components as

$$\vec{E} = \vec{E}_t + \vec{E}_n, \qquad \vec{D} = \vec{D}_t + \vec{D}_n,$$

where $\vec{E}_t$ and $\vec{D}_t$ are tangential components and $\vec{E}_n$ and $\vec{D}_n$ are normal components respectively.

2.2.7 (a) Dielectric-Dielectric Boundary

Consider the electric field $\vec{E}$ existing in a region consisting of two different dielectrics ε_{r1} and ε_{r2}.

Electric field in media 1 and 2 can be written as

$$E_1 = E_{1t} + E_{1n}$$

$$E_2 = E_{2t} + E_{2n}.$$

Applying eqn. (2.127) to the closed path *abcda* of Fig. (2.42a) and assume that the path is very small with respect to variation of *E*.

Then,

$$E_{1t}\,\Delta w - E_{1n}\frac{\Delta h}{2} - E_{2n}\frac{\Delta h}{2} - E_{2t}\,\Delta w + E_{2n}\frac{\Delta h}{2} + E_{1n}\frac{\Delta h}{2} = 0, \quad ...(2.129)$$

where $E_t = |\vec{E}_t|,\ E_n = |\vec{E}_n|$

As $\Delta h \to 0$, eqn. (2.129) becomes

$$E_{1t} = E_{2t}. \quad ...(2.130)$$

Thus the tangential components of $\vec{E}$ are the same on the two sides of the boundary. In other words, E_t undergoes no change on the boundary and it is said to be continuous across the boundary.

Since $D = \varepsilon E$

$$\frac{D_{1t}}{\varepsilon_1} = E_{1t}$$

$$\frac{D_{2t}}{\varepsilon_2} = E_{2t}.$$

As $E_{1t} = E_{2t}$

$$\frac{D_{1t}}{\varepsilon_1} = \frac{D_{2t}}{\varepsilon_2}$$

$$\frac{D_{1t}}{D_{2t}} = \frac{\varepsilon_1}{\varepsilon_2}. \qquad ...(2.131)$$

The tangential components of electric flux density is discontinuous across the surface. Similarly eqn. (2.128) is applied to the pill box and allowing $\Delta h \to 0$

$$\underset{(\text{top})}{D_{1n}\,\vec{a}_z \cdot \Delta s\vec{a}_z} + \underset{(\text{bottom})}{D_{2n}\,\vec{a}_z \cdot \Delta s(-\vec{a}_z)} = \Delta Q. \qquad ...(2.132)$$

As $\Delta Q = \rho_s\,\Delta s$, where ρ_s is the charge density placed at the boundary.

Then $D_{1n}\,\Delta s - D_{2n}\,\Delta s = \rho_s\,\Delta s$

$$D_{1n} - D_{2n} = \rho.$$

Note that $\vec{D}$ is directed from region 2 to 1.

If no free charges exist at the interface, $\rho_s = 0$ and so

$$D_{1n} = D_{2n}. \qquad ...(2.133)$$

Thus normal component of $\vec{D}$ is continuous across the surface, *i.e.* D_n undergoes no change at the boundary.

Since $\vec{D} = \varepsilon\vec{E}$

$$\varepsilon_1 E_{1n} = \varepsilon_2 E_{2n}$$

$$\frac{E_{1n}}{E_{2n}} = \frac{\varepsilon_{r_2}}{\varepsilon_{r_1}}. \qquad ...(2.134)$$

The normal component of $\vec{E}$ is discontinuous at the boundary. This boundary conditions are used to determine the 'refraction' of the electric field across the surface.

Consider D_1 or E_1 and D_2 or E_2 making angles θ_1 and θ_2 with the normal to the interface as in Fig. 2.43.

$$E_1 \sin\theta_1 = E_{1t} = E_{2t} = E_2 \sin\theta_2.$$

(or) $E_1 \sin\theta_2 = E_2 \sin\theta_2$.

Similarly, by applying other condition,

$$D_{1n} = D_{2n}.$$

$$\varepsilon_1 E_1 \cos\theta_1 = \varepsilon_2 E_2 \cos\theta_2$$

$$\frac{\tan\theta_1}{\varepsilon_1} = \frac{\tan\theta_2}{\varepsilon_2}$$

$$\frac{\tan\theta_1}{\tan\theta_2} = \frac{\varepsilon_1}{\varepsilon_2} = \frac{\varepsilon_{r1}}{\varepsilon_{r2}}. \qquad ...(2.135)$$

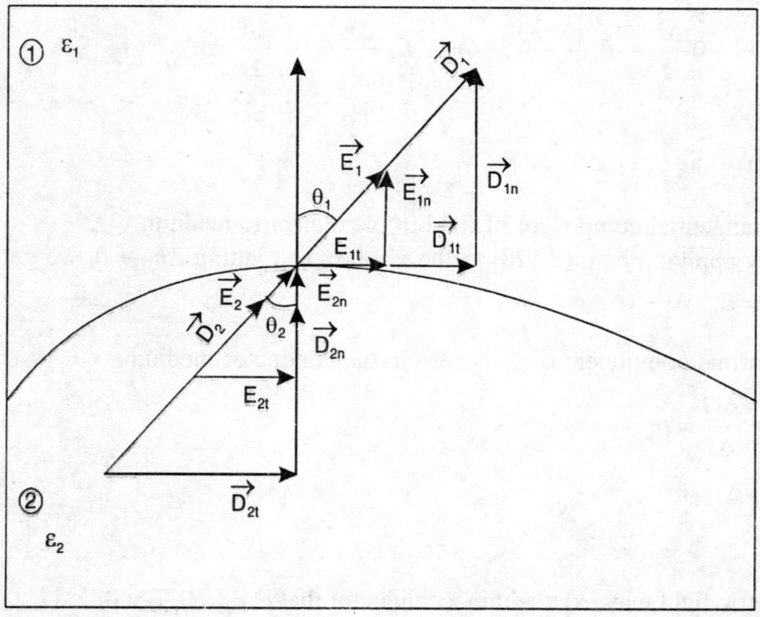

Fig. 2.43 Refraction of $\vec{D}$ or $\vec{E}$ at a dielectric-dielectric boundary

2.2.7 (b) Conductor-dielectric Boundary

The same procedure is followed here also to find the boundary conditions for a conductor–dielectric interface. It should be remembered that electric field intensity is zero inside the conductor.

Applying eqn. (2.127) to the closed path *abcda* of Fig. (2.44a) as in the previous section (2.2.7a).

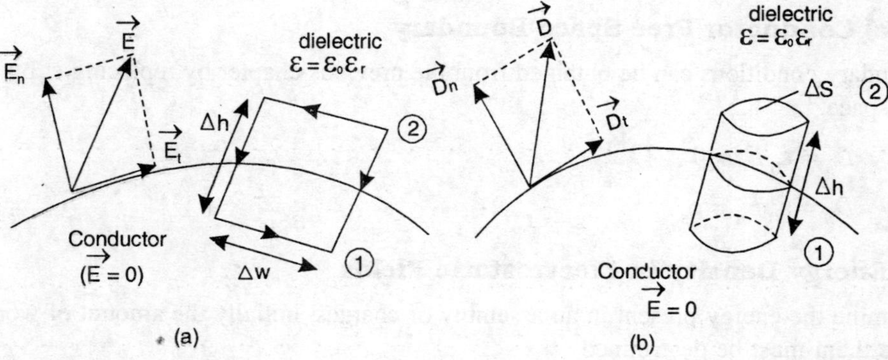

Fig. 2.44 Conductor-dielectric boundary.

As the electric field intensity $\vec{E}$ in the medium ① is zero, its tangential and normal components of E are also zero. So, eqn. (2.129) becomes

$$0 \cdot \Delta w. - 0\frac{\Delta h}{2} - E_n\frac{\Delta h}{2} - E_t\Delta w + E_n\frac{\Delta h}{2} + 0 \cdot \frac{\Delta h}{2} = 0$$

$$0 \cdot \Delta w - 0\frac{\Delta h}{2} - E_n\frac{\Delta h}{2} - E_t\Delta w + E_n\frac{\Delta h}{2} + 0\cdot\frac{\Delta h}{2} = 0$$

As $\quad \Delta h \to 0$

$$E_t = 0, \quad ...(2.136)$$

where E_t is the tangential component of field in the dielectric medium ②.

Similarly, by applying eqn. (2.128) to the pill box and letting $\Delta h \to 0$, we get

$$\Delta Q = D_n \cdot \Delta s - 0 \cdot \Delta s$$

Since the normal component of $\vec{D}$ is zero in the conductor medium.

$$\therefore \quad D_n = \frac{\Delta Q}{\Delta s} = \rho_s$$

$$D_n = \rho_s. \quad ...(2.137)$$

Conclusions

1. No electric field may exist within a conductor that is $\rho_v = 0, E = 0$.
2. Since $\vec{E} = -\nabla V = 0$, there can be no potential difference between any two points in the conductor *i.e.* a conductor is an equipotential body.
3. The electric field $\vec{E}$ can only be external to the conductor and normal to its surface *i.e.*

$$\vec{D}_t = \varepsilon_0\,\varepsilon_r\,\vec{E}_t = 0 \quad ...(2.138)$$

$$\vec{D}_n = \varepsilon_0\,\varepsilon_r\,\vec{E}_n = \rho_s \quad ...(2.139)$$

2.2.7 (c) Conductor Free Space Boundary

The boundary conditions can be obtained from the previous chapter by replacing ε_r by 1. Since $\varepsilon_r = 1$ for free space.

$$\therefore \quad D_t = \varepsilon_0\,E_t = 0 \quad ...(2.140)$$

$$D_n = \varepsilon_0\,E_n = \rho_s \quad ...(2.141)$$

2.2.8 Energy Density in Electrostatic Fields

To determine the energy present in an assembly of charges, initially the amount of work necessary to assemble them must be determined.

Let us assume that three point charges are to be positioned in an initially empty space.

No work is required to transfer Q_1 from infinity to P_1, because the space is initially charge free and there is no electric field so $W_1 = 0$

The work done in transferring Q_2 from infinity to P_2 is equal to the product of Q_2 and potential V_{21} at P_2 due to Q_1. Similarly, the work done in positioning Q_3 at P_3 is equal to $Q_3\,(V_{32} + V_{31})$ where V_{32} and V_{31} are the potentials at P_3 due to Q_2 and Q_1 respectively. Hence the total work done in positioning the three charges is

$$W_E = W_1 + W_2 + W_3$$
$$= 0 + Q_2\, V_{21} + Q_3\,(V_{31} + V_{32}). \qquad ...(2.142)$$

If the charges were positioned in reverse order,

$$W_E = W_3 + W_2 + W_1$$
$$= 0 + Q_2\, V_{23} + Q_1\,(V_{12} + V_{13}), \qquad ...(2.143)$$

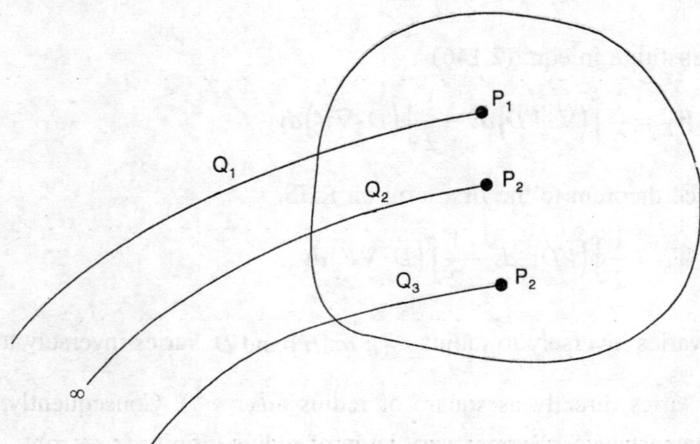

Fig. 2.45 Assembling of charges

where V_{23} is the potential at P_2 due to Q_3; V_{12} and V_{13} are respectively the potentials at P_1 due to Q_2 and Q_3.

Adding the two eqns. (2.141) and (2.142)

$$2W_E = Q_1\,(V_{12} + V_{13}) + Q_2\,(V_{21} + V_{23}) + Q_3\,(V_{31} + V_{32})$$
$$= Q_1V_1 + Q_2V_2 + Q_3V_3.$$
$$W_E = \frac{1}{2}\,[Q_1V_1 + Q_2V_2 + Q_3V_3], \qquad ...(2.144)$$

where V_1, V_2 and V_3 are total potentials at P_1, P_2 and P_3 respectively. In general, if there are n point charges, then

$$W_E = \frac{1}{2}\sum_{k=1}^{n} Q_k\, V_k \text{ joules} \qquad ...(2.145)$$

Instead of point charges, if the region has a continuous charge distribution of ρ_v, then the summation becomes integration

$$W_E = \frac{1}{2}\int_v (\rho_v\, dv)V \quad Q = \int_v \rho_v dv.$$

From eqn. (2.103)

$$\rho_v = \nabla \cdot \vec{D}$$

$$W_E = \frac{1}{2}\int(\nabla\cdot\vec{D})V\,dv \qquad ...(2.146)$$

Using vector identity

$$\nabla\cdot(V\vec{A}) = \vec{A}\cdot\nabla V + V(\nabla\cdot\vec{A})$$

$$\therefore \quad (\nabla\cdot\vec{A})V = (\nabla\cdot V\vec{A}) - (\vec{A}\cdot\nabla V) \qquad ...(2.147)$$

Let $\vec{A} = \vec{D}$, and substitute in eqn. (2.146)

$$W_E = \frac{1}{2}\int(\nabla\cdot V\vec{D})dv - \frac{1}{2}\int_v(\vec{D}\cdot\nabla V)dv \qquad ...(2.148)$$

Applying divergence theorem to the first term on RHS.

$$W_E = \frac{1}{2}\oint_s(V\vec{D})\cdot ds - \frac{1}{2}\int_v(\vec{D}\cdot\nabla V)dv. \qquad ...(2.149)$$

We know that 'V' varies inversely to radius r $(V \propto 1/r)$ and $\vec{D}$ varies inversely to square of radius r $(\vec{D} \propto 1/r^2)$ while ds varies directly as square of radius $(ds \propto r^2)$. Consequently, the first integral must tend to zero as the surface S becomes very large of radius infinity ($r \to \infty$).

$$\therefore \quad \oint_{\substack{s \\ r\to\infty}} (V\vec{D})\cdot d\vec{s} \propto \frac{1}{r} = 0$$

Hence

$$W_E = -\frac{1}{2}\int_v(\vec{D}\cdot\nabla V)dV$$

$$= \frac{1}{2}\int_v(\vec{D}\cdot\vec{E})dV \qquad [\because \vec{E} = -\nabla V]$$

Since $\vec{D} = \varepsilon_0\vec{E}$,

$$W_E = \frac{1}{2}\int \varepsilon_0\vec{E}^2\,dv \text{ joules.} \qquad ...(2.150)$$

Electrostatic energy density W in (J/m^3) is

$$w_E = \frac{dW_E}{dv} = \frac{1}{2}\varepsilon_0 E^2 = \frac{D^2}{2\varepsilon_0} \qquad ...(2.151)$$

2.2.9 Electric Field Intensity due to Dipole

Let $+Q$ and $-Q$ be the two charges separated by a small distance 'l' m which is small compared to the distance of the point 'P' at which the electric field intensity is to be determined.

Assume the point '*P*' is described by spherical co-ordinate $P(r, \theta, \phi)$

Let '*P*' be at R_1, R_2 and r m respectively from $+Q$, $-Q$ and from mid point '*O*', as in Fig. 2.46a.

The potential at '*P*' due to $+Q$ alone is $V_1 = \dfrac{Q}{4\pi\varepsilon_0 R_1}$ and due to $-Q$ alone is $V_2 = \dfrac{-Q}{4\pi\varepsilon_0 R_2}$.

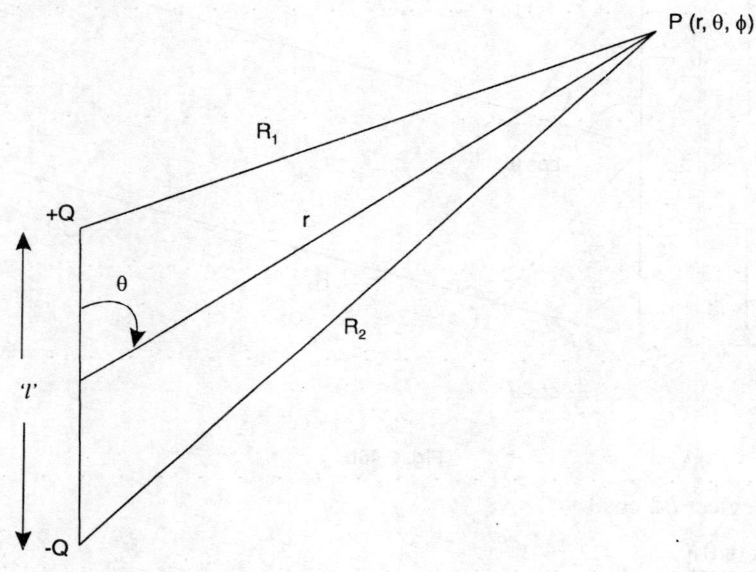

Fig. 2.46a

The potential at P due to the dipole is

$$V = V_1 + V_2$$

$$= \frac{Q}{4\pi\varepsilon_0}\left[\frac{1}{R_1} - \frac{1}{R_2}\right].$$

From the Fig. 2.46(*b*),

$$R_1 = r - l/2 \cos\theta$$

$$R_2 = r + l/2 \cos\theta$$

$$R_2 - R_1 = l \cos\theta.$$

$$V = \frac{Q}{4\pi\varepsilon_0}\left[\frac{R_2 - R_1}{R_2 R_1}\right]$$

$$= \frac{Q}{4\pi\varepsilon_0}\left[\frac{l\cos\theta}{\left(\frac{l}{2}\cos\theta + r\right)\left(r - \frac{l}{2}\cos\theta\right)}\right]$$

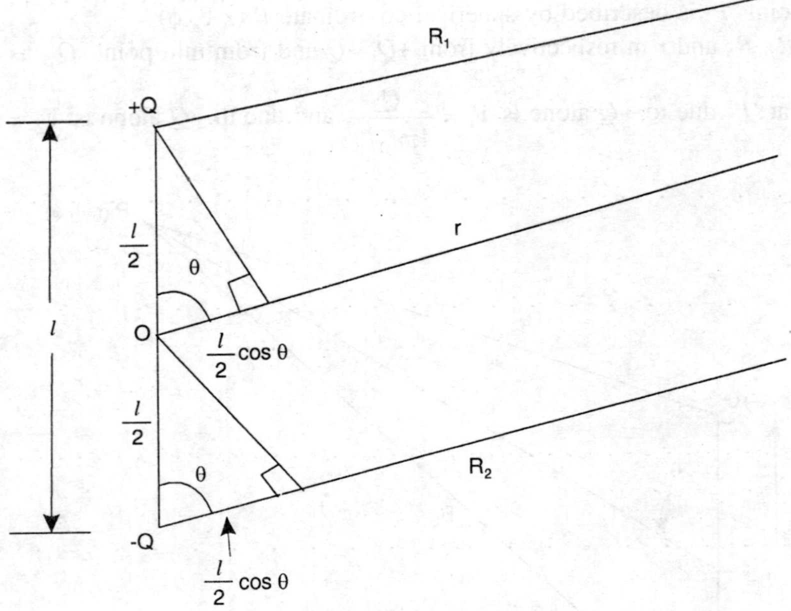

Fig. 2.46b

Since $l << r$, neglect $l/2\ \cos\theta$.

$$V = \frac{Q}{4\pi\varepsilon_0}\left(\frac{l\cos\theta}{r^2}\right)$$

$$V = \frac{Ql\cos\theta}{4\pi\varepsilon_0 r^2}.$$

Dipolement $p = Ql$

$$V = \frac{p\cos\theta}{4\pi\varepsilon_0 r^2}\ \text{volt.} \qquad ...(2.152)$$

Potential gradient (or) electric field intensity is $E = -\nabla V$.
In spherical co-ordinate

$$\vec{E} = -\nabla V = -\left(\frac{\partial V}{\partial r}\vec{a}_r + \frac{1}{r}\frac{\partial V}{\partial\theta}\vec{a}_\theta + \frac{1}{r\sin\theta}\frac{\partial V}{\partial\phi}\vec{a}_\phi\right).$$

From eqn. (2.151), $\frac{\partial V}{\partial r} = -\frac{Qr\cos\theta}{2\pi\varepsilon_0 r^3}$

$$\frac{1}{r}\frac{\partial V}{\partial\theta} = -\frac{Qr\sin\theta}{4\pi\varepsilon_0 r^3}$$

$$\frac{1}{r\sin\theta}\frac{\partial V}{\partial\phi} = 0$$

$$\therefore \quad \vec{E} = \frac{Ql}{4\pi\varepsilon_0 r^3}\left[2\cos\theta\,\vec{a}_r + \sin\theta\,\vec{a}_\theta\right] \text{ V/m} \qquad \text{...(2.153)}$$

The electric field lines are shown in Fig. 2.47.

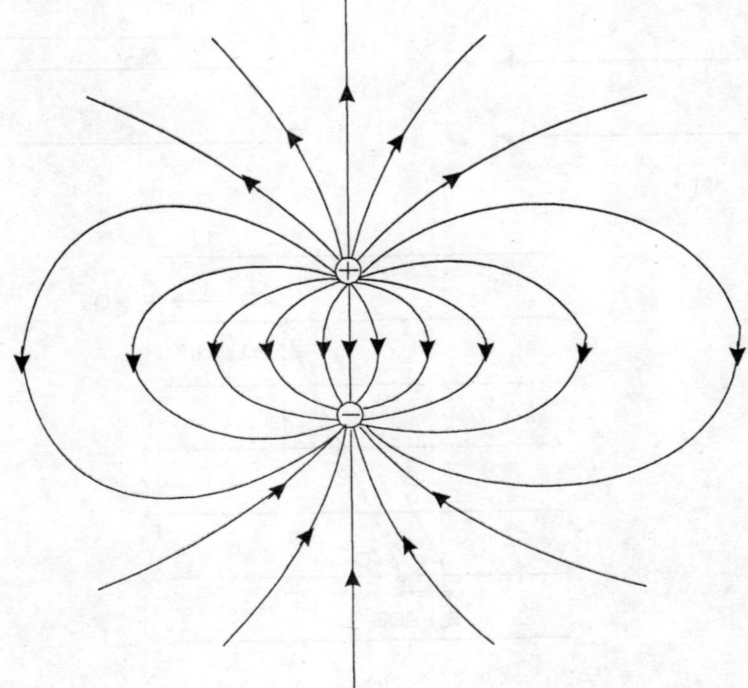

Fig. 2.47 Electric field lines of an electric dipole

2.2.10 Torque on Dipole

We have already defined the electric dipole moment $\vec{P}$ as a vector of magnitude Qd directed from $-Q$ to $+Q$.

In a field $\vec{E}$, the dipole experiences a torque aligning it so that $\vec{P}$ is parallel to $\vec{E}$. Torque is the product of force and perpendicular distance from the centre of the dipole. The forces acting on the two charges ($\pm Q$) are $+EQ$ and $-EQ$, so the net force is zero. However, there is a resultant torque. Torque = Force × Perpendicular distance.

$$T = F_+\frac{d}{2}\sin\theta - F_-\frac{d}{2}\sin\theta$$

$$= QEd\sin\theta$$

$$T = PE\sin\theta \qquad \text{...(2.154)}$$

Torque is a vector quantity whose direction gives the axis of rotation which is normal to the plane containing $\vec{P}$ and $\vec{E}$. Hence

$$\vec{T} = \vec{P}\times\vec{E} \qquad \text{...(2.155)}$$

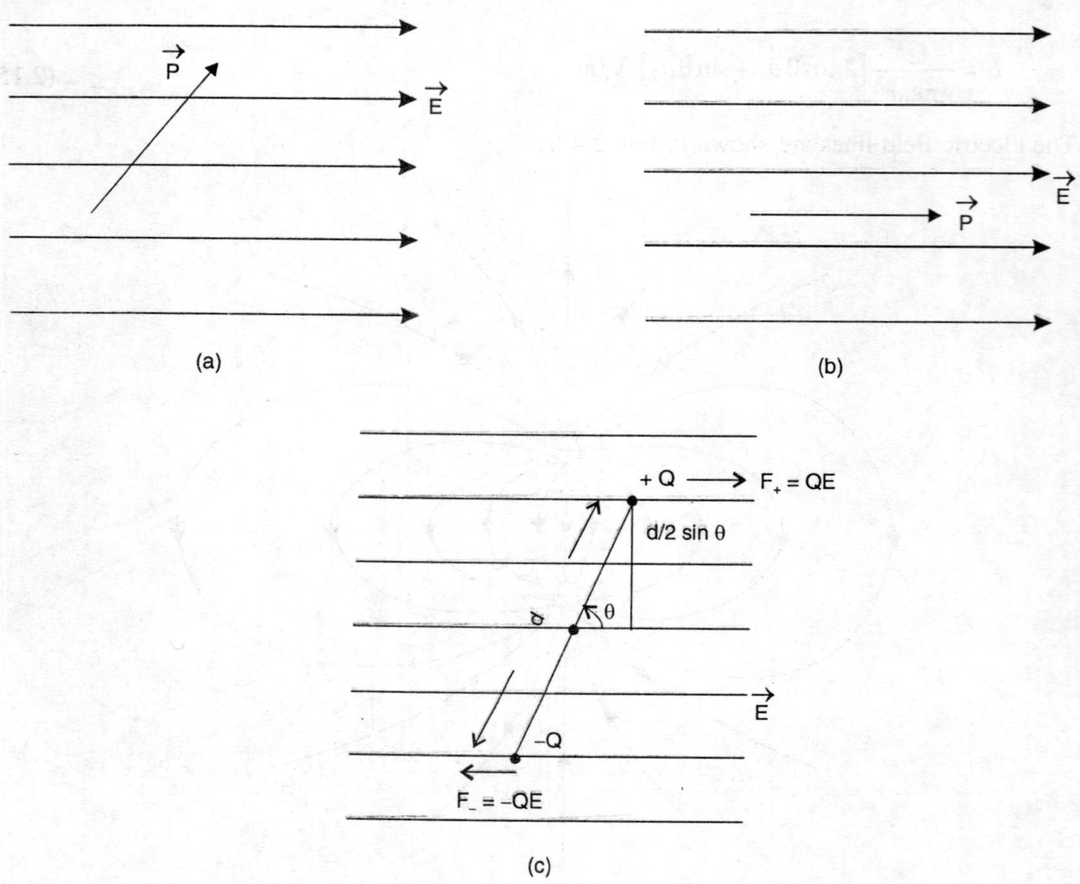

Fig. 2.48 Torque on dipole

The potential energy U of the dipole is minimum when $\vec{P}$ is parallel to $\vec{E}$ and maximum when $\vec{P}$ is perpendicular to $\vec{E}$. The work done to rotate the dipole from θ_i (initial) to θ_f (final) equals the change in U:

$$U_f - U_i = \text{Work done} = \int_{\theta_i}^{\theta_f} |T| \, d\theta$$

$$= \int_{\theta_i}^{\theta_f} PE \sin\theta \, d\theta = -PE\left(\cos\theta_f - \cos\theta_i\right)$$

By convention, assume $U_i = 0$ at $\theta_i = 90°$ so that $U_f = -PE\cos\theta_f$

In vector notation, potential energy

$$U = \vec{P} \cdot \vec{E} \qquad \text{...(2.156)}$$

2.3 CAPACITANCE-CAPACITORS

Introduction

A capacitor is an electrical device composed of two conductors which are separated through a dielectric medium and which can store equal and opposite charges. In this section, capacitance of various types of capacitors are derived using Gauss's law.

2.3.1 Theory of Capacitors

Generally, in order to have a capacitor, two (or more) conductors carrying equal but opposite charges are required. This states that all flux lines leaving one conductor must necessarily terminate at the surface of the other conductor. The conductors are referred to as the plates of the capacitor. The plates may be separated by free space or a dielectric.

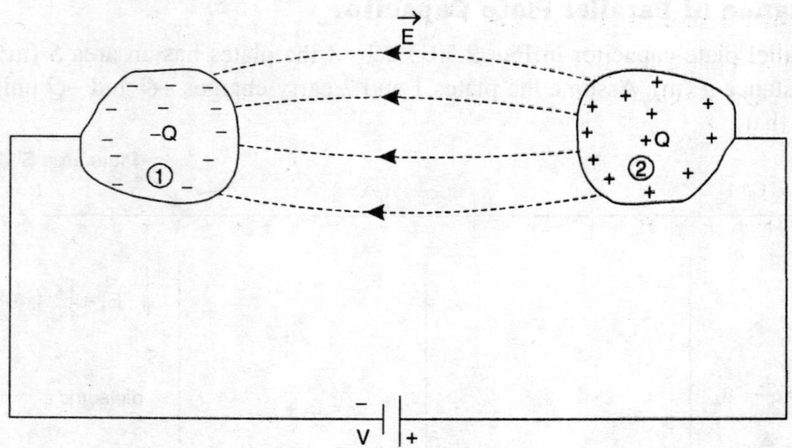

Fig. 2.49 Two-conductor capacitor

Consider the two-conductor capacitor of Fig. 2.49. The conductors are maintained at a potential difference V given by

$$V = V_1 - V_2 = -\int_2^1 \vec{E} \cdot d\vec{l},$$

where $\vec{E}$ is electric field existing between the conductors and conductor 1 is assumed to carry positive charge.

The capacitance 'C' of the capacitor is the ratio of the magnitude of the charge on one of the plates to the potential difference between them. That is,

$$C = \frac{Q}{V} = \frac{\varepsilon \oint \vec{E} \cdot d\vec{s}}{\int_l \vec{E} \cdot d\vec{l}} \qquad \text{...(2.157)}$$

The negative sign before $V = -\int \vec{E} \cdot d\vec{l}$ has been dropped because only absolute value of V is to be considered. The capacitance of the capacitor is measured in farad.

The capacitance 'C' can be obtained for any given two-conductor capacitance by assuming Q and determining V in terms of Q. This method involves Gauss's law.

The following steps are used to find the capacitance of any two-conductor capacitance.

1. Choose a suitable coordinate system.
2. Let the conducting plates carry charges $+Q$ and $-Q$.
3. Determine $\vec{E}$ using Coulomb's law or Gauss's law and find V from $V = -\int \vec{E} \cdot d\vec{l}$. The negative sign may be ignored in this case since only absolute value of V is considered.
4. Finally obtain C from $C = \dfrac{Q}{V}$.

2.3.2 Capacitance of Parallel Plate Capacitor

Consider the parallel plate capacitor in Fig. 2.50. Each of the plates has an area S (m^2) and they are separated by a distance d (m). Assume the plates 1 and 2 carry charges $+Q$ and $-Q$ uniformly distributed on them, so that

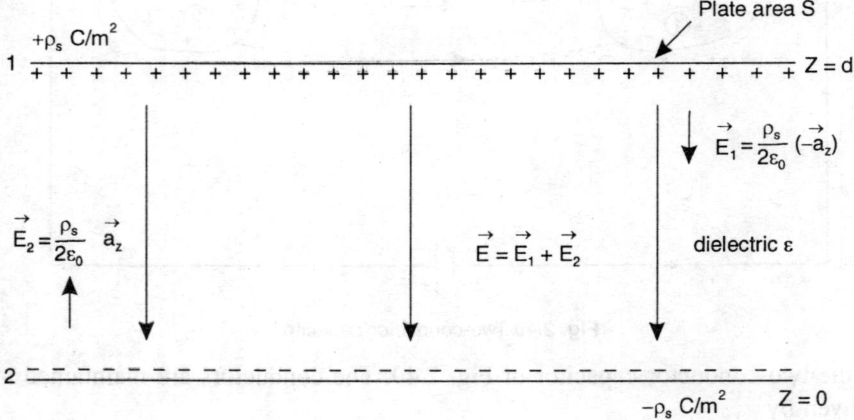

Fig. 2.50 Parallel plate capacitor

$$\rho_s = \frac{Q}{S}\ \text{C/m}^2. \qquad \text{...(2.158)}$$

An ideal parallel plate capacitor is one in which the plate separation d is very small compared with the dimensions of the plate. Assuming such an ideal case, the fringing field at the edge of the plates can be ignored so that the field between them is considered uniform. If the space between the plates is filled with homogeneous dielectric with permittivity ε.

Electric field intensity at any point in between two plates is found by Gauss's law.

Electric field intensity $\vec{E}$ at a point below the surface charge sheet of $+\rho_s$ is given by equation (2.39).

$$\vec{E} = \frac{\rho_s}{2\varepsilon}(-\vec{a}_z) \qquad \text{(Here, the medium is not free space)}$$

Similarly, electric field intensity at a point above the surface charge sheet of $-\rho_s$ is given by equation $\vec{E}_2 = \frac{-\rho_s}{2\varepsilon}(\vec{a}_z)$.

Total electric field intensity between the two plates $\vec{E} = \vec{E}_1 + \vec{E}_2 = \frac{\rho_s}{\varepsilon}(-\vec{a}_z)$.

From eqn. (2.158) $\rho_s = \frac{Q}{S}$

$$\vec{E} = \frac{Q}{s\varepsilon}(-\vec{a}_z)\ \text{V/m}.$$

Hence $\quad V = -\int_2^1 \vec{E}\cdot d\vec{l} \qquad$ ($d\vec{l}$ is along $d\vec{z}$)

$$= -\int_{z=0}^{z=d} \frac{Q}{S\varepsilon}(-\vec{a}_z)\cdot dz\,\vec{a}_z$$

$$V = \frac{Qd}{S\varepsilon}$$

$$C = \frac{Q}{V} = \frac{\varepsilon S}{d}\ \text{farad.} \qquad ...(2.159)$$

2.3.3 Energy Stored in the Parallel Plate Capacitor

From eqn. (2.149) the energy stored in electrostatic field is

$$W_E = \frac{1}{2}\int_v \varepsilon E^2 dv$$

From eqn. (2.156), $E = |\vec{E}| = \frac{Q}{\varepsilon s}$

Sub in eqn. (2.149),

$$W_E = \frac{1}{2}\int_v \varepsilon\left(\frac{Q}{\varepsilon s}\right)^2 dv$$

$$= \frac{1}{2}\frac{Q^2}{\varepsilon s^2}\int_v dv.$$

Volume of the parallel plate capacitor is

= Area × length

= $S \times d$.

$$\therefore \qquad W_E = \frac{1}{2}\frac{Q^2}{\varepsilon s^2}(sd)$$

$$= \frac{1}{2}\frac{Q^2 d}{\varepsilon s}$$

$$= \frac{1}{2}\frac{Q^2}{\left(\frac{\varepsilon s}{d}\right)} = \frac{1}{2}\frac{Q^2}{C}.$$

$$\left.\begin{aligned} W_E &= \frac{1}{2}\ \frac{Q^2}{C}\ \text{joules (J)} \\ W_E &= \frac{1}{2}\ QV\ \text{J}\ [\because Q = CV] \\ W_E &= \frac{1}{2}\ CV^2\ \text{J} \end{aligned}\right] \qquad ...(2.160)$$

2.3.4 Capacitance of Co-axial Capacitor

Consider two co-axial conductors of length 'L' m having inner radius 'a' and outer radius b as shown in Fig. 2.51.

Let the space between the conductors be filled with a homogeneous dielectric with permittivity ε. Assume the conductors 1 and 2 carry $+Q$ and $-Q$ coulomb of charges uniformly distributed on them. For convenience, assume cylindrical co-ordinates system.

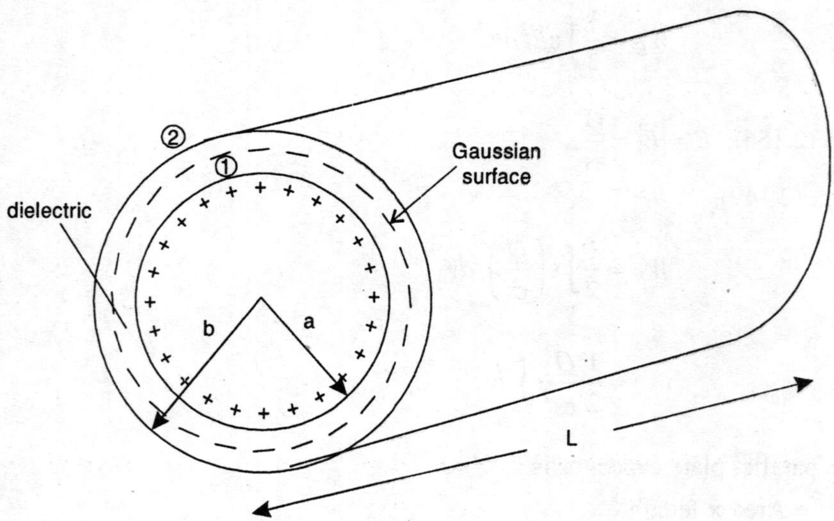

Fig. 2.51 Co-axial capacitor

The linear charge density ρ is given as

$$\rho_L = \frac{Q}{L} \text{ C/m} \qquad \text{...(2.161)}$$

Electric field intensity between two co-axial conductors is obtained by applying Gauss's law to an arbitrary Gaussian cylindrical surface of radius ρ.

From eqn. (2.63),

$$\vec{E} = \frac{\rho_L}{2\pi\varepsilon_0\rho}\vec{a}_\rho \qquad (a < \rho < b)$$

where ρ is the distance of point from the linear charge.

Substituting eqn. (2.161) in eqn. (2.63)

$$\vec{E} = \frac{Q}{2\pi\varepsilon\rho L}\vec{a}_\rho \qquad \begin{bmatrix}\text{Here the medium of dielectric is not}\\ \text{free space so } \varepsilon_0 \text{ is replaced by } \varepsilon\end{bmatrix}$$

Neglecting flux fringing at the cylindrical ends.

$$V = -\int_2^1 \vec{E}\cdot d\vec{l} = -\int_{\rho=b}^{\rho=a} E\vec{a}_\rho \cdot d\rho\,\vec{a}_\rho$$

$$= -\int_b^a \frac{Q}{2\pi\varepsilon\rho L}d\rho$$

$$= \frac{Q}{2\pi\varepsilon L}\ln\,[b/a]\cdot$$

Thus the capacitance of co-axial cylinder is

$$C = \frac{Q}{V} = \frac{2\pi\varepsilon L}{\ln\dfrac{b}{a}} \text{ farad.} \qquad \text{...(2.162)}$$

2.3.5 Capacitance of a Spherical Capacitor

Consider two concentric spherical conductors of inner radius a and outer radius b separated by a dielectric medium with permittivity ε as shown in Fig. 2.52.

Assume charges $+Q$ and $-Q$ are on the inner and outer spheres respectively.

Applying Gauss's law to an arbitary Gaussian spherical surface of radius r $(a < r < b)$

$$Q = \varepsilon\oint_S E\cdot ds$$

$$= \varepsilon\oint_S E\vec{a}_r \cdot ds\,\vec{a}_r$$

$$= \varepsilon E\oint_S ds$$

$$Q = \varepsilon E \times 4\pi r^2$$

$$\vec{E} = \frac{Q}{4\pi\varepsilon r^2}\vec{a}_r.$$

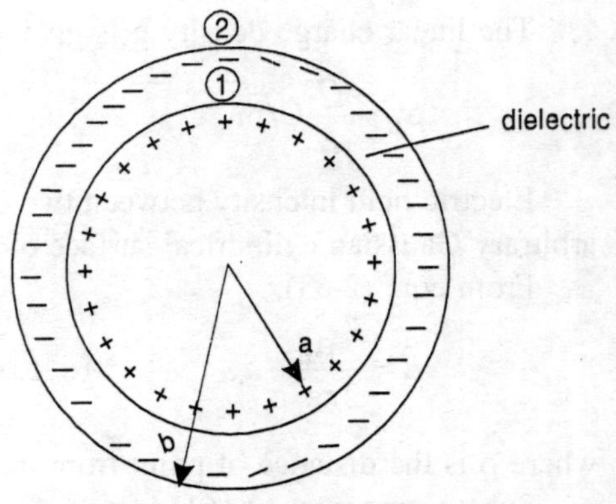

Fig. 2.52 Spherical capacitor

Hence

$$V = -\int_2^1 \vec{E}\cdot d\vec{l} = -\int_{r=b}^{r=a} \frac{Q}{4\pi\varepsilon r^2}\vec{a}_r \cdot dr\,\vec{a}_r$$

$$V = \frac{Q}{4\pi\varepsilon}\left[\frac{1}{a} - \frac{1}{b}\right].$$

Thus the capacitance of spherical capacitor is

$$C = \frac{Q}{V} = \frac{4\pi\varepsilon}{\left(\frac{1}{a} - \frac{1}{b}\right)} \text{ farad.} \qquad ...(2.163)$$

Letting $b \to \propto$, $c = 4\pi\varepsilon a$, which is the capacitance of a spherical capacitor whose outer plate is infinitely large. This is the case of an isolated sphere.

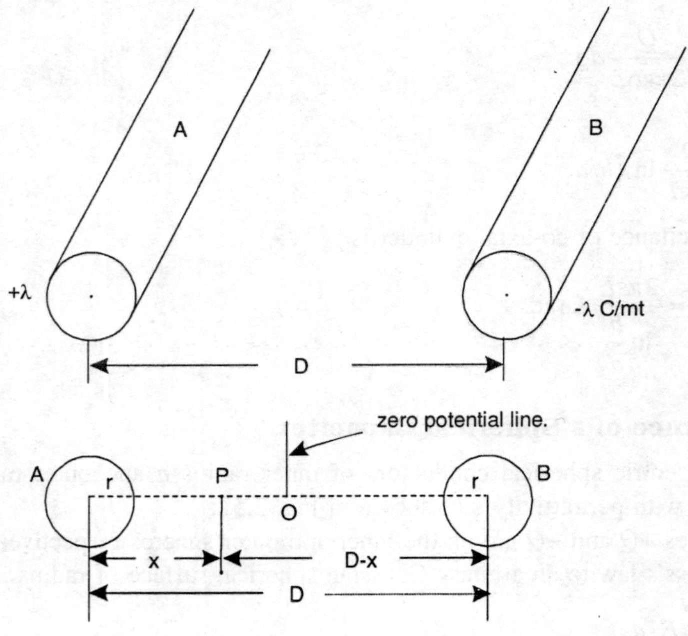

Fig. 2.53 Capacitance of two-wire transmission line.

2.3.6 Capacitance of Two-Wire Transmission Line

In Fig. 2.53, $+\lambda$ and $-\lambda$ are the charges in coulomb/m length of wires A and B spaced D m apart and radius of each wire in r m, remembering that $D >> r$.

To determine the capacitance between A and B, the potential difference between them should be derived initially. This potential difference can be evaluated by considering the straight line path of integration from B to A.

If $\vec{E}_x$ is the field intensity at any point P on this line at distant x from A and $(D-x)$ from B, then

$$V_{BA} = \text{potential rise from } B \text{ to } A$$

$$= -\int_B^A E_x \, dx \qquad \text{...(2.164)}$$

The limits being the equipotential surfaces of wires B and A will be $x = D - r$ at B and $x = r$ at A respectively.

The field at 'P' is the vector sum of the fields due to A and B. The field due to charge of $+\lambda$ per meter length at a distance x is $\dfrac{\lambda}{2\pi\varepsilon_0 x}$ directed from A to B, and that due to $-\lambda$, at a distance $(D-x)$, the field is also directed from A to B. So the resultant field at P is

$$\vec{E}_x = \frac{\lambda}{2\pi\varepsilon_0 x} + \frac{\lambda}{2\pi\varepsilon_0 D - x}. \qquad \text{...(2.165)}$$

Substitute eqn. (2.165) in eqn. (2.164)

$$V = -\frac{\lambda}{2\pi\varepsilon_0}\int_{D-r}^{r}\left(\frac{1}{x} + \frac{1}{D-x}\right)dx$$

$$= -\frac{\lambda}{2\pi\varepsilon_0}\left[\ln\frac{r}{D-r} + \ln\frac{r}{D-r}\right]$$

$$= \frac{\lambda}{2\pi\varepsilon_0} 2\ln\frac{D-r}{r}$$

$$= \frac{\lambda}{\pi\varepsilon_0}\ln\frac{D-r}{r}.$$

As D is too large in comparison with r, $\dfrac{D-r}{r}$ can be written as $\dfrac{D}{r}$.

$$\therefore \quad V = \frac{\lambda}{\pi\varepsilon_0}\ln\frac{D}{r} \text{ volt} \qquad \text{...(2.166)}$$

The capacitance between the two wires A and B is given by

$$C_{AB} = \frac{\lambda}{V} = \frac{\pi\varepsilon_0}{\ln D/r} \text{ farad/m.} \qquad \text{...(2.167)}$$

Consider any point on the perpendicular bisector of the line AB. As 'C' is equidistant from equal and opposite charges, $\pm\lambda$, the resultant potential at C is zero. Therefore, the perpendicular bisector is an equipotential line of zero potential. As the potential difference between A and B is V, the potential difference between A and O and between O and B is equal.

$$V_{AO} = V_{OB} = \frac{V}{2}.$$

So, the capacitance of each conductor (A or B) per meter length is

$$C_A \text{ or } C_B = \frac{\lambda}{V/2}$$

$$= 2\frac{\lambda}{V} \text{ farad/m.}$$

$$\therefore \quad C_A \text{ or } C_B = 2C_{AB}$$

$$C_A = C_B = \frac{2\pi\varepsilon_0}{\ln D/r} \text{ F/m.} \qquad \text{...(2.168)}$$

Thus, capacitance of infinite single-wire transmission line is two times of two-wire transmission line.

Example 2.21

Deduce an expression for the capacitance of a parallel plate capacitor having two dielectric media.

A parallel plate capacitor has a plate separation t. The capacitance with air only between the plates is C farads. When a slab of thickness t' and relative permittivity ε_r is placed on one of the plates, the capacitance is C. Show that

$$\frac{C'}{C} = \frac{\varepsilon_0 t}{t' + \varepsilon_r (t - t')}.$$

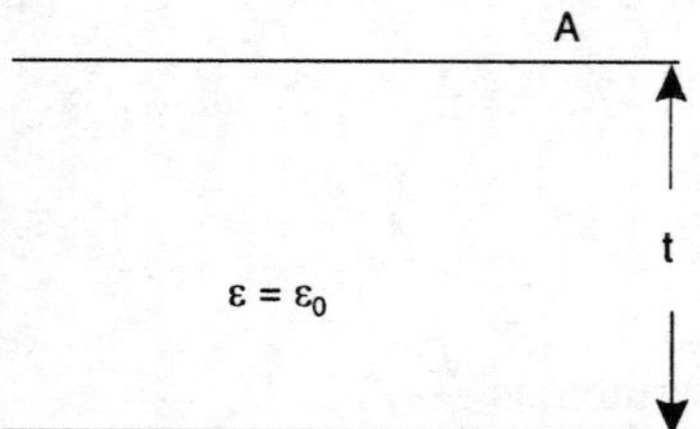

Fig. 2.54 (a) Capacitor with air alone as di-electric

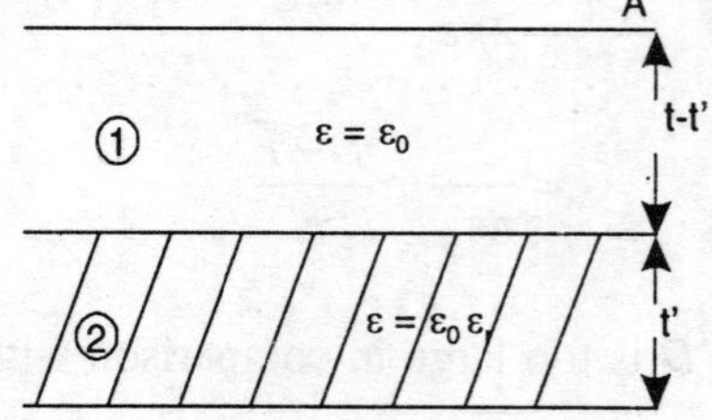

Fig. 2.54 (b) Capacitor having two di-electric

With air alone as dielectric as in Fig. 2.52 (a).

$$\text{Capacitance } C = \frac{\varepsilon_0 A}{t} \left(\text{as } \varepsilon_r = 1\right). \qquad \text{...(2.169)}$$

With the introduction of a slab of thickness t', the thickness of air film is reduced to $(t - t')$. Let E_1 and E_2 be the field intensities in the air film and slab respectively.

Let V_1 and V_2 be the potential difference across them.

If V is the total potential difference across the capacitor,

$$V = V_1 + V_2$$

$$V_1 = E_1(t - t')$$

$$V_2 = E_2 t'.$$

If Q is the charge on the plates, then from the boundary conditions, the normal component of electric flux densities $\vec{D}$ being equal in two dielectrics:

$$\vec{D}_1 = \vec{D}_2 = \vec{D},$$

where $$\vec{D} = \frac{Q}{A}.$$

Also from eqn. (2.120),

$$D = \varepsilon E$$

$$\therefore \quad E_1 = \frac{D}{\varepsilon_0}, \quad E_2 = \frac{D}{\varepsilon} = \frac{D}{\varepsilon_0 \varepsilon_r}.$$

Hence,

$$V = \frac{Q}{A}\left[\frac{(t - t')}{\varepsilon_0} + \frac{t'}{\varepsilon_0 \varepsilon_r}\right] \quad \text{...(2.170)}$$

$$= \frac{Q}{A\varepsilon_0}\left[\frac{t'}{\varepsilon_r} + (t - t')\right]$$

$$C' = \frac{Q}{V} = \frac{A\varepsilon_0}{\dfrac{t'}{\varepsilon_r} + (t - t')} \quad \text{...(2.171)}$$

$$C' = \varepsilon_0 \cdot \frac{A\varepsilon_r}{t' + \varepsilon_r(t - t')}. \quad \text{...(2.172)}$$

Dividing eqn. (2.172) by eqn. (2.169)

$$\frac{C'}{C} = \frac{\varepsilon_0 t}{t' + \varepsilon_r(t - t')}.$$

Example 2.22

A parallel plate condenser has the following specifications.

Cross-sectional area of plates: $A = 1$ sq. m.

Separation between the plates: $d = 0.01$ m

The condenser has two layers of dielectric. One has permittivity ε_1 for a thickness of $d_1 = 0.008$ m, and other has air as dielectric for the remaining thickness $(d - d_1)$. Calculate the capacitance of the system. If the voltage 5000 V is applied across the capacitor, determine the potential gradient in the two dielectrics $\varepsilon_1 = 6\varepsilon_0$.

From eqn. (2.170)

$$V = \frac{Q}{A}\left[\frac{d-d_1}{\varepsilon_0} + \frac{d_1}{\varepsilon_0\varepsilon_r}\right]$$

$$C = \frac{Q}{V} = \left[\frac{A}{\dfrac{d-d_1}{\varepsilon_0} + \dfrac{d_1}{\varepsilon_0\varepsilon_r}}\right].$$

So, the capacitance C = 2660 pF.

If $\pm Q$ is the charge on plates of the capacitor and A is the area of plate, then $D = Q/A$.

As in the previous Example (2.21),

$$E_1 = \frac{Q}{A\varepsilon_1}, \quad E_2 = \frac{Q}{A\varepsilon_2} \qquad \text{...(2.173)}$$

$$V_1 = E_1 d_1, \quad V_2 = E_2(d-d_1). \qquad \text{...(2.174)}$$

From eqns. (2.173) and (2.174),

$$\frac{V_1}{V_2} = \frac{\dfrac{Q}{A\varepsilon_1}d_1}{\dfrac{Q}{A\varepsilon_2}(d-d_1)} \qquad \text{...(2.175)}$$

$$\frac{V_1}{V_2} = \frac{d_1}{\varepsilon_1} \times \frac{\varepsilon_2}{d-d_1}$$

$$\varepsilon_1 = 6\varepsilon_0, \quad \varepsilon_2 = \varepsilon_0$$

$$\frac{V_1}{V_2} = \frac{.008}{6\varepsilon_0} \times \frac{\varepsilon_0}{.002}$$

$$\frac{V_1}{V_2} = \frac{2}{3}. \qquad \text{...(2.176)}$$

Given $$V_1 + V_2 = 5000 \qquad \text{...(2.177)}$$

Using the above two equations

V_1 = 2000 volt V_2 = 3000 volt

Thus, $$E_1 = \frac{V_1}{d_1} = 2.5 \text{ kV/cm}.$$

$$E_2 = \frac{V_2}{d-d_1} = 15 \text{ kV/cm}.$$

2.3.7 Dielectric Strength of a Co-axial Cable

Consider the co-axial cable of inner and outer radius '*a*' and '*b*' respectively having uniform dielectric of permittivity ε.

From the Section 2.3.4, the field intensity and potential at any point of this co-axial cable are given as

$$\vec{E} = \frac{Q}{2\pi\varepsilon\rho l}\vec{a}_\rho, \quad V = \frac{Q}{2\pi\varepsilon L}\ln b/a.$$

From the above two equations

$$E = \frac{V}{\rho \ln b/a}. \qquad ...(2.178)$$

From this eqn. (2.178), it is known that dielectric strength (or) electric stress (or) field intensity varies inversely proportional to radius *r*.

The electric stress is maximum at the surface of inner conductor and minimum at the surface of outer conductor.

$$E_{max} = \frac{V}{a\ln(b/a)}\text{V/m} \qquad ...(2.179)$$

$$E_{min} = \frac{V}{b\ln(b/a)}\text{V/m} \qquad ...(2.180)$$

2.3.8 Capacitance of Co-axial Cable with Two Dielectrics

Let us consider a cable with two dielectrics with permittivities ε_1 and ε_2.

If E_1 is the field intensity at any radial distance *r* in dielectric medium (1) and E_2 is in (2), then from eqn. (2.63),

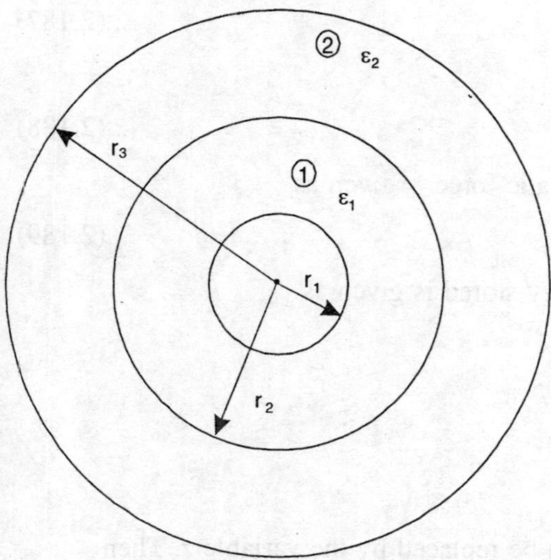

Fig. 2.55 Cross-section of co-axial cable with two di-electric

$$E_1 = \frac{\lambda}{2\pi\varepsilon_1 r}\ (r_1 \le r \le r_2) \quad [\text{Here } \rho = r] \qquad ..(2.181)$$

$$E_2 = \frac{\lambda}{2\pi\varepsilon_2 r}\ (r_2 \le r < r_3). \qquad ...(2.182)$$

Potential difference across (1) and (2) are V_1 and V_2.

$$V_1 = -\int_{r_2}^{r_1} E_1 dr, \quad V_2 = -\int_{r_3}^{r_2} E_2 dr. \qquad ...(2.183)$$

Substituting eqns. (2.181) and (2.182) in eqn. (2.183) and integrating

$$V_1 = \frac{\lambda}{2\pi\varepsilon_0\varepsilon_{r_1}}\ln\left(\frac{r_2}{r_1}\right)$$

$$V_2 = \frac{\lambda}{2\pi\varepsilon_0\varepsilon_{r_2}} \ln\left(\frac{r_3}{r_2}\right)$$

$$V = V_1 + V_2$$

$$V = \frac{\lambda}{2\pi\varepsilon_0}\left[\frac{1}{\varepsilon_{r_1}}\ln\left(\frac{r_2}{r_1}\right) + \frac{1}{\varepsilon_{r_2}}\ln\left(\frac{r_3}{r_2}\right)\right]. \quad ...(2.184)$$

Capacitance per mt length

$$C = \frac{\lambda}{V} = \frac{2\pi\varepsilon_0\varepsilon_1\varepsilon_2}{\varepsilon_{r_2}\ln\left(\frac{r_2}{r_1}\right) + \varepsilon_{r_1}\ln\left(\frac{r_3}{r_2}\right)}. \quad ...(2.185)$$

2.3.9 Electrostatic Forces

The force $(\vec{F})$ acting on any system of charges may be obtained from energy changes in the system. Force $(\vec{F})$ can be determined from W_E, the electrostatic energy stored in the system by virtue of the presence of electric charges.

Let us consider a parallel plate capacitor to derive the electrostatic force. If one of the plates of the parallel plate capacitor is displaced by an amount $d\vec{l}$ under the influence of the electric force $\vec{F}$, then the mechanical work done by the system (capacitor) is

$$dW = \vec{F} \cdot d\vec{l}. \quad ...(2.186)$$

Here the assumption is that the charges on the plates are maintained constant.

For an isolated system, this mechanical work is provided by expending electrostatic energy. Hence, dW is equal to the loss of energy stored in the dielectric material of the capacitor, *i.e.*,

$$dW = -dW_E. \quad ...(2.187)$$

dW_E may be written in terms of the gradient of W_E as

$$dW_E = -\nabla W_E \cdot d\vec{l}. \quad ...(2.188)$$

By comparing eqns. (2.186) and (2.188), the electrostatic force is given as

$$\vec{F} = -\nabla W_E. \quad ...(2.189)$$

For a parallel plate capacitor, from eqn. (2.159), energy stored is given as

$$W_E = \frac{1}{2}\frac{Q^2}{C} \text{ joule,}$$

where $$C = \frac{\varepsilon A}{d} \text{ farad.}$$

As the distance d between the plates is varying, d can be replaced by the variable z. Then

$$W_E = \frac{1}{2}\frac{Q^2 Z}{\varepsilon A}.$$

So, $$\vec{F} = -\nabla W_E = -\frac{\partial}{\partial z}\left(\frac{Q^2 Z}{2\varepsilon A}\right)\vec{a}_z$$

$$\vec{F} = -\frac{Q^2}{2\varepsilon A}\vec{a}_z \quad \text{...(2.190)}$$

As $$Q = \varepsilon AE \qquad [\because Q = D \times A = \varepsilon E \times A] \quad \text{...(2.191)}$$

$$\vec{F} = -\frac{\varepsilon AE^2}{2}\vec{a}_z. \quad \text{...(2.192)}$$

2.4 ELECTROSTATIC BOUNDARY VALUE PROBLEMS

Introduction

In the preceding chapters, the determination of electric field $\vec{E}$ is done by using either Coulomb's law or Gauss's law when the charge distribution is known or using $\vec{E} = -\nabla V$ when the potential V is known throughout the region.

In this chapter, the electrostatic problems where only electrostatic conditions (charge and potential) at some boundaries are known are to be considered, and it is desired to find $\vec{E}$ and V throughout the region. Such problems are usually solved by using Poisson's or Laplace referred to as boundary-value problems.

2.4.1 Solutions of Poisson's and Laplace's Equations

The Laplace equation in cartesian, cylindrical or spherical co-ordinates respectively is given by

$$\frac{\partial^2 V}{\partial x^2} + \frac{\partial^2 V}{\partial y^2} + \frac{\partial^2 V}{\partial z^2} = 0 \quad \text{...(2.193)}$$

$$\frac{1}{\rho}\frac{\partial}{\partial \rho}\left(\rho\frac{\partial V}{\partial \rho}\right) + \frac{1}{\rho^2}\frac{\partial^2 V}{\partial \phi^2} + \frac{\partial^2 V}{\partial z^2} = 0 \quad \text{...(2.194)}$$

$$\frac{1}{r^2}\frac{\partial}{\partial r}\left(r^2\frac{\partial r}{\partial r}\right) + \frac{1}{r^2 \sin\theta}\frac{\partial}{\partial \theta}\left(\sin\theta\frac{\partial V}{\partial \theta}\right) + \frac{1}{r^2 \sin^2\theta}\frac{\partial^2 V}{\partial \phi^2} = 0 \quad \text{...(2.195)}$$

depending on whether the potential is $V(x, y, z)$, $V(\rho, \phi, z)$ or $V(r, \theta, \phi)$. Poisson's equation in those co-ordinate systems may be obtained by simply replacing zero on the right hand side of equations with $\frac{-\rho_v}{\varepsilon}$.

2.4.2 Uniqueness Theorem

Since there are several methods of solving a given problem, there may be a question that whether solving Laplace's equation in different ways gives different solutions. But the answer is, if a solution of Laplace's equation satisfies a given set of boundary conditions, then, that is the only possible solution and that solution is said to be unique.

Thus any solution of Laplace's equation which satisfies the same boundary conditions must be the only solution regardless of the method used. This is known as the '*Uniqueness theorem*'. This theorem applies to any solution of Poission's and Laplace's equation in a given region or closed surface.

General procedure for solving Poission's or Laplace equations (One dimensional only).

The following general procedure may be taken in solving a given boundary value problem involving Poission's and Laplace's equation.

1. Solve Laplace's (if $\rho_v = 0$) or Poission's $(\text{if } \rho_v \neq 0)$ equation using direct integration when V is a function of one variable.
2. Apply the boundary conditions to determine a unique solution for V. Imposing the given boundary conditions makes the solution unique.
3. Having obtained V, find $\vec{E}$ using $\vec{E} = -\nabla V$ and $\vec{D}$ from $\vec{D} = \varepsilon \vec{E}$.
4. If desired, find the charge Q induced on a conductor using $Q = \int \rho_s ds$ where $\rho_s = D_n$, and D_n is the component of $\vec{D}$ normal to the conductor. If necessary, the capacitance between two conductors can be found by using $C = Q/V$.

Example 2.23

Semi-infinite conducting planes $\phi = 0$ and $\phi = \pi/6$ are separated by an infinitesimal insulating gap. If $V(\phi = 0) = 0$ and $V(\phi = \pi/6) = 100V$, calculate r and $\vec{E}$ in the region between the planes.

Solution:

As V depends only on ϕ, Laplace's equation in cylindrical co-ordinates becomes

$$\nabla^2 V = \frac{1}{\rho^2}\frac{\partial^2 V}{\partial \phi^2} = 0$$

$$\frac{\partial^2 V}{\partial \phi^2} = 0,$$

which is integrated twice to give $V = A\phi + B$.

We apply the boundary conditions to determine constants A and B, where $\phi = 0$, $V = 0$.

$$B = 0$$

when $\quad \phi = \phi_0, \; V = V_0$

$$V_0 = A\phi_0$$

$$A = \frac{V_0}{\phi_0}.$$

Hence,

$$V = \frac{V_0}{\phi_0}\phi \text{ and}$$

$$\vec{E} = -\nabla V = \frac{-1}{\rho}\frac{\partial V}{\partial \phi}\vec{a}_\phi$$

$$\frac{\partial V}{\partial \phi} = \frac{V_0}{\phi_0}.$$

$$\therefore \quad \vec{E} = -\frac{V_0}{\rho\phi_0}\vec{a}_\phi.$$

Substituting $V_0 = 100,\ \phi_0 = \frac{\pi}{6}$ gives

$$V = \frac{600}{\pi}\phi \text{ and } \vec{E} = \frac{600}{\pi\rho}\vec{a}_\phi.$$

Example 2.24

Conducting spherical shells with radii $a = 10$ cm and $b = 30$ cm are maintained at a potential difference of 100 V such that $V(r = b) = 0$ and $V(r = a) = 100$ V.

Determine V and $\vec{E}$ in the region between the shells.

Solution:

As V depends only on r, the Laplace's equation becomes

$$\nabla^2 V = \frac{1}{r^2}\frac{d}{dr}\left(r^2\frac{dv}{dr}\right) = 0$$

$$\frac{d}{dr}\left(r^2\frac{dv}{dr}\right) = 0.$$

Integrating once

$$r^2\frac{dV}{dr} = A.$$

$$\frac{dV}{dr} = \frac{A}{r^2}.$$

Again integrating, $V = -\frac{A}{r} + B.$

The constants A and B are determined from the boundary conditions

when $r = b$, $V = 0$

$$0 = -\frac{A}{b} + B$$

$$B = \frac{A}{b}.$$

Hence,
$$V = -\frac{A}{r} + \frac{A}{b}$$

$$V = A\left[\frac{1}{b} - \frac{1}{r}\right],$$

when
$$r = a,\ V = V_0.$$

$$V_0 = A\left[\frac{1}{b} - \frac{1}{a}\right]$$

$$A = \frac{V_0}{\left[\frac{1}{b} - \frac{1}{a}\right]}.$$

Thus,

$$V = V_0\left\{\frac{\left[\frac{1}{r} - \frac{1}{b}\right]}{\left(\frac{1}{a} - \frac{1}{b}\right)}\right\}$$

$$\vec{E} = -\nabla V = -\frac{dV}{dr}\vec{a}_r$$

$$\vec{E} = -\frac{A}{r^2}\vec{a}_r$$

$$\vec{E} = \frac{V_0}{r^2\left[\frac{1}{a} - \frac{1}{b}\right]}\vec{a}_r.$$

Q can also be found out using

$$Q = \int_s \vec{D}\cdot d\vec{s} = \int \varepsilon\vec{E}\cdot d\vec{s}$$

$$= \int_{\theta=0}^{\pi} \int_{\phi=0}^{2\pi} \frac{\varepsilon_0 \varepsilon_r V_0}{r^2 \left[\frac{1}{a} - \frac{1}{b}\right]} r^2 \sin\theta \, d\phi \, d\theta$$

$$Q = \frac{4\pi\varepsilon_0\varepsilon_r V_0}{\frac{1}{a} - \frac{1}{b}},$$

when $\dfrac{V_0}{\frac{1}{a} - \frac{1}{b}}$ is replaced by $Er^2 \, \vec{a}_r$.

Hence $\vec{E} = \dfrac{Q}{4\pi\varepsilon_0\varepsilon_r r^2} \; \vec{a}_r$,

which is same as in eqn. (2.97) which is determined by using Gauss's law.

Also, capacitance can be determined by

$$C = \frac{Q}{V_0} = \frac{4\pi\varepsilon_0\varepsilon_r}{\frac{1}{a} - \frac{1}{b}}.$$

Substituting $a = 0.1$ m, $b = 0.3$ m, $V_0 = 100$ V.

$$V = \frac{100\left[\frac{1}{r} - \frac{10}{3}\right]}{\left(10 - \frac{10}{3}\right)} = 15\left[\frac{1}{r} - \frac{10}{3}\right] \text{ volt.}$$

$$\vec{E} = \frac{100}{r^2\left[10 - \frac{10}{3}\right]} = \frac{15}{r^2}\vec{a}_r \text{ V/m}$$

Example 2.25

The two plates of a parallel plate capacitor are separated by a distance 'd' and maintained at potentials 0 and V_0 as shown in Fig. 2.56. Assuming negligible fringing effect at the edges, determine

(*a*) the potential at any point between the plates and

(*b*) the surface charge densities on the plates.

Solution:

As V depends only on z, the Laplace's equation becomes

$$\nabla^2 V = \frac{d^2 V}{dz^2} = 0.$$

Integrating once,

$$\frac{dV}{dz} = A.$$

Again integrating

$$V = Az + B$$

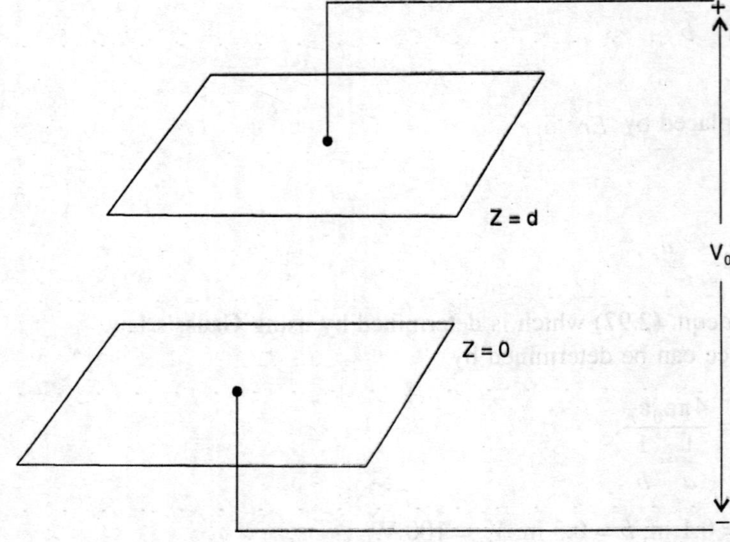

Fig. 2.56

The constants A and B are determined from the boundary conditions,

when $Z = 0,\ V = 0$

$B = 0,\ V = Az$

when $Z = d,\ V = V_0$

$\therefore \quad V_0 = Ad$

$$A = \frac{V_0}{d}.$$

Hence, $A = \frac{V_0}{d}$.

To find $\vec{E}$,

$$\vec{E} = -\nabla V$$

$$= -\frac{dV}{dz}\vec{a}_z$$

$$\vec{E} = -A \cdot \vec{a}_z$$

$$\vec{E} = -\frac{V_0}{d}\vec{a}_z.$$

To find charge Q and charge density ρ_s,

$$Q = \int \varepsilon \vec{E} \cdot d\vec{s}$$

$$= \varepsilon E \cdot S\text{, where surface area of plates be } S \text{ m}^2.$$

$$Q = -\varepsilon \frac{V_0 S}{d} \text{ coulomb}$$

and $$E = \frac{Q}{\varepsilon s}\ \vec{a}_z$$

$$\vec{E} = -\frac{\rho_s}{\varepsilon}\vec{a}_z \text{ (upper plate)}.$$

The equation is same as derived from Gauss's law.
Electric field lines begin from positive charges and end in negative charges.
Capacitance is also derived as

$$C = \frac{Q}{v_0} = \frac{\varepsilon s}{d} \text{ farad} \quad \text{[Taking only the magnitude]}$$

Example 2.26

Determine the $\vec{E}$ field inside and outside a spherical cloud of electrons with a uniform volume charge density $\rho_s = -\rho_0$ (where ρ_0 is a positive quantity) for $0 \le r \le b$ and $\rho = 0$ for $r > b$ by solving Laplace's and Poission's equations for V.

Solution:

Since there are no variations in θ and ϕ directions, the Laplace's and Poission's equations will be

$$\nabla^2 V = \frac{\rho_s}{\varepsilon} = \frac{1}{r^2}\frac{d}{dr}\left(r^2\frac{dV}{dr}\right) = \frac{-\rho_s}{\varepsilon}$$

$$\nabla^2 V = 0 = \frac{1}{r^2}\frac{d}{dr}\left(r^2\frac{dV}{dr}\right) = 0.$$

(*a*) *Inside the cloud* $0 \le r \le \rho;\quad \rho = -\rho_0.$

The charge density exists inside the cloud, so, the Poission's equation is used.

$$\frac{1}{r^2}\frac{d}{dr}\left(r^2\frac{dV_i}{dr}\right) = -\frac{\rho_s}{\varepsilon}$$

$$\frac{d}{dr}\left(r^2\frac{dV_i}{dr}\right) = -\frac{\rho_s}{\varepsilon}r^2.$$

Integrating once,

$$r^2 \frac{dV_i}{dr} = -\frac{\rho_s r^3}{3\varepsilon} + A$$

$$\frac{dV_i}{dr} = -\frac{\rho_s}{3\varepsilon} r + \frac{A}{r^2}.$$

The electric field intensity inside the electron cloud is

$$\vec{E}_i = -\nabla V_i = \frac{\nabla V_i}{dr}(-\vec{a}_r).$$

Since $\vec{E}_i$ cannot be infinite at $r = 0$, the integration constant A must vanish.

Hence, $$\vec{E}_i = \frac{\rho_0}{3\varepsilon_0} r \vec{a}_r. \qquad [\because \ \rho_s = -\rho_0] \qquad 0 \le r \le b$$

Since A is zero,

$$\frac{dV_i}{dr} = \frac{\rho_0}{3\varepsilon_0} r.$$

Again integrating,

$$V_i = \frac{\rho_0 r^2}{6\varepsilon_0} + A_1.$$

(*b*) *Outside the cloud.*

$r \ge b,\ \rho_s = 0.$

Here, Laplace's equation holds

$$\frac{1}{r^2}\frac{d}{dr}\left(r^2 \frac{dv}{dr}\right) = 0.$$

Integrating once

$$\frac{dV_0}{dr} = \frac{B}{r^2}$$

$$\vec{E}_0 = -\nabla V_0 = \frac{dV_0}{dr}(-\vec{a}_r) = \frac{B}{r^2}(-\vec{a}_r).$$

The integration constant is found out by equating $\vec{E}_0$ and $\vec{E}_i$ at $r = b$, where there is no discontinuity in medium characteristics.

$$\frac{B}{b^2} = \frac{\rho_0}{3\varepsilon_0} b$$

$$B = \frac{\rho_0 b^3}{3\varepsilon_0}$$

$$\therefore \qquad \vec{E}_0 = \frac{\rho_0 b^3}{3\varepsilon_0 r^2}(-\vec{a}_r) \qquad r \geq b.$$

The total charge contained in the electron cloud is $Q = -\rho_0\left(\frac{4\pi}{3}b^3\right)$.

$\vec{E}_0$ can also be written as $\vec{E}_0 = \frac{Q}{4\pi\varepsilon_0 r^2}\vec{a}_r$.

The above equation is same as derived from Gauss's law.
Again, to find V_0, integrating again the equation, after substituting B

$$\frac{dV_0}{dr} = \frac{\rho_0 b^3}{3\varepsilon_0 r^2}$$

$$V_0 = -\frac{\rho_0 b^3}{3\varepsilon_0 r} + B_1.$$

Here, B_1 must vanish, since V_0 is zero at $r \to \infty$. As electrostatic potential is continuous at a boundary, A_1 is found out by equating V_i and V_0 at $r = b$.

$$\frac{\rho_0 b^2}{6\varepsilon_0} + A_1 = -\frac{\rho_0 b^2}{3\varepsilon_0}$$

$$A_1 = -\frac{\rho_0 b^2}{2\varepsilon_0}.$$

So,

$$V_i = -\frac{\rho_0}{3\varepsilon_0}\left(\frac{3b^2}{2} - \frac{R^2}{2}\right).$$

2.5 METHOD OF IMAGES

Introduction

The method of images commonly used to determine V, $\vec{E}$, $\vec{D}$ and ρ_s due to charges in the presence of conductors. By this method, solving the problems by Poisson's or Laplace's equation was avoided. But the fact that a conducting surface is an equipotential was utilised to obtain the solution.

Poisson's or Laplace's equation is feasible in principle, but the solution is quite complicated mathematically. Alternatively, the problem can be solved by using image theory.

2.5.1 Image Theory

The image theory states that "given a charge configuration above an infinite grounded perfect conducting plane may be replaced by the charge configuration itself, its image, and an equipotential surface in place of the conducting plane.

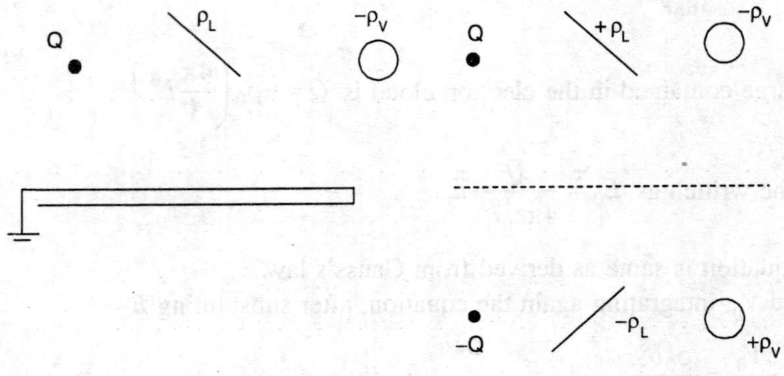

Fig. 2.57

In applying the image method, two conditions must be always satisfied:

1. The image charges must be located in the conducting region.
2. The image changes must be located such that on conducting surfaces, the potential is zero.

2.5.2 A Point Charge Above Grounded Conducting Plane

Consider a point charge Q placed at a distance h from a perfect conducting plane as shown in Fig. 2.58(a). The image configuration is shown in Fig. 2.57. The electric field at point p (x, y, z) is given by

$$\vec{E} = \vec{E}_+ + \vec{E}_-$$

$$\vec{E}_+ = \frac{Q}{4\pi\varepsilon_0 r_1^3}\vec{r}_1, \quad \vec{E}_- = \frac{-Q}{4\pi\varepsilon_0 r_2^3}\vec{r}_2$$

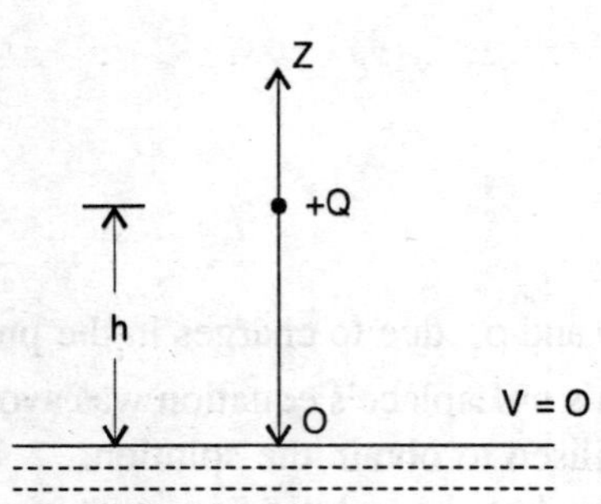

Fig. 2.58 (a) Point charge and grounded conducting plane

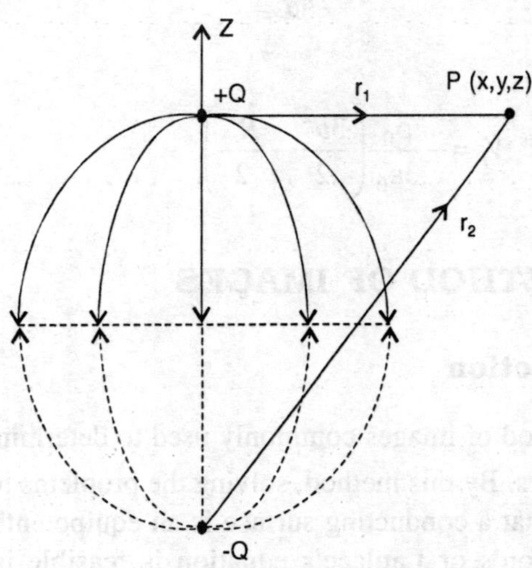

(b) Image configuration

$$\vec{r}_1 = (x, y, z) - (0, 0, h) = (x, y, z - h)$$

$$\vec{r}_2 = (x, y, z) - (0, 0, -h) = (x, y, z + h)$$

$$\vec{E} = \frac{Q}{4\pi\varepsilon_0}\left[\frac{x\vec{a}_x + y\vec{a}_y + (z-h)\vec{a}_z}{\left[x^2 + y^2 + (z-h)^2\right]^{3/2}} - \frac{x\vec{a}_x + y\vec{a}_y + (z+h)\vec{a}_z}{\left[x^2 + y^2 + \left(z^2 + h^2\right)\right]^{3/2}}\right].$$

It should be noted that when $z = 0$, $\vec{E}$ has only the z component *i.e.*, if $z = 0$, then

$$\vec{E} = \frac{-Qh}{2\pi\varepsilon_0\left(x^2 + y^2 + h^2\right)^{3/2}} \cdot \vec{a}_z. \qquad ...(2.196)$$

The potential at point P is given by

$$V = V_+ + V_-$$

$$= \frac{Q}{4\pi\varepsilon_0 r_1} + \frac{-Q}{4\pi\varepsilon_0 r_2}$$

$$= \frac{Q}{4\pi\varepsilon_0}\left[\frac{1}{\left(x^2 + y^2 + (z-h)^2\right)^{1/2}} - \frac{1}{\left(x^2 + y^2 + (z+h)^2\right)^{1/2}}\right].$$

If $z = 0$, then $V = 0$.

The surface charge density of the induced charge can be obtained by the equation

$$\rho_s = D_n = \varepsilon_0 E_n \quad \text{at } z = 0.$$

$$= \frac{-Qh}{2\pi\left(x^2 + y^2 + h^2\right)^{3/2}}. \qquad ...(2.197)$$

The total induced charge on the conducting plane is given by

$$Q_i = \int \rho_s \cdot ds = \int_{-\infty}^{\infty}\int_{-\infty}^{\infty} \frac{-Qh}{2\pi\left(x^2 + y^2 + h^2\right)^{3/2}}\, dx\, dy. \qquad ...(2.198)$$

To convert cartesian coordinates into cylindrical co-ordinates:

Let $\quad x^2 + y^2 = \rho^2$

$$dx\, dy = \rho d\rho\, d\phi$$

$$Q_i = -\frac{Qh}{2\pi}\int_0^{2\pi}\int_0^{\infty} \frac{\rho d\rho d\phi}{\left(\rho^2 + h^2\right)^{3/2}}$$

$$= -\frac{Qh}{2\pi}\int_0^{\infty}\frac{\frac{1}{2}d\left(\rho^2\right)}{\left(\rho^2+h^2\right)^{3/2}}\cdot(\phi)_0^{2\pi}$$

$$= -Qh\int_0^{\infty}\left[\rho^2+h^2\right]^{-3/2}\cdot\frac{1}{2}d\left(\rho^2\right)$$

$$= \left[\frac{Qh}{\left(\rho^2+h^2\right)^{1/2}}\right]_0^{\infty}$$

$$Q_i = -Q.$$

All flux lines terminating on the conductor would have terminated on the image charge if the conductor were absent.

2.5.3 Capacitance of Two-Wire Transmission Line — Including Earth Effect

Consider a two-wire transmission circuit. The capacitance of the transmission system may be evaluated by considering the proximity effect of earth.

Figure 2.59 shows a pair of transmission wires. If A and B are two wires of transmission line with the charge/m length of $+\lambda$ and $-\lambda$, their images A' and B' will have $-\lambda$ and $+\lambda$.

Potential difference between the conductor A' and B' can be evaluated by considering the contribution made by all the charges.

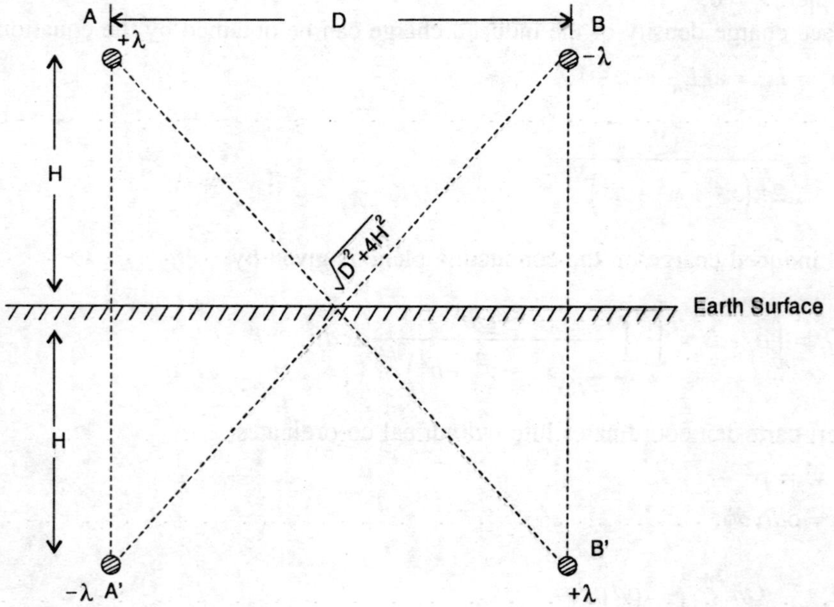

Fig. 2.59

V_{BA}^{A} is the potential difference between A and B due to A

$$V_{BA}^{A} = -\int_{D}^{r} \frac{\lambda}{2\pi\varepsilon_0 x} \cdot dx$$

$$= \frac{\lambda}{2\pi\varepsilon_0} \ln \frac{D}{r}$$

$$V_{BA}^{B} = \frac{\lambda}{2\pi\varepsilon_0} \ln \frac{D}{r}$$

$$V_{BA}^{A'} = -\int_{\sqrt{D^2+4H^2}}^{2H} -\frac{\lambda}{2\pi\varepsilon_0 x} dx$$

$$= \frac{\lambda}{2\pi\varepsilon_0} \ln \frac{2H}{\sqrt{D^2+4H^2}}$$

$$V_{BA}^{B'} = \frac{\lambda}{2\pi\varepsilon_0} \ln \frac{2H}{\sqrt{D^2+4H^2}}.$$

V_{BA} by superposition principle,

$$V_{BA} = \frac{\lambda}{2\pi\varepsilon_0} \left[2\ln\frac{D}{r} + 2 \cdot \ln \frac{2H}{\sqrt{D^2+4H^2}} \right]$$

$$= \frac{\lambda}{2\pi\varepsilon_0} \left[\ln \frac{2HD}{r\sqrt{D^2+4H^2}} \right]$$

$$C_{AB} = \frac{\lambda}{V_{BA}} = \frac{\pi\varepsilon_0}{\ln \frac{D}{r} \frac{2H}{\sqrt{D^2+4H^2}}} \text{F/m}.$$

Proximity of the earth has the effect in measuring the capacitance of the transmission line.

SOLVED PROBLEMS

Problem 31

A dielectric slab of flat surface with relative permittivity 4 is disposed with its surface normal to a uniform field with flux density of 1.5 C/m^2. The slab occupies a volume of 0.08 m^3 and is uniformly polarised. Determine

(1) The polarisation of the slab.

(2) The total dipole moment of the slab.

Solution:

$$\varepsilon_r = 4$$

$$D = 1.5 \text{ C/m}^2$$

$$\text{Volume} = 0.08 \text{ m}^3$$

From equation, $\vec{D} = \varepsilon\vec{E}$

$$= \varepsilon_0\varepsilon_r\vec{E}$$

$$\vec{E} = \frac{\vec{D}}{\varepsilon_0\varepsilon_r} = \frac{1.5}{4\varepsilon_0}$$

$$\therefore \quad \varepsilon_0\vec{E} = \frac{1.5}{4}$$

We know that $\vec{P} = \vec{D} - \varepsilon_0\vec{E}$

$$= 1.5 - \frac{1.5}{4}$$

Polarisation $\vec{P} = 1.125 \text{ C/m}^2$

$\therefore$ Dipole moment $= \vec{P} \times \text{volume}$

$$= 1.125 \times 0.08$$

$$= 0.09 \text{ C–m}$$

Problem 32

At the boundary between glass ($\varepsilon_r = 4$) and air, the lines of electric induction make an angle of 32° with the normal to the interface. If the density of such lines in air is 0.25 μC/m², determine

(*a*) the angle of $\vec{D}$ in the glass with the normal to the surface.

(*b*) determine the value of $\vec{D}$ in the glass.

Solution:

From the relation of boundary condition,

$$\frac{\tan\theta_1}{\tan\theta_2} = \frac{\varepsilon_1}{\varepsilon_2} \text{ and } E_2 = E_1 \cdot \frac{\sin\theta_1}{\sin\theta_2}$$

$$\frac{E_2}{E_1} = \frac{\sin\theta_1}{\sin\theta_2}$$

$$\frac{D_2}{\varepsilon_2}\cdot\frac{\varepsilon_1}{D_1}=\frac{\sin\theta_1}{\sin\theta_2}$$

(or) $$D_2=\frac{\varepsilon_2}{\varepsilon_1}\cdot D_1\cdot\frac{\sin\theta_1}{\sin\theta_2}$$

Let ε_1 = permittivity of air = $\varepsilon_0\varepsilon r_1 = \varepsilon_0$.

ε_2 = permittivity of glass = $\varepsilon_0\varepsilon r_2 = 4\varepsilon_0$.

Given $\theta_1 = 32°$

$$\tan\theta_2=\frac{\varepsilon_2}{\varepsilon_1}\cdot\tan\theta_1$$

$$= 4\cdot\tan 32$$

$$= 2.499$$

$\therefore \quad \theta_2 = 68.2°$.

Substituting for θ_1, θ_2, ε_1, ε_2 and D_1, we get

$$D_2 = 4\times 0.25\mu\times\frac{\sin 32}{\sin 68.2}$$

$$D_2 = 0.57\,\mu\text{C/m}^2.$$

Problem 33

A 2 μF capacitor is charged by connecting it across a 100 V d.c. supply. Determine the energy stored. The supply is now disconnected and the capacitor is connected across another 2 μF capacitor. Assuming no leakage of charge, determine the potential difference between the plates of each capacitor and energy stored.

Solution:

(*i*) Energy stored $=\frac{1}{2}C_1V_1^2$

$$=\frac{1}{2}\times 2\times 10^{-6}\times 100^2$$

$$= 0.01 \text{ joules.}$$

(*ii*) Charge on first capacitor

$$Q_1 = C_1V_1$$

$$= 2\times 10^{-6}\times 100$$

$$= 2\times 10^{-4}\text{ C.}$$

When another capacitor of value C_2 is connected across C_1, the charge 2×10^{-4} is equally divided.

$\therefore$ Charge/capacitor $= 1 \times 10^{-4}$ C

$$\text{P.d. across each } \frac{Q_1}{C_1} = \frac{Q_2}{C_2} = \frac{1 \times 10^{-4}}{2 \times 10^{-6}}$$

$$= 50 \text{ V}$$

Effective capacitance $C = C_1 + C_2 = 4\mu F$

$$\therefore \quad \text{Energy stored } = \frac{1}{2} \times 4 \times 10^{-6} \times 50^2$$

$$= 0.005 \text{ joule.}$$

Problem 34

(*a*) What is the capacitance/kilometre of length of air-filled co-axial cable with an inner conductor of 3 mm diameter and an outer conductor of inside diameter of 1 cm.

(*b*) What will be the capacitance of co-axial cable if the inner conductor is covered with rubber insulation ($\varepsilon_r = 3$) to a diameter of 7mm and the remainder space is filled with a dielectric of permittivity $\varepsilon_r = 6$.

Solution:

The capacitance/metre of length of the line will be given by

$$C = \frac{2\pi\varepsilon_0}{\ln b/a} \text{ F/m}$$

Radius of the conductor $= a = 1.5 \times 10^{-3}$ m

Radius of the cable $= b = 0.5 \times 10^{-2}$ m

$$C = \frac{2\pi \times 8.854 \times 10^{-12}}{\ln\left(\dfrac{0.5 \times 10^{-2}}{1.5 \times 10^{-3}}\right)} \times 1000$$

$$= 0.4 \ \mu\text{F/km}.$$

We know from eqn. (2.185), capacitance of co-axial cable with two dielectrics is

$$C = \frac{2\pi\varepsilon_0\varepsilon_{r_1}\varepsilon_{r_2}}{\varepsilon_{r_2} \ln\dfrac{r_2}{r_1} + \varepsilon_{r_1} \ln\dfrac{r_3}{r_2}} \text{ F/m}$$

$$= \frac{2\pi \times 8.854 \times 10^{-12} \times 3 \times 6}{6 \ln\dfrac{3.5}{1.5} + 3 \ln\dfrac{5}{3.5}} \times 1000$$

$$C = 0.16 \ \mu\text{F/km}.$$

Problem 35

A capacitor consists of squared two metal plates each 100 cm side placed parallel and 2 mm apart. The space between the plates is filled with a dielectric having a relative permittivity of 3.5. A potential drop of 500 V is maintained between the plates. Calculate.

(*a*) the capacitance
(*b*) the charge of capacitor
(*c*) the electric flux density
(*d*) the potential gradient

Solution:

Area of the plate = (100 × 100) = $1 m^2$
Distance between the plate d = 2mm = 2×10^{-3}m

$$\varepsilon_r = 3.5$$

$$V = 500 \text{ V}.$$

(*i*) Capacitance $C = \dfrac{\varepsilon A}{d}$

$$= \frac{\varepsilon_0 \varepsilon_r A}{d}$$

$$= \frac{8.854 \times 10^{-12} \times 3.5 \times 1}{2 \times 10^{-3}}$$

$$= 15.49 \times 10^{-9} \text{ F/m}$$

$$= 15.49 \, \mu\text{F/m}.$$

(*ii*) Charge on the capacitor

$$Q = CV$$

$$= 15.5 \times 10^{-9} \times 500$$

$$= 7.75 \times 10^{-6} \text{ C}$$

(*iii*) Flux density $\vec{D} = \dfrac{Q}{A} \text{ C/m}^2$

$$= 7.75 \times 10^{-6}$$

$$= 7.75 \times 10^{-6} \text{ C/m}^2$$

(*iv*) Potential gradient $\vec{E} = \dfrac{\vec{D}}{\varepsilon}$

$$\vec{E} = \frac{\vec{D}}{\varepsilon_0 \varepsilon_r}$$

$$= \frac{7.75 \times 10^{-6}}{8.85 \times 10^{-12} \times 3.5}$$

$= 250.2\,\text{kV/m}$

$= 2.5$ kV/cm.

Problem 36

An air condenser consisting of parallel plates of 50 cm side is charged to a potential difference of 250 volt, when the plates are 1 mm apart. Find the work done in separating the plates from 1 mm to 3 mm. Assume perfect insulation.

Solution:

Area of the plate $= \left(50 \times 10^{-2}\right)^2 = 0.25\ \text{m}^2$

$V = 250\ \text{V}$

$d = 1\ \text{mm}$

Work done in separating the plate = Energy stored when $d = 3$ mm − Energy stored when $d = 1$ mm

$$C_1 = \frac{\varepsilon_0 \varepsilon_r A}{d}$$

$$= \frac{8.854 \times 10^{-12} \times 1 \times 0.25}{1 \times 10^{-3}}$$

$$= 2.2 \times 10^{-9}\ \text{F/m}.$$

$$C_2 = \frac{\varepsilon_0 \varepsilon_r A}{d}$$

$$= \frac{8.854 \times 10^{-12} \times 0.25}{3 \times 10^{-3}}$$

$$= 7.378 \times 10^{-10}\ \text{F/m}.$$

$$\text{Work done} = \left(\frac{1}{2} C_1 - \frac{1}{2} C_2\right) V^2$$

$$= \frac{1}{2} \times 250^2 \left(2.2 \times 10^{-9}\right)^2 - \left(7.378 \times 10^{-10}\right)^2$$

$= 1.342 \times 10^{-13}$ joule.

Problem 37

The electric field between two concentric cylindrical conductors at $r = 0.01$ m and $r = 0.05$ m is given by $\vec{E} = \dfrac{10^6}{r}\vec{a}_r$ V/m . Find the energy stored in 1.0 m length of conductor. Assume free space.

(Anna. Univ. Oct 2002)

Solution:

Energy stored in electric field $W = \int_V \frac{1}{2}\varepsilon E^2 dv$. Differential volume for cylinder is $dV = 2\pi rl\,dr$.

$$W = \int_{r=0.01}^{r=0.05} \frac{1}{2}\left(\frac{10^6}{r}\right)\varepsilon \times 2\pi r\,ldr$$

$$= \frac{1}{2} \times 10^{12} \cdot \varepsilon \times 2\pi r \int_{r=0.01}^{0.05} \frac{1}{r}dr$$

$l = 1$ m, $\varepsilon_r = 1$

$$= \frac{1}{2} \times 10^{12} \times 8.85 \times 10^{-12} \times 2\pi(\ln r)_{0.01}^{0.05}$$

$$= 8.85 \times \pi \ln\left(\frac{0.05}{0.01}\right)$$

= 44.7 joule.

Problem 38

A condenser is composed of two plates separated by a sheet of insulating material 3 mm thick and of relative permittivity 4. The distance between the plates is increased to allow the insertion of a second sheet 5 mm thick and of relative permittivity ε_r. If the capacitance of the condenser so formed is one third of the original capacitance, find ε_r.

Solution:

Let the area of the plate = A m^2

Capacitance with single dielectric = C_1

$$C_1 = \frac{A \cdot \varepsilon_0 \varepsilon_r}{d}$$

$$= \frac{A \cdot \varepsilon_0 \times 4}{3 \times 10^{-3}}.$$

After the insertion of the second sheet,

$$C_2 = \frac{A\varepsilon_0}{\dfrac{t_1}{\varepsilon_{r_1}} + \dfrac{t_2}{\varepsilon_{r_2}}}$$

$$= \frac{A\varepsilon_0}{\dfrac{3\times10^{-3}}{4} + \dfrac{5\times10^{-3}}{\varepsilon_r}}$$

$$= \frac{1}{3} \times \frac{A\varepsilon_0 \times 4}{3\times10^{-3}}.$$

$$\therefore \quad \frac{1}{3} \times \frac{A\varepsilon_0}{3\times10^{-3}} = \frac{A\varepsilon_0}{\dfrac{3\times10^{-3}}{4} + \dfrac{5\times10^{-3}}{\varepsilon_r}}$$

$$9 = 4\left(\frac{3}{4} + \frac{5}{\varepsilon_r}\right)$$

$$\varepsilon_r = 3.33.$$

Problem 39

A single core lead-sheathed cable, with a conductor diameter of 2 cm is designed to withstand 66 kV. The dielectic consists of 2 layers *A* and *B* having relative permittivities of 3.5 and 3 respectively. The corresponding permissible electrostatic stresses are 72 kV/cm and 60 kV/cm. Find the thickness of two layers.

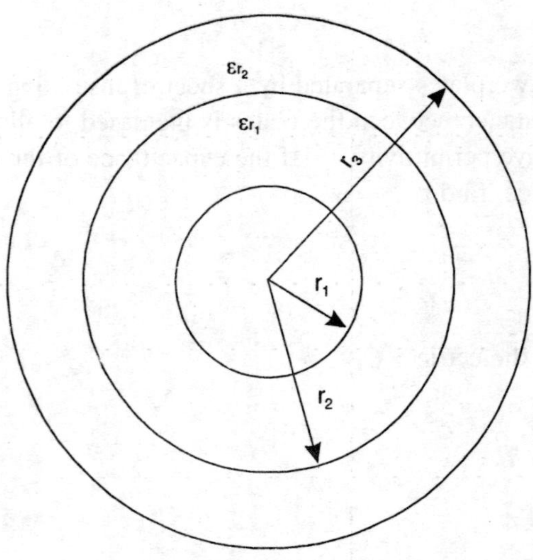

Fig. P-39

Solution:

Let E_{1max} and E_{2max} be the maximum dielectric stress. If V_1 and V_2 are potential differences across the respective dielectrics, then from eqn. (2.179)

$$E_{1\,max} = \frac{V_1}{r_1 \ln \frac{r_2}{r_1}} \text{ V/m} \qquad ...(i)$$

$$E_{2\,max} = \frac{V_2}{r_2 \ln \frac{r_3}{r_2}} \text{ V/m} \qquad ...(ii)$$

$$r_1 = 1\text{cm} \qquad E_{1\,max} = 72 \text{ kV/cm}$$

$$E_{2\,max} = 60 \text{ kV/cm}$$

$$E_{1\,max} = \frac{\lambda}{2\pi\varepsilon_0\varepsilon_{r_1} r_1}$$

$$E_{2\,max} = \frac{\lambda}{2\pi\varepsilon_0\varepsilon_{r_2} r_2}$$

$$\frac{E_{1\,max}}{E_{2\,max}} = \frac{\varepsilon_{r_2} r_2}{\varepsilon_{r_1} \cdot r_1}$$

$$\frac{72}{60} = \frac{3 \cdot r_2}{3.5 \times 1}$$

$$r_2 = 1.4 \text{ cm.}$$

Substituting r_2 = 1.4 cm in (*i*)

$$72 = \frac{V_1}{1 \times \ln \frac{1.4}{1}}$$

$$V_1 = 24.2 \text{ kV}$$

and $$V_2 = 66 - 24.2$$

$$= 41.8 \text{kV.}$$

Substituting V_2 = 41.8 kV in (*ii*)

$$60 = \frac{41.8}{1.4 \ln\left(\frac{r_3}{1.4}\right)}$$

$$r_3 = 2.3 \text{ cm}$$

Thickness of the 1st layer = 1.4 − 1.0 = 0.4 cm.

Thickness of the 2nd layer = 2.3 − 1.4 = 0.9 cm.

Problem 40

The electric field in air above a block of paraffin ($\varepsilon_r = 2.1$) is at an angle of 45° with respect to the plane surface of the block. Find the angle between $\vec{E}$ and the surface in the paraffin.

Solution:

According to the boundary condition,

$$\frac{\tan\theta_1}{\tan\theta_2} = \frac{\varepsilon_{r_1}}{\varepsilon_{r_2}}$$

where

ε_{r_1} = relative permittivity of air

ε_{r_2} = relative permitivity of paraffin

θ_1 = angle made with the normal to the boundary.

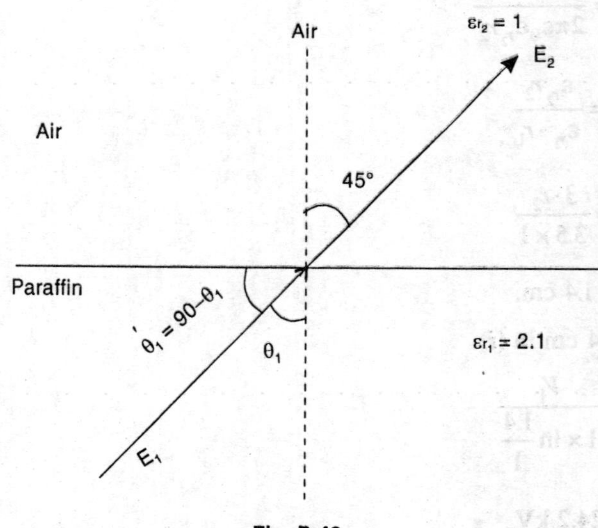

Fig. P-40

$$\frac{\tan\theta_1}{\tan\theta_2} = \frac{2\cdot 1}{1}$$

$$\frac{\tan\theta_1}{\tan 45} = 2.1$$

$\therefore$ $\theta_1 = 64.53$

θ_1' = angle between $\vec{E}$ and surface of paraffin

$\theta_1' = 90° - 64.53° = 25.4°$

Problem 41

A parallel plate capacitor consists of two plates each of 30 × 30 cm, separated 2 mm apart and dielectrics each 1 mm thick having relative permittivities of 3 and 5. If the potential difference between the plates is 5 kV, calculate the voltage gradient in each dielectric.

Solution:

Let V_1 be the p.d. across first medium

V_2 be the p.d. across second medium.

$$V_1 = E_1\, t_1 \text{ and } V_2 = E_2\, t_2$$

We know that $\vec{D} = \varepsilon \vec{E}$ and by boundary conditions, normal component of electric flux density D in two medium is same.

$$D_1 = D_2 = D$$

So,

$$\left|\vec{E}_1\right| = \frac{\left|\vec{D}\right|}{\varepsilon_{r_1}}$$

$$\left|\vec{E}_2\right| = \frac{\left|\vec{D}\right|}{\varepsilon_{r_2}}$$

5000 V

$\varepsilon_2 = 5$ $t_2 = 1\text{mm}$

$\varepsilon_1 = 3$ $t_1 = 1\text{mm}$

Fig. P-41

$$\frac{V_1}{V_2} = \frac{\dfrac{\vec{D}}{\varepsilon_{r_1}} \cdot t_1}{\dfrac{\vec{D}}{\varepsilon_{r_2}} \cdot t_2}$$

$$\frac{V_1}{V_2} = \frac{\varepsilon_{r_2}}{\varepsilon_{r_1}} \cdot \frac{t_1}{t_2} = \frac{5}{3}$$

$$V_1 = \frac{5}{3} V_2.$$

$$V_1 + V_2 = 5000.$$

Substituting $V_1 = \frac{5}{3} V_2$ and solving for V_1 and V_2

$$V_1 = 3125\,V; \quad V_2 = 1875\text{ V}$$

$$\therefore \quad E_1 = 3125\text{ kV/mm}$$

$$E_2 = 1875\text{ kV/mm}.$$

Problem 42

A cable has core diameter of 2.5 cm, impregnated paper insulation thickness of 2.13 cm whose relative permittivity is 3.5. Determine the capacitance/km length of the cable.

Solution:

Radius of conductor $a = \frac{2.5}{2} = 1.25\,\text{cm}$

Radius of cable $= b = a + t$

$= 1.25 + 2.13$

$= 3.38$ cm.

Capacitance $C = \dfrac{2\pi\varepsilon_0\varepsilon_r}{\ln\left(\dfrac{b}{a}\right)}$ F/m

$$= \frac{2\pi \times 8.854 \times 10^{-2} \times 3.5}{\ln\left(\dfrac{3.38}{1.25}\right)}$$

$$= 1.95 \times 10^{-10} \text{ F/m}$$

$C = 0.195$ µF/km.

Problem 43

A capacitor has two large horizontal plates spaced 't' m apart. A dielectric slab of relative permittivity ε_r and thickness 't' is now placed on the lower plate. A potential drop of V is applied across the capacitor plates. Show that the electric field E_d in the dielectric is $E_d = \dfrac{V}{\varepsilon_r t - t'(\varepsilon_r - 1)}$ and the electric field in the air space between the top dielectric slab and the upper capacitor plate is

$E_g = \dfrac{\varepsilon_r V}{\varepsilon_r t - t'(\varepsilon_r - 1)}$.

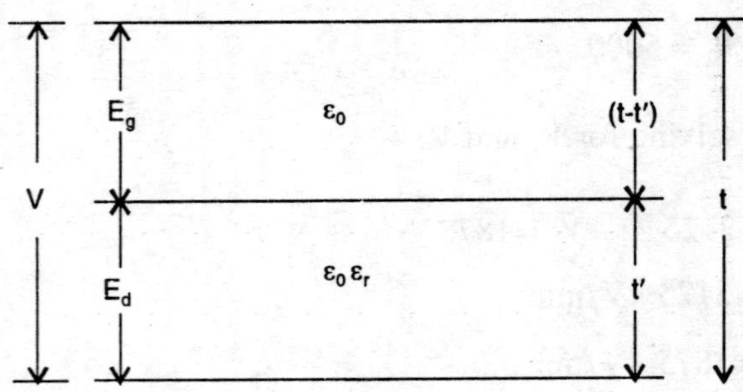

Fig. P-43

Solution:

The total potential (V) is the sum of potential across each dielectric material.

$$V = V_g + V_d \quad ...(1)$$

where $V_g \rightarrow$ potential across air gap

$V_d \rightarrow$ potential across dielectric slab.

As we know

$$V = E \times t$$

$$V_g = E_g t_g \text{ and } V_d = E_d t_d$$

t_g – thickness of air gap

$$t_g = t - t'$$

t_d – thickness of dielectric slab

$$t_d = t',$$

where E_g, E_d are potential gradient of air gap and dielectric slab respectively.

$$V_g = E_g(t - t') \quad ...(2)$$

$$V_d = E\ t'. \quad ...(3)$$

As $$E = \frac{D}{\varepsilon_0 \varepsilon_r}$$

$$E_g = \frac{D_g}{\varepsilon_0}(\varepsilon_r = 1), \quad E_d = \frac{D_d}{\varepsilon_0 \varepsilon_r}.$$

As per boundary conditions, normal components of electric flux density between two different medium is same.

$$\therefore \quad D_g = D_d = D$$

$$E_g = \frac{D}{\varepsilon_0} \quad ...(4)$$

$$E_d = \frac{D}{\varepsilon_0 \varepsilon_r}. \quad ...(5)$$

Substituting (4) and (5) in (2) and (3)

$$V_g = \frac{D}{\varepsilon_0}(t - t'), \quad V_d = \frac{D}{\varepsilon_0 \varepsilon_r} t'$$

$$\frac{V_g}{V_d} = \frac{t-t'}{t'}\varepsilon_r$$

$$V_g = \frac{t-t'}{t'}\varepsilon_r \cdot V_d. \qquad ...(6)$$

Substituting (6) in (1)

$$V = \left[\frac{(t-t')\varepsilon_r}{t'} + 1\right]V_d$$

$$V = \frac{V_d}{t'}\left[(t-t')\varepsilon_r + t'\right]$$

or

$$V_d = \frac{Vt'}{(t-t')\varepsilon_r + t'}$$

$$\frac{V_d}{t'} = E_d = \frac{V}{\varepsilon_r t - t'(\varepsilon_r - 1)}.$$

Similarly, from (6)

$$V_d = \frac{t'}{(t-t')\varepsilon_r}V_g \qquad ...(7)$$

Substituting (7) in (1)

$$V = V_g\left[1 + \frac{t'}{(t-t')\varepsilon_r}\right]$$

$$V_g = \frac{V(t-t')\varepsilon_r}{(t-t')\varepsilon_r + t'}$$

$$\frac{V_g}{t-t'} = E_g = \frac{\varepsilon_r V}{\varepsilon_r(t-t') + t'}$$

$$E_g = \frac{\varepsilon_r V}{\varepsilon_r t - t'(\varepsilon_r - 1)}.$$

Problem 44

Determine the maximum working voltage of a single core lead-sheathed cable with conductor 1 cm diameter and sheath 5 cm inside diameter. Two dielectrics are used:

inner layer: maximum working stress 72 kV/cm, $\varepsilon_{r_1} = 4$

outer layer: maximum working stress 60 kV/cm, $\varepsilon_{r_2} = 2.5$

Solution:

$$r_1 = \frac{1}{2} = 0.5 \text{ cm}$$

$$r_3 = \frac{5}{2} = 2.5 \text{ cm}$$

$$\varepsilon_{r_1} = 4 \quad \varepsilon_{r_2} = 2.5$$

$$E_{1\max} = 72 \text{ kV/cm}$$

$$E_{2\max} = 60 \text{ kV/cm.}$$

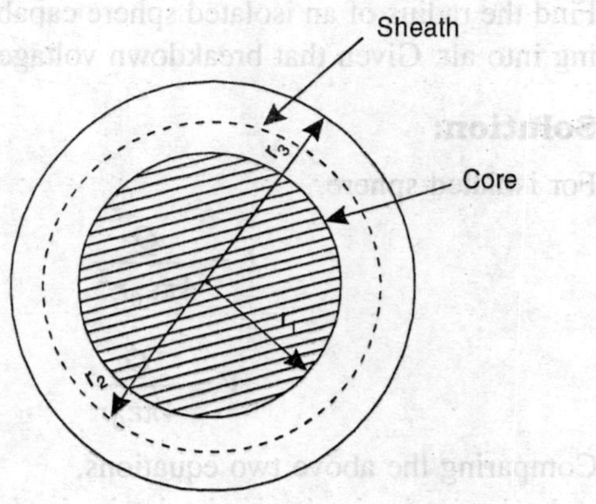

Fig. P-44

We know that $E_{1\max} = \dfrac{\lambda}{2\pi\varepsilon_0\varepsilon_{r_1} r_1}$

$$E_{2\max} = \frac{\lambda}{2\pi\varepsilon_0\varepsilon_{r_2} r_2}.$$

$$\frac{E_{1\max}}{E_{2\max}} = \frac{72}{60} = \frac{\varepsilon_{r_2} \cdot r_2}{\varepsilon_{r_1} \cdot r_1} = \frac{2.5}{4} \times \frac{r_2}{0.5}$$

$$\therefore \quad r_2 = 0.96 \text{ cm.}$$

$$E_{1\max} = \frac{V_1}{r_1 \ln \dfrac{r_2}{r_1}}$$

$$72 = \frac{V_1}{0.5 \times \ln \dfrac{0.96}{0.5}}$$

$$V_1 = 22.83 \text{ kV.}$$

$$E_{2\max} = \frac{V_2}{r_2 \ln \dfrac{r_3}{r_2}}$$

$$60 = \frac{V_2}{0.9b \ln \dfrac{2.5}{0.9b}}$$

$$V_2 = 55.13 \text{ kV.}$$

$$V = V_1 + V_2 = 22.83 + 55.13$$

$$= 78 \text{ kV.}$$

Problem 45

Find the radius of an isolated sphere capable of being charged to 1 million volt potential before sparking into air. Given that breakdown voltage of air is 30 kV/cm.

Solution:

For isolated sphere,

$$E = \frac{Q}{4\pi\varepsilon_0 r^2}$$

$$V = \frac{Q}{4\pi\varepsilon_0 r}.$$

Comparing the above two equations,

$$E = \frac{V}{r}$$

$$V = 1 \times 10^6$$

$$E_{max} = 30 \text{ kV/cm}$$

$\therefore$ Radius of the sphere $= \frac{V}{E}$

$$= \frac{1 \times 10^6}{30 \times 10^3}$$

$$= \frac{1}{3} \times 10^2 \text{ cm}$$

Radius of the sphere $= \frac{1}{3}$ m.

Problem 46

A single core co-axial cable is designed for 7.5 kV with rubber dielectric having an allowable field strength of 50 kV/cm. If the inner conductor is a solid wire of radius 0.4 cm, determine the minimum thickness of the dielectric.

Solution:

Potential of the cable = 7.5 kV

E_{max} = 50 kV/cm.

Radius of the conductor r_1 = 0.4 cm.

We know that

$$E_{max} = \frac{V}{r_1 \ln \frac{r_2}{r_1}}$$

$$50 = \frac{7.5}{0.4 \ln \frac{r_2}{0.4}}.$$

Solving for r_2,

$$r_2 = 0.582 \text{ cm}$$

$\therefore$ Thickness of the dielectric = 0.582 – 0.4

= 0.182 cm.

Problem 47

The radii of spherical capacitor are 20 cm and 5 cm and the space between two spheres is filled with an insulator of $\varepsilon_r = 2$. Find the greatest electric force (E_{max}) in the dielectric when the p.d. between the sphere is 20 kV.

Solution:

Radius of inner conductor = 5 cm

Radius of outer conductor = 20 cm

$\varepsilon_r = 2$

$V = 20$ kV.

From the equation $E = \dfrac{Q}{4\pi\varepsilon_0 r^2}$

$$V = \frac{Q}{4\pi\varepsilon_0}\left[\frac{1}{a} - \frac{1}{b}\right]$$

$$E = \frac{V}{r^2\left[\frac{1}{a} - \frac{1}{b}\right]}$$

$$E_{max} = \frac{V}{a^2\left[\frac{1}{a} - \frac{1}{b}\right]} \quad (r = a)$$

$$= \frac{20\,\text{kV}}{5^2\left[\frac{1}{5} - \frac{1}{20}\right]}$$

$$E_{max} = 5.33 \text{ kV/cm}.$$

Problem 48

Determine the capacitance/km of a pair of parallel wires 5 mm in diameter spaced uniformly 25 cm apart in air. What is the capacitance/km of each wire with respect to an equipotential plane interposed between parallel wires.

Solution:

We know that capacitance $= \dfrac{\pi\varepsilon_0\varepsilon_r}{\ln\dfrac{D}{r}}$

Given $D = 25$ cm

$r = 2.5$ mm $= 0.25$ mm

$$= \frac{\pi \times 8.854 \times 10^{-12} \times 1}{\ln\dfrac{25}{0.25}}$$

$= 6.037 \times 10^{-12}$ F/m

$= 6.037$ pf/km.

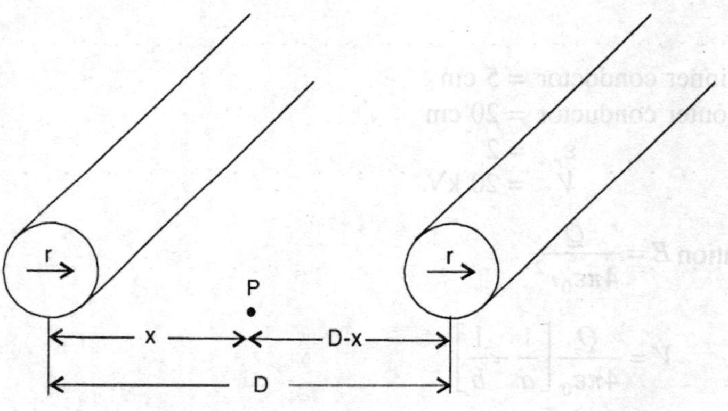

Fig. P-48

(*b*) Consider point *P* at a distance *x* m from conductor *A*.

Electric field intensity at $P = \vec{E} = \dfrac{\lambda}{2\pi\varepsilon_0}\left[\dfrac{1}{x} - \dfrac{1}{D-x}\right]\vec{a}_x.$

Electric field intensity will be maximum at the surface of the conductor. At the surface $x = r$.

$$\left|\vec{E}_{\max}\right| = \frac{\lambda}{2\pi\varepsilon_0}\left[\frac{1}{r} - \frac{1}{D-r}\right]$$

$$= \frac{\lambda}{2\pi\varepsilon_0}\left[\frac{D-2r}{D\cdot r}\right].$$

As $D >> r$, neglect r.

$$\vec{E}_{\max} = \frac{\lambda}{2\pi\varepsilon_0}\left[\frac{1}{r}\right]$$

$$\vec{E}_{\max} = \frac{\lambda}{2\pi\varepsilon_0 r}.$$

We know that the potential at any point is

$$V = \frac{\lambda}{2\pi\varepsilon_0} \ln \frac{D}{r} \text{ V/m}$$

$$\therefore \quad \vec{E}_{\max} = \frac{V}{2r \ln \cdot D/r}.$$

Given

$$E_{\max} = 3 \times 10^6 \text{ V/m} = 30 \text{ kV/cm}$$

$$\therefore \quad V = 30 \times 2 \times 0.75 \times \ln \frac{25}{0.025}$$

$$V = 207 \text{ kV}.$$

Problem 49

A condenser consists of two parallel plates each having an area of 0.15 m^2 and separated by a layer of 5mm thick sheet of ebonite. Again a sheet of ebonite 1.8 mm thick is introduced between the plates and their distance is increased to 6.1 mm in order to keep the capacitance intact. Determine the capacity of the condenser with ebonite in position and spacing of 5 mm between the plates.

Solution:

We know that

$$C = \frac{\varepsilon_0 A}{d}$$

$$= \frac{\varepsilon_0 \times 0.15}{5 \times 10^{-3}}$$

$$= 2.655 \times 10^{-10} \text{ F.}$$

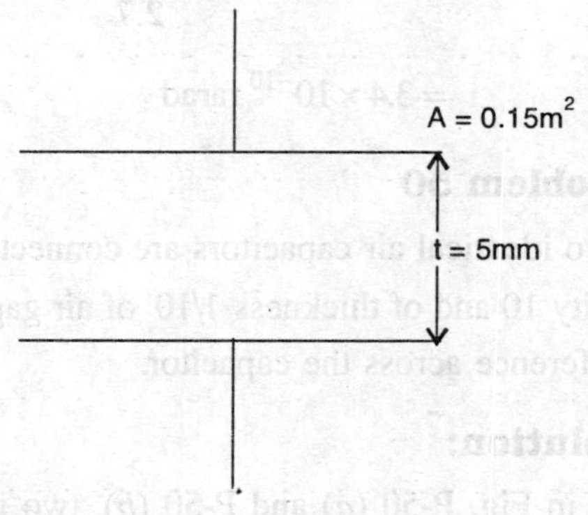

Fig. P-49 (a)

As in Fig. P-49 (*b*), another dielectric of relative permittivity of ε_{r_2} is introduced so that the capacitance remains unaltered.

To retain the capacitance at the same value, the distance between the plate is increased from 5mm to 6.1 mm.

If it has two dielectrics, the

$$\text{capacitance} = \frac{\varepsilon_0 A}{\dfrac{d - d'}{\varepsilon_{r_1}} + \dfrac{d'}{\varepsilon_{r_2}}}$$

$$\varepsilon_{r_1} = 1$$

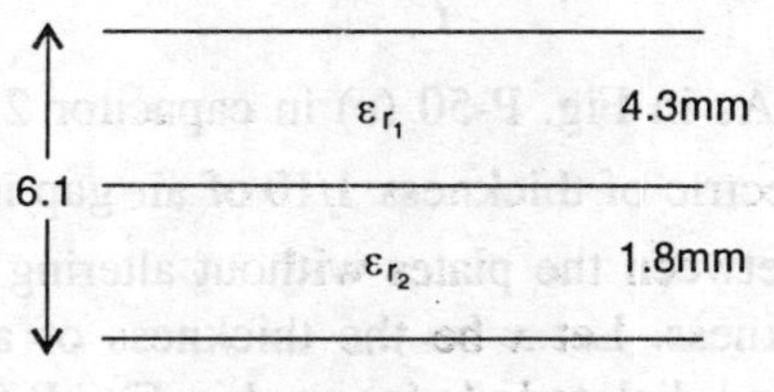

Fig. P-49 (b)

$$C = \frac{\varepsilon_0 \times 0.15}{\dfrac{4.3 \times 10^{-3}}{1} + \dfrac{1.8 \times 10^{-3}}{\varepsilon_{r_2}}}.$$

In both the cases, capacitance is equal.

$$2.655 \times 10^{-10} = \frac{8.854 \times 10^{-12}}{\dfrac{4.3 \times 10^{-3}}{1} + \dfrac{1.8 \times 10^{-3}}{\varepsilon_{r_2}}}$$

$$\varepsilon_{r_2} = 2.7.$$

(*c*) If the distance between the plates remains 5 mm with the ebonite thickness of 1.8 mm.

$$\text{Capacitance } C = \frac{\varepsilon_0 A}{\dfrac{d_1}{\varepsilon_{r_1}} + \dfrac{d_2}{\varepsilon_{r_2}}}$$

$$= \frac{8.854 \times 10^{-12} \times 0.15}{3.2 \times 10^{-3} + \dfrac{1.8 \times 10^{-3}}{2.7}}$$

$$= 3.4 \times 10^{-10} \text{ farad}.$$

$\varepsilon_{r_1} = 1$ 3.2

5mm

$\varepsilon_{r_2} = 2.7$ 1.8

Fig. P-49 (c)

Problem 50

Two identical air capacitors are connected in series across a constant 100 V. If a dielectric of permittivity 10 and of thickness 1/10 of air gap is inserted into one of the capacitors, determine the potential difference across the capacitor.

Solution:

As in Fig. P-50 (*a*) and P-50 (*b*), two identical capacitors are connected in series. If another dielectric is introduced in the second capacitor, it is equivalent of creating a third capacitor.

C_1 has air alone as the dielectric.

$$C_1 = \frac{\varepsilon_0 A}{t}$$

As in Fig. P-50 (*c*) in capacitor 2, the second dielectric of thickness 1/10 of air gap is introduced in between the plates without altering the original thickness. Let x be the thickness of air gap after second dielectric is inserted in Fig. P-50 (*c*)

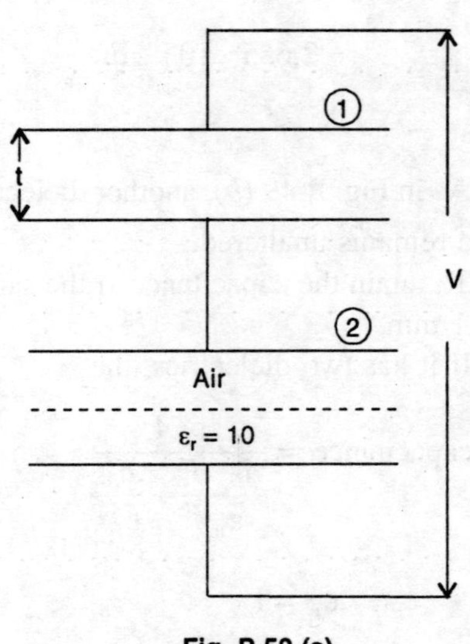

Fig. P-50 (a)

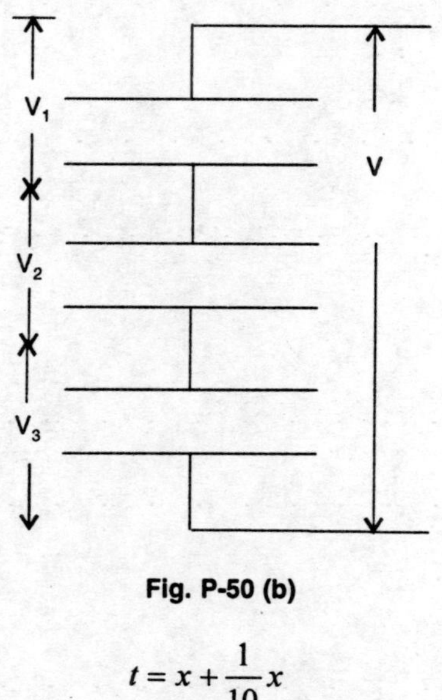

Fig. P-50 (b)

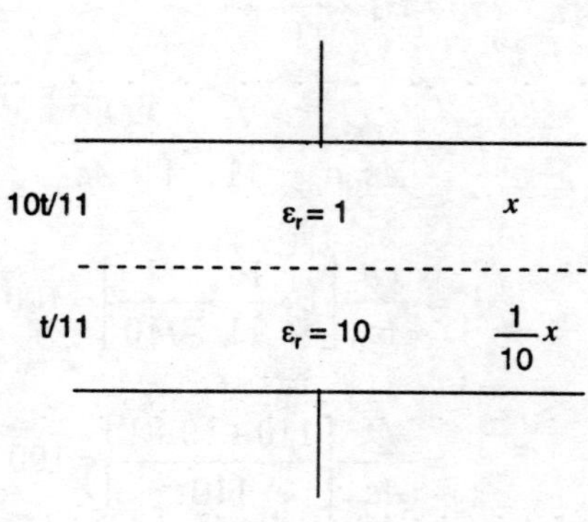

Fig. P-50 (c)

$$t = x + \frac{1}{10}x$$

$$= \frac{11x}{10}$$

$$x = \frac{10t}{11}.$$

Let C_2 having thickness of $\frac{10t}{11}$ and of permittivity $\varepsilon_r = 1$.

C_3 is having thickness of $\frac{t}{11}$ and of permittivity 10.

It is given that

$V = V_1 + V_2 + V_3 = 100$ V.

V_1 = Voltage across the first capacitor

$$= E_1 \times t$$

$$= \frac{Q}{A\varepsilon_0}t.$$

V_2 = Voltage across the second capacitor

$$= E_2 \times \frac{10t}{11}$$

$$= \frac{Q}{A\varepsilon_0} \times \frac{10t}{11} = \frac{10Qt}{11\varepsilon_0 A}.$$

V_3 = Voltage across the third capacitor

$$= E_3 \times \frac{t}{11}$$

$$= \frac{Q}{A\varepsilon_0\varepsilon_r} \times \frac{t}{11} = \frac{Qt}{110A\varepsilon_0}$$

$$V = \frac{Qt}{A\varepsilon_0}\left[1 + \frac{10}{11} + \frac{1}{110}\right] = 100$$

$$= \frac{Qt}{A\varepsilon_0}\left[\frac{110 + 10 + 1}{110}\right] = 100$$

$$\therefore \quad \frac{Qt}{A\varepsilon_0} = V_1 = \frac{100 \times 110}{211} = 52.13.$$

$$V_2 = V_1 \times \frac{10}{11} = 47.39$$

$$V_3 = V_1 \times \frac{1}{110} = 0.474$$

∴ Voltage across the first capacitor = 52.13

Voltage across the second capacitor having two dielectrics = 47.39 + 0.474

= 47.864 V.

Problem 51

A parallel plate capacitor has 3 similar plates, the outside two being joined together. The inner plate is movable so that it can be used as a variable capacitor. If C_1 is the capacitance when the inner plate is exactly midway between the outers and C_2 is the capacitance when the inner plate is 3 times as near one plate as the other. Find C_1/C_2.

Solution:

Since the two plates have been joined together and the centre plate is at midway, this is equivalent to two capacitors connected in parallel each having equal thickness of $t/2$ as in Fig. P-51.

$$\therefore \quad C_1 = \frac{\varepsilon_0 A}{t/2} + \frac{\varepsilon_0 A}{t/2}$$

$$= \frac{4\varepsilon_0 A}{t}$$

When the centre plate is 3 times nearer to one plate, it has a thickness of $\frac{t}{4}$ and $\frac{3}{4}t$ as in Fig. P-51 (*b*).

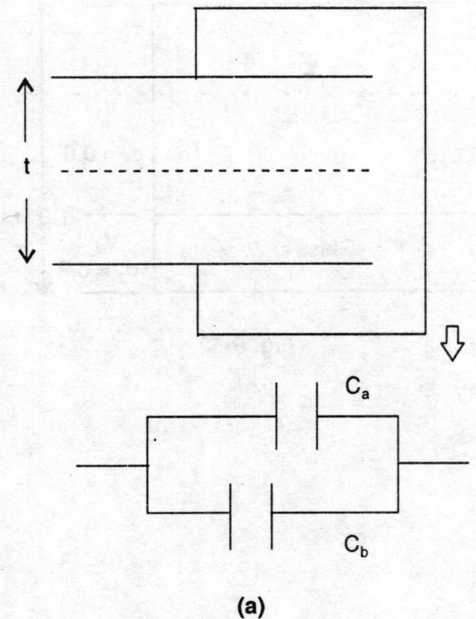

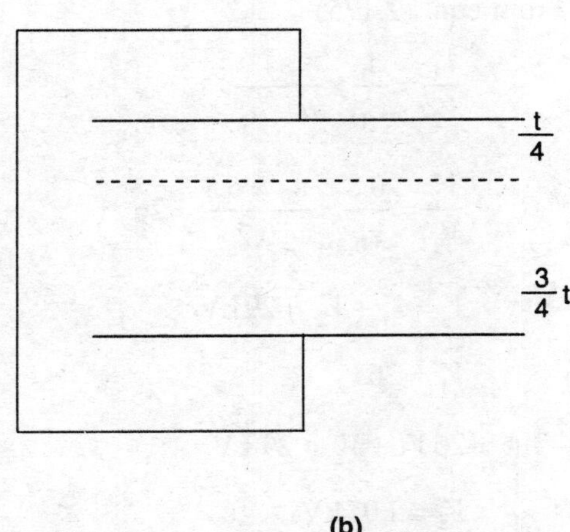

Fig. P-51

$$\therefore \quad C_2 = \frac{\varepsilon_0 A}{t/4} + \frac{\varepsilon_0 A}{3t/4}$$

$$= \frac{4\varepsilon_0 A}{t}\left(\frac{4}{3}\right)$$

$$= \frac{4}{3} C_1$$

$$\therefore \quad \frac{C_1}{C_2} = \frac{3}{4}.$$

Problem 52

A parallel plate capacitor with a separation of $d = 1$ cm has 29 KV applied when free space is the only dielectric. Assume air has a dielectric strength of 30 KV/cm. Show why the air breaks down when a thin piece of glass ($\varepsilon_r = 6.5$) with a dielectric strength of 290 KV/cm and thickness $d_2 = 0.2$ cm is inserted as shown in Fig. P-52. **(Anna Univ. Oct. 2002)**

Solution:

Potential across the capacitor is $V = 29$ KV

When a glass dielectric is inserted,

$V = V_1 + V_2$

V_1 = Potential across air.

V_2 = Potential across glass.

$$V_1 = E_1 d_1 \qquad V_1 = E_1 \times 0.8$$

$$V_2 = E_2 (d - d_1), \; V_2 = E_2 \times 0.2.$$

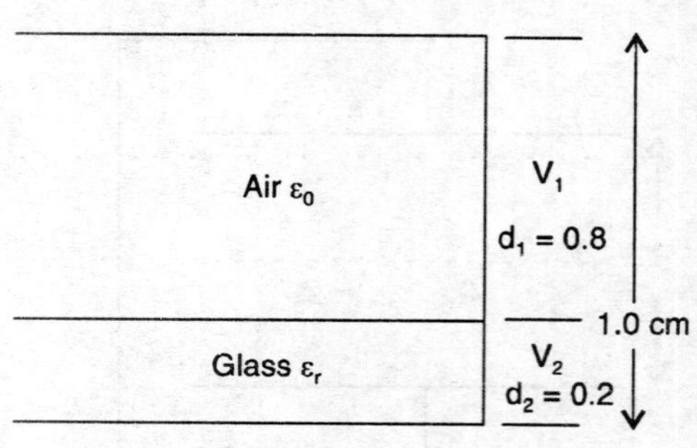

Fig. P-52

From eqn. (2.175)

$$\frac{V_1}{V_2} = \frac{d_1}{\varepsilon_1} \times \frac{\varepsilon_2}{d - d_1}$$

$$\frac{V_1}{V_2} = \frac{0.8}{\varepsilon_0} \times \frac{\varepsilon_0 \times 6.5}{0.2} = 26$$

$$V = V_1 + V_2 = 29\,\text{kV}$$

$$V_1 = 26\,V_2$$

$$26\,V_2 + V_2 = 29\text{ kV}$$

$$V_2 = 1.07\,\text{kV}$$

$$V_1 = 27.93\,\text{kV}$$

when $V_1 = 27.93$ applied across the air dielectric, dielectric stress or potential gradient.

$$E = \frac{V_1}{d_1} = \frac{27.93}{0.8} = 34.91\text{ kV/cm}.$$

This stress exceeds the maximum dielectric strength of air (30 kV/cm). So, the air breaks down.

Problem 53

A transmission line of conductors of diameter 5 mm are 1.5 m apart. The line is situated 7 m above the ground level. Calculate the capacitance.

Solution:

Radius of the conductor $= \frac{5}{2}$ mm

$= 2.5 \times 10^{-3}$ m.

Distance between the two conductors (D) = 1.5 m.
Height above the ground level (H) = 7 m.

$$C_{AB} = \frac{\pi\varepsilon_0}{\ln\dfrac{D}{r}\dfrac{2H}{\sqrt{4H^2 + D^2}}}$$

$$= \frac{\pi \times 8.85 \times 10^{-12}}{\ln\dfrac{1.5}{2.5 \times 10^{-3}} \times \dfrac{2 \times 7}{\sqrt{4 \times 7^2 + 1.5^2}}}$$

$$= 4.352 \times 10^{-12} \text{ farad/m}$$

$$= 4.352 \text{ pf/m}$$

SUMMARY

1. Coulomb's law of electrostatic force between two point charges Q_A and Q_B is:

$$\vec{F}_{AB} = \frac{Q_A Q_B}{4\pi\varepsilon_0 R^2}\vec{a}_{AB} \qquad \text{(if distance between } Q_A \text{ and } Q_B \text{ is given)}$$

$$\vec{F}_{AB} = \frac{Q_A Q_B}{4\pi\varepsilon_0 R^3}\vec{R}_{AB} \qquad \text{(if position vectors of } Q_A \text{ and } Q_B \text{ is given)}$$

$$\vec{R}_{AB} = \vec{r}_B - \vec{r}_A$$

$$R = \left|\vec{R}_{AB}\right|.$$

2. Electric field intensity $\vec{E}$ is defined as force per unit charge. For point charge,

$$\vec{E} = \frac{Q}{4\pi\varepsilon_0 R^2}\vec{a}_R \quad \text{N/coulomb (or) volt/m.}$$

3. For continuous charge distribution, the electric field intensity formula is modified as

$$\vec{E} = \int_l \frac{\lambda_L\, dl}{4\pi\varepsilon_0 R^2}\,\vec{a}_R \qquad \text{line charge with density } \lambda_L \text{ c/m}$$

$$= \int_s \frac{\sigma_s\, ds}{4\pi\varepsilon_0 R^2}\vec{a}_R \qquad \text{surface charge with density } \sigma_s \text{ c/m}^2$$

$$= \int_l \frac{\rho_v\, dv}{4\pi\varepsilon_0 R^2}\vec{a}_R \qquad \text{volume charge with density } \rho_v \text{ c/m}^3$$

4. For a finite line charge

$$\vec{E}_x = \frac{\lambda_L}{4\pi\varepsilon_0 h}[\sin\alpha_2 - \sin\alpha_1]\;\vec{a}_x \text{ V/m}$$

$$\vec{E}_y = \frac{\lambda_L}{4\pi\varepsilon_0 h}[\cos\alpha_1 + \cos\alpha_2]\;\vec{a}_y \text{ V/m.}$$

5. For an infinite line charge

$$\vec{E} = \frac{\lambda_L}{2\pi\varepsilon_0 h}\;\vec{a}_y \text{ V/m}$$

(or) $$= \frac{\lambda_L}{2\pi\varepsilon_0 h}\,\vec{a}_\rho \text{ V/m.}$$

6. For an finite disc surface charge

$$\vec{E} = \frac{\sigma_s}{2\varepsilon_0}(1-\cos\alpha)\vec{a}_z.$$

7. For an infinite surface charge

$$\vec{E} = \frac{\sigma_s}{2\varepsilon_0}\,\vec{a}_z.$$

8. The electric flux density $\vec{D}$ is related to the electric field intensity (E) in free space as

$$\vec{D} = \varepsilon_0 \vec{E}.$$

9. The electric flux through the surface S is

$$\Psi = \int_s \vec{D}\cdot d\vec{s}.$$

10. Gauss's law states that the net electric flux coming out from a closed surface is equal to the total charge enclosed.

$$\Psi = Q_{enc}.$$

$$\Psi = \oint_s \vec{D}\cdot d\vec{s} = Q_{enc}.$$

(or) $$\oint_s \vec{E}\cdot d\vec{s} = Q_{enc}/\varepsilon_0.$$

11. Integral form of Gauss's law is

$$\oint_s \vec{D}\cdot d\vec{s} = \int_v \rho_v dv.$$

Differential form is

$$\nabla\cdot\vec{D} = \rho_v.$$

12. Gauss's law is applied only for symmetrical charge distribution.
13. The charge distribution is said to be symmetrical if the resultant electric field intensity done to the charge distribution is along only one direction e.g.

Infinite line charge distribution – cylindrical symmetry
Infinite surface charge distribution – cartesian (or) rectangular symmetry
Spherical volume charge distribution – spherical symmetry.

14. For a spherical volume charge distribution

$$\vec{E} = \begin{cases} \rho_v r/3\varepsilon_0 \; \vec{a}_r & 0 < r < a \\ \dfrac{\rho_v a^3}{3\varepsilon_0 r^2} \; \vec{a}_r & r > a. \end{cases}$$

$$\text{(or)} \quad \vec{E} = \begin{cases} \dfrac{Qr}{4\pi\varepsilon_0 a^3} \; \vec{a}_r & 0 < r < a \\ \dfrac{Q}{4\pi\varepsilon_0 r^2} \; \vec{a}_r & r > a. \end{cases}$$

15. Total work done to move a point charge Q from point A to B in an electric field $(\vec{E})$ is

$$W = -Q\int_A^B E \cdot d\vec{l}\,.$$

16. Potential difference between two points A and B is the work done in taking a test charge from A to B, in the electric field.

$$V_{AB} = -\frac{W}{Q} = -\int_A \vec{E} \cdot dl\,.$$

17. Potential at a point is the work done in bringing a test charge from infinity to that point in the electric field.

$$V = -\int_\infty^r \vec{E} \cdot d\vec{l}\,.$$

18. Potential due to a point charge

$$V = \frac{Q}{4\pi E_0 r} \text{ volt.}$$

For a line charge distribution

$$V = \int_L \frac{\lambda_L \, dl}{4\pi\varepsilon_0 r} \text{ volt.}$$

For a surface charge distribution

$$V = \int_S \frac{\sigma_s \, ds}{4\pi\varepsilon_0 r} \text{ volt.}$$

For a volume charge distribution

$$V = \int_s \frac{\rho_v \, dv}{4\pi\varepsilon_0 r} \text{ volt.}$$

19. Electrostatic field is conservative (or) irrotational. No work is done in moving a charge along a closed path in electrostatic field.

$$\oint \vec{E}\cdot d\vec{l} = 0 \quad \text{(or)} \quad \nabla\times\vec{E} = 0.$$

20. Relation between $\vec{E}$ and V is

$$\vec{E} = -\nabla V.$$

21. The direction of electric field intensity $\vec{E}$ (or) electric flux lines is always normal to the equipotential surfaces.
22. The Laplace's equation is

$$\nabla^2 V = 0.$$

The Poisson's equation is

$$\nabla^2 V = -\rho_v/\varepsilon.$$

Electric Field in Material Medium

23. Conduction current density $\vec{J} = \sigma\vec{E}$ (Ohm's law in point form).
24. Convection current density $\vec{J} = \rho_v u$.
25. In dielectric medium, $\vec{D}$ and $\vec{E}$ fields are related as $\vec{D} = \varepsilon\vec{E}$ where $\varepsilon = \varepsilon_0\varepsilon_r$ is the permittivity of the medium.
26. The electric susceptibility $X_e.(= \varepsilon_r - 1)$ of a dielectric, measures the sensitivity of the material in dielectric field.
27. For a dielectric region, if ε does not vary, then it is homogeneous. If ε depends on the space co-ordinate, then it is inhomogeneous.
28. If $\vec{D}$ and $\vec{E}$ along the same direction, in a dielectric it is isotropic, otherwise it is anisotropic.
29. A dielectric material is said to be linear, if $\vec{D}$ varies linearly with $\vec{E}$.
30. The principle of charge conservation is stated in continuity equation

$$\nabla\cdot\vec{J} + \frac{\partial\rho_v}{\partial t} = 0.$$

For electrostatic field $\nabla\cdot\vec{J} = 0$.

31. Electric field inside the conductor is zero, $\vec{E} = 0$.
32. Boundary conditions must be satisfied by an electric field existing in two differential media separated by an interface.

For a dielectric-dielectric interface

$$E_{1t} = E_{2t}.$$

$$D_{1n} - D_{2n} = \rho_s \text{ (or) } D_{1n} = D_{2n} \text{ if } \rho_s = 0.$$

$$\boxed{\frac{\tan\theta_1}{\tan\theta_2} = \frac{\varepsilon_{r_1}}{\varepsilon_{r_2}}.}$$

For a dielectric-conductor interface

$$E_t = 0$$

$$D_n = \varepsilon E_n = \rho_s.$$

33. The electrostatic energy due to a continuous volume charge distribution

$$W_E = \frac{1}{2}\int_v \varepsilon_0 E^2 dv \text{ joule.}$$

Energy density $w = \frac{1}{2}\varepsilon_0 E^2 \text{ (or)} = \frac{1}{2}\frac{D^2}{\varepsilon_0} \text{ J/m}^3.$

34. Two opposite charges with same magnitude separated by small distance constitute a dipole.

35. Dipole moment $\vec{p} = Q\vec{d}$ C/m.

Polarisation density (or) vector $\vec{P} = \frac{\text{dipole moment}}{\text{volume}} = \frac{Qd}{v}$ C/m^2

36. Process of creating dipoles is polarisation.

37. Electric potential due to dipole is $V = \frac{p\cos\theta}{4\pi\varepsilon_0 r^2}$ volt.

Electric field intensity $\vec{E} = \frac{Qr}{4\pi\varepsilon_0 r^3}\left[2\cos\theta\,\vec{a}_r + \sin\theta\,\vec{a}_\theta\right]$ V/m.

38. Torque on a dipole is $\vec{T} = \vec{P} \times \vec{E}.$

39. Potential energy of dipole is $U = -\vec{P}\cdot\vec{E}.$

40. The capacitance 'C' of the capacitor is

$$C = \frac{Q}{V} = \frac{\varepsilon \oint_s \vec{E}\cdot d\vec{s}}{\oint_r \vec{E}\cdot d\vec{l}} \text{ farad.}$$

41. Capacitance:

For a parallel plate capacitor

$$C = \frac{\varepsilon s}{d} \text{ farad.}$$

For a cylindrical capacitor

$$C = \frac{2\pi\varepsilon l}{\ln b/a} \text{ farad.}$$

For a spherical capacitor

$$C = \frac{4\pi\varepsilon}{(1/a - 1/b)} \text{ farad.}$$

For an isolated sphere

$$C = 4\pi\varepsilon_0 a \text{ farad}$$

For a two-wire transmission line

$$C_{AB} = \frac{\pi\varepsilon_0}{\ln D/l} \text{ F/m.}$$

Capacitance of single wire is 2 times that of two-wire transmission line.

42. Capacitance of a parallel plate capacitor with two dielectric medium

$$C = \frac{\varepsilon_0 A}{\frac{t'}{\varepsilon r_1} + \frac{t - t'}{\varepsilon r_2}}.$$

43. Capacitance of a co-axial cable with two dielectric is

$$C = \frac{2\pi\varepsilon_0 \varepsilon_{r_1} \varepsilon_{r_2}}{\varepsilon r_1 \ln\left(\frac{r_2}{r_1}\right) + \varepsilon r_1 \ln\left(\frac{r_3}{r_2}\right)}.$$

44. The maximum stress above which the dielectric breaks down to conductor is called dielectric strength (KV/cm). The corresponding potential is breakdown potential in KV.

45. The dielectric strength of co-axial cable is

$$E_{\max} = \frac{V}{a \ln b/a} \text{ KV/cm.}$$

46. The energy stored in parallel plate capacitor is

$$W = \frac{1}{2} CV^2 \text{ joule (or)}$$

$$W = \frac{1}{2} QV \text{ joule (or)}$$

$$W = \frac{1}{2}\frac{Q^2}{C} \text{ joule.}$$

47. Electrostatic force $\vec{F} = -\nabla W_E$ N.

For parallel plate capacitor $\vec{F} = \dfrac{-\varepsilon A E^2}{2} \vec{a}_z$ N.

48. Boundary value problems are those in which potentials at the boundaries of a region are specified and the potential field within the region is to be determined using Laplace's and Poisson's equations.
49. The uniqueness theorem states that if V satisfies Poisson's or Laplace's equation and the prescribed boundary condition, then V is the only possible solution for that given problem.
50. Method of images is used to solve a boundary value problem involving an infinite conducting plane.
51. Capacitance of two-wire transmission line including earth effect using image theory is

$$C_{AB} = \frac{\pi \varepsilon_0}{\ln D'/r} \quad \text{F/m}$$

where $D' = D \cdot \dfrac{2H}{\sqrt{4H^2 + D^2}}$.

52. Grading is the process of distributing the electric stress uniform in the dielectric medium by inserting more than one dielectric material.

EXERCISE

1. Two point charges with $q_1 = 3 \times 10^{-5}$ C and $q_2 = -4 \times 10^{-5}$ C are located in free space at (1, 3, –1) and (–3, 1, 0) respectively in cartesian system.
   Find (*a*) Force on charge located at (4, –1, 2) and (*b*) Electric field at (4, –1, 2).
2. Two identical charges are located on the X-axis at $x = 3$ and $x = 7$. At what point in space is the net electric field zero?
3. Four charges of each are located in free space at (–3, 0, 0), (3, 0, 0), (0, –3, 0), and (0, 3, 0) is a cartesian co-ordinate system. Find the force on a 20 μC charge located at (0, 0, 4). All distances are in meters. $[F = \hat{z}\, 0.23(\text{N})]$
4. Two long parallel conductors of a dc transmission line separated by 2 m have charges of $\rho_L = 5\,\mu\text{C/m}$ opposite sign. Both lines are 8 m above ground. What is the magnitude of the electric field 4 m directly below one of the wires? $(\varepsilon_r = 1)$. $[|E| = 1 + 16 \text{ kV/m}]$
5. Four 3 pc charges are at the corners of a 1m square. Two charges on the left side of the square are positive. Two charges on the right side are negative. Find the field $\vec{E}$ at the center of the square. $(\varepsilon_r = 1)$. [153 mV/m]
6. The plane $y = 3$ m contains a uniform charge distribution of density $\rho_s = 10^{-8}/6\pi \text{ C/m}^2$. Determine $\vec{E}$ at a point above the plane and below the plane. $[\pm\vec{a}_y \text{ V/m}]$

7. A uniform sheet of charge with $\rho_s = (-1/3\pi)$ nC/m² is located at $z = 5$ m and a uniform line of charge with $\rho_s = (-25/9)$ nC/m is located at $z = -3$ m, $y = 3$ m. Find the electric field at (0, −1, 0) m. $\left[8\vec{a}_y \text{ V/m}\right]$

8. Point charges Q_1 and Q_2 are respectively located at (4, 0, −3) and (2, 0, 1). If $Q_2 = 4$ nC, find Q_1 such that

   (*a*) The $\vec{E}$ at (5, 0, 6) has no *z*-component.

   (*b*) The force on a total charge at (5, 0, 6) has no *x*-component.

9. Given that

$$\rho_v = \begin{cases} 12\rho \text{ nC/m}^3 & 1 < \rho < 2 \\ 0 & \text{otherwise.} \end{cases}$$

Determine $\vec{D}$ everywhere.

$$\left[D_\rho = \begin{cases} 0 & \rho < 1 \\ \dfrac{8(\rho^3 - 1)}{2\rho} & 1 < \rho < 2 \\ 28/\rho & \text{otherwise} \end{cases} \right]$$

10. Point charges 30 nC, −20 nC and 10 nC are located at (−1, 0, 2), (0, 0, 0) and (1, 5, −1) respectively. What is the total flux leaving a cube of side 6 m centered at the origin? [10 nC]

11. Determine the total charge within the sphere $r = 4$ m if $\rho_v = \dfrac{10}{r\sin\theta}$ C/m³. [157.9 nC]

12. Planes $X = 2$, $Y = -3$, respectively, carry charges 10 nC/m² and 15 nC/m². If the line $x = 0$, $Z = 2$ carries charge 10 π nC/m. Calculate $\vec{E}$ at (1, 1, −1) due to three charge distributions.

$$\left[\begin{aligned} E_0 &= -162\pi\,\vec{a}_x + 270\pi\,\vec{a}_y - 54\pi\,\vec{a}_z \\ E_1 &= -180\pi\,\vec{a}_x \\ E_2 &= 270\pi\,\vec{a}_y \\ E_3 &= 18\pi\vec{a}_x - 54\pi\,\vec{a}_z \end{aligned}\right]$$

Potential

13. Determine the electric potential at the origin in free space due to four charges of 20 μC each located at the corners of a square in the *X-Y* plane and whose center is at the origin. The square has sides of 2 m each. $\left[V = \sqrt{2} \times 10^{-5}/\pi\varepsilon_0 \cdot V\right]$

14. A spherical shell of radius *R* has a uniform surface charge density ρ_s. Determine the electric potential at the center of the shell. $\left[V = \rho_s R/\varepsilon\right]$

15. A circular ring of charge of radius a lies in the X-Y plane and is centered at the origin. Assume also that the ring is in air and carries a uniform density ρ_L. Show that the electric potential at (0, 0, z) is given by $V = \dfrac{\rho_l a}{2\varepsilon_0\left[a^2+z^2\right]^{1/2}}$.

16. Three point charges $-4\mu C, 5\mu C$ and $3\mu C$ are located at (2, –1, 3), (0, 4, –2) and (–1, 5, 2) respectively. Find the potential at (1, 0, 1) assuming zero potential at infinity. [10.23 kV]

17. Given the potential $V = \dfrac{10}{r^2}\sin\theta\cos\phi$, find the electric flux density D at $(2, \pi/2, 0)$.

$\left[22.1\,\vec{a}_x \text{ pC/m}^2\right]$

18. Two dipoles, with dipole moments $-5\vec{a}_z$ nC/m and $9\vec{a}_z$ nC/m, are located at point (0, 0, –2) and (0, 0, 3) respectively. Find the potential at the origin. $\left[V = \sum_{k=1}^{2} \dfrac{\vec{P}_k\cdot\vec{r}_k}{4\pi\varepsilon_0 rk^3} = -20.25 \text{ V}\right]$

19. Find the work done in carrying a 5 C charge from P (1, 2, –4) to R (3, –5, 6) in an electric field $\vec{E} = \vec{a}_x + z^2\,\vec{a}_y + 2yz\,\vec{a}_z$ V/m. [1050 J]

20. A circular ring 2.4 m in diameter has a charge density of $8\,\mu$ C/m. Find V and E at the center of ring. $[V = 452 \text{ kV},\ E = 0]$

21. At what distance is the absolute potential due to a 120 nC point charge equal to 1kV (*a*) in vacuum and (*b*) in a medium with $\varepsilon_r = 4$. [(*a*) 1.08 m, (*b*) 0.27 m]

22. A point charge $Q = 0.4$ nC is located at (2, 3, 3) m in cartesian co-ordinates. Find the potential difference V_{AB} where point A is (2, 2, 3) m and B is (–2, 3, 3) m. [2.70 V]

23. Find the polarization in a dielectric material with $\varepsilon_r = 2.8$, if $\vec{D} = 3.0\times10^{-7}\,\vec{a}_x\,\text{C/m}^2$.

$\left[1.93\times10^{-7}\,\vec{a}_x \text{ C/m}\right]$

24. Region $y \le 0$ consists of a perfect conductor while region $y \ge 0$ is a dielectric medium of $\varepsilon_r = 2$. If there is a surface charge of 2 nC m^2 on the conductor, determine $\vec{E}$ and $\vec{D}$ at (*a*) A (3, –2, 2) and (*b*) B (–4, 1, 5). [(*a*) $\vec{E} = \vec{D} = 0$ (*b*) $D = 2\vec{a}_y$ nC/m^2 $E = 113.1\,\vec{a}_y$ V/m]

25. Find the capacitance of a parallel plate capacitor containing two dielectrics $\varepsilon_{r_1} = 1.5$ and $e_{v_2} = 3.5$ and $A = 2\text{ m}^2$ and $d = 10^{-3}$m and for the two cases.
(*i*) if each dielectric comprises one half of the volume and
(*ii*) if each dielectric occupies one half of the space between the plates, but interface is parallel to the plates.

26. Find the voltage across each dielectric in the capacitor shown in Fig. 4, when applied voltage is 200 V.

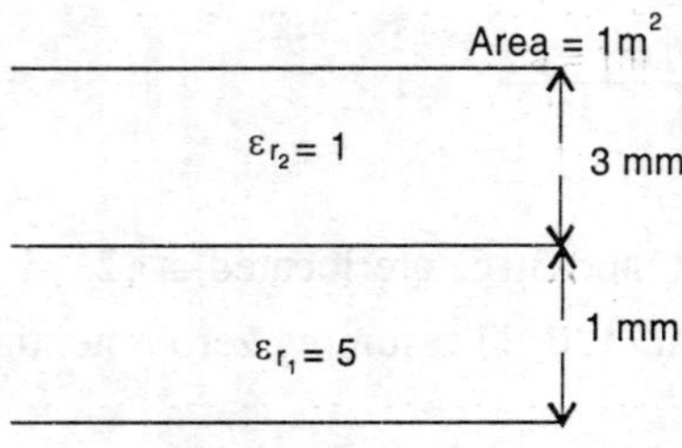

$$\begin{bmatrix} V_1 = 12.5 \text{ V} \\ V_2 = 187.5 \text{ V} \end{bmatrix}$$

27. A parallel plate capacitor with a separation $d = 1.0$ cm has 29 kV applied when free space is the only dielectric. Assume that air has a dielectric strength of 30 kV/cm. Show why the air breaks down when a thin piece of glass $(\varepsilon_r = 6.5)$ with a dielectric strength of 290 kV/cm and thickness $d_2 = 0.20$ cm is inserted in the dielectric space between the two plates.
[As E_1 across air is 34.9 kV/cm, air breaks down]

28. Find the capacitance per unit length of a co-axial conductor with outer radius 4 mm and inner radius 0.5 mm if the dielectric has $\varepsilon_r = 5.2$. [139 pF/m]

29. Prove that the force of attraction between the two plates of a parallel plate capacitor is $\vec{F} = \dfrac{\varepsilon A E^2}{2}(-\vec{a}_z)$.

30. A parallel plate capacitor has plate area 200 cm^2 and plate separation 3 mm. The charge density is 1 μC/m^2 with air as dielectric. Find (*a*) the capacitance of the capacitor (*b*) the voltage between the plates (*c*) the force with which the plates attract each other.

31. Electrostatically, a typical thundercloud may be represented by a capacitor model with horizontal plates 10 km^2 in area separated by a vertical distance of 1 km. The upper plate has a positive charge of 200 C and the lower plate has equal negative charge. Find (*a*) the electrostatic energy stored in the cloud (*b*) the potential difference 'V' between top and bottom of the cloud (*c*) the average electric field *E* in the cloud. [2.26×10^{11} J, 2.26×10^{9} V, 2.26×10^{6} V/m]

32. In a dielectric medium with $\varepsilon_r = 4$, the electric field is given by $\vec{E} = \left(x^2 + 2z\right)\vec{a}_x + x^2\vec{a}_y - (y+z)\vec{a}_z$. Calculate the electrostatic energy stored in the region $-1\,\text{m} \le x \le 1\,\text{m},\ 0 \le y \le 2\,\text{m}$ and $0 \le z \le 3\,\text{m}$. [4.62×10^{-9} J]

3

Magnetostatic Fields

INTRODUCTION

In the previous chapter, the study of electrostatic field which is produced by static or stationary charges has been dealt with. If the charges are moving with constant velocity, a static magnetic field (or magnetostatic field) is produced. The movement of charges (or current flow) is due to magnetisation currents as in permanent magnets, electron beam currents as in vacuum tubes or conduction currents as in current-carrying wires. In this chapter, magnetic fields due to direct current in free space and in material space is considered.

There are two major laws governing magnetostatic fields: (*i*) Biot-Savart's law and (*ii*) Ampere's circuital law. Like Coulomb's law, Biot-Savart's law is the general law of magnetostatics. Just as Gauss's law is a special case of Coulomb's law, Ampere's law is a special case of Biot-Savart's law and is easily applied to problems involving symmetrical current distribution.

3.1 BIOT-SAVART'S LAW

The law states that the magnetic field intensity $\left(d\vec{H}\right)$ produced at a point P, as shown in Fig. 3.1, by the differential current element Idl is proportional to the product of Idl and the sine of the angle α between the element and the line joining P to the element and is inversely proportional to the square of the distance R between the point P and the differential current element. The direction of magnetic field intensity is normal to the plane containing the differential element and the line drawn from the differential element to the point P. That is,

$$d\vec{H} \propto \frac{Idl \sin\alpha}{R^2}\vec{a}_n \qquad ...(3.1)$$

$$d\vec{H} = \frac{kIdl \sin\alpha}{R^2}\vec{a}_n, \qquad ...(3.2)$$

where k is the constant of proportionality in SI units, $(k = 1/4\pi)$. So, eqn. (3.2) becomes

$$d\vec{H} = \frac{Idl \sin\alpha}{4\pi R^2}\vec{a}_n \qquad ...(3.3)$$

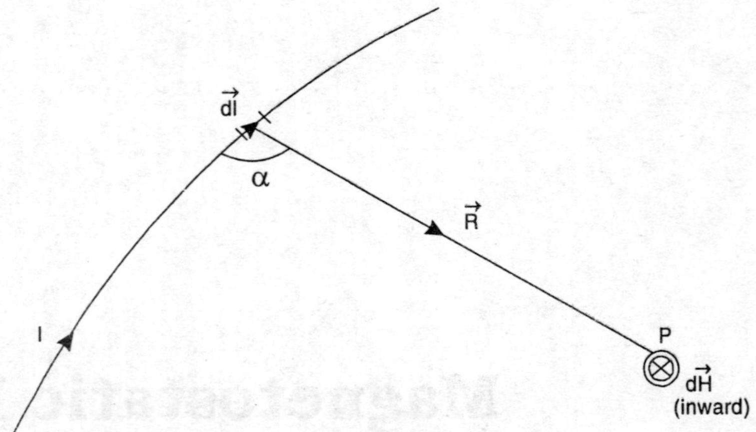

Fig. 3.1 Magnetic field $d\vec{H}$ at P due to current element $Id\vec{l}$

Equation (3.3) can also be written in vector form as

$$d\vec{H} = \frac{Id\vec{l} \times \vec{a}_R}{4\pi R^2}, \qquad ...(3.4)$$

where $\vec{a}_R$ is the unit vector of the line joining differential element to point 'P'.

$$\vec{a}_R = \frac{\vec{R}}{|\vec{R}|} \text{ and } R = |\vec{R}|.$$

So, eqn. (3.4) becomes

$$d\vec{H} = \frac{Id\vec{l} \times \vec{R}}{4\pi R^3}. \qquad ...(3.5)$$

Thus the direction of $d\vec{H}$ can be determined by the right-hand rule with the right-hand thumb pointing in the direction of the current, the right-hand fingers encircling the wire in the direction of $d\vec{H}$. Alternatively, the direction of $d\vec{H}$ can be determined also by the right handed screw rule, with the screw placed along the wire and pointing in the direction of current flow, the direction of rotation of the screw in the direction of $d\vec{H}$ as in Fig. 3.2.

The direction of the magnetic field intensity (H or current I) is represented by a small circle with dot ⊙ if H or I is out of page or by a small circle with cross sign ⊗ if H or I is into the page.

Similar to different charge distributions as in Fig. 2.11, there are different current distributions such as line current, surface current and volume current as in Fig. 3.3.

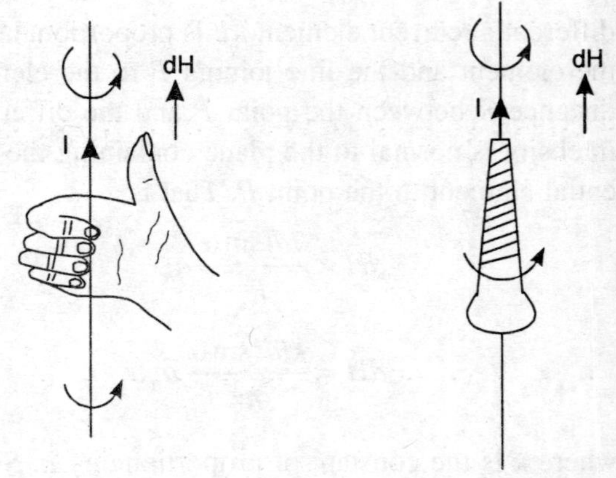

Fig. 3.2

If $\vec{K}$ is defined as the surface current density (in ampere/meter) and $\vec{J}$ as the volume current density (in ampere/m^2), the source elements are related as

$$I d\vec{l} = \vec{K} ds = \vec{J} dv. \qquad ...(3.6)$$

Thus, in terms of the distributed current sources and the Biot-Savart law, eqn. (3.4) becomes

$$\vec{H} = \int_L \frac{I d\vec{l} \times \vec{a}_R}{4\pi R^2} \qquad \text{(line current)} \qquad ...(3.7)$$

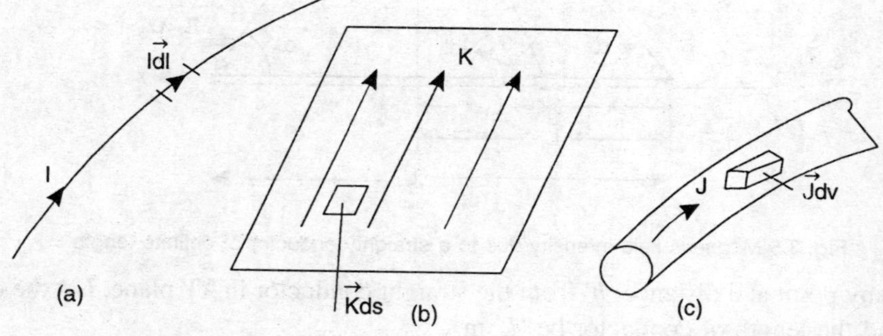

Fig. 3.3 Current distributions (*a*) Line current (*b*) Surface (*c*) Volume current

$$\vec{H} = \int_S \frac{\vec{K} ds \times \vec{a}_R}{4\pi R^2} \qquad \text{(surface current)} \qquad ...(3.8)$$

$$\vec{H} = \int_v \frac{\vec{J} dv \times \vec{a}_R}{4\pi R^2} \qquad \text{(volume control)} \qquad ...(3.9)$$

Example 3.1

Find the field intensity at a point P due to a straight conductor carrying current I as shown in Fig. 3.4.

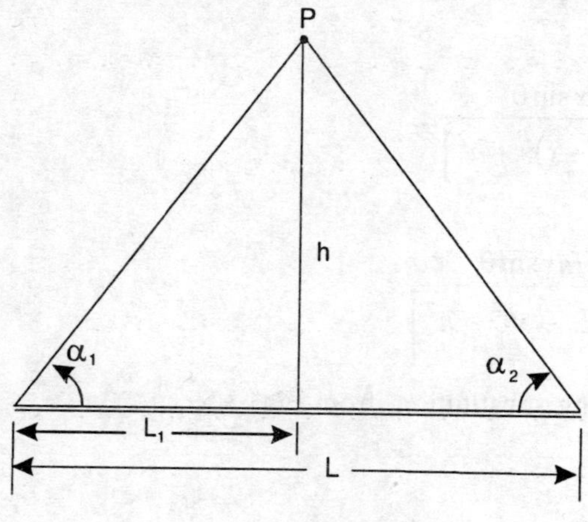

Fig. 3.4

Solution:

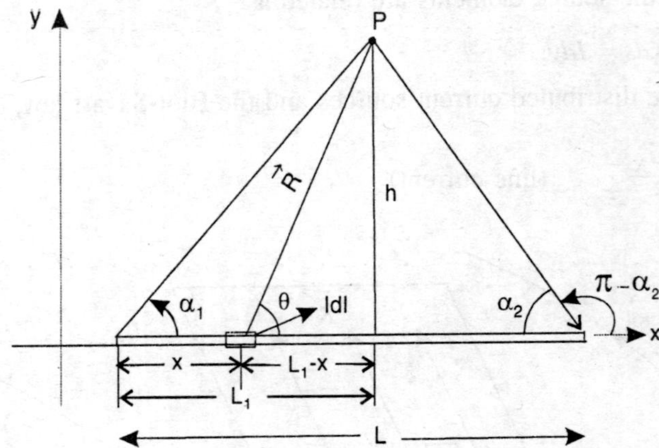

Fig. 3.5 Magnetic field intensity due to a straight conductor of definite length.

Let 'P' be any point at a distance 'h' from the straight conductor in XY plane. Let the conductor be along X-axis and the length of conductor be 'L' m.

Consider any differential current element $I\vec{dx}$ (since it is along X-axis $\vec{dl} = \vec{dx}$) at a distance x from one end of the conductor. Let the distance between this differential current element and the point 'P' be 'R' m. The angle between the line joining Idl to P and the conductor be θ (which is equal to α in eqn. 3.3).

From eqn. (3.3)

$$d\vec{H} = \frac{Idl \sin\theta}{4\pi R^2}\vec{a}_n. \qquad \text{...(3.10)}$$

Here the direction of magnetic field intensity which is normal to the plane (XY) containing the current element and the point P is along Z direction. Hence $\vec{a}_n = \vec{a}_z$.

From Fig 3.5, $R^2 = \left(L - x^2\right) + h^2$ and limits of x are from 0 to L.

Eqn (3.10) becomes

$$d\vec{H} = \frac{Idx \sin\theta}{4\pi\left[\left(L_1 - x\right)^2 + h^2\right]}\vec{a}_z$$

$$\vec{H} = \int_{x=0}^{x=L} \frac{Idx \sin\theta}{4\pi\left[\left(L_1 - x\right)^2 + h^2\right]}\vec{a}_z. \qquad \text{...(3.11)}$$

Integrating eqn. (3.11) using substitution, from Fig. 3.5

$$L_1 - x = h\cot\theta.$$

So, $\quad -dx = -h\,\text{cosec}^2\theta\, d\theta$

Also, change the limits as θ varies from α_1 to $(\pi - \alpha_2)$

$$\vec{H} = \int_{\theta=\alpha_1}^{\theta=(\pi-\alpha_2)} \frac{I \sin\theta h \operatorname{cosec}^2\theta \, d\theta}{4\pi\left[h^2 \operatorname{cosec}^2\theta\right]} \vec{a}_z$$

$$\vec{H} = \frac{I}{4\pi h} \int_{\alpha_1}^{\pi-\alpha_2} \sin\theta \, d\theta \, \vec{a}_z$$

$$\vec{H} = \frac{I}{4\pi h} [-\cos\theta]_{\alpha_1}^{\pi-\alpha_2} \vec{a}_z$$

$$\vec{H} = \frac{I}{4\pi h} \left[\cos\alpha_1 + \cos\alpha_2\right] \vec{a}_z. \quad \text{...(3.12)}$$

If the conductor is infinitely long, $\alpha_1 = \alpha_2 = 0$.

Hence substituting the above condition in eqn. (3.12)

$$\vec{H} = \frac{I}{4\pi h}(2) = \frac{I}{2\pi h} \vec{a}_z$$

$$\vec{H} = \frac{I}{2\pi h} \vec{a}_z \quad \text{...(3.13)}$$

We have already known that $\vec{E}$, due to a charged long wire along *X*-axis, is directed away and along *Y*-axis, but here we know from eqn. (3.13), $\vec{H}$ is along *Z*-axis. Hence, these two fields $\vec{H}$ and $\vec{E}$ are mutually perpendicular.

Example 3.2

Find the magnetic field intensity $\left(\vec{H}\right)$ due to an infinitely long straight conductor carrying a current *I* along *Z*-axis.

The cylindrical co-ordinates system is chosen for solving the problem.

From the Fig. 3.6, it is seen that, because of symmetry, field will not vary with *Z* or ϕ. Therefore the point *P* at which the field is to be determined is chosen in $Z = 0$ plane.

Let the point '*P*' be at a distance of 'ρ' m from the line conductor. Position vector of this point '*P*' is $\vec{r} = \rho\vec{a}_\rho$. Similarly, differential current element *dz* is chosen along *Z*-axis at a distance of '*Z*' m from 0. The position vector of this current element is $\vec{r}\,' = z\vec{a}_z$.

The line joining the differential element and the point '*P*' is $\vec{R}_{12} = \vec{r} - \vec{r}\,' = \rho_{a\bar{\rho}} - z_{a\bar{z}}$.

From Biot-Savart's law and eqn. (3.4)

$$d\vec{H} = \frac{I d\vec{l} \times \vec{a}_r}{4\pi R^2}.$$

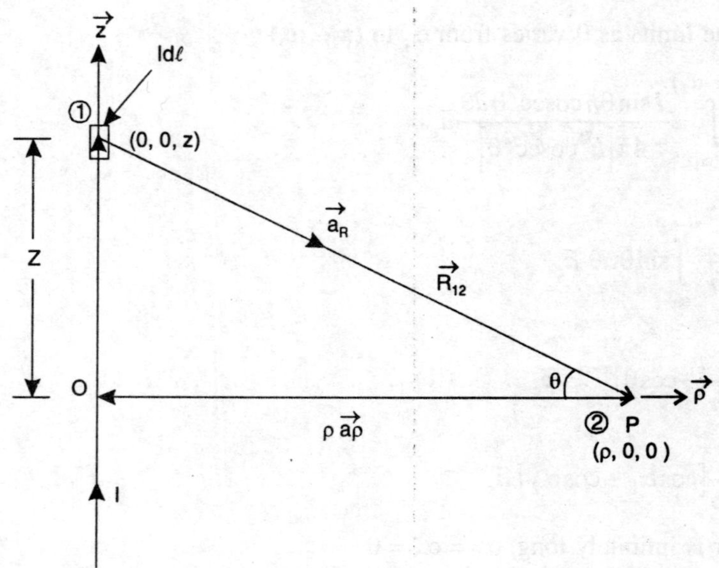

Fig. 3.6 Magnetic field due to infinite length of current firmament

Here

$$d\vec{l} = dz\vec{a}_z,\ |\vec{R}| = |\vec{R}_{12}| = R$$

and $\vec{a}_r$ is the unit vector of $\vec{R}_{12}$

$$\vec{a}_r = \frac{\vec{R}_{12}}{|\vec{R}_{12}|} = \frac{\rho\vec{a}_\rho - z\vec{a}_z}{\sqrt{\rho^2 + z^2}}.$$

Hence, $d\vec{H} = \dfrac{Idz\,\vec{a}_z}{4\pi R^2} \times \dfrac{(\rho\vec{a}_\rho - z\vec{a}_z)}{\sqrt{\rho^2 + z^2}}.$

From Fig. 3.6, $R = \sqrt{\rho^2 + z^2}$, also limits of Z from $-\infty$ to ∞.

Hence,

$$\vec{H} = \int_{z=-\infty}^{z=\infty} \frac{I\vec{a}_z \times (\rho\vec{a}_\rho - z\vec{a}_z)}{4\pi(\rho^2 + z^2)^{3/2}} dz$$

$$\vec{H} = \int_{-\infty}^{\infty} \frac{I\rho\, dz}{4\pi(\rho^2 + z^2)^{3/2}} \vec{a}_\phi \cdot \begin{bmatrix} \vec{a}_z \times \vec{a}_\rho = \vec{a}_\phi \\ \vec{a}_z \times \vec{a}_z = 0 \end{bmatrix}.$$

Integrating by substituting,

$z = \rho \tan\theta$

$$dz = \rho \sec^2 \theta \, d\theta$$

$$\vec{H} = \int_{z=-\infty}^{z=\infty} \frac{I\rho\left(\rho \sec^2 \theta \, d\theta\right)}{4\pi \rho^3 \sec^3 \theta} \vec{a}_\phi$$

$$= \frac{I}{4\pi\rho} \int_{z=-\infty}^{z=\infty} \cos\theta \, d\theta \, \vec{a}_\phi$$

$$= \frac{I}{4\pi\rho} [\sin\theta]_{z=-\infty}^{z=\infty} \, \vec{a}_\phi$$

$$= \frac{I}{4\pi\rho} \left[\frac{Z}{\sqrt{Z^2 + \rho^2}} \right]_{z=-\infty}^{z=\infty} \vec{a}_\phi$$

$$\vec{H} = \frac{I}{2\pi\rho} \vec{a}_\phi \qquad \text{...(3.14)}$$

From eqn. (3.14), the magnitude of field is not a function of ϕ or Z and it varies inversely as the distance from the current filament. The direction of the magnetic field intensity vector is circumferential. The streamlines are therefore circles about the filament and the field may be mapped in cross-section as in Fig. 3.7.

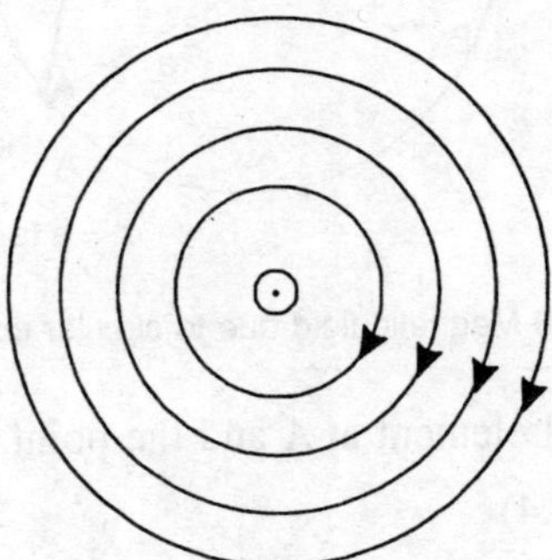

Fig. 3.7 Streamlines of the magnetic field intensity about an infinite length of conductor carrying current I into the Page.

Example 3.3

Find the magnetic field intensity on the axis of a circular current loop of radius '*a*' m. Specialize the result to the centre of the loop.

The cylindrical co-ordinates system is chosen for solving the problem.

Let the point '*P*' at which the magnetic field intensity is to be determined is at a height '*h*' m along the *Z*-axis. The *Z*-axis is passing through the centre of the current loop '*O*'. The position vector of *P* is $\vec{r} = h\vec{a}_z$.

A differential current element Idl along the circular loop at a point A is selected which is represented by $Id\vec{l} = Iad\phi\,\vec{a}_\phi$ [since differential length varying only about the circumference is given in cylindrical co-ordinate as $d\vec{l} = a\,d\phi\vec{a}_\phi$]. The position vector of A is $\vec{r}' = a\vec{a}_\rho$, where a is the radius of the circular loop.

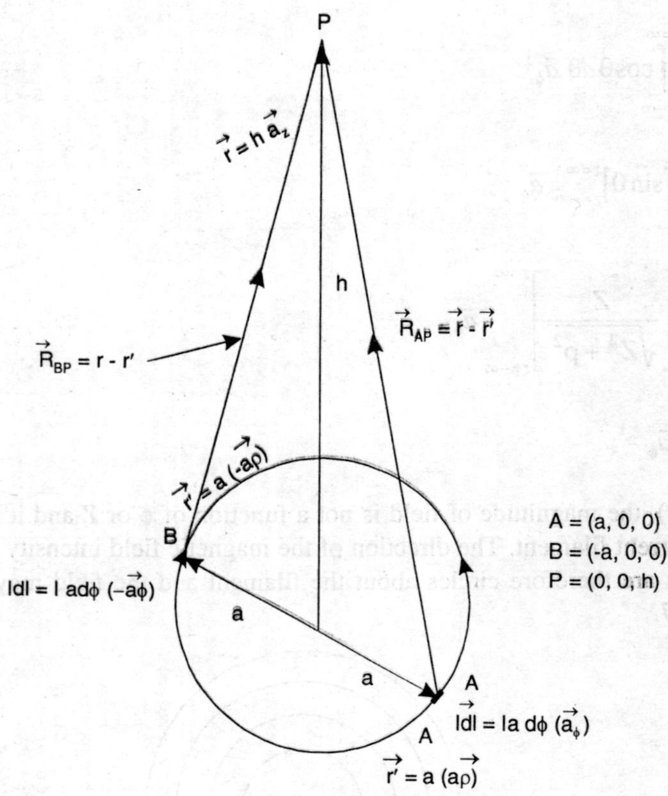

Fig. 3.8 Magnetic field due to circular current loop

The line joining the differential element at A and the point P is $\vec{R}_{AP} = \vec{r} - \vec{r}' = h\vec{a}_z - a\vec{a}_\rho$.
From Biot-Savart's law eqn. (3.4)

$$d\vec{H} = \frac{Id\vec{l} \times \vec{a}_r}{4\pi R^2},$$ where $\vec{a}_r$ is the unit vector of $\vec{R}_{AP}$.

$$\vec{a}_l = \frac{\vec{R}_{AP}}{\left|\vec{R}_{AP}\right|} \quad \text{and} \quad \left|\vec{R}_{AP}\right| = R.$$

Substitute $Id\vec{l}$ and $\vec{a}_r$ in eqn. (3.4)

$$d\vec{H}_A = \frac{Iad\phi\,\vec{a}_\phi}{4\pi R^2} \times \frac{\left[h\vec{a}_z - a\,\vec{a}_\rho\right]}{\sqrt{h^2 + a^2}}.$$

From the Fig 3.8, $R = \sqrt{h^2 + a^2}$

$$d\vec{H}_A = \frac{I a\, d\phi}{4\pi\left[a^2 + h^2\right]^{3/2}}\left[a\vec{a}_z + h\vec{a}_\rho\right] \quad \begin{bmatrix} \text{Since } \vec{a}_\phi \times \vec{a}_z = \vec{a}_\rho \\ \vec{a}_\phi \times \left(-a_\rho\right) = \vec{a}_z \end{bmatrix} \qquad \text{...(3.15)}$$

Similarly, another differential current element at B which is diametrically opposite to A is represented as $Id\vec{l} = Iad\phi\left(-\vec{a}_\phi\right)$. The position vector of B is $\vec{r}\,' = a\left(-a\vec{a}_\rho\right)$.

Now the line joining B and P is $\vec{R}_{BP} = \vec{r} - \vec{r}' = h\vec{a}_z - \left(-a\vec{a}_\rho\right)$

$$\vec{R}_{BP} = h\vec{a}_z + a\vec{a}_\rho.$$

So, $\vec{a}_r$ in eqn. (3.4) becomes

$$\vec{a}_l = \frac{\vec{R}_{BP}}{\left|\vec{R}_{BP}\right|} \quad \text{and} \quad \left|\vec{R}_{BP}\right| = R$$

$$d\vec{H}_B = \frac{I\, a\, d\phi\left(-\vec{a}_\phi\right)}{4\pi R^2} \times \frac{\left[h\vec{a}_z + \vec{a}_\rho\right]}{\sqrt{h^2 + a^2}}$$

$$d\vec{H}_B = \frac{I\, a\, d\phi\left[a\vec{a}_z - h\vec{a}_\rho\right]}{4\pi\left[a^2 + h^2\right]^{3/2}}. \qquad \text{...(3.16)}$$

The total field $d\vec{H}$ due to A and B at P is $d\vec{H} = d\vec{H}_A + d\vec{H}_B$.

Adding eqn. (3.15) and eqn. (3.16)

$$d\vec{H} = \frac{Ia^2\, d\phi}{2\pi\left(a^2 + h^2\right)^{3/2}}\vec{a}_z$$

$$\vec{H} = \int_{\phi=0}^{\phi=\pi} \frac{Ia^2}{2\pi\left[a^2 + h^2\right]^{3/2}}\,\vec{a}_z$$

$$\vec{H} = \frac{Ia^2}{2\left[a^2 + h^2\right]^{3/2}}\vec{a}_z, \qquad \text{...(3.17)}$$

when $h = 0$, that is, if the point P is at the centre of the loop, then

$$\vec{H} = \frac{I}{2a}\vec{a}_z. \qquad \text{...(3.18)}$$

Note:

If the direction of current I in the current loop is taken along the counter clockwise direction $\left(\vec{a}_\phi\right)$, then $\vec{H}$ is in the positive Z direction $\left(\vec{a}_z\right)$. Alternatively, if I is along the clockwise direction $\left(-\vec{a}_\phi\right)$, $\vec{H}$ is in the negative Z direction $\left(-\vec{a}_z\right)$.

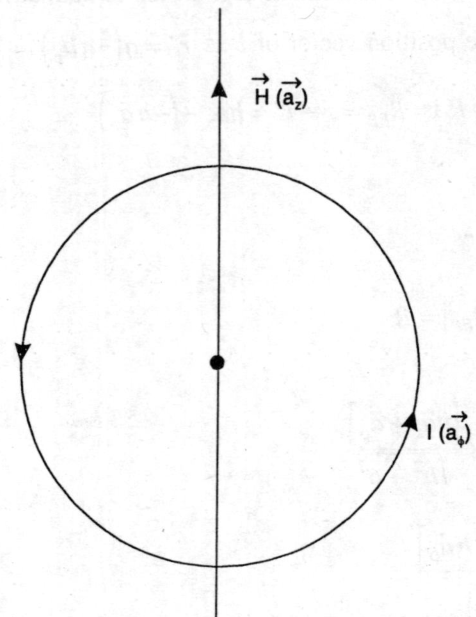

Fig. 3.9 Magnetic flux line due to circular current loop.

Example 3.4

Derive an expression for magnetic field intensity $\vec{H}$ at the centre of a circular wire carrying current I in the counter-clockwise direction. The radius of the circle is 'a' m and the wire is in XY plane.

Let the point 'P' be at the centre of the circular loop. A differential current element $\vec{dl} = ad\phi\,\vec{a}_\phi$ is selected at a point 'A' along the circular loop. The line joining the point A (a, 0, 0) and P (0, 0, 0) is $\vec{R} = -a\,\vec{a}_\rho$.

The unit vector of $\vec{R}$ is $\vec{a}_r = \dfrac{\vec{R}}{\left|\vec{R}\right|}$, where $\left|\vec{R}\right| = a$

$$\vec{a}_r = \frac{-a\,\vec{a}_\rho}{a}$$

$$d\vec{H} = \frac{Id\vec{l} \times \vec{a}_r}{4\pi R^2}$$

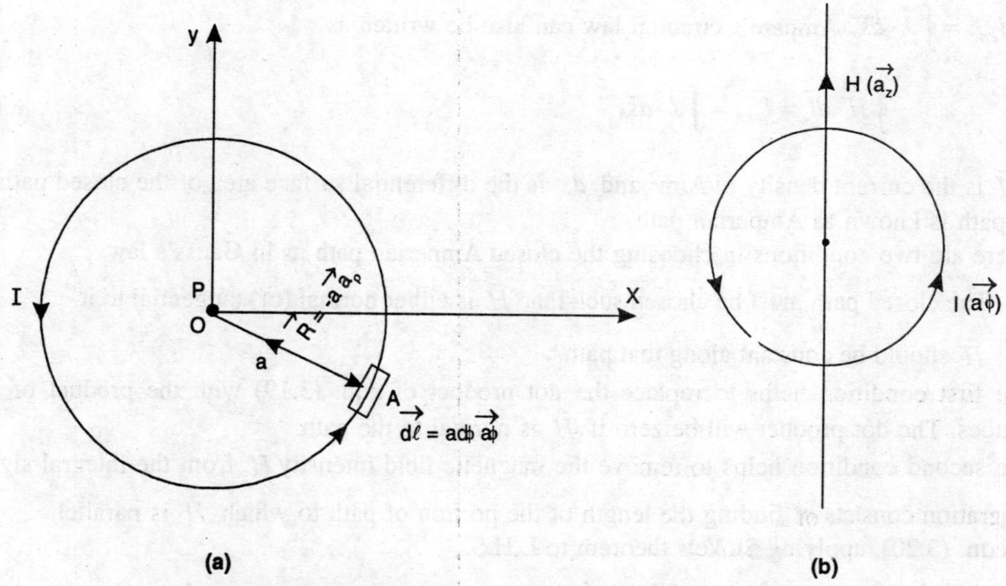

Fig. 3.10 Magnetic field at the centre due to a circular current loop.

$$= \frac{Ia\,d\phi\,\vec{a}_\phi}{4\pi a^2} \times \frac{\left(-a\,\vec{a}_\rho\right)}{a}$$

$$d\vec{H} = \frac{Ia^2\,d\phi\,\vec{a}_z}{4\pi a^3}$$

$$\vec{H} = \int_{\phi=0}^{\phi=2\pi} \frac{I}{4\pi a}\,d\phi\;\vec{a}_z$$

$$\vec{H} = \frac{I}{4\pi a}[\phi]_0^{2\pi}\;\vec{a}_z$$

$$\vec{H} = \frac{I}{2a}\vec{a}_z.$$

3.2 AMPERE'S CIRCUITAL LAW

Ampere's circuital law helps to determine the magnetic field intensity more easily than Biot-Savart's law. This law is also known as Ampere's work law and it is derived from Biot-Savart's law. This law is applicable only for the symmetrical current distribution systems.

Ampere's circuital law states that the line integral of $\vec{H}$ about any closed path is exactly equal to the direct current enclosed by that path. That is,

$$\oint \vec{H} \cdot d\vec{l} = I_{enc}. \qquad ...(3.19)$$

Since $I_{enc} = \int \vec{J} \cdot d\vec{s}$, Ampere's circuital law can also be written as

$$\oint \vec{H} \cdot d\vec{l} = I_{enc} = \int \vec{J} \cdot d\vec{s}, \qquad ...(3.20)$$

where $\vec{J}$ is the current density in A/m^2 and $\vec{ds}$ is the differential surface area of the closed path. This closed path is known as Amperian path.

There are two conditions in choosing the closed Amperian path as in Gauss's law.

1. The closed path must be chosen such that $\vec{H}$ is either normal (or) tangential to it
2. $\vec{H}$ should be constant along that path.

The first condition helps to replace the dot product of eqn. (3.19) with the product of scalar magnitudes. The dot product will be zero if $\vec{H}$ is normal to the path.

The second condition helps to remove the magnetic field intensity $\vec{H}$ from the integral sign and the integration consists of finding the length of the portion of path to which $\vec{H}$ is parallel.

In eqn. (3.20), applying Stoke's theorem to L.H.S.

$$\oint_r \vec{H} \cdot d\vec{l} = \int_S \left(\nabla \times \vec{H}\right) \cdot d\vec{s} = \int_S \vec{J} \cdot d\vec{s}$$

$$\nabla \times \vec{H} = \vec{J} \qquad ...(3.21)$$

The above equation is differential form of Ampere's circuital law.

Example 3.5

Derive an expression for the magnetic field intensity $\vec{H}$ due to an infinitely long current carrying filament placed along Z-axis using Ampere's circuital law.

The filament lies on the Z-axis in free space and the current flows along the Z-direction $(\vec{a}_z)$. As in Example 3.2, symmetry shows that there is no variation with Z or ϕ. It is also known from Biot-Savart's law that direction of dH is normal to the plane containing $d\vec{l}$ and $\vec{R}$ and so in the direction of $\vec{a}_\phi$. Hence, the only component of $\vec{H}$ is H_ϕ and it is function of only ρ, where ρ is the distance of point at which the magnetic field is to be determined. So, the closed Amperian path must be a circle of radius ρ.

From the Ampere circuital law

$$\oint H \cdot dl = I_{enc}$$

where I_{enc} is the current enclosed by the Amperian path which is equivalent to 'I'.

$$\oint \vec{H}_\phi \vec{a}_\phi \cdot \rho d\phi \vec{a}_\phi = I \qquad ...(3.22)$$

$$\oint_{\phi=0}^{\phi=2\pi} \vec{H}_\phi \rho d\phi = I$$

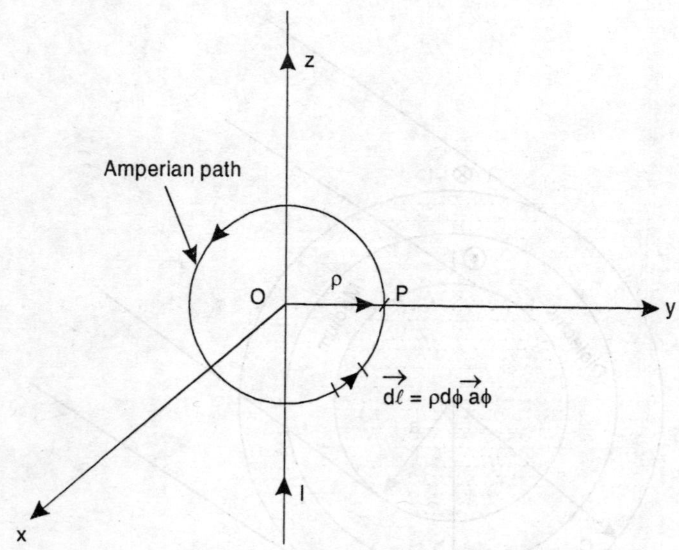

Fig. 3.11 Ampere's law applied for an infinite filament.

$$\vec{H}_\phi \, \rho(2\pi) = I \quad \text{...(3.23)}$$

$$\vec{H}_\phi = \frac{I}{2\pi\rho} \quad \text{...(3.24)}$$

$$\vec{H} = \frac{I}{2\pi\rho}\vec{a}_\phi . \quad \text{...(3.25)}$$

Example 3.6

Currents in the inner and outer conductor of a co-axial cable shown in Fig. 3.12 are uniformly distributed. Use Ampere's law to show that for $b \le \rho \le C$.

$$\vec{H} = \frac{I}{2\pi\rho}\left[\frac{c^2 - \rho^2}{c^2 - b^2}\right]\vec{a}_\phi .$$

Assume the co-axial cable is placed along Z axis and it is carrying $+I$ amp in the inner conductor of radius 'a' and $-I$ amp in the outer conductor having inner radius 'b' and outer radius 'c'.

The qualitative analysis made in the Example (3.5) is also applicable for this example. Hence, H_ϕ is the only component of $\vec{H}$ and it varies with ρ.

The Ampere's circuital law is applied for the four possible cases.

Case i: $\rho < a$ (Inside the inner conductor)

In this case the selected closed Amperian path (L_1) is of radius 'ρ', which is smaller than the conductor radius 'a'.

The current enclosed by this path (I_{enc}) can be determined using the eqn. (3.20)

$$\oint_l H \cdot dl = I_{enc} = \int_S \vec{J} \cdot d\vec{s} .$$

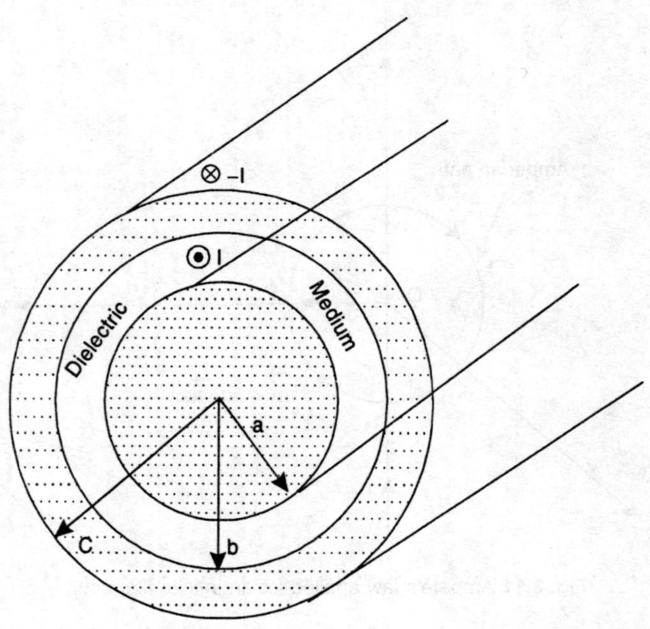

Fig. 3.12 Co-axial cable.

From the Fig. 3.13,

$$I_{enc} = \int_S J\vec{a}_z \cdot ds\,\vec{a}_z \qquad \text{...(3.26)}$$

$$= J\int ds. \qquad \text{...(3.27)}$$

The surface area enclosed by the Amperian path is $\pi\rho^2$ and $J = \dfrac{I}{\pi a^2}$ A/m^2.

$$\therefore \qquad I_{enc} = \frac{I}{\pi a^2}\cdot \pi\rho^2$$

$$I_{enc} = \frac{I\rho^2}{a^2}. \qquad \text{...(3.28)}$$

As from Example 3.5,

$$\oint \vec{H}\cdot dl = H_\phi(2\pi\rho) \qquad \text{...(3.29)}$$

Substitute eqn. (3.28) and eqn. (3.29) in eqn. (3.20)

$$\vec{H}_\phi(2\pi\rho) = \frac{I\rho^2}{a^2}$$

$$\vec{H}_\phi = \frac{I\rho}{2\pi a^2}$$

$$\vec{H} = \frac{I\rho}{2\pi a^2}\vec{a}_\phi. \qquad ...(3.30)$$

Case ii: $a < \rho < b$ (outside the inner conductor).

For the selected Amperian path L_2, the result is same as in Example 3.5.

$$\vec{H} = \frac{I}{2\pi\rho}\vec{a}_\phi$$

Case iii: $b < \rho < c$ (At a point in the outer conductor).

Here the selected Amperian path (L_3) is of radius ρ which is greater than b and less than c as in Fig. (3.13).

Using Ampere's circuital law, $\oint \vec{H}\cdot d\vec{l} = I_{enc}$.

As in previous cases, the left hand side expression $\oint \vec{H}\cdot d\vec{l} = H_\phi(2\pi\rho)$

$$I_{enc} = I + \int_S \vec{J}\cdot d\vec{s}, \qquad ...(3.31)$$

where I is the current in the inner conductor and $\vec{J}$ is the current density of the outer conductor which is along $(-\vec{a}_z)$.

Current density $\vec{J}$ for outer conductor is given by $\dfrac{I}{(\pi c^2 - \pi b^2)}(-\vec{a}_z)$.

$d\vec{s}$ is the differential surface area of Amperian path of radius ρ. [along $\vec{a}_z$].

So, $$\int_S \vec{J}\cdot d\vec{s} = \int J(-\vec{a}_z)\cdot ds\,\vec{a}_z$$

$$= -\vec{J}\int_S d\vec{s}$$

The total surface area of the Amperian path is $(\pi\rho^2 - \pi b^2)$

$$\int_S \vec{J}\cdot d\vec{s} = -J(\pi\rho^2 - \pi b^2)$$

$$= -\frac{I}{(\pi c^2 - \pi b^2)}\cdot(\pi\rho^2 - \pi b^2)$$

$$\int_S \vec{J}\cdot d\vec{s} = -\frac{I(\rho^2 - b^2)}{(c^2 - b^2)}. \qquad ...(3.32)$$

Eqn. (3.32) is substituted in eqn. (3.31)

$$I_{enc} = I - \frac{I\left(\rho^2 - b^2\right)}{\left(c^2 - b^2\right)}$$

$$= \frac{I\left(c^2 - \rho^2\right)}{c^2 - b^2}$$

$$H_\phi(2\pi\rho) = \frac{I\left(c^2 - \rho^2\right)}{\left(c^2 - b^2\right)}$$

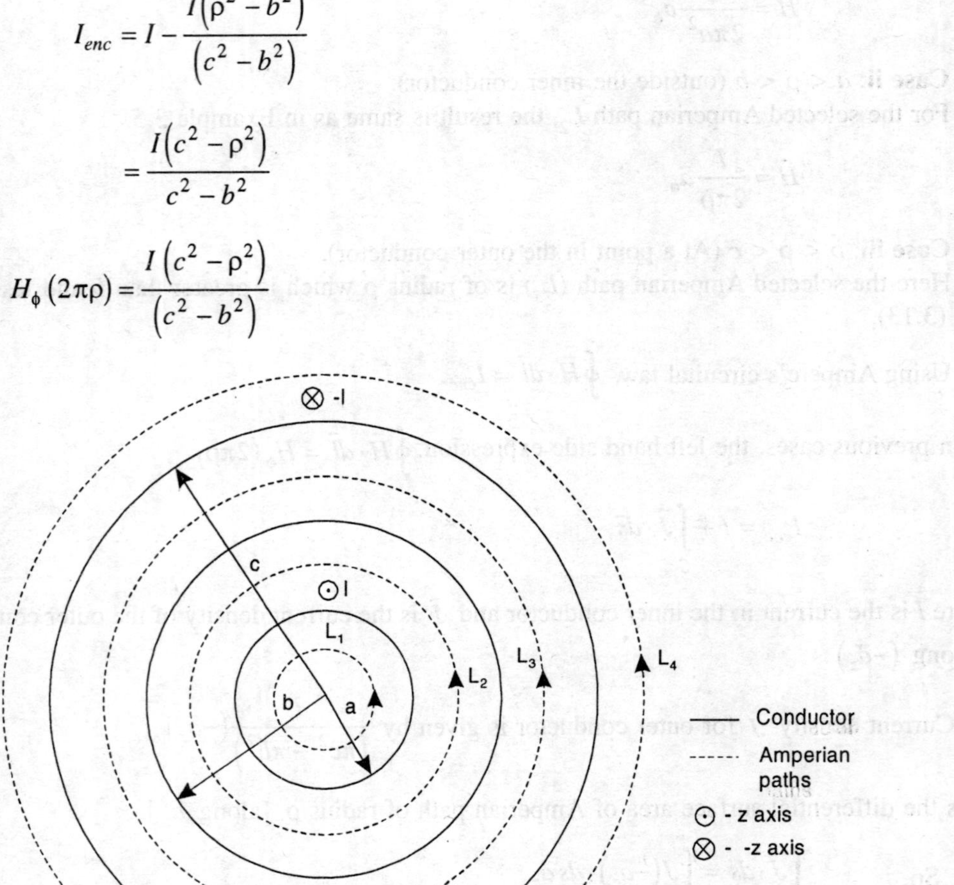

Fig. 3.13 Cross-section of co-axial cable

$$H_\phi = \frac{I}{2\pi\rho}\frac{\left(c^2 - \rho^2\right)}{\left(c^2 - b^2\right)}$$

$$\vec{H} = \frac{I}{2\pi\rho}\frac{\left(c^2 - \rho^2\right)}{\left(c^2 - b^2\right)}\vec{a}_\phi \qquad ...(3.33)$$

Case iv: $\rho > c$.

For this region of Amperian path (L_4) $I_{enc} = I - I$

$= 0$

$$\therefore \quad \oint H \cdot dl = I_{enc} = 0 \qquad \text{...(3.34)}$$

$$\vec{H} = 0.$$

The variation of $\vec{H}$ with the distance 'ρ' is plotted in Fig. 3.14.

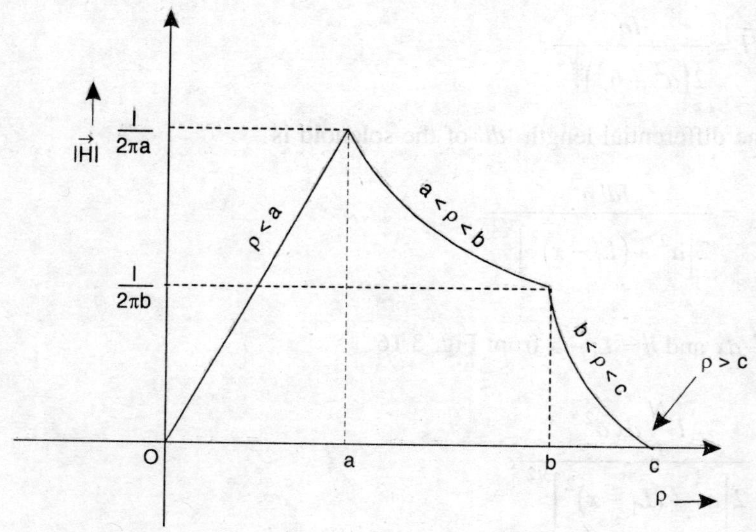

Fig. 3.14 Variation of $\vec{H}$ with ρ

Example 3.7

Derive the general expression for the field B at any point along the axis of a solenoid (uniform cylindrical coil wound on a non-magnetic frame). The length of the solenoid is L m and radius is 'a' m and it consists of N turns of wire carrying a current of I amp.

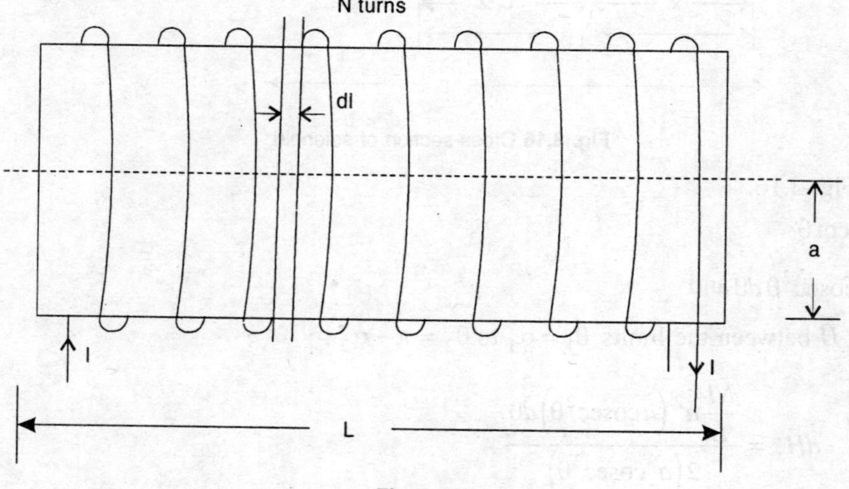

Fig. 3.15

Consider the cross-section of the solenoid as in Fig. 3.15. Since the solenoid consists of circular loops, the result of the Example 3.3 can be applied here. The contribution to the magnetic field *H* at *P* by an element of the solenoid of length *dx* is obtained as follows.

Consider a small length *dx* at a distance *x* m from one end of the solenoid. Let the point '*P*' be at a distance of L_1 m and the angles made by the two ends of the solenoid at *P* is α_1 and α_2 as in Fig. 3.16.

From the eqn. (3.17), the magnetic field due to one circular loop of radius '*a*' is

$$\vec{H} = \frac{Ia}{2\left(a^2 + h^2\right)^{3/2}}.$$

Now due to the differential length '*dl*' of the solenoid is

$$d\vec{H}_x = \frac{Idl\, a^2}{2\left[a^2 + (L_1 - x)^2\right]^{3/2}}, \qquad \text{...(3.35)}$$

where $d\vec{l} = \frac{N}{L}dx$ and $h = L_1 - x$ from Fig. 3.16

$$d\vec{H}_x = \frac{I\frac{N}{L}dx\, a^2}{2\left[a^2 + (L_1 - x)^2\right]^{3/2}}. \qquad \text{...(3.36)}$$

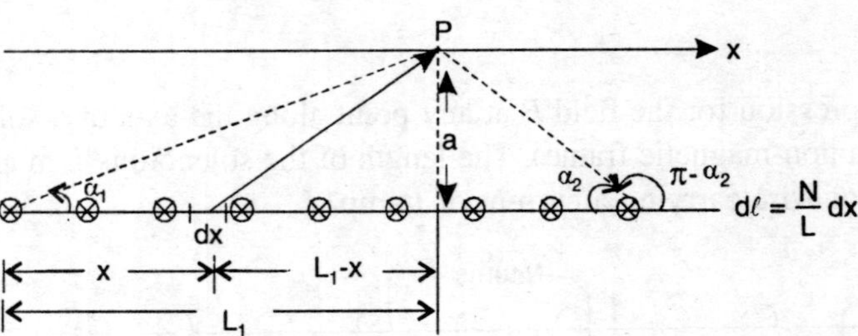

Fig. 3.16 Cross-section of solenoid

From the Fig. 3.16,

$L_1 - x = a\cot\theta$

$-dx = -a\,\text{cosec}^2\theta\, d\theta$ and

integrating *H* between the limits $\theta_1 = \alpha_1$ to $\theta_2 = \pi - \alpha_2$.

$$d\vec{H}_x = \frac{\frac{NI}{L}a^2\left(a\,\text{cosec}^2\theta\right)d\theta}{2\left(a^3\text{cosec}^3\theta\right)}$$

$$d\vec{H}_x = \frac{NI}{2L}\sin\theta\, d\theta$$

$$\vec{H}_x = \int_{\theta=\alpha_1}^{\theta=\pi-\alpha_2} \frac{NI}{2L}\sin\theta\, d\theta$$

$$= \frac{NI}{2L}[-\cos\theta]_{\alpha_1}^{\pi-\alpha_2}$$

$$\vec{H} = \frac{NI}{2L}[\cos\alpha_1 + \cos\alpha_2]\vec{a}_x. \qquad ...(3.37)$$

Case I

Let the point P be at one of the ends of the solenoid.

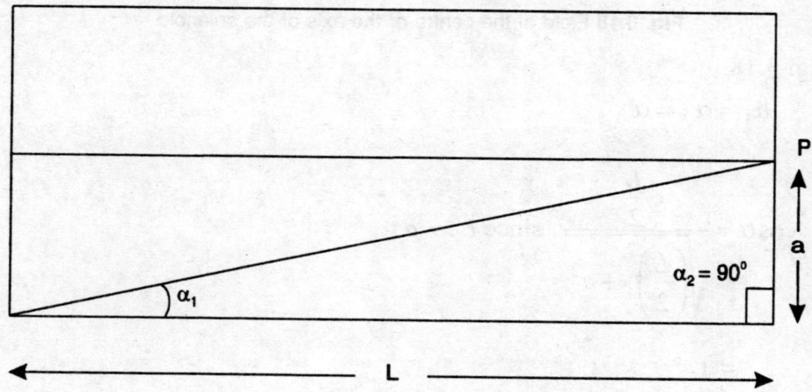

Fig. 3.17 Field at one of end of the solenoid.

From the Fig. 3.17,

$$\alpha_2 = 90° \text{ and}$$

$$\cos\alpha_1 = \frac{L}{\sqrt{L^2 + a^2}}.$$

Eqn. (3.37) becomes

$$\vec{H} = \frac{NI}{2L}\cos\alpha_1$$

$$= \frac{NI}{2L}\cdot\frac{L}{\sqrt{L^2 + a^2}}.$$

Since $L >> a$

$$\vec{H} = \frac{NI}{2L}\vec{a}_x \qquad ...(3.38)$$

Case II

Let '*P*' be at the centre.

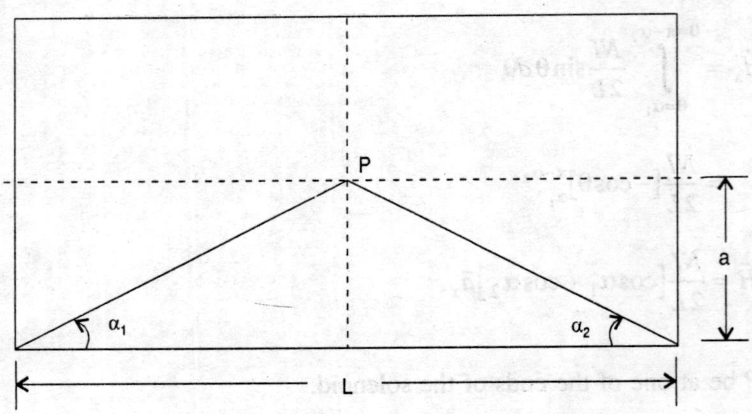

Fig. 3.18 Field at the centre of the axis of the solenoid.

From the Fig. 3.18,

$$\alpha_1 = \alpha_2 = \alpha$$

$$\cos\alpha = \frac{\frac{L}{2}}{\sqrt{\left(\frac{L}{2}\right)^2 + a^2}}, \text{ since } L >> a$$

$$= 1$$

$$\therefore \quad \vec{H} = \frac{NI}{L}\vec{a}_x. \qquad ...(3.39)$$

Example 3.8

A toroid whose dimensions are shown in Fig. 3.19 has *N* turns and carries current of *I* amp. Determine $\vec{H}$ inside and outside the toroid.

The Ampere's circuital law is applied to the Amperian path which is a circle of radius ρ shown dotted in Fig. 3.19. Since *N* wires cut through this path each carrying current *I*, the net current enclosed by the Amperian path is *NI*. Hence

$$\oint_l \vec{H} \cdot d\vec{l} = I_{enc}$$

$$\vec{H} 2\pi\rho = NI$$

$$\vec{H} = \frac{NI}{2\pi\rho}\vec{a}_\phi$$

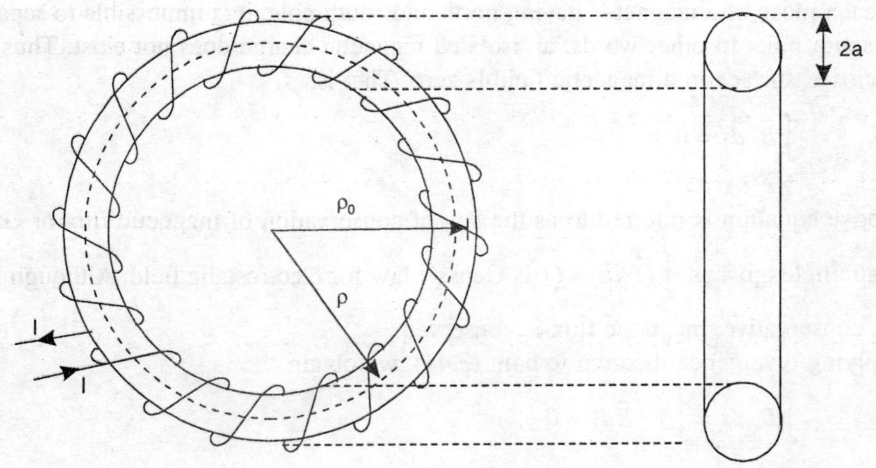

Fig. 3.19

Approximate value of $\vec{H} = \frac{NI}{2\pi\rho_0} = \frac{NI}{l_m}\vec{a}_\phi$, ...(3.40)

where ρ_0 is the mean radius of toroid and l_m is mean circumference.

At any point outside the toroid, the current enclosed by an Amperian path is zero. Hence $\vec{H} = 0$.

3.3 MAGNETIC FLUX DENSITY

The magnetic flux density $\vec{B}$ which is similar to electric flux density $\vec{D}$ is related to magnetic field intensity $\vec{H}$ as

$$\vec{B} = \mu_0 \vec{H}, \quad \text{...(3.41)}$$

where μ_0 is a constant known as the permeability of free space. The constant is in henry/meter and has the value of

$$\mu_0 = 4\pi \times 10^{-7} \text{ H/m}. \quad \text{...(3.42)}$$

The magnetic flux through a surface S is given by

$$\psi = \int_s \vec{B} \cdot d\vec{s}, \quad \text{...(3.43)}$$

where the magnetic flux ψ is in wb and the magnetic flux density is in wb/m² or tesla.

In electrostatic field, the flux passing through a closed surface is the same as the charge enclosed, that is,

$$\psi = \oint \vec{D} \cdot d\vec{s} = Q.$$

Thus it is possible to have an isolated electric charge and also the electric flux lines which are not closed. But the magnetic flux lines always close upon themselves. This is due to the fact that it is not possible to have isolated magnetic poles (or magnetic charges).

Since each piece of a magnet is having north and south pole, it is impossible to separate the north pole from south pole. In other words, an isolated magnetic charge does not exist. Thus the total flux through a closed surface in a magnetic field is zero. That is,

$$\oint_s \vec{B} \cdot d\vec{s} = 0. \qquad \text{...(3.44)}$$

The above equation is referred to as the law of conservation of magnetic flux or Gauss's law for magnetostatic fields just as $\oint \vec{D} \cdot d\vec{s} = Q$ is Gauss's law for electrostatic field. Although magnetostatic field is not conservative, magnetic flux is conserved.

By applying divergence theorem to eqn. (3.44), we obtain

$$\oint_s \vec{B} \cdot d\vec{s} = \int_v \nabla \cdot \vec{B} dv = 0$$

(or) $$\nabla \cdot \vec{B} = 0 \qquad \text{...(3.45)}$$

The above eqn. (3.45) shows that the magnetostatic fields have no sources or links and it is solenoidal.

3.4 MAGNETIC POTENTIALS

3.4.1 Magnetic Vector Potential:

In vector analysis, two vector identities are particularly significant in understanding and deriving electromagnetic potential. The identities are from eqn. (1.67) and eqn. (1.68),

$$\text{curl grad } \phi = \nabla \times \nabla \phi = 0 \qquad \text{...(3.46)}$$

where ϕ is a scalar function

and $$\text{div curl } \vec{A} = \nabla \cdot \left(\nabla \times \vec{A}\right) = 0 \qquad \text{...(3.47)}$$

where $\vec{A}$ is a vector function.

From the first vector identity eqn. (3.46), we can state that, if $\nabla \times \vec{F} = 0$ (F is irrotational), there exists a scalar ϕ such that $\vec{F} = \nabla \phi$.

So, the identity guarantees that an irrotational field can always be derived from a scalar field usually called a scalar potential.

For example, the irrotational electrostatic field $\vec{E}$ can be derived from a scalar field or scalar potential V is volts.

From eqn. (2.94) $\vec{E} = -\nabla V$.

The scalar sources for this field are given by taking the divergence of $\vec{E}$.

$$\text{Scalar scources} = \nabla \cdot \vec{E} = \nabla \cdot (-\nabla V) = -\nabla^2 V = \frac{\rho}{\varepsilon}. \qquad \text{...(3.48)}$$

The above relation between $\vec{E}$ and charge density ρ is given by Gauss's law in point form.

We know that eqn. (3.48) is Poisson's eqn. as in eqn. (2.104).

Similarly, from the second identity (3.47), we can state that if $\nabla \cdot \vec{F} = 0$ ($\vec{F}$ is solenoidal), there exists a vector $\vec{A}$ such that $\vec{F} = \nabla \times \vec{A}$.

This identity implies that a solenoidal field can be derived from a vector field, usually called a vector potential.

As magnetic field is solenoidal, from eqn. (3.45), a similar procedure can be followed to define the vector potential $\vec{A}$.

The vector sources for $\vec{B}$ are obtained by taking the curl of $\vec{B}$.

$$\text{Vector source} = \nabla \times \vec{B} = \nabla \times \nabla \times \vec{A} = \mu_0 \vec{J}, \quad ...(3.49)$$

where the relationship between $\vec{B}$ and $\vec{J}$ is given by Ampere's circuital law in point form.

From the vector identity

$\nabla \times (\nabla \times \vec{A}) = \nabla \times (\nabla \cdot \vec{A}) - \nabla^2 \vec{A}$, eqn. (3.49) becomes

$$\nabla(\nabla \cdot \vec{A}) - \nabla^2 \vec{A} = \mu_0 \vec{J}.$$

With the purpose of simplifying the above equation, the divergence of $\vec{A}$ is chosen as zero *i.e.*,

$$\nabla \cdot \vec{A} = 0. \quad ...(3.50)$$

The above condition eqn. (3.50) is called Coulomb's condition or Coulomb gauge.

Hence, eqn. (3.49) becomes

$$\nabla^2 \vec{A} = \mu_0 \vec{J}. \quad ...(3.51)$$

The above equation is vector Poisson's equation.

In cartesian co-ordinates, eqn. (3.51) is equivalent to three scalar Poisson's equations

$$\nabla^2 \vec{A}_x = -\mu_0 \vec{J}_x$$

$$\nabla^2 \vec{A}_y = -\mu_0 \vec{J}_y$$

$$\nabla^2 \vec{A}_z = -\mu_0 \vec{J}_z.$$

Each of these equations is similar to Poisson's eqn. (2.104) for the electrostatic potential V,

$$\nabla^2 V = \frac{-\rho_0}{\varepsilon} \quad ...(3.52)$$

which has the general solution in eqn. (2.89).

$$V = \frac{1}{4\pi\varepsilon} \int_V \frac{\rho\, dv}{r}, \quad ...(3.53)$$

where ρ is volume charge density.

Note: If surface charge or line charge distribution is given, eqn. (2.87) or eqn. (2.88) can be taken as solution.

By comparison to eqns. (3.52) and (3.53), the solutions for the component equations of eqn. (3.51) are:

$$A_x = \frac{\mu_0}{4\pi}\int_V \frac{J_x dv}{r}$$

$$A_y = \frac{\mu_0}{4\pi}\int_V \frac{J_y dv}{r}$$

$$A_z = \frac{\mu_0}{4\pi}\int_V \frac{J_z dv}{r} \qquad \text{...(3.54)}$$

Combining the three solutions, the vector potential

$$\vec{A} = A_x\vec{a}_x + A_y\vec{a}_y + A_z\vec{a}_z$$

$$= \frac{\mu_0}{4\pi}\int_V \frac{\vec{J}\,dv}{r} \text{ wb/m.} \qquad \text{...(3.55)}$$

$\vec{A}$ can also be defined as

$$\vec{A} = \frac{\mu_0}{4\pi}\int_S \frac{\vec{K}\,ds}{r} \text{ wb/m for surface current} \qquad \text{...(3.56)}$$

$$\vec{A} = \frac{\mu_0}{4\pi}\int_r \frac{\vec{I}\,dl}{r} \text{ wb/m for line current.}$$

The magnetic vector potential will serve a pivotal role in the development of magnetic material effects. It will also serve ancillary roles in the computation of the flux B through an open surface and stored magnetostatic energy. However, computing $\vec{A}$ first and then $\vec{B}$ is generally as cumbersome as computing $\vec{B}$ directly from the current.

3.4.2 Magnetic Scalar Potential:

Just as the electric potential V is related to electric field intensity $\vec{E}$, $\vec{E} = -\nabla V$, it is possible to define a scalar potential associated with magnetostatic field $\vec{B}$.

The scalar magnetic potential is designated as V_m (ampere) and is related with magnetic field $\vec{H}$ as $\vec{H} = -\nabla V_m$.

This definition must not conflict with our previous results for magnetostatic field.

But from the identity (3.46),

$$\nabla \times (-\nabla V_m) = \nabla \times \vec{H} = 0$$

which is contradictory to the Ampere's circuital law eqn. (3.21), $\nabla \times \vec{H} = \vec{J}$.

Hence, magnetic field intensity can be related to scalar magnetic potential, only in the region where current density is zero or current free region *i.e.*,

$$\vec{H} = -\nabla V_m \ \left(\vec{J} = 0\right). \quad ...(3.57)$$

Combining eqns. (3.57) and (3.21)

$$\nabla \times \vec{H} = \nabla \times \left(-\nabla V_m\right) = J = 0$$

$$\nabla^2 V_m = 0 \ (J = 0). \quad ...(3.58)$$

We can see that V_m satisfies Laplace's equation in current-free region.

Solution of this Laplace's equation is unique, only if the region is current free.

As already mentioned, in a region where current exists, the magnetic field is not conservative and the scalar magnetic potential V_m is not a single-valued function.

It will be observed from eqn. (3.57) that the magnetic scalar potential difference between two points in free space can be expressed as

$$V_{m_2} - V_{m_1} = -\int_1^2 \vec{H} \cdot d\vec{l}. \quad ...(3.59)$$

The magnetic potential difference evaluated by the eqn. (3.59) depends on the path of integration.

Thus if the path of integration encircles a current, V_m does not have a unique value. In eqn. (3.59), if 1 and 2 are the same points in space and the path of integration encloses a current I, then two values of V_m differing by I will be assigned to the point. To make the scalar magnetic potential unique, only the regions which do not entirely encircle currents are to be considered. The regions are called 'simply connected', because any two paths connecting a pair of points in the region form a loop which does not enclose any exterior points.

An example of a simply-connected region between co-axial conductors is shown in Fig. 3.19.

Example 3.9

Prove that scalar magnetic potential V_m is a multivalued function of ϕ, the given co-axial line conductor in the region $a < \rho < b$.

The scalar magnetic potential can be determined for the given co-axial line conductor, in the region $a < J < b$, where $J = 0$.

From Example 3.6, $\vec{H} = \dfrac{I}{2\pi\rho}\vec{a}_\phi$,

where I is the total current flowing in $\vec{a}_z$ direction in the inner conductor.

$$\vec{H} = -\nabla V_m = \frac{I}{2\pi\rho}.$$

As V_m is given as a function of ϕ, the gradient of V_m can be taken in one-dimensional as

$$-\frac{1}{\rho}\frac{\partial V_m}{\partial \phi} = -\frac{I}{2\pi\rho}$$

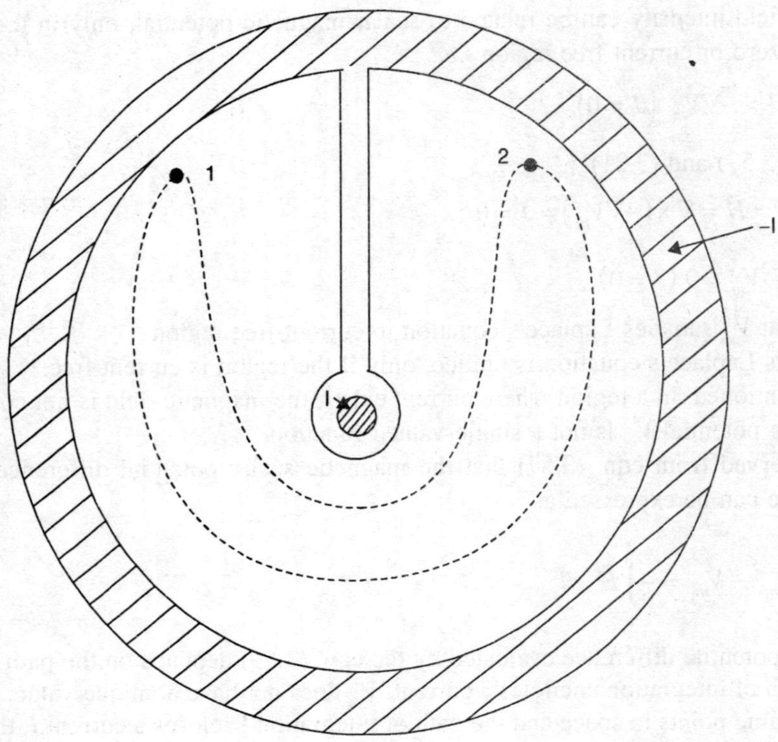

Fig. 3.20 Simply connected region between coaxial cylinders

$$\frac{\partial V_m}{\partial \phi} = -\frac{I}{2\pi}$$

$$V_m = -\frac{I}{2\pi}\phi,$$

where the constant of integration has been set equal to zero.

Let V_m be zero at P_1 where $\phi = 0$ and as we proceed counterclockwise around the circle, the magnetic potential goes negative linearly. After completing one revolution, at point P_1 ($\phi = 2\pi$), now V_m becomes $-I$ which was originally assigned to zero.

Similarly at P_2 $\left(\phi = \frac{\pi}{4}, \frac{9\pi}{4}, \frac{17\pi}{4}, \ldots \text{(or)} \frac{-7\pi}{4}, \frac{-15\pi}{4}, \frac{-23\pi}{4} \ldots\right)$

$$V_{mP_2} = \frac{I}{2\pi}\left(2n - \frac{1}{4}\right)\pi \quad n = 0, \pm 1, \pm 2.$$

Thus the magnetic scalar potential is multivalued function of ϕ in the region $a < J < b$.

The important reason for the multiple solution is that, the current I in the inner conductor is completely enclosed for every counterclockwise encirclement around it.

Let us consider a barrier (branch cut) at $\phi = \pi$ in Fig. 3.21.

Now the selected path will not cross the barrier and so the current in the inner conductor will not be encircled *i.e.*, $-\pi < \phi < \pi$.

Scalar magnetic potential $V_m = -\frac{I}{2\pi}\phi$

$$-\pi < \phi < \pi$$

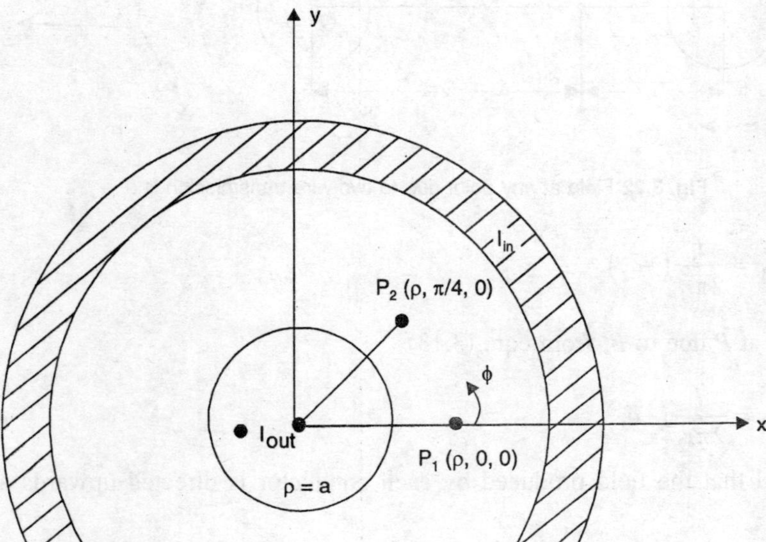

Fig. 3.21 Cross-section of coaxial line

At P_1,

$$V_m P_1 = 0 \ (\phi = 0).$$

At P_2,

$$V_m P_2 = -\frac{I}{8}\left(\phi = \frac{\pi}{4}\right).$$

Thus the magnetic scalar potential has unique value for $-\pi < \phi < \pi$.

Example 3.10

A single phase, two-wire transmission line circuit comprises two parallel conductors A and B, 1 cm diameter and spaced 1m apart. The conductors carry current of +100 A and −100 A respectively. Determine $\vec{H}$ at the surface of each conductor and also in the space exactly midway between A and B.

The field at any point P between A and B is $\vec{H} = \vec{H}_A + \vec{H}_B$.

$\vec{H}_A$ is the field at P due to A. From eqn. (3.13)

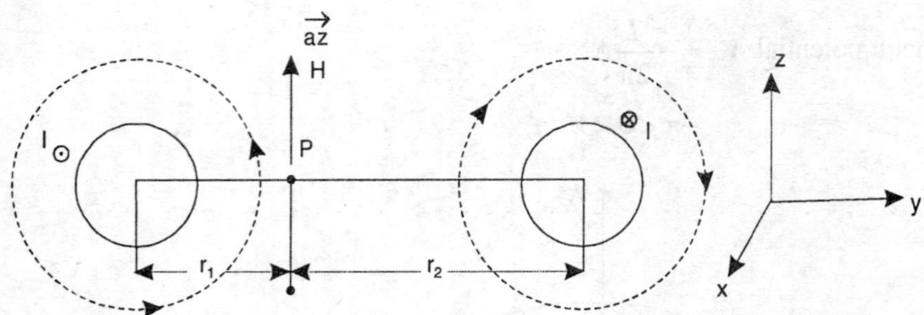

Fig. 3.22 Field at any point due to two-wire transmission line

$$\vec{H}_A = \frac{I}{2\pi r_1}(+\vec{a}_z).$$

$\vec{H}_B$ is the field at P due to B, from eqn. (3.13)

$$\vec{H}_B = \frac{I}{2\pi r_2}(+\vec{a}_z).$$

It is to be noted that the field produced by each conductor is directed upwards at any point P between A and B.

$$\vec{H} = \frac{I}{2\pi}\left[\frac{1}{r_1}+\frac{1}{r_2}\right][+\vec{a}_z].$$

Taking only the magnitudes

$$H = \frac{I}{2\pi}\left[\frac{1}{r_1}+\frac{1}{r_2}\right]. \qquad ...(3.60)$$

To determine the field at the conductor surface, consider the point at the surface of A.
In eqn. (3.56), substitute $r_1 = 0.5$ cm

$$= 0.5 \times 10^{-2} \text{ m}$$

$$r_2 = 1 \text{ m}.$$

$$H = \frac{100}{2\pi}\left[\frac{10^2}{0.5}+1\right]$$

$$= \frac{10^4}{\pi} \text{ A/m}.$$

Field at the midway between A and B:
Substitute $r_1 = r_2 = 0.5$ m

$$H = \frac{100}{2\pi}\left[\frac{1}{0.5}+\frac{1}{0.5}\right]$$

$$= \frac{200}{\pi} \text{ A/mt}.$$

3.5 MAGNETIC DIPOLE

The distant magnetic field produced by a small current loop is similar to the electric field from electric dipole. A bar magnet or a small filamentary current loop is usually referred to as magnetic dipole.

Let us determine the magnetic field $\vec{B}$ at a point $P(r, \theta, \phi)$ due to circular loop carrying current as in Fig. 3.23.

The distance (r) of the point P from the centre of the loop satisfies the relation that $r >> a$ and also certain simplifying approximations are made.

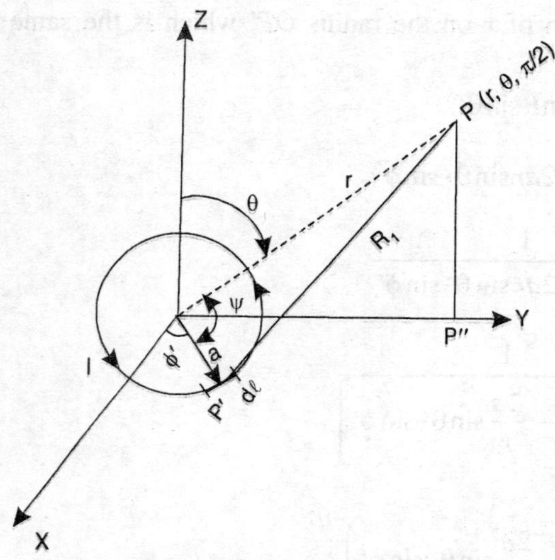

Fig. 3.23 Magnetic field due to dipole

Assume the point P is located directly above the Y-axis as shown in Fig. 3.23. The magnetic vector potential $\vec{A}$ at P is

$$\vec{A} = \int_l \frac{\mu_0 I d\vec{l}}{4\pi R}, \quad \text{...(3.61)}$$

where $I \cdot d\vec{l} = I_a d_\phi \vec{a}_\phi$. The variable direction of $\vec{a}_\phi$ in the integrand is handled by pairing the effects of the current elements Idl_1 and Idl_2 at the symmetrical locations about the Y-axis.

From the geometry,

$$Idl_1 = I \cdot a \cdot d\phi \cdot \vec{a}_\phi$$

$$= \left(-\vec{a}_x \sin\phi + \vec{a}_y \cos\phi\right) Ia\,d\phi \quad \text{...(3.62)}$$

and

$$= -\left(\vec{a}_x \sin\phi - \vec{a}_y \cos\phi\right) Ia\,d\phi$$

$$Idl_2 = -\left(\vec{a}_x \sin\phi - \vec{a}_y \cos\phi\right) Ia\,d\phi. \quad \text{...(3.63)}$$

The net dA at P due to these pain of element is along $-\vec{a}_x$

$$-\vec{A}_x = \vec{A}_\phi = 2\int_{\phi'=-\frac{\pi}{2}}^{\phi'=\frac{\pi}{2}} \frac{\mu_0 Ia\sin\phi'}{4\pi R}\,d\phi'. \qquad \text{...(3.64)}$$

From the law of cosines applied to *PoP′*

$$R_1^2 = a^2 + r^2 - 2ar\cos\psi,$$

where rcosψ is the projection of *r* on the radius *OP′* which is the same as the projection of *OP″* (= rsinθ) on *OP′*. Hence

$$r\cos\psi = r\sin\theta\sin\theta'$$

$$R_1^2 = a^2 + r^2 - 2ar\sin\theta\cdot\sin\phi' \qquad \text{...(3.65)}$$

$$\frac{1}{R_1^2} = \frac{1}{r^2 + a^2 - 2ar\sin\theta\cdot\sin\phi'}$$

$$\frac{1}{R_1^2} = \frac{1}{r^2\left[1 + \frac{a^2}{r^2} - \frac{2a}{r}\sin\theta\cdot\sin\phi'\right]}$$

$$\frac{1}{R_1} = \frac{1}{r}\left(1 + \frac{a^2}{r^2} - \frac{2a}{r}\sin\theta\cdot\sin\phi'\right)^{-1/2}$$

$r >> a$, $\frac{a^2}{r^2}$ is neglected in comparison with 1.

$$\frac{1}{R_1} \approx \frac{1}{r}\left(1 + \frac{a}{r}\sin\theta\cdot\sin\phi'\right). \qquad \text{...(3.66)}$$

Substituting,

$$\vec{A}_\phi = \frac{\mu_0 Ia}{2\pi r}\int_{\phi'=-\pi/2}^{\phi'=\pi/2}\left(1 + \frac{a}{r}\sin\theta\cdot\sin\phi'\right)\sin\phi' d\phi'$$

$$= \frac{\mu_0 Ia}{2\pi r}\left[\int_{\phi'=-\pi/2}^{\phi'=\pi/2}\sin\phi' d\phi' + \int_{\phi'=-\pi/2}^{\phi'=\pi/2}\frac{a}{r}\sin\theta\cdot\sin^2\phi\, d\phi'\right]$$

$$= \frac{\mu_0 Ia}{2\pi r}\left[(-\cos\phi')_{-\pi/2}^{\pi/2} + \frac{a}{r}\sin\theta\cdot\int_{-\pi/2}^{\pi/2}\left(\frac{1-\cos 2\phi'}{2}\right)\cdot d\phi'\right].$$

$$\vec{A}_\phi = \frac{\mu_0 Ia}{2\pi r}\left[\frac{a}{r}\sin\theta\left(\frac{1}{2}\phi' - \sin\frac{2\phi'}{4}\right)\right]_{-\pi/2}^{\pi/2}$$

$$= \frac{\mu_0 Ia^2 \sin\theta}{4\pi r^2}\left[\phi' - \frac{\sin 2\phi'}{2}\right]_{-\pi/2}^{\pi/2} \vec{a}_\phi$$

$$= \frac{\mu_0 Ia^2 \sin\theta}{4\pi r^2}\left[\left(\frac{\pi}{2}+\frac{\pi}{2}\right) - \frac{\sin 2\pi/2}{2} - \frac{\sin(-2\pi/2)}{2}\right]$$

$$= \frac{\mu_0 Ia^2 \pi \sin\theta}{4\pi r^2}\vec{a}_\phi.$$

$$\vec{A} = \frac{\mu_0 m \sin\theta}{4\pi r^2}\cdot\vec{a}_\phi, \qquad \text{...(3.67)}$$

where m = magnetic moment = $I\pi a^2$.

The magnetic flux density $\vec{B} = \nabla \times \vec{A}$

$$\vec{B} = \frac{1}{r^2 \sin\theta}\begin{vmatrix} \vec{a}_r & r\cdot\vec{a}_\theta & r\sin\theta\cdot\vec{a}_\phi \\ \partial/\partial r & \partial/\partial\theta & \partial/\partial\phi \\ A_r & r\cdot A_\theta & r\sin\theta\cdot A_\phi \end{vmatrix}$$

$$= \frac{1}{r^2 \sin\theta}\begin{vmatrix} \vec{a}_r & r\cdot\vec{a}_\theta & r\sin\theta\cdot\vec{a}_\phi \\ \partial/\partial r & \partial/\partial\theta & \partial/\partial\phi \\ 0 & 0 & r\sin\theta\cdot A_\phi \end{vmatrix}$$

$$= \frac{1}{r^2 \sin\theta}\left[\vec{a}_r \frac{\partial}{\partial\theta}(r\sin\theta\cdot A\phi) - r\cdot\vec{a}_\theta \frac{\partial}{\partial r}(r\sin\theta\cdot A\phi)\right]$$

$$= \frac{1}{r^2 \sin\theta}\left[\vec{a}_r\left\{\frac{\partial}{\partial\theta}\left(r\sin\theta\cdot\frac{\mu_0 m}{4\pi r^2}\sin\theta\right)\right\} - r\vec{a}_\theta.\frac{\partial}{\partial r}\left[r\sin\theta\cdot\frac{\mu_0 m}{4\pi r^2}\sin\theta\right]\right] \qquad \text{...(3.68)}$$

$$= \frac{1}{r^2 \sin\theta}\left[\frac{\mu_0 m}{4\pi r}\frac{\partial}{\partial\theta}\left(\sin^2\theta\right)\vec{a}_r - r\vec{a}_\theta\frac{\mu_0 m \sin^2\theta}{4\pi}\frac{\partial}{\partial r}\left[\frac{1}{r}\right]\right]$$

$$= \frac{1}{r^2 \sin\theta}\left[\frac{\mu_0 m}{4\pi r}2\sin\theta\cos\theta\vec{a}_r - r\cdot\frac{\mu_0 m \sin^2\theta}{4\pi}\left(\frac{-1}{r^2}\right)\vec{a}_\theta\right]$$

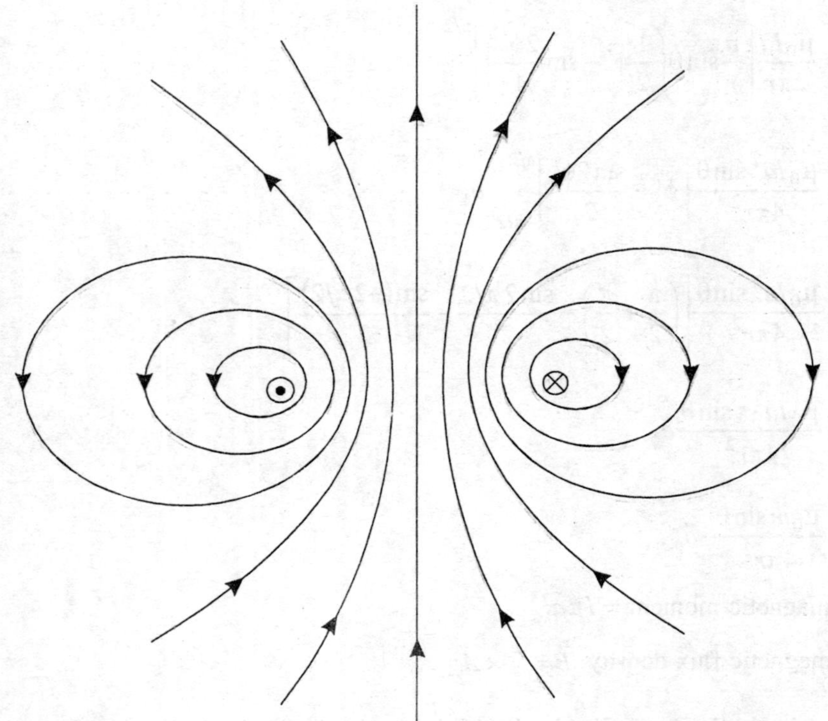

Fig. 3.24 Magnetic field lines due to magnetic dipole

$$= \frac{1}{r^2 \sin\theta}\left[\frac{\mu_0 m}{4\pi r} 2\sin\theta\cos\theta\,\vec{a}_r + \frac{\mu_0 m \sin^2\theta}{4\pi r}\,\vec{a}_\theta\right]$$

$$= \frac{\mu_0 m \sin\theta}{4\pi r^3 \sin\theta}\left[2\cos\theta\,\vec{a}_r + \sin\theta\,\vec{a}_\theta\right]$$

$$\vec{B} = \frac{\mu_0 m}{4\pi r^3}\left[2\cos\theta\,\vec{a}_r + \sin\theta\,\vec{a}_\theta\right] \text{ wb/m}^2. \qquad ...(3.69)$$

The magnetic field lines due to magnetic dipole are shown in Fig. 3.24.

In order to understand the significance of magnetic dipole moment, we will compare it to electric dipole moment.

Electric dipole:

$$\vec{P} = Q\vec{d}$$

+Q

d

–Q

where $\vec{P}$ is electric dipole moment. $\vec{d}$ is vector pointing from –ve to +ve charge.

Magnetic dipole:

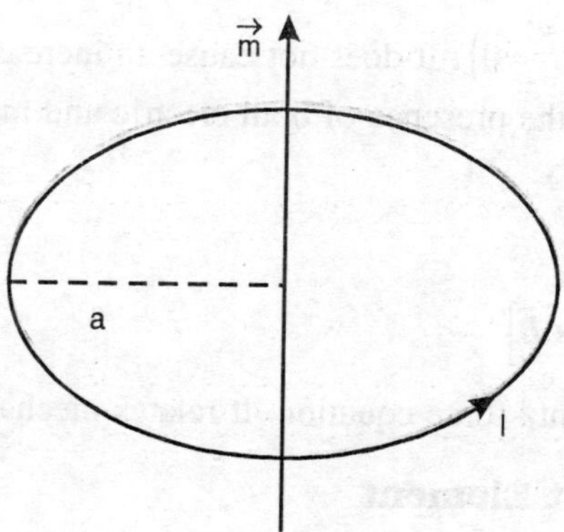

$$\vec{m} = I\pi a^2 \vec{a}_n$$

where $\vec{m}$ is magnetic dipole moment.

$\vec{a}_n$ is the direction of $\vec{m}$ and is given by right hand rule.

3.6 MAGNETIC FORCES

The study of force exerted by a magnetic field on charged particles, on current elements, and on loops is important to solve problems in electronic devices such as ammeters, voltmeters, galvanometers, cyclotrons, etc.

There are three ways in which force due to magnetic fields can be experienced. The force can be (*a*) due to a moving charged particle in a magnetic field (B) (*b*) on a current element in an external magnetic field (B) (*c*) between two current elements.

3.6.1 Force on a Charged Particle

According to the discussion in Chapter 2, the electric force $\vec{F}_e$ on a stationary or moving electric charge Q in an electric field is given by Coulomb's experimental law and is related to the electric field intensity $\vec{E}$ as

$$\vec{F}_e = Q\vec{E}. \qquad \text{...(3.70)}$$

This shows that if Q is positive, $\vec{F}_e$ and $\vec{E}$ have the same direction.

A magnetic field can only exert force on a moving charge. From experiments, it is found that the magnetic force $\vec{F}_m$ experienced by a charge Q moving with a velocity $\vec{u}$ in a magnetic field $\vec{B}$ is

$$\vec{F}_m = Q\vec{u} \times \vec{B}. \qquad \text{...(3.71)}$$

This clearly shows that $\vec{F}_m$ is perpendicular to both $\vec{u}$ and $\vec{B}$.

From eqn. (3.70), it can be seen that $\vec{F}_e$ is independent of the velocity of the charge and can perform work on the charge and change its kinetic energy. From eqn. (3.71), $\vec{F}_m$ depends on the charge velocity and is normal to it. $\vec{F}_m$ cannot perform work because it is at right angles to the direction of motion of the charge $\left(\vec{F}_m \cdot d\vec{l} = 0\right)$. It does not cause an increase in kinetic energy of the charge.

For a moving charge Q in the presence of both electric and magnetic fields, the total force on the charge is given by

$$\vec{F} = \vec{F}_e + \vec{F}_m$$

(or) $$\vec{F} = Q\left[\vec{E} + \vec{u} \times \vec{B}\right] \quad ...(3.72)$$

This is known as the Lorentz force equation. It relates mechanical force to electrical force.

3.6.2 Force on a Current Element

The force on a current element $Id\vec{l}$ of a current-carrying conductor due to the magnetic field $\vec{B}$ can be determined by modifying the eqn. (3.67) using the equation for convection current.

$$\vec{J} = \rho_v \vec{u}.$$

From eqn. (3.6), the relationship between current elements is

$$Id\vec{l} = \vec{K}ds = \vec{J}dv$$

$$Id\vec{l} = \rho_v \vec{u}dv = \rho_v dv\, \vec{u}$$

$$= dQ\vec{u}. \quad ...(3.73)$$

Alternatively

$$Id\vec{l} = \frac{dQ}{dt} d\vec{l} = dQ \cdot \frac{d\vec{l}}{dt} = dQ\vec{u}.$$

Hence $$Id\vec{l} = dQ\vec{u}. \quad ...(3.74)$$

This shows that a charge dQ moving with velocity $\vec{u}$ (thereby producing convection current element $dQ\vec{u}$) is equivalent to a conduction current element $Id\vec{l}$. Thus the force on a current element $Id\vec{l}$ in a magnetic field $\vec{B}$ can be obtained by replacing $Q\vec{u}$ by $Id\vec{l}$ in eqn. (3.71). That is,

$$d\vec{F} = Id\vec{l} \times \vec{B}. \quad ...(3.75)$$

If the current I is through a closed path L or circuit, the force on the circuit is given by

$$\vec{F} = \oint_L Id\vec{l} \times \vec{B} \quad ...(3.76)$$

While using eqns. (3.75) and (3.76) it should be kept in mind that the magnetic field produced by the current element $Id\vec{l}$ does not exert force on the element itself just as a point charge does not exert force on itself. The $\vec{B}$ field that exerts force on $Id\vec{l}$ must be due to another element. In other words, the $\vec{B}$ field is external to the current element Idl in eqns. (3.75) and (3.76). So,

$$\left.\begin{aligned} d\vec{F}_2 &= I_1 d\vec{l} \times \vec{B}_2 \quad \text{or} \\ d\vec{F}_{21} &= I_2 d\vec{l}_2 \times \vec{B}_1 \end{aligned}\right] . \qquad \text{...(3.77)}$$

From eqn. (3.7), force can also be written in terms of current elements as

$$d\vec{F}_{12} = I_1 d\vec{l} \times \mu \int \frac{I_2 d\vec{l} \times \vec{a}_R}{4\pi R^2} . \qquad \text{...(3.78)}$$

Since the current elements must necessarily be part of closed loops of current, a more physically meaningful force would be that on an entire current loop due to another current loop as in Fig. 3.25.

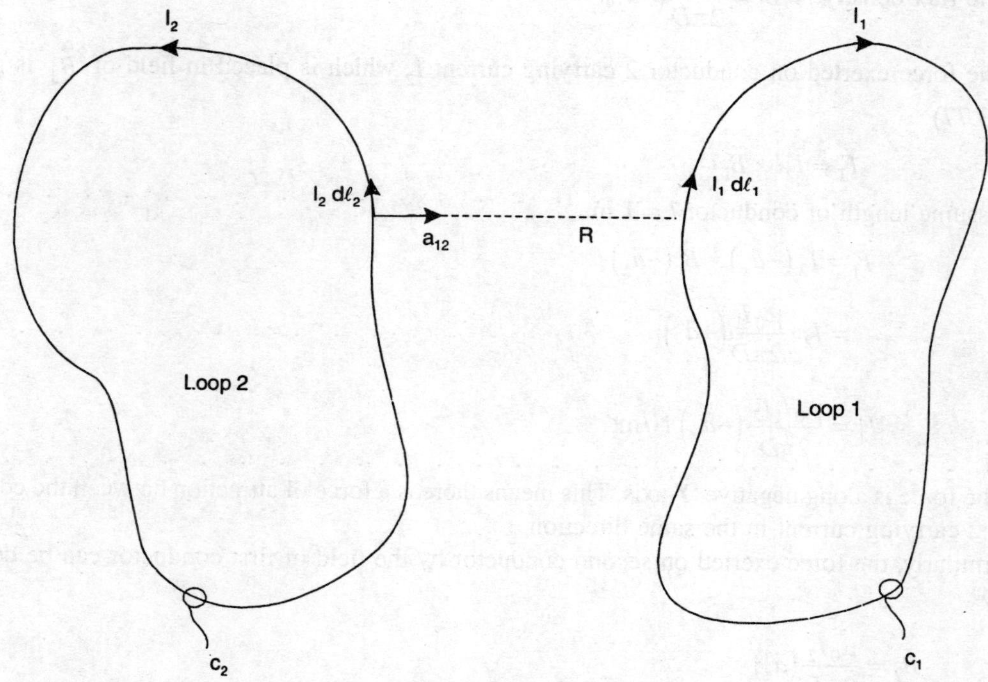

Fig. 3.25 Two closed loops of filamentary current

Hence by summing up all contributions of current elements in loop 2 to the force on every current element in loop 1 using eqn. (3.74) and taking the limit as the lengths of the current segments tend to zero, the resulting net force experienced by loop 1 due to loop 2 is

$$\vec{F}_{12} = \frac{\mu I_1 I_2}{4\pi} \oint_{c_1} \oint_{c_2} \frac{d\vec{l}_1 \times \left(d\vec{l}_2 \times \vec{a}_{12}\right)}{R^2} \text{N}. \qquad \text{...(3.79)}$$

This equation is referred to as Ampere's force law for two current loops (or) Ampere's work law.

3.6.3 Force Between Two Current Elements

Consider two straight, long, parallel conductors situated in space and spaced D m apart, carrying current I_1 amp and I_2 amp in the same direction.

Assume that the conductors are placed in XY-plane and they carry currents along $-X$ direction.

From eqn. (3.13), the field produced by the conductor at any point is $\vec{H} = \frac{I}{2\pi h}\vec{a}_z$, where h is the distance of point from the conductor.

From Fig. 3.19, the field $\vec{H}_1$ produced by conductor 1 at the location of conductor 2 is

$$\vec{H}_1 = \frac{I_1}{2\pi D}(-\vec{a}_z).$$

The flux density is $\vec{B}_1 = \frac{\mu_0 I_1}{2\pi D}(-\vec{a}_z)$.

The force exerted on conductor 2 carrying current I_2 which is placed in field of $\vec{B}_1$ is given by eqn. (3.77)

$$\vec{F}_1 = \vec{I}_2 l \times \vec{B}_1 .$$

Assume length of conductor $l = 1$ m.

$$\vec{F}_1 = I_2(-\vec{a}_x) \times B_1(-\vec{a}_z)$$

$$= I_2 \cdot \frac{\mu_0 I_1}{2\pi D}(-\vec{a}_y)$$

$$\vec{F}_1 = \frac{\mu_0 I_1 I_2}{2\pi D}(-\vec{a}_y) \text{ N/mt.} \quad ...(3.80)$$

The force is along negative Y-axis. This means there is a force of attraction between the conductors 1 and 2 carrying current in the same direction.

Similarly, the force exerted on second conductor by the field of first conductor can be derived as follows.

$$\vec{B}_2 = \frac{\mu_0 I_2}{2\pi D}(\vec{a}_z)$$

$$\vec{F}_2 = I_1 l \times B_2$$

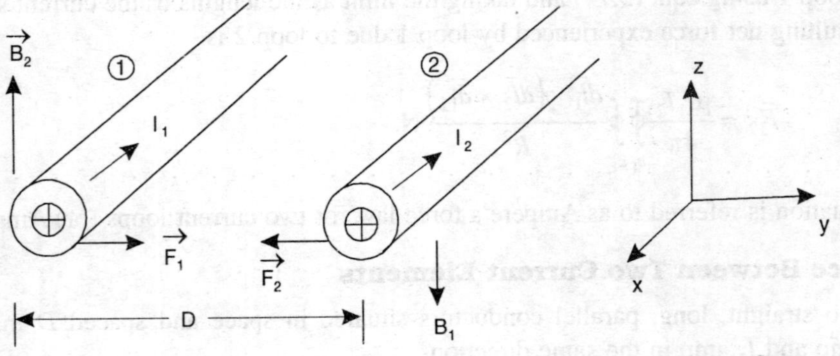

Fig. 3.26 Force between two current-carrying conductors

$$= I_1(-\vec{a}_x)\times B_2(\vec{a}_z)$$

$$= I_1 \cdot \frac{\mu_0 I_2}{2\pi D}(\vec{a}_y)$$

$$\vec{F}_2 = \frac{\mu_0 I_1 I_2}{2\pi D}\vec{a}_y \qquad ...(3.81)$$

$$|\vec{F}_1| = |\vec{F}_2|.$$

If the direction of current I_2 is reversed, the direction of force is also reversed. Consequently, there will be a force of repulsion.

3.7 MAGNETISATION ON MATERIALS

The magnetisation on materials is equivalent to the concept of polarisation of materials in an electric field.

In an atom, the internal magnetic field is produced by electrons orbiting around the nucleus or electrons spinning. Both of these electron motions produce internal magnetic fields that are similar to the magnetic field produced by a current loop. The equivalent current loop has a magnetic moment of $\vec{m} = I_b S\vec{a}_n$, where S is the area of the loop and I_b is the bound current (bound to the atom). The magnetisation $\vec{M}$ (in amp/m) is defined as the magnetic dipole moment per unit volume. This vector $\vec{M}$ is analogous to the polarisation $\vec{P}$ in dielectrics and is sometimes called the magnetic polarisation density of the medium. Both have same units.

In material medium, $\vec{M} \neq 0$ and as a result $\vec{B}$ in eqn. (3.41) changes into

$$\vec{B} = \mu_0(\vec{H} + \vec{M}). \qquad ...(3.82)$$

The above relationship holds good for all materials whether they are linear or not. The concepts of linearity, isotropy and homogeneity discussed in previous section for dielectric media equally apply here for magnetic media. For linear materials, $\vec{M}$ (in A/m) depends linearly on $\vec{H}$, such that

$$\vec{M} = \chi_m \vec{H} \qquad ...(3.83)$$

where χ_m is a dimensionless quantity called magnetic susceptability (ratio of $\vec{M}$ to $\vec{H}$) of the medium. It is the measure of susceptability (sensitivity) of the material to a magnetic field. Substituting eqn. (3.82) in eqn. (3.83)

$$\vec{B} = \mu_0(1+\chi_m)\vec{H} = \mu\vec{H} \qquad ...(3.84)$$

(or)

$$\vec{B} = \mu_0\mu_r\vec{H}, \qquad ...(3.85)$$

where

$$\mu_r = 1 + \chi_m = \frac{\mu}{\mu_0}. \qquad ...(3.86)$$

The quantity $\mu = \mu_0 \mu_r$ is called the permeability of the material and is measured in henry/metre. The dimensionless quantity μ_r is the ratio of the permeability of a given material to that of free space and is known as the relative permeability of the material.

3.8 CLASSIFICATION OF MAGNETIC MATERIALS

In general, the magnetic susceptibility (χ_m) or the relative permeability (μ_r) is useful in classifying the materials in terms of their magnetic property or behaviour. A material is said to be non-magnetic if $\chi_m = 0$ (or) $\mu_r = 1$. Free space and materials with $\chi_m = 0$ (or) $\mu_r = 1$ are regarded as non-magnetic.

Magnetic materials may be grouped into three major classes. They are diamagnetic, paramagnetic and ferromagnetic.

Diamagnetism occurs in materials where the magnetic fields due to electron motions of orbiting and spinning completely cancel each other. Thus the permanent (or intrinsic) magnetic moment of each atom is zero and the materials are weakly affected by a magnetic field. Examples for diamagnetic materials are bismuth, lead and copper. For diamagnetic materials, the value of $\mu_r \leq 1$ and χ_m is of the order of -10^{-5}. Perfect diamagnetism occurs in superconductors at temperatures near absolute zero.

Materials whose atoms have non-zero permanent magnetic moment may be paramagnetic or ferromagnetic. Paramagnetism occurs in materials where the magnetic fields produced by orbital or spinning electrons do not cancel completely. Examples for para-magnetic materials are platinum and tungsten. For paramagnetic materials, χ_m is of the order $+10^{-5}$ to $+10^{-3}$ and $\mu_r \geq 1$. Paramagnetic materials find application in masers.

Ferromagnetism occur in materials whose atoms have relatively large permanent magnetic moment. Cobalt, iron, nickel and their alloys are ferromagnetic materials. They are capable of being magnetised very strongly by a magnetic field. They retain a considerable amount of their magnetisation when removed from the field.

3.9 MAGNETISATION CURVE (OR) B-H CURVE

The relationship between $|\vec{B}|$ and $|\vec{H}|$ is called magnetisation curve or *B-H* curve.

For ferromagnetic materials, the relationship between *B* and *H* is non-linear and at any point on the curve, μ is given by the ratio of *B* to *H* or by dB/dH.

A typical *B-H* curve for a ferromagnetic material is shown in Fig. 3.27.

Initially, the ferromagnetic material is unmagnetised. As *H* increases due to increase in current from '0' to maximum applied field intensity H_{max}, curve *OP* is produced. This curve is referred to as the initial magnetisation curve. After reaching saturation at *P*, if *H* is decreased, *B* does not follow the initial curve but lags behind *H*. This phenomenon of *B* lagging behind *H* is called hysteresis.

If *H* is reduced to zero, *B* is not reduced to zero but to B_r, which is referred to as permanent flux density. The value of B_r depends on H_{max}, the maximum applied field intensity. If *H* increases negatively by reversing the direction of current, *B* becomes zero, when *H* becomes H_c, which is known as the coercive field intensity.

Further increase in *H* in the negative direction to reach *Q* and a reverse in its direction to reach *P* gives a closed curve called a hysteresis loop. The shape of hysteresis loops varies from one material to another. For example, some ferrites have an almost rectangular hysteresis loop and are used in digital computers as magnetic information storage devices. The area of a hysteresis loop gives the energy loss (hysteresis loss) per unit volume during one cycle of the periodic magnetisation of the ferromagnetic

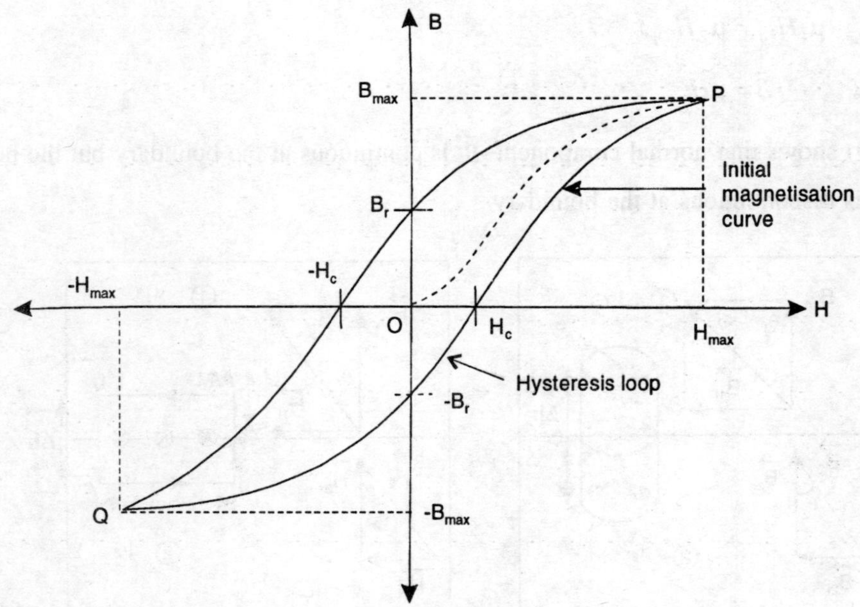

Fig. 3.27 Hysteresis loop

material. This energy loss is in the form of heat. It is therefore desirable that materials used in electric generators, motors and transformers should have tall but narrow hysteresis loops, so that hysteresis losses are minimum.

3.10 MAGNETIC BOUNDARY CONDITIONS

Magnetic boundary conditions are the conditions that $|\vec{B}|$ or $|\vec{H}|$ field must satisfy at the boundary between two different media. The derivations are similar to electrostatic boundary conditions.

The two laws used for deriving the boundary conditions are

1. Gauss's law for magnetic fields

$$\oint_s \vec{B} \cdot d\vec{s} = 0 \qquad ...(3.87)$$

2. Ampere's circuital law

$$\oint_l \vec{H} \cdot d\vec{l} = I. \qquad ...(3.88)$$

Consider the boundary between two magnetic media 1 and 2, characterised by μ_1 and μ_2 as in Fig. 3.28. Applying eqn. (3.87) to the pill box (Gaussian surface) of Fig. 3.28a and allowing $\Delta h \to 0$, then

$$B_{1n}\Delta s - \vec{B}_{2n}\Delta s = 0.$$

Thus,

$$\vec{B}_{1n} = \vec{B}_{2n} \text{ (or)} \qquad ...(3.89a)$$

$$\mu_1 \vec{H}_{1n} = \mu_2 \vec{H}_{2n}, \quad ...(3.89b)$$

since $\qquad \vec{B} = \mu \vec{H}.$

Eqn. (3.89) shows that normal component $\vec{B}$ is continuous at the boundary but the normal component of $\vec{H}$ is discontinuous at the boundary.

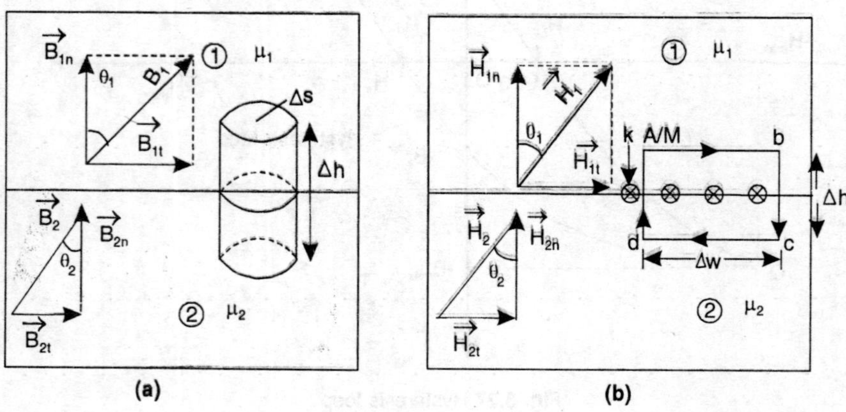

Fig. 3.28 Boundary conditions between two magnetic media.

Similarly, by applying eqn. (3.82) to the closed path adcda of Fig. 3.28b, the conditions for tangential component can be obtained. The surface current k A/M on the boundary is assumed to be normal to the path. Then from the Fig. 3.28b,

$$H_{1t} \cdot \Delta w + H_{1n} \cdot \frac{\Delta h}{2} + H_{2n} \cdot \frac{\Delta h}{2}$$

$$- H_{2t} \cdot \Delta w - H_{2n} \frac{\Delta h}{2} - H_{1n} \frac{\Delta h}{2} = k \Delta w. \quad ...(3.90)$$

As $\Delta h \to 0$, eqn. (3.86) becomes

$$\vec{H}_{1t} - \vec{H}_{2t} = k \quad ...(3.91)$$

This shows that the tangential component of H is also discontinuous. Eqn. (3.87) can also be written in terms of B as

$$\frac{\vec{B}_{1t}}{\mu_1} - \frac{\vec{B}_{2t}}{\mu_2} = k. \quad ...(3.92)$$

If the boundary is free of current or the media are not conductors (for K is free current density), $K = 0$ and eqn. (3.87) becomes

$$\vec{H}_{1t} = \vec{H}_{2t} \quad ...(3.93a)$$

(or)

$$\frac{\vec{B}_{1t}}{\mu_1} = \frac{\vec{B}_{2t}}{\mu_2}. \quad ...(3.93b)$$

Thus the tangential component of $\vec{H}$ is continuous while that of $\vec{B}$ is discontinuous at the boundary.

If the fields make an angle θ with the normal to the interface, eqn. (3.84) results in

$$B_1 \cos\theta_1 = B_{1n} = B_{2n} = B_2 \cos\theta_2, \quad ...(3.94)$$

while eqn. (3.80) becomes

$$\frac{B_1}{\mu_1}\sin\theta_1 = H_{1t} = H_{2t} = \frac{B_2}{\mu_2}\sin\theta_2. \quad ...(3.95)$$

Dividing eqn. (3.92) by eqn. (3.91) gives

$$\frac{\tan\theta_1}{\tan\theta_2} = \frac{\mu_1}{\mu_2}. \quad ...(3.96)$$

The above eqn. (3.96) is the law of refraction for magnetic flux lines at a boundary with no surface current.

3.11 TORQUE ON CLOSED CIRCUIT

The concept of current loop experiencing a torque in a magnetic field is important in understanding the principle of d.c. motor. The torque (or mechanical moment of force) is defined as the vector product of force $\vec{F}$ and the moment arm $\vec{r}$. That is, the torque about the axis is

$$\vec{T} = \vec{r} \times \vec{F} \quad ...(3.97)$$

$\vec{T}$ has the unit of newton-metres.

Consider the single turn coil in the $Z = 0$ plane shown in Fig. (3.29) of width w in the x-direction and length l along y-direction. The field is uniform and in the $+x$ direction.

The force acting on a current element placed in the field is given by eqn. (3.71).

$$\vec{F} = I\vec{l} \times \vec{B}$$

$$\vec{F} = BIl \sin\alpha, \quad ...(3.98)$$

where α is the angle made by the current element with B.

Now, the force exerted on the side ab carrying current from a to b is

$$\vec{F} = Il(\vec{a}_y) \times B\vec{a}_x$$

$$= -BIl\vec{a}_z.$$

Similarly, for side cd

$$\vec{F} = Il(-\vec{a}_y) \times B(\vec{a}_x)$$

$$= BIl\,\vec{a}_z.$$

Since the sides bc and da are parallel with B, no force is exerted on them. Equal forces act on length l only at different points in opposite directions.

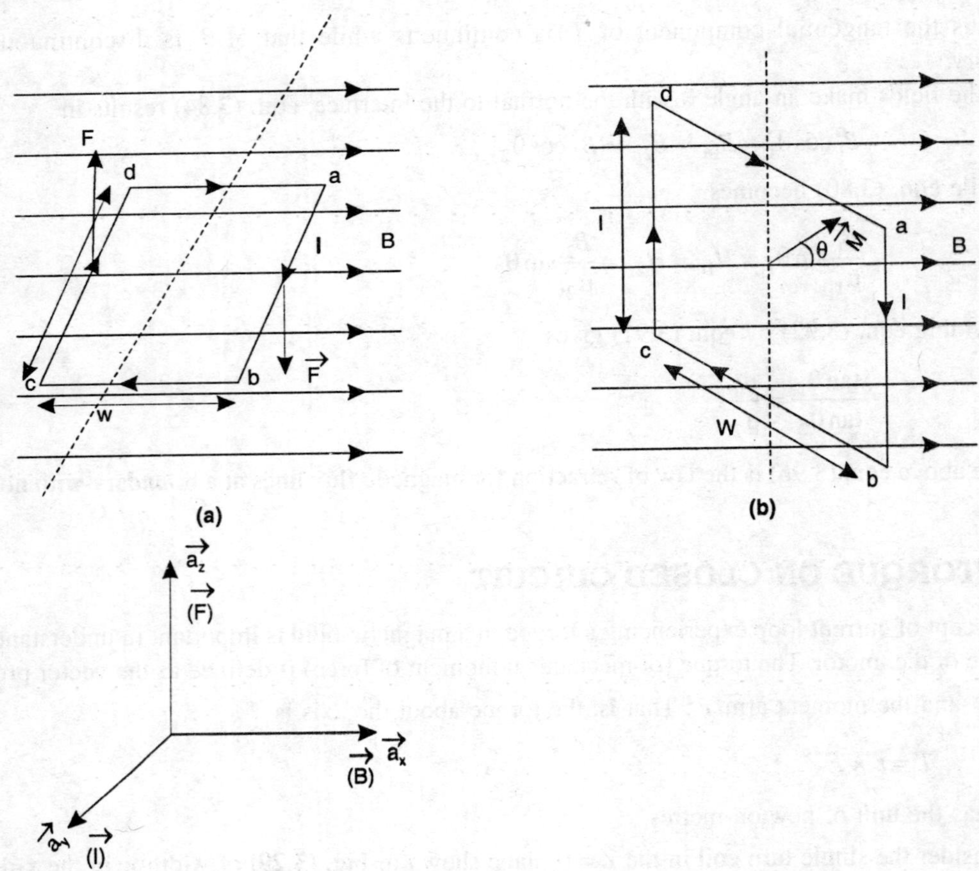

Fig. 3.29 Torque on a closed loop

The direction of force on *ab* is such that it will tend to move downwards whereas *cd* will tend to move upwards.

The torque about *Y*-axis from the left current element (*cd*) requires a lever arm $r = (-w/2)\vec{a}_x$. The sign will change for the lever arm to the right side (ab) $r = w/2(\vec{a}_x)$.

The torque from both elements is

$$\vec{T} = \left(\frac{-w}{2}\right)\vec{a}_x \times (-BIl\,\vec{a}_z) + \left(\frac{w}{2}\right)\vec{a}_x \times BIl\,\vec{a}_z$$

$$\vec{T} = BIl\,w\left(-\vec{a}_y\right)$$

$$\vec{T} = BIA\left(-\vec{a}_y\right) \text{ N-mt,} \qquad \text{...(3.99)}$$

where *A* is the area of the coil.

If the coil is not parallel with $\vec{B}$ but is normal and makes an angle α as in Fig. 3.28 (*b*), then force will be

$$\left|\vec{F}\right| = B\,Il\sin\alpha, \qquad ...(3.100)$$

so
$$\left|\vec{T}\right| = BIA\sin\alpha \qquad ...(3.101)$$

$$\vec{T} = I\vec{A}\times\vec{B} \text{ or } IA\vec{a}_n\times\vec{B},$$

where $\vec{m} = IA\vec{a}_n$ is the magnetic dipole moment of the current loop (or) coil and $\vec{a}_n$ is the unit normal of the loop.

$$\therefore \qquad \vec{T} = \vec{m}\times\vec{B}. \qquad ...(3.102)$$

SOLVED PROBLEMS

Problem 1

Two narrow coils A and B have a common axis and are placed 10 cm apart. Coil A has 10 turns of radius 5 cm with a current of 1A passing through it. Coil B has a single turn of radius 7.5 cm. If the magnetic field at the center of the coil A is zero, what current should be passed through B?

Solution:

Magnetic field intensity $\vec{H}$ at centre of coil A due to current in coil A is

$$\vec{H}_A = \frac{NI}{2a}(-\vec{a}_z)$$

[As current is in CW direction, $\vec{H}$ is along $(-\vec{a}_z)$ as per thumb's rule].

$$\vec{H}_A = \frac{10\times 1}{2\times 5\times 10^{-2}}(-\vec{a}_z) = 100\frac{AT}{\text{m}}(-\vec{a}_z).$$

Field intensity $\vec{H}$ at the centre of coil A due to current in coil B is

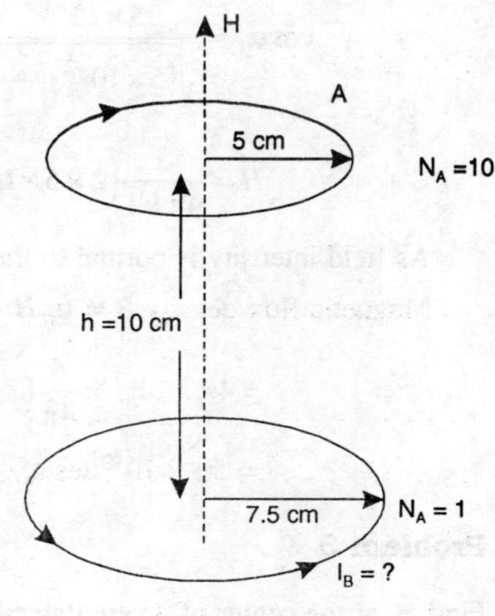

Fig. P. 3.1

$$\vec{H}_B = \frac{NIa^2}{2\left(a^2+h^2\right)^{3/2}}\vec{a}_z$$

$$= \frac{1\times I_B\times(7.5)^2\times 10^{-4}}{2\left[(7.5)^2+10^2\right]^{3/2}\times 10^{-6}}$$

$$\vec{H}_B = 1.44\,I_B$$

To make the field intensity at the centre of coil A to be zero.

$$|\vec{H}_A| = |\vec{H}_B|$$

$$100 = 1.44 I_B$$

$$I_B = 69.44 \text{ amp.}$$

Problem 2

A straight conductor in air carries a current of 5A. Calculate the flux density produced by 10 cm length of the conductor at a point 1m distance along the perpendicular bisector drawn to the conductor length.

Solution:

Assume the conductor along X-axis and the point along Y-axis.

Magnetic field intensity H at point 'P' due to the finite length of conductor is

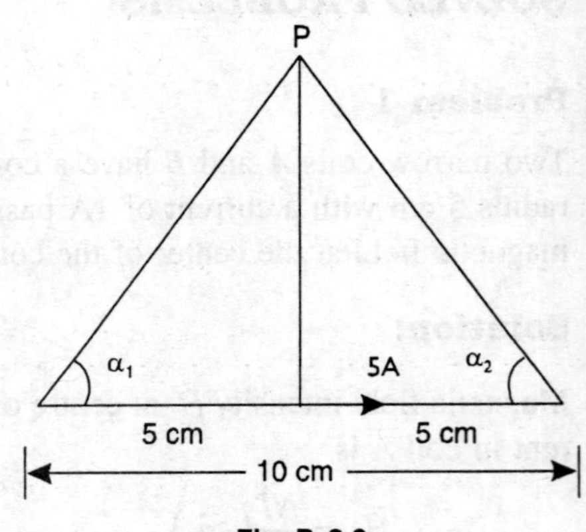

Fig. P. 3.2

$$\vec{H} = \frac{I}{4\pi h}(\cos\alpha_1 + \cos\alpha_2)\vec{a}_z$$

$$\cos\alpha_1 = \cos\alpha_2$$

$$\cos\alpha_1 = \frac{5\times10^{-2}}{\sqrt{(5\times10^{-2})^2+1}} = 5\times10^{-2}$$

$$\vec{H} = \frac{5}{4\pi\times1}(2\times5\times10^{-2})\vec{a}_z.$$

As field intensity is normal to the plane XY, it is along Z-axis.

Magnetic flux density $B = \mu_0 H$

$$= 4\pi\times10^{-7}\times\frac{5}{4\pi}(2\times5\times10^{-2})$$

$$= 50\times10^{-9} \text{ tesla.}$$

Problem 3

Find $\vec{H}$ at the center of an equilateral triangular loop of side 4 m carrying current of 5A.

Solution:

Assume that the triangular loop is in XY plane. Consider one side (horizontal) of the equilateral triangle.

Magnetic field intensity at a point 'P' due to finite length of conductor is

$$\vec{H} = \frac{I}{4\pi h}[\cos\alpha_1 + \cos\alpha_2]\vec{a}_z.$$

From Fig. P. 3.3,

$$\tan 30° = \frac{h}{2}, \cos\alpha_1 = \cos 30° = \cos\alpha_2$$

$$h = 1.54 \text{ m}$$

$$\vec{H}_p = \frac{5}{4\pi \times 1.54} 2\cos 30° = 0.596\,\vec{a}_z$$

$\vec{H}$ due to all three sides = 3 × 0.596 = 1.79 $\vec{a}_z$ AT/m.

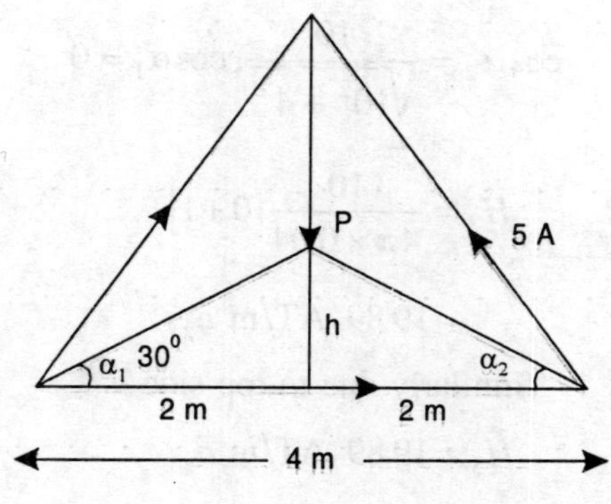

Fig. P. 3.3

Problem 4

For the given filamentary loop, find H at '0'.

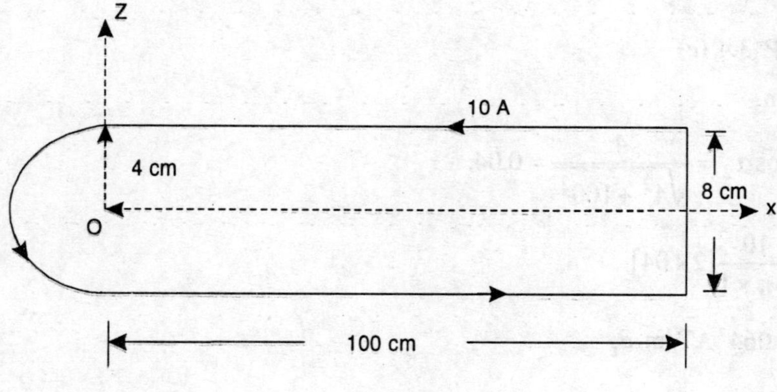

Fig. P. 3.4 (a)

Solution:

Let H_1 be magnetic field intensity due to semi-circular loop at O.

$$a = 0.04 \text{ m}$$

$$\vec{H}_1 = \frac{1}{2}\frac{I}{2a}\vec{a}_z$$

$$= \frac{10}{2 \times 2 \times 0.04} = 62.5 \text{ AT/m}.$$

Let H_2 be magnetic field intensity due to bottom side of loop.

$$\vec{H}_2 = \frac{I}{4\pi h}[\cos\alpha_1 + \cos\alpha_2]\vec{a}_z$$

as in Fig. P. 3.4 (*b*)

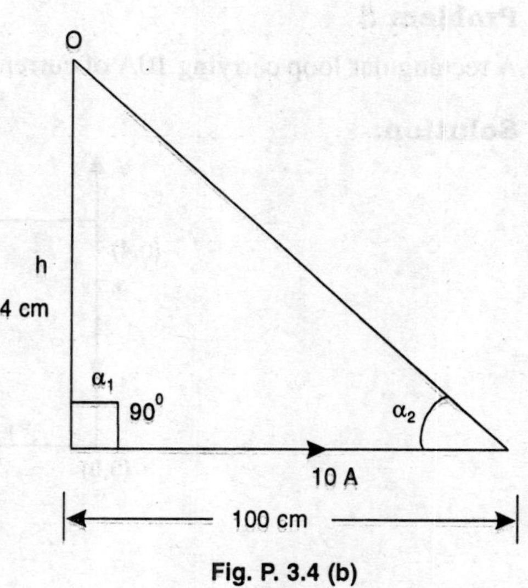

Fig. P. 3.4 (b)

$$\cos\alpha_2 = \frac{10}{\sqrt{10^2 + 4^2}}, \cos\alpha_1 = 0$$

$$\vec{H}_2 = \frac{10}{4\pi \times 0.04}[0+1]$$

$$= 19.89 \text{ AT/m } \vec{a}_z$$

Similarly due to top side is

$$\vec{H}_3 = 19.89 \text{ AT/m } \vec{a}_z$$

Due to side (vertical) of 8 cm is

$$\vec{H}_4 = \frac{I}{4\pi h}[\cos\alpha_1 + \cos\alpha_2]\vec{a}_z$$

as in Fig. P. 3.4 (*c*)

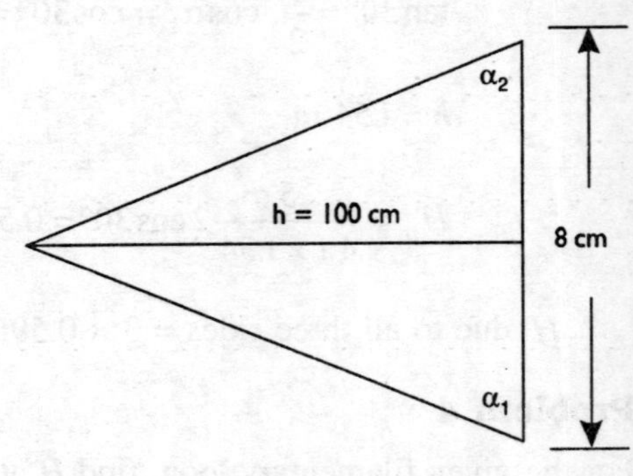

Fig. P. 3.4 (c)

$$h = 1\text{m}$$

$$\cos\alpha_1 = \cos\alpha_2 = \frac{4}{\sqrt{4^2 + 100^2}} = 0.04.$$

$$= \frac{10}{4\pi \times 1}[2 \times .04]$$

$$= 0.064 \text{ AT/m } \vec{a}_z$$

Total $\vec{H}$ at '0' is

$$= 62.5 + 2(19.89) + 0.064 = 102.32 \text{ AT/m } \vec{a}_z.$$

Problem 5

A rectangular loop carrying 10A of current is placed on $Z = 0$ plane. Find the field intensity at (4, 2, 0).

Solution:

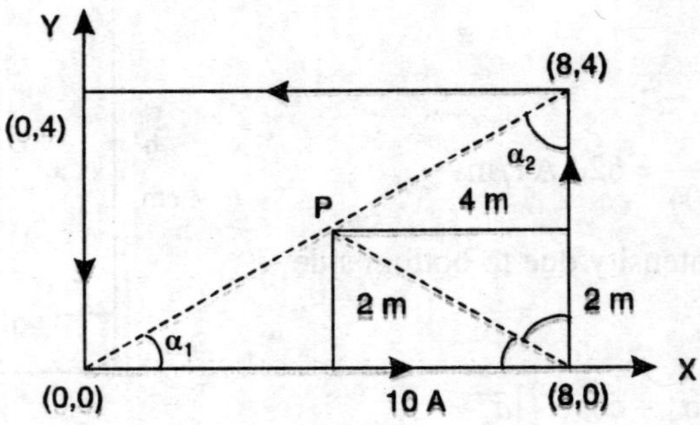

Fig. P. 3.5

(*i*) $\vec{H}_1$ at *P* due to horizontal side of 8 m is

$$\vec{H}_1 = \frac{I}{4\pi h}[\cos\alpha_1 + \cos\alpha_2]$$

$$h = 2\,\text{m}$$

$$\cos\alpha_1 = \cos\alpha_2 = \frac{4}{\sqrt{4^2 + 2^2}} = 0.894$$

$$H_1 = \frac{10}{4\pi \times 2} 2 \times 0.894$$

$$= 0.72 \text{ AT/m } \vec{a}_z$$

(*ii*) $\vec{H}_2$ at *P* due to vertical side of 4 m is

$$h = 4\,\text{m},\ \cos\alpha_1 = \cos\alpha_2 = \frac{2}{\sqrt{2^2 + 4^2}} = 0.447$$

$$\vec{H}_2 = \frac{10}{4\pi \times 4}(2 \times 0.447)$$

$$= 0.178 \text{ AT/m } \vec{a}_z$$

Total $\vec{H}$ at *P*

$$= 2 \times 0.172 + 2 \times 0.178$$

$$= 1.779\ \vec{a}_z \text{ AT/m}$$

Problem 6

An infinitely long conductor is bent into an '*L*' shape as in Fig. P-3.6. If 5A current flows, find $\vec{H}$ at (0, 0, 2) and (0, –2, 0).

Solution:

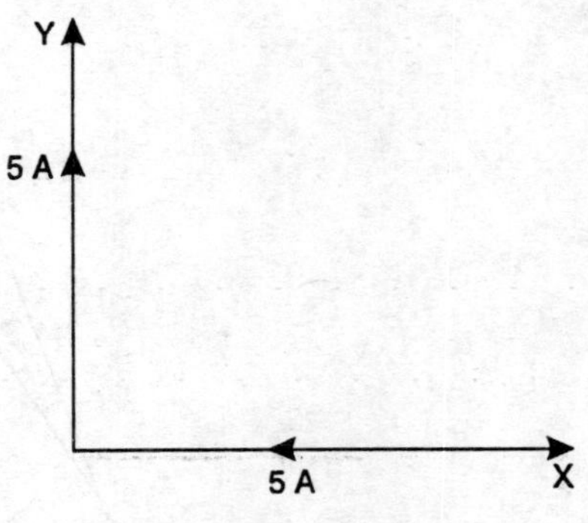

Fig. P. 3.6

(*i*) At (0, –2, 0):

Consider the horizontal line of '*L*' shape. As one end is towards infinity, the angle subtended at '*P*' is zero ($\alpha_2 = 0$).

As the point is below the *X*-axis, the magnetic field intensity is along $(-\vec{a}_z)$ axis.

$$\vec{H}_x = \frac{I}{4\pi h}[\cos\alpha_1 + \cos\alpha_2](-\vec{a}_z)$$

$$h = 2,\ \alpha_1 = 90°,\ \alpha_2 = 0$$

$$= \frac{5}{4\pi \times 2}[1] = 0.1990(-\vec{a}_z).$$

Magnetic field intensity due to conductor portion along Y-axis is zero, as the point is on the Y-axis.

$$\therefore \qquad \vec{H} = 0.199\,(-\vec{a}_z)\cdot$$

(*ii*) $\vec{H}$ at (0, 0, 2).

As the point (0, 0, 2) is on Z-axis, draw the 'L' shaped conductor as in Fig. P. 3.6 (*b*).

Now $\vec{H}$ at P due to conductor along X-axis is

$$\vec{H}_x = \frac{5}{4\pi \times 2}[\cos 90° + \cos 0°]\vec{a}_y$$

$$= 0.199\ \vec{a}_y.$$

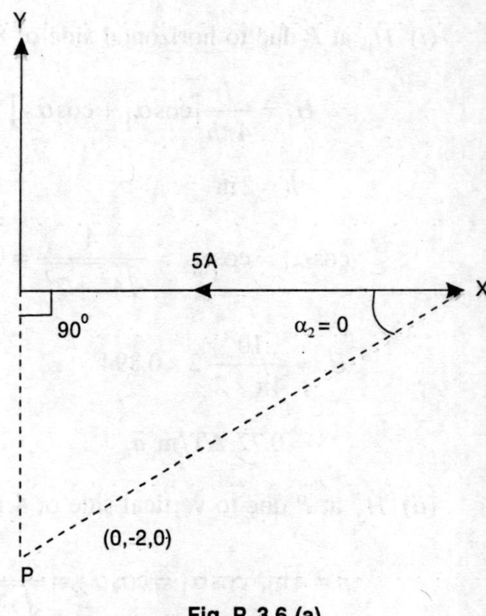

Fig. P. 3.6 (a)

As the plane involves X and Z-axes, the direction of $\vec{H}$ is normal to the plane. So, along 'Y' axis, $\vec{H}$ at P due to conductor along Y axis is

$$\vec{H}_y = \frac{5}{4\pi \times 2}[\cos 0° + \cos 90°]\vec{a}_x$$

$= 0.199\ \vec{a}_x$ (same argument holds here to find the direction)

Total $\vec{H} = 0.199\ \vec{a}_x + 0.199\vec{a}_y$

Fig. P. 3.6 (b)

Problem 7

Show that the field at the center of circular conductor is about 1.047 times as strong as that at the center of hexagonal conductor carrying the same current. The distance between the opposite sides is equal to the diameter of circle.

Solution:

Assume both circular loop and hexagonal loop are in the *XY* plane. Consider the circular loop, field intensity at the center *P* is

$$\vec{H}_c = \frac{I}{2a}\vec{a}_z,$$

where radius $a = \frac{d}{2}$, $H_c = \frac{I}{2 \times d/2} = \frac{I}{d}\vec{a}_z$.

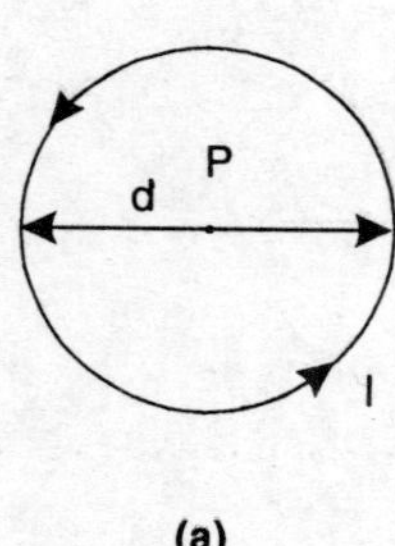

(a)

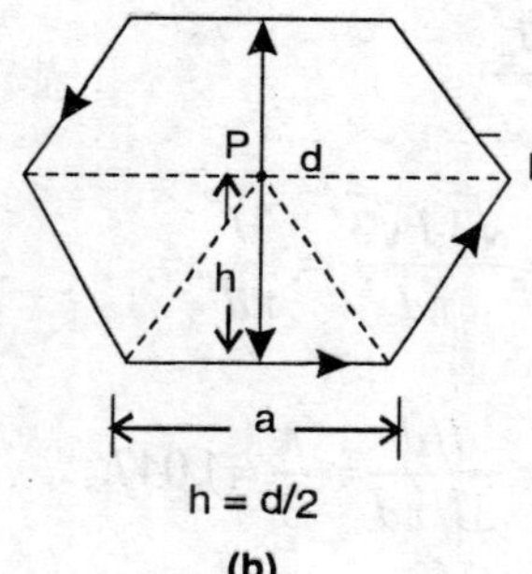

(b)

Fig. P. 3.7

For the hexagonal loop, to find the field intensity at *P*, first find out the field intensity due to one side and then multiply by 6.

Consider the bottom horizontal line, as in Fig. P. 3.7 (*c*)

[Given that distance between opposite sides of hexagon is equal to diameter of circular loop].

From the Fig. P. 3.7 (*b*) and (*c*)

$$\tan 60° = \frac{h}{\frac{a}{2}}$$

$$h = \frac{a\sqrt{3}}{2},$$

as $$h = \frac{d}{2}$$

$$d = \sqrt{3}\,a.$$

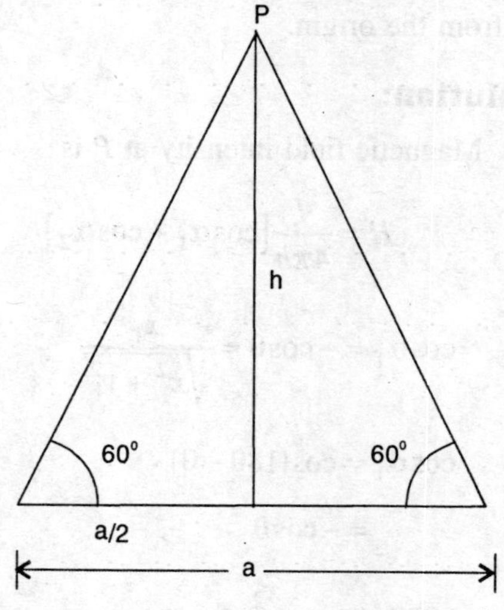

Fig. P. 3.7 (c)

Magnetic field intensity due to bottom side

$$|\vec{H}| = \frac{I}{4\pi h}[\cos\alpha_1 + \cos\alpha_2]$$

$$|\vec{H}| = \frac{I}{4\pi\left(\frac{a\sqrt{3}}{2}\right)} 2 \cdot \cos 60°$$

$$|\vec{H}| = \frac{I}{2\sqrt{3}\,\pi a}$$

$$|\vec{H}| \text{ due to all 6 sides } |\vec{H}_h| = 6 \times \frac{I}{2\sqrt{3}\,\pi a} = \frac{\sqrt{3}\,I}{\pi a}.$$

As $a = \frac{d}{\sqrt{3}}$,

$$|\vec{H}_h| = \frac{\sqrt{3}\,I\sqrt{3}}{\pi d} = \frac{3I}{\pi d}.$$

Hence $$\frac{|\vec{H}_c|}{|\vec{H}_h|} = \frac{I/d}{3I/\pi d} = \frac{\pi}{3} = 1.047.$$

Problem 8

A thin straight conductor of length '*l*' carrying current '*I*' is situated along the *X*-axis such that one end of the conductor is at a distance x_1 from the origin and the other at a distance x_2, the direction of the current is away from the origin. Find an expression for flux density at a point on the *Y*-axis at a distance y_1 from the origin.

Solution:

Magnetic field intensity at *P* is

$$H = \frac{I}{4\pi h}[\cos\alpha_1 + \cos\alpha_2]$$

$$\cos\alpha_1 = -\cos\theta = \frac{x_1}{\sqrt{x_1^2 + y_1^2}}$$

$$\cos\alpha_1 = \cos(180 - \theta)$$

$$= -\cos\theta$$

$$\cos\alpha_2 = \frac{x_2}{\sqrt{x_2^2 + y_1^2}}$$

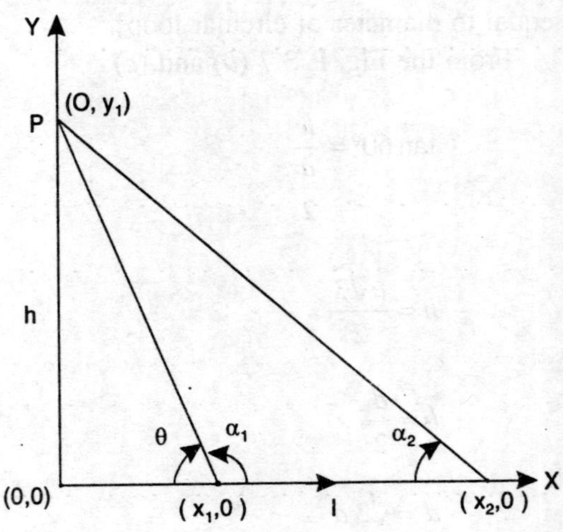

Fig. P. 3.8

$$\vec{H} = \frac{I}{4\pi y_1}\left[\frac{-x_1}{\sqrt{x_1^2 + y_1^2}} + \frac{x_2}{\sqrt{x_2^2 + y_1^2}}\right]\vec{a}_z .$$

Problem 9

A solenoid with radius 2 cm is wound with 20 turns/cm and carries 10 mA. Find $\vec{H}$ at the center of solenoid if the length is 10 cm. If all the turns of the solenoid were compressed into a ring of radius 2 cm. What would be the magnetic field intensity at the center of the ring?

Solution:

From the Example 3.7:

$$\vec{H} = \frac{NI}{2l}(\cos\alpha_1 + \cos\alpha_2)\vec{a}_x .$$

From the Fig. P. 3.9,

$$\cos\alpha_1 = \frac{5}{\sqrt{5^2 + 2^2}} = 0.928 = \cos\alpha_2 .$$

No. of turns per cm = 20.

$\therefore$ Total turns = 20 × 10 = 200

$$\vec{H} = \frac{200 \times 10 \times 10^{-3}}{2 \times 10 \times 10^{-2}} \times 2 \times 0.928 = 18.57 \text{ AT/m}.$$

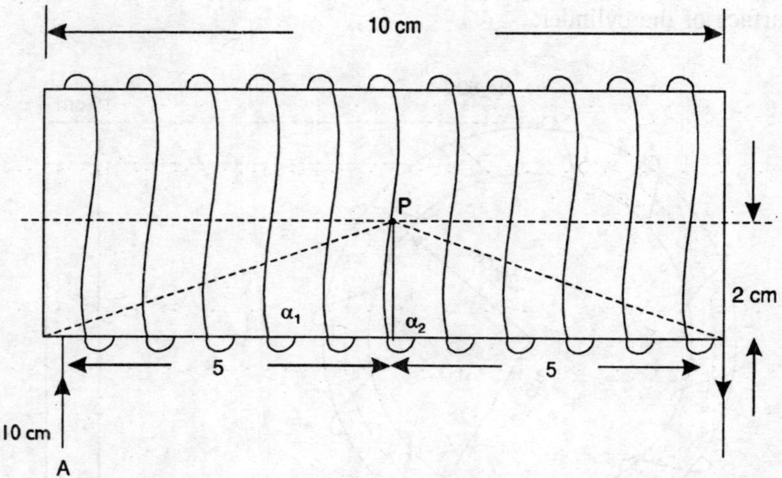

Fig. P. 3.9

If all the turns are compressed to form a ring of radius 2 cm, then number of turns in a single ring is $N = 200$.

$\therefore$ $\vec{H}$ at the centre of the ring is

$$\vec{H} = \frac{NI}{2a}$$

$$= \frac{200 \times 10 \times 10^{-3}}{2 \times 2 \times 10^{-2}} = 50 \text{ AT/m}.$$

Problem 10

A wooden rail with internal diameter 50 cm and of square cross-section with 10 cm side is uniformly wound with 500 turns. If the coil carries a current of 1 amp, find the magnetic flux density on the inner and outer cylindrical surface of the wood.

From Example 3.8,

Magnetic field intensity due to a toroid at a point of radius ρ is

$$\vec{H} = \frac{NI}{2\pi\rho} \vec{a}_\phi.$$

$\vec{H}$ at inner surface of the cylinder:

$$\vec{H} = \frac{500 \times 1}{2\pi \times 0.35}$$

$$\vec{B} = \mu_0 H \quad [\text{For wood, } \mu_r = 1]$$

$$\vec{B} = \frac{4\pi \times 10^{-7} \times 500}{2\pi \times 0.35} = 4 \times 10^{-4} \vec{a}_\phi \text{ tesla}.$$

$\vec{H}$ at outer surface of the cylinder:

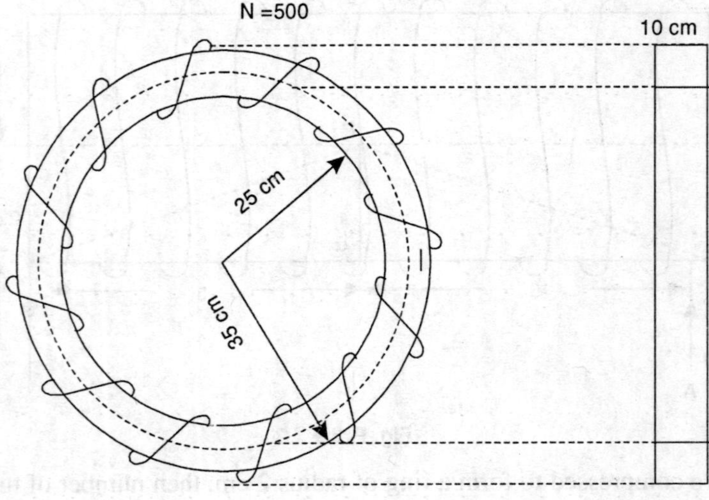

Fig. P. 3.10

$$\left|\vec{H}\right| = \frac{500 \times 1}{2\pi \times 0.35}$$

$$\left|\vec{B}\right| = \frac{4\pi \times 10^{-7} \times 500}{2\pi \times 0.35}$$

$$\vec{B} = 2.86 \times 10^{-4}\, \vec{a}_\phi \text{ tesla.}$$

Problem 11

If the vector potential $\vec{A}$ is given as $\vec{A} = 5\left(x^2 + y^2 + z^2\right)\vec{a}_x$. Find out flux density.

From the equation

$$\nabla \times \vec{A} = \vec{B}$$

$\vec{A}$ is in cartesian co-ordinate.

So,
$$\nabla \times \vec{A} = \begin{vmatrix} \vec{a}_x & \vec{a}_y & \vec{a}_z \\ \dfrac{\partial}{\partial x} & \dfrac{\partial}{\partial y} & \dfrac{\partial}{\partial z} \\ 5x^2 + 5y^2 + 5z^2 & 0 & 0 \end{vmatrix}$$

$$= -\vec{a}_y(-10z) + \vec{a}_z(-10y)$$

$$= 10z\,\vec{a}_y - 10y\,\vec{a}_z$$

$$\vec{B} = 10\left(z\vec{a}_y - y\vec{a}_z\right).$$

Problem 12

A straight conductor placed along *X*-axis, carries a current of 200 A in positive *x* direction. If a uniform field with $B = 1$ tesla exist everywhere in direction parallel to the *XY* plane and at an angle 45° with *X*-axis, determine the magnitude and direction of force per unit length of conductor.

Solution:

Force on a conductor placed in a magnetic field is

$$\vec{F} = BIl \sin\theta\, \vec{a}_n \left(I\vec{l} \times \vec{B}\right)$$

Given

$B = 1$ tesla

$I = 200$ A

$\theta = 45°$

$$\frac{F}{l} = 1 \times 200 \times 45° = 141.4 \text{ N}$$

Direction of the force is normal to *XY*-plane (i.e.) along *z* direction.

Problem 13

What is the maximum torque on a square loop of 100 turns in a field of flux density '*B*' tesla. The loop has 10 cm side and carries a current of 3 amp. What is the magnetic moment of the loop?

Maximum torque on a single loop = BIA N–m

Maximum torque on 1000 turns of a square loop = $1000 \times B \times I \times A$ N–m

Given I = 3amp, $A = (10 \times 10) \times 10^{-4}$, B = 1 tesla

$T_m = 1000 \times 3 \times 100 \times 10^{-4} = 30$ N

Magnetic moment (m) = IA

$= 0.3 \times 1 \times 10^{-4}\ \vec{a}_n$

$$= 30 \text{mA} - \text{m}^2\ \vec{a}_n$$

3.12 INDUCTORS AND INDUCTANCES

A circuit (or closed conducting path) carrying current I produces a magnetic field $\vec{B}$ which causes a flux $\psi = \int_s \vec{B} \cdot d\vec{s}$ to pass through each turn of the circuit as shown in Fig. 3.30. If the circuit has N identical turns, then flux linkage λ is defined as $\lambda = N\psi$ ...(3.103)

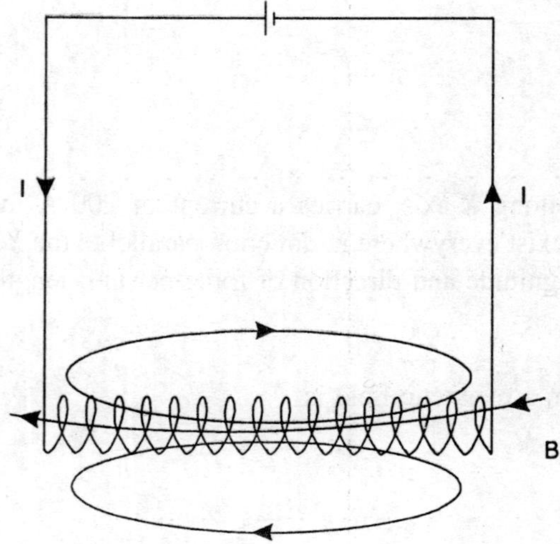

Fig. 3.30 Magnetic field produced by a circuit

Also, if the medium surrounding the circuit is linear, the flux linkage λ is proportional to the current I producing it, that is,

$$\lambda \alpha I$$

$$\lambda = LI, \qquad ...(3.104)$$

where L is a constant of proportionality called the inductance of the circuit. The inductance of an inductor circuit is defined as the ratio of magnetic flux linkage λ to the current I through the conductor. That is,

$$L = \frac{\lambda}{I} = \frac{N\psi}{I}. \qquad ...(3.105)$$

The unit of inductance is henry (H) which is same as wb/amp.

Inductances are usually expressed in millihenrys.

The inductance defined by eqn. (3.105) is commonly referred to as self-inductance, since the linkages are produced by the inductor itself.

Instead of having a single circuit, if there are two circuits carrying current I_1 and I_2 as shown in Fig. 3.31, then a magnetic interaction exists between the circuits. Four component fluxes ψ_{11}, ψ_{12}, ψ_{21} and ψ_{22} are produced. The flux ψ_{12} is the flux passing through circuit 1 due to current I_2 in circuit 2. If $\vec{B}_2$ is the field due to I_2 and S_1 is the area of circuit 1, then

$$\psi_{12} = \int_{S_1} \vec{B}_2 \cdot d\vec{s}. \qquad ...(3.106)$$

The mutual inductance M_{12} is defined as the ratio of the flux linkage $\lambda_{12} = N_1\psi_{12}$ on circuit 1 due to current I_2, that is

$$M_{12} = \frac{\lambda_{12}}{I_2} = \frac{N_1\psi_{12}}{I_2}. \qquad ...(3.107)$$

Similarly, the mutual inductance M_{21} is defined as the flux linkages of circuit 2 per unit current I_1, that is,

$$M_{21} = \frac{\lambda_{21}}{I_1} = \frac{N_2\psi_{21}}{I_1}. \qquad ...(3.108)$$

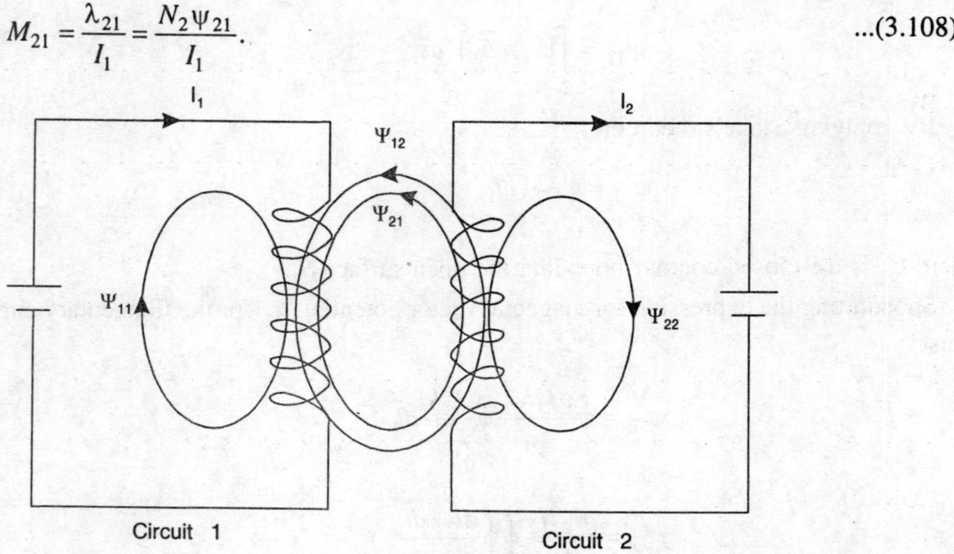

Fig. 3.31 Magnetic interaction between two circuits

Self-inductance of circuits 1 and 2 can be defined as

$$L_1 = \frac{\lambda_{11}}{I_1} = \frac{N_1\psi_1}{I_1} \quad ...(3.109)$$

$$L_2 = \frac{\lambda_{22}}{I_2} = \frac{N_2\psi_2}{I_2}, \quad ...(3.110)$$

where $\psi_1 = \psi_{11} + \psi_{12}$ and $\psi_2 = \psi_{21} + \psi_{22}$.

As already explained, an inductor is a conductor arranged in proper shape to store magnetic energy. Typical examples of inductors are toroids, solenoids, co-axial transmission lines and parallel-wire transmission lines. The inductance of each of these inductors can be determined by following a procedure similar to that taken in determining the capacitance of a capacitor. For a given inductor, the self-inductance L can be found by following the steps.

1. Choose a suitable co-ordinate system
2. Let the conductor carry current I
3. Determine $\vec{B}$ from Biot-Savart's law (or Ampere's circuital law) and calculate ψ from $\psi = \int \vec{B} \cdot d\vec{s}$.
4. Finally, find inductance L from $L = \dfrac{\lambda}{I} = \dfrac{N\psi}{I}$.

The mutual inductance between two circuits may be calculated by taking a similar procedure.

In co-axial and parallel wire transmission line, the inductance produced by the flux, internal to the inductor, is called the internal inductance L_{in}, while that produced by the external flux is called external inductance L_{ext}. The total inductance is $L = L_{in} + L_{ext}$

An alternative formula for directly computing mututal inductance can be derived using magnetic vector potential $\vec{A}$. In eqn. (3.102), $\vec{B}_2$ is replaced by $\vec{B}_2 = \nabla \times \vec{A}_2$ such that

$$\psi_{12} = \int_{S_1} \left(\nabla \times \vec{A}_2\right) \cdot d\vec{s}. \quad ...(3.111)$$

By applying Stoke's theorem

$$\psi_{12} = \oint_{C_1} \vec{A}_2 \cdot d\vec{l}_1, \quad ...(3.112)$$

where C_1 is the closed contour bounding the open surface S_1.

Substituting the expression for magnetic vector potential $\vec{A}_2$ for the filamentary current loop of N_2 turns

$$\psi_{12} = \oint_{C_1} \frac{\mu I_2 N_2}{4\pi} \oint_{C_2} \frac{d\vec{l}_2}{r_{12}} \cdot d\vec{l}_1$$

$$= \frac{\mu I_2 N_2}{4\pi} \oint_{C_1}\oint_{C_2} \frac{d\vec{l}_2 \cdot d\vec{l}_1}{r_{12}}, \quad ...(3.113)$$

where r_{12} is the distance from any point on loop 2 to any point on loop 1.

From the eqn. (3.103),

Mutual inductance $M_{12} = \dfrac{N_1\Psi_{12}}{I_2}$ can be written as

$$M_{12} = \frac{\mu N_1 N_2}{4\pi} \oint_{c_1} \oint_{c_2} \frac{d\vec{l}_2 \cdot d\vec{l}_1}{r_{12}} \qquad \text{...(3.114)}$$

This equation is referred to as Neumann formula for filamentary conductors.

Example 3.10

Calculate the self-inductance of an infinitely long solenoid.

From Example 3.7, the field intensity of an infinitely long solenoid is

$$H = \left|\vec{H}\right| = \frac{NI}{l}, \text{ so } B = \mu H = \frac{\mu NI}{l}.$$

If A is the cross-sectional area in m^2 of the solenoid, the total flux through the cross-section is $\psi = BA = \dfrac{\mu NIA}{l}$.

Flux linkage in the solenoid having N turns $\lambda = N\psi = \dfrac{\mu N^2 IA}{l}$.

Thus, inductance is

$$L = \frac{\lambda}{I} = \frac{\mu N^2 A}{l} \text{ henry} \qquad \text{...(3.115)}$$

Example 3.10 (a)

Determine the self-inductance of a co-axial cable of inner radius a and outer radius b.

Internal Inductance:

Consider the cross-section of the cable as shown in Fig. 3.32(a).

From Example 3.6, the field intensity H at any point inside the conductor $\rho < a$ is given by eqn. (3.30) as

$$\vec{H}_1 = \frac{I\rho}{2\pi a^2}\vec{a}_\phi.$$

The internal inductance L_{in} is found out by considering the flux linkages due to the inner conductor. From Fig. 3.32(a), the flux leaving a differential shell of thickness $d\rho$ is

$$d\psi_1 = \vec{B}_1 \cdot d\vec{s} = B_1\vec{a}_\phi \cdot ds\,\vec{a}_\phi$$

$$d\psi_1 = B_1 d\rho\, dz \qquad \left(\because d\vec{s} = d\rho\, dz\, \vec{a}_\phi\right),$$

where $\quad B_1 = \mu H_1$

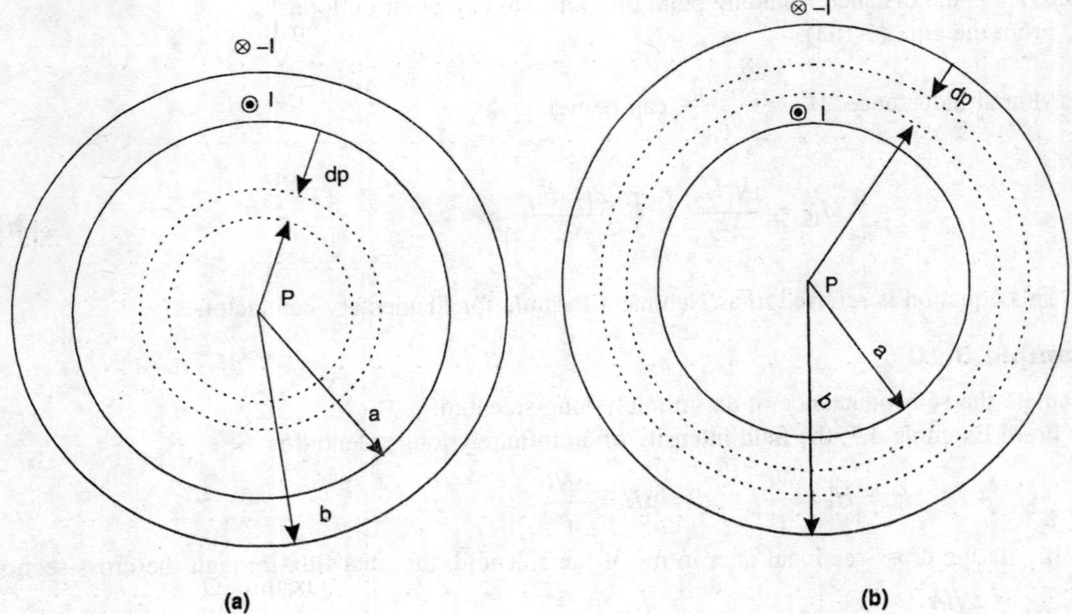

Fig. 3.32

$$d\psi_1 = \frac{\mu I \rho}{2\pi a^2} d\rho dz.$$

The flux linkage $d\lambda_1$ is $d\psi_1$ multiplied by the ratio of the area within the path enclosing the flux to the total area *i.e.*

$$d\lambda_1 = d\psi_1 \cdot \frac{I_{enc}}{I} = d\psi_1 \cdot \frac{\pi\rho^2}{\pi a^2}.$$

Thus the total flux linkages with in the differential flux element are

$$d\lambda_1 = \frac{\mu I \rho\, d\rho\, dz}{2\pi a^2} \cdot \frac{\rho^2}{a^2}.$$

For length l of the cable

$$\lambda_1 = \int_{\rho=0}^{\rho=a} \int_{z=0}^{z=l} \frac{\mu I \rho^3}{2\pi a^4} d\rho dz$$

$$\lambda_1 = \frac{\mu I l}{8\pi}.$$

Internal inductance = $L_1 = \dfrac{\lambda_1}{I} = \dfrac{\mu l}{8\pi}$ henry. ...(3.116)

Internal inductance per unit length

$$L_{in} = \frac{L_1}{l} = \frac{\mu}{8\pi} \text{ H/m}. \quad ...(3.117)$$

External Inductance

From Example 3.6, the field intensity at any point ρ between inner and outer conductor, ($a < \rho < b$) is given by eqn. (3.25)

$$\vec{H}_2 = \frac{I}{2\pi\rho}\vec{a}_\phi$$

$$\vec{B}_2 = \frac{\mu I}{2\pi\rho}\vec{a}_\phi.$$

The external inductance is determined by considering the flux linkages between inner and outer conductor as in Fig. 3.32 (*b*).

For a differential shell of thickness $d\rho$,

$$d\psi_2 = \vec{B}_2 \cdot d\vec{s}$$

$$= B_2\vec{a}_\phi \cdot ds\,\vec{a}_\phi$$

$$= \frac{\mu I}{2\pi\rho}d\rho\, dz$$

$$\psi_2 = \int_{\rho=a}^{\rho=b}\int_{z=0}^{z=l}\frac{\mu I}{2\pi\rho}d\rho\, dz$$

$$= \frac{\mu I l}{2\pi}\ln\frac{b}{a}$$

$$L_2 = \frac{\lambda_2}{I} = \frac{\mu l}{2\pi}\ln\frac{b}{a}\ \text{H} \qquad ...(3.118)$$

$$L_{ext} = \frac{L_2}{l} = \frac{\mu}{2\pi}\ln\frac{b}{a}\ \text{H/m} \qquad ...(3.119)$$

Thus total inductance per unit length is $L = L_{in} + L_{ext}$

$$= \frac{\mu}{2\pi}\left[\frac{1}{4} + \ln\frac{b}{a}\right]\text{H/m} \qquad ...(3.120)$$

Example 3.11

Determine the inductance per unit length of two-wire transmission line with separation distance *d*. Each wire has radius *a*.

Consider the first wire carrying current *I* amp.

The internal inductance is determined as in the previous example. Thus for region $0 < \rho < a$, the flux linkages $\lambda_1 = \frac{\mu I l}{8\pi}$.

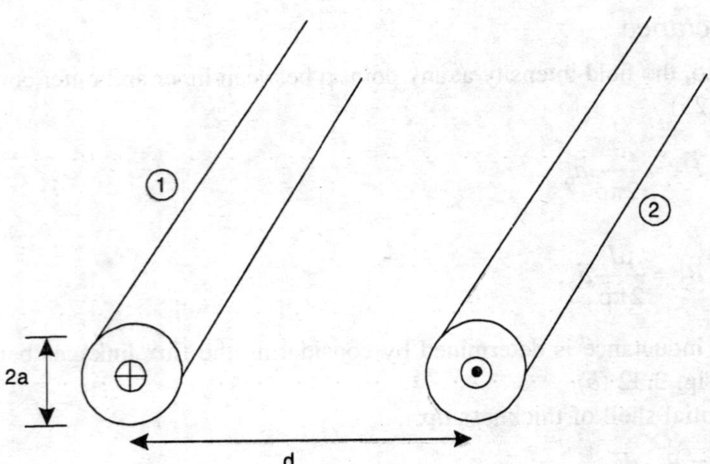

Fig. 3.33 Two-wire transmission line

To find external inductance for the same wire, consider the region $a \le \rho \le d - a$. The flux linkages between the two wires are

$$\lambda_2 = \int_{\rho=a}^{\rho=d-a} \int_{z=0}^{l} \frac{\mu I}{2\pi\rho} d\rho\, dz$$

$$= \frac{\mu I l}{2\pi} \ln \frac{d-a}{a}.$$

The total flux linkages produced by wire 1 is $\lambda_1 + \lambda_2 = \frac{\mu I l}{8\pi} + \frac{\mu I l}{2\pi} \ln \frac{d-a}{a}$.

By symmetry, the same amount of flux is produced by current-I amp in wire 2. Hence, total flux linkages are

$$\lambda = 2(\lambda_1 + \lambda_2)$$

$$= \frac{\mu I l}{\pi}\left[\frac{1}{4} + \ln \frac{d-a}{a}\right].$$

If $d >> a$, the self-inductance per unit length is

$$L' = \frac{L}{l} = \frac{\lambda}{Il} = \frac{\mu}{\pi}\left[\frac{1}{4} - \ln \frac{d}{a}\right] \text{H/m}. \quad ...(3.121)$$

Example 3.12

Show that the mutual inductance of a co-axial solenoid in either direction is same.

Consider two co-axial solenoids of radius R_1 and R_2, where $R_2 > R_1$, carrying current I_1 and I_2 respectively. Let the number of turns in each solenoid be N_1 and N_2 and their length is d m.

The mutual inductance M_{12} is given by eqn. (3.107)

$$M_{12} = \frac{N_1 \psi_{12}}{I_2}$$

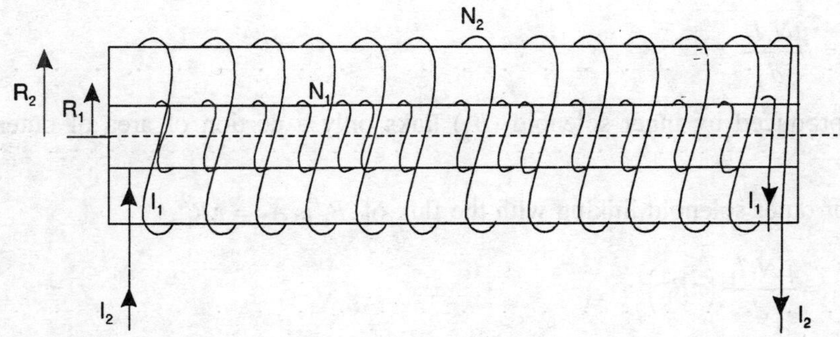

Fig. 3.34 Co-axial solenoid of same length different radius

$$\psi_{12} = \int_{S_1} \vec{B}_2 \cdot d\vec{s} \;\; i.e.,$$

$$= B_2 \times A_1$$

where A_1 is the area of circuit 1 linking with the flux of B_2.

B_2 is the flux density produced in circuit 2 (outer solenoid).

For a solenoid, $\vec{H}$ is given as

$$\vec{H} = N\frac{I}{l} \text{ and } \vec{B} = \mu\vec{H}.$$

Hence,

$$B_2 = \frac{\mu N_2 I_2}{d}. \qquad [\because r = d]$$

The flux produced by B_2 links with entire area of circuit 1 (inner solenoid).

$$\therefore \quad A_1 = \pi R_1^2.$$

Hence, $\psi_{12} = \dfrac{\mu N_2 I_2}{d} \cdot \pi R_1^2$

$$M_{12} = \frac{N_1}{I_2}\left(\frac{\mu N_2 I_2 \pi R_1^2}{d}\right)$$

$$M_{12} = \frac{\mu N_1 N_2 \pi R_1^2}{d} \text{ henry.} \qquad \text{...(3.122)}$$

Similarly, from eqn. (3.104)

$$M_{21} = \frac{N_2 \psi_{21}}{I_1}$$

$$\psi_{21} = B_1 \times A_2$$

i.e. flux produced by the inner solenoid which links with the outer solenoid.

$$B_1 = \frac{\mu N_1 I_1}{d}.$$

The flux produced by inner solenoid (B_1) links only a portion of area of outer solenoid since ($R_2 > R_1$).

The area of outer solenoid linking with the flux of B_1 is $A_2 = \pi R_1^2$.

$$\therefore \quad \psi_{21} = \frac{\mu N_1 I_1}{d}\left(\pi R_1^2\right)$$

$$M_{21} = \frac{N_2}{I_1}\left(\frac{\mu N_1 I_1}{d}\pi R_1^2\right)$$

$$M_{21} = \frac{\mu N_1 N_2 \pi R_1^2}{d}. \qquad ...(3.123)$$

Comparing eqn. (3.122) and eqn. (3.123), it is proved that mutual inductance in both direction of a co-axial solenoid is same.

Example 3.13

Derive the expression for co-efficient of coupling in terms of mutual and self-inductances.

Let us take a general case where two coils may be associated with different magnetic circuits, so that their reluctances are different.

If L_1 and L_2 are self-inductances of coil 1 and 2 having N_1 and N_2 turns, with the currents passing through them are I_1 and I_2, then from eqns. (3.109) and (3.110)

$$L_1 = \frac{N_1 \psi_1}{I_1} \text{ and } L_2 = \frac{N_2 \psi_2}{I_2},$$

where ψ_1 and ψ_2 are flux produced in coil 1 and coil 2 by the currents I_1 and I_2 respectively.

If a fraction k_1 of flux ψ_1 is linking with the coil 2 and similarly a fraction k_2 of ψ_2 is linking with coil 1, then from eqn. (3.103), mutual inductance M_{12} is

$$M_{12} = \frac{N_1 \psi_{12}}{I_2},$$

where ψ_{12} is the flux linking with coil 1 produced by coil 2 = $k_2\, \psi_2$.

$$\text{Thus,} \quad M_{12} = \frac{N_1 k_2 \psi_2}{I_2}. \qquad ...(3.124)$$

Similarly,

$$M_{21} = \frac{N_2 \psi_{21}}{I_1},$$

where ψ_{21} is the flux linking with coil 2 produced by coil 1 = $k_1 \psi_1$.

$$\therefore \quad M_{21} = \frac{k_1 N_2 \psi_1}{I_1}. \qquad ...(3.125)$$

Multiply eqns. (3.117) and (3.118)

$$M_{12}\,M_{21}=\frac{k_1k_2N_1N_2\psi_1\psi_2}{I_1I_2}. \qquad ...(3.126)$$

Substitute eqns. (3.109) and (3.110) in eqn. (3.126)

$$M_{12}\,M_{21}=k_1k_2\left(\frac{N_1\psi_1}{I_1}\right)\left(\frac{N_2\psi_2}{I_2}\right)$$

$$M_{12}M_{21}=K_1K_2L_1L_2. \qquad ...(3.127)$$

Assume $k_1 = k_2 = k$ and $M_{21} = M_{12} = M$.

Then eqn. (3.127) becomes

$$M^2=k^2L_1L_2$$

$$M=k\sqrt{L_1L_2}, \qquad ...(3.128)$$

where k is called coupling coefficient.

3.13 ENERGY STORED IN MAGNETIC FIELD

When a current flowing through an inductance L changes, an e.m.f. induced in it opposing the change in the current and the applied voltage must overcome this induced e.m.f., if the change in current is to be maintained. Now, let the current increases from 0 to I amps in t sec, then the work done in establishing this current in the inductive circuit is

$$W=-\int_0^t e\,i\,dt \qquad \text{watt-second or joule.}$$

where e and i are induced e.m.f. and current at any instant.

But, $$e=-L\frac{di}{dt}$$

$$W=\int_{i=0}^{I} Li\frac{di}{dt}dt \text{ joule}$$

$$=\int_{i=0}^{I} Li\,di \text{ joule}$$

$$W=\frac{1}{2}LI^2 \text{ joule.} \qquad ...(3.129)$$

This work is stored as energy of the magnetic field. This energy is released when the current is brought down to zero again.

3.14 SELF INDUCTANCE OF A TOROID (OR) TOROIDAL SOLENOID

A toroid or toroidal solenoid is a long slender solenoid which is bent into a form of a ring and thereby closed on itself as in Fig. 3.18.

From Example 3.8, the field at any point inside the toroid is from eqn. (3.40)

$$H = \frac{NI}{l_m}$$

$$B = \frac{\mu NI}{l_m}.$$

From Fig. 3.18, mean radius of coil is 'a' m. Area of coil links with the field is $A = \pi a^2$.

So, $\psi = B \times A$

$$= \frac{\mu NIA}{l_m} \, wb.$$

Flux linkages $\lambda = N\psi$

$$= \frac{\mu N^2 IA}{l_m}$$

So, the inductance of toroid is

$$L = \frac{\lambda}{I} = \frac{\mu N^2 A}{l_m} \text{ henry.} \qquad \text{...(3.130)}$$

where l_m is the mean circumference.

3.15 ENERGY DENSITY IN THE MAGNETIC FIELD

Consider a toroidal solenoid. The inductance of it is given as

$$L = \frac{\mu N^2 A}{l_m}$$

where N = number of turns

A = area of the solenoid section

l_m = mean circumference, let it be l.

From eqn. (3.112), energy stored (W) $= \frac{1}{2} LI^2$

$$= \frac{1}{2}\left(\frac{\mu N^2 A}{l}\right) I^2$$

$$= \frac{1}{2}\mu \frac{N^2 I^2}{l^2} \cdot (Al).$$

From eqn. (3.40) $H = NI/l$.

$$W = \frac{1}{2}\mu\left(H^2\right)Al$$

$$= \frac{1}{2}\mu H^2 (Al),$$

where Al = volume of the space inside the coil.

Energy stored per unit volume = Energy density (w)

$$w = \frac{W}{Al} = \frac{1}{2}\mu H^2 \text{ J/m}^3$$

$$w = \frac{1}{2}\mu H^2 \text{ J/m}^3. \qquad ...(3.131)$$

Substitute $B = \mu H$ in eqn. (3.131).

So, $$w = \frac{1}{2}BH \text{ J/m}^3 \qquad ...(3.132)$$

Substitute $H = B/\mu$ in eqn. (3.132).

Then $$w = \frac{1}{2}\frac{B^2}{\mu} \text{ J/m}^3 \qquad ...(3.133)$$

3.16 LIFTING POWER OF AN ELECTROMAGNET

Consider two magnetic poles arranged opposite to each other as in Fig. 3.35.

Let each have an area 'a' m^2 and let F newton be the force of attraction between them.

If one pole is moved through a short distance dx, then the work done against the force of attraction is Fdx joule.

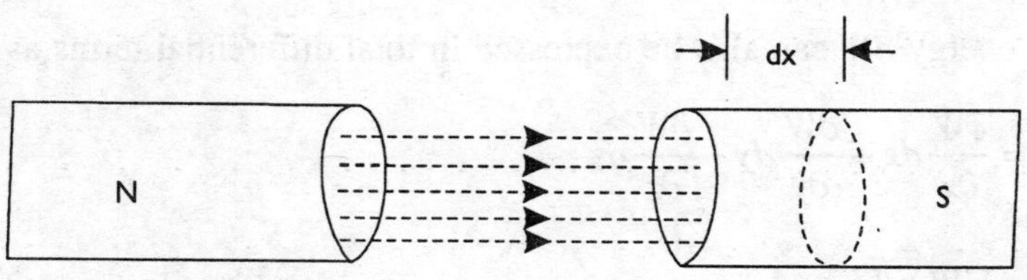

Fig. 3.35 Lifting or pulling power of magnet

This work input must manifest itself in some other form of energy in the system. Due to the movement of one pole with respect to the other, the volume of the magnetic field in between them has increased by adx m^3.

So, the energy stored is also increased. Let us assume that the flux density remains unchanged due to the movement of the pole.

From eqn. (3.133), the energy density $w = \frac{1}{2}\frac{B^2}{\mu} \text{ J/m}^3$

Increase in stored energy $W = w \times$ increased volume

$$= \frac{1}{2}\frac{B^2}{\mu} \times (adx) \text{ joule}$$

where $\mu = \mu_o \mu_r$.

This increase in the stored energy in the magnetic field must be equal to the work input.

$$Fdx = \frac{1}{2}\frac{B^2}{\mu_0 \mu_r} adx$$

$$F = \frac{1}{2}\frac{B^2 a}{\mu_0 \mu_r} \text{ newton}. \qquad ...(3.134)$$

3.17 FORCE AND TORQUE IN TERMS OF STORED ENERGY

From eqn. (3.129), the energy stored within the magnetic field is

$$W = \frac{1}{2} L I^2 \text{ J}$$

As already mentioned, change in stored energy is equal to the work done in physically moving the magnetic materials.

Change in stored energy $dW = \frac{1}{2} I^2 dL.$ ...(3.135)

Since the current is constant, change in energy is related to the change in inductance in (3.135).

Work done $dW_m = \vec{F}_m \cdot d\vec{l}$ ...(3.136)

where $d\vec{l}$ is differential displacement.

The change in stored energy dW can also be expressed in total differential terms as

$$dW = \frac{\partial W}{\partial x} dx + \frac{\partial W}{\partial y} dy + \frac{\partial W}{\partial z} dz$$

$$dW = \nabla W \cdot d\vec{l} \qquad ...(3.137)$$

Equating eqns. (3.136) and (3.137)

$$\vec{F}_m = \nabla W \qquad ...(3.138)$$

Thus the magnetostatic force can be computed as simply the gradient of stored magnetostatic energy.

If the displacement is angular (ϕ) about Z-axis, then the torque can be expressed as

$$T_z = \frac{dW_m}{d\phi} \text{ N-m} \qquad ...(3.139)$$

Example 3.14

A toroid of length l with rectangular cross-section of side h is shown in Fig. 3.36. Derive the expression for the inductance in terms of r_1 and r_2.

Consider a circular loop of Amperian path at a distance r from the centre and let there be a small element of length dl. By applying Ampere's law

$$\oint \vec{H}\cdot d\vec{l} = NI$$

$$\vec{H}\int d\vec{l} = NI$$

$$\vec{H} = \frac{NI}{2\pi r}\vec{a}_\phi, \quad \text{where } \int dl = 2\pi r.$$

The solution for $\vec{H}$ is same as given in eqn. (3.40).

$$\vec{B} = \mu\vec{H},$$

hence $$\vec{B} = \frac{\mu NI}{2\pi r}\cdot\vec{a}_\phi.$$

The flux inside the toroid is $\psi = B \times A$ or $\psi = \int_s \vec{B}\cdot d\vec{s}$,

where A is the area of the solenoidal section.

$$\psi = \int_s \vec{B}\vec{a}_\phi \cdot ds\,\vec{a}_\phi$$

$$ds = dz\,d\rho\,\vec{a}_\phi \qquad \text{(here let } d\rho = dr\text{)}.$$

From Fig. 3.36,

$$\psi = \int_s \vec{B}\,dz\,dr$$

$$\psi = \int_{r=r_1}^{r=r_2}\int_{z=0}^{z=h} \frac{\mu NI}{2\pi r}dz\,dr$$

$$\phi = \frac{\mu NIh}{2\pi}\ln\frac{r_2}{r_1}$$

$$\lambda = N\psi = \frac{\mu N^2 Ih}{2\pi}\ln\frac{r_2}{r_1}$$

$$l = \frac{\lambda}{I} = \frac{\mu N^2 h}{2\pi}\ln\frac{r_2}{r_1} \text{ henry.}$$

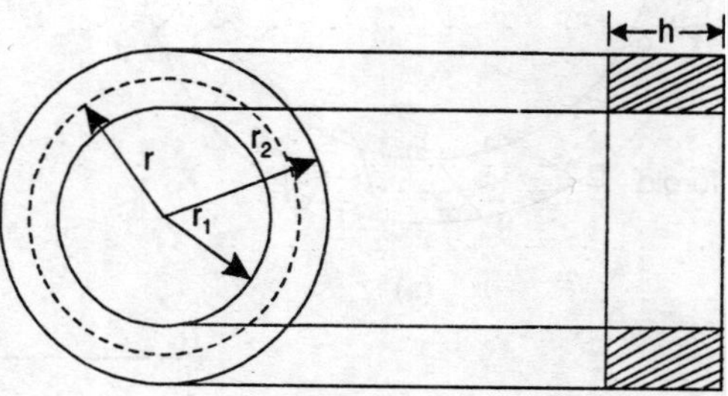

Fig. 3.36 Toroid with rectangular cross-section.

Example 3.15

A toroidal solenoid having N turns and carrying current of I is shown in Fig. 3.37. Its cross-section is a square of side a and its mean radius is ρ_0. Show that the self-inductance is

$$L = \frac{\mu_0 N^2 a}{2\pi} \ln\left(\frac{2\rho_0 + a}{2\rho_0 - a}\right).$$

The field intensity at any point inside the toroid is given by eqn. (3.40)

$$\vec{H} = \frac{NI}{2\pi\rho_0}\vec{a}_\phi.$$

$$\text{Flux density of toroid } = \vec{B} = \frac{\mu NI}{2\pi\rho_0}\vec{a}_\phi.$$

Flux surrounding the solenoidal section of the toroid is

$$\psi = \int_s \vec{B}\cdot d\vec{s},$$

where $d\vec{s}$ is the area of the solenoidal section.
From Fig. 3.37 (*c*)

$$d\vec{s} = dz\, d\rho\, \vec{a}_\phi.$$

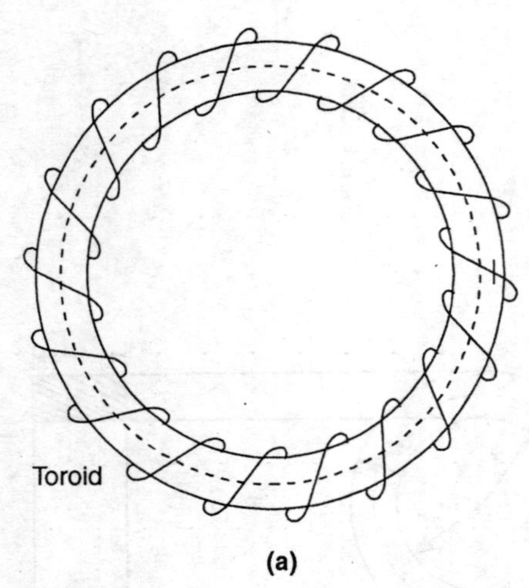

(a)

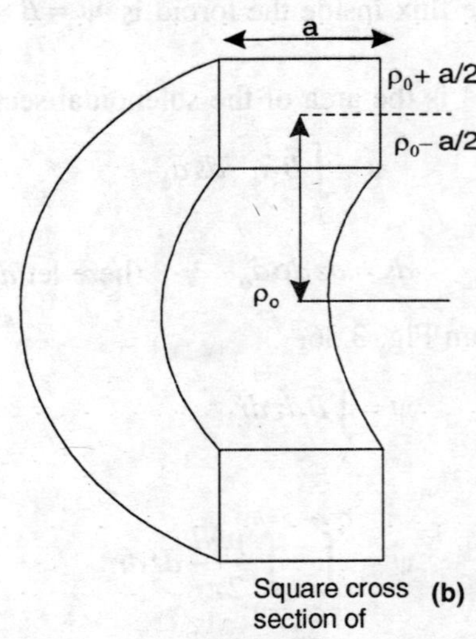

Square cross section of toroid (b)

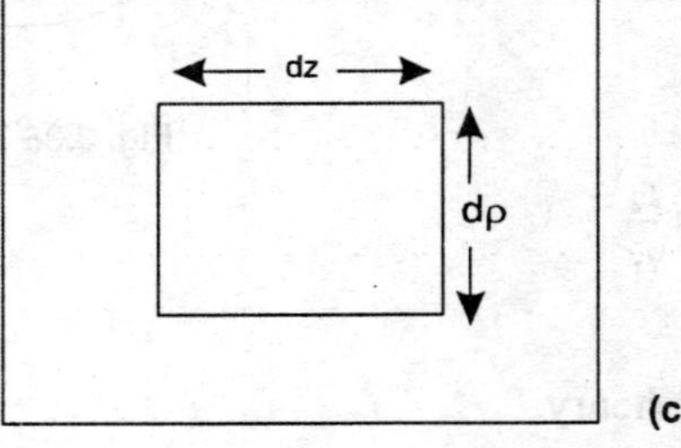

(c)

Fig. 3.37

$$\therefore \qquad \psi = \iint B\vec{a}_\phi \cdot dz\, d\rho\, \vec{a}_\phi$$

$$= \int_{\rho=\rho_1}^{\rho=\rho_2} \int_{z=0}^{z=a} \frac{\mu NI}{2\pi\rho} dz\, d\rho$$

$$= \check{\rho}_1$$

where, from Fig. 3.36 (*b*)

$$\rho_1 = \rho_0 - \frac{a}{2}, \; \rho_2 = \rho_0 + \frac{a}{2}.$$

$$\therefore \qquad \phi = \frac{\mu NIa}{2\pi} \ln\left(\frac{\rho_0 + a/2}{\rho_0 - a/2}\right)$$

$$\lambda = N\psi = \frac{\mu N^2 Ia}{2\pi} \ln\left(\frac{2\rho_0 + a}{2\rho_0 - a}\right)$$

$$L = \frac{\lambda}{I} = \frac{\mu N^2 a}{2\pi} \ln\left(\frac{2\rho_0 + a}{2\rho_0 - a}\right) \quad \text{henry.}$$

Example 3.16

Show that the mutual inductance in either direction of a co-axial solenoid having same diameter but of different length as in Fig. 3.38 is same.

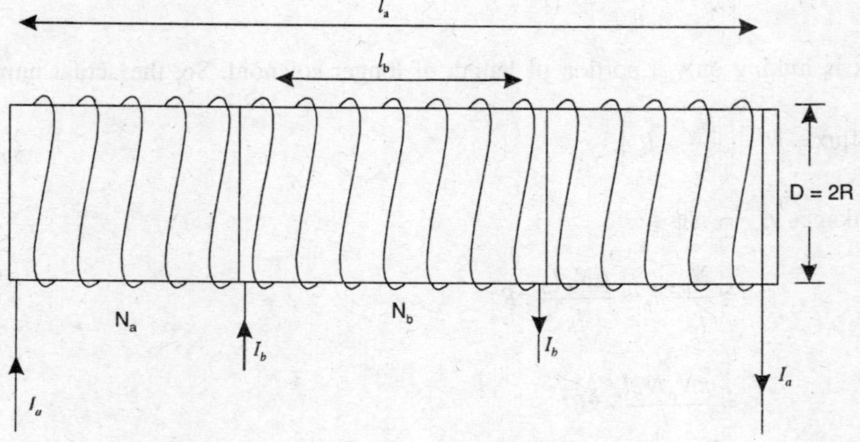

Fig. 3.38 Co-axial solenoid with same radius but different length

Assume the radius of the solenoid is '*R*'. The longer solenoid has N_a turns and its length is 'l_a' and smaller solenoid has N_b turns over its length 'l_b'.

From eqn. (3.39), the field intensity in the longer solenoid is $H = |\vec{H}| = \frac{N_a I_a}{l_a}$,

where I_a is the current through the longer solenoid.

$$B = |\vec{B}| = \frac{\mu N_a I_a}{l_a}.$$

Flux surrounding it, $\psi_{ab} = B \times A$

$$\psi_{ab} = \frac{\mu N_a I_a}{l_a} \times \pi R^2$$

Since the flux is linking with entire length of smaller solenoid, flux linkages

$$\lambda_{ab} = N_b \psi_{ab}$$

$$= \frac{\mu N_a N_b \pi R^2 I_a}{l_a}.$$

So, mutual inductance $M_{ab} = \frac{\lambda_{ab}}{I_a} = \frac{\mu N_a N_b \pi R^2}{l_a}.$

Similarly, the flux density of the smaller solenoid $B_b = \frac{\mu N_b I_b}{l_b}.$

The flux surrounding it, $\psi_{ab} = B_b \times A$ $\left(A \text{ is same} = \pi R^2\right)$

$$\psi_{ab} = \frac{\mu N_b I_b}{l_b} \pi R^2.$$

This flux is linking only a portion of length of longer solenoid. So, the actual number of turns linking with flux $= N = \left(\frac{N_a}{l_a} \times l_b\right).$

$\therefore$ Flux linkages $\lambda_{ab} = N\psi_{ab}$

$$= \frac{N_a}{l_a} \cdot l_b \times \frac{\mu N_b I_b}{l_b} \pi R^2$$

$$= \frac{\pi N_a N_b I_b}{l_a} \pi R^2$$

$$M_{ab} = \frac{\lambda_{ab}}{I_b} = \frac{\mu N_a N_b \pi R^2}{l_a}.$$

SOLVED PROBLEMS

Problem 14

Find magnetic field intensity around infinitely long straight conductor by magnetic vector potential.

Solution:

Magnetic vector potential is given by

$$\vec{A} = \int \frac{\mu_0 I d\vec{l}}{4\pi r}$$

$$\vec{A}_z = \frac{\mu_0 I}{4\pi} \int_{-L}^{L} \frac{dz}{r}$$

$$= \frac{\mu_0 I}{4\pi} \int_{-L}^{+L} \frac{dz}{\sqrt{Z^2 + R^2}}$$

$$= \frac{\mu_0 I}{2\pi} \left[\ln\left(Z + \sqrt{Z^2 + R^2}\right) \right]_0^L$$

$$= \frac{\mu_0 I}{2\pi} \left[\ln\left(L + \sqrt{L^2 + R^2}\right) - \ln R \right]$$

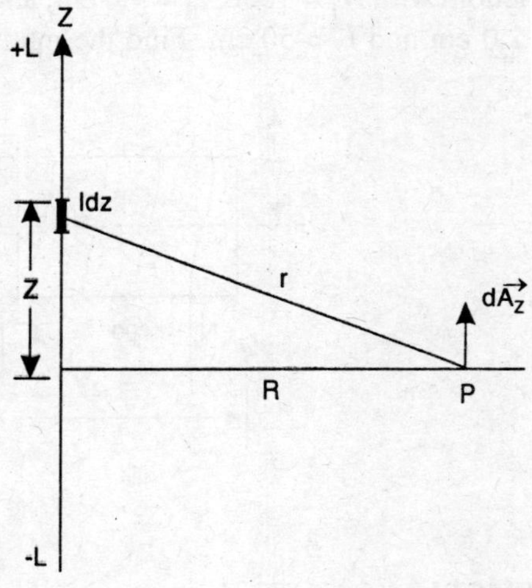

Fig. P. 3.14

$$\vec{A}_z = \frac{\mu_0 I}{2\pi} [\ln(2L) - \ln R] \vec{a}_z$$

$(\because L >> R,\ L^2 + R^2 \approx L^2)$

$\vec{B} = \nabla \times \vec{A}$, in cylindrical co-ordinator

$$\nabla \times \vec{A} = \frac{1}{\rho} \begin{vmatrix} \vec{a}_\rho & \rho \vec{a}_\phi & \vec{a}_z \\ \partial/\partial\rho & \partial/\partial\phi & \partial/\partial z \\ A_\rho & \rho A_\phi & A_z \end{vmatrix}.$$

Let $\rho = R$

$$= \frac{1}{R} \begin{vmatrix} \vec{a}_\rho & R \cdot \vec{a}_\phi & \vec{a}_z \\ \partial/\partial R & \partial/\partial\phi & \partial/\partial z \\ 0 & 0 & \frac{\mu_0 I}{2\pi} \ln\left(\frac{2L}{R}\right) \end{vmatrix}$$

$$= \frac{1}{R}\left(-Ra_\phi\right)\frac{\partial}{\partial R}\left(\frac{\mu_0 I}{2\pi}\ln\left(\frac{2L}{R}\right)\right)$$

$$\vec{B} = \frac{\mu_0 I}{2\pi R}\vec{a}_\phi \text{ A/m}$$

Problem 15

A solenoid with $N_1 = 1000$, $r_1 = 1.0$ cm, and $l_1 = 50$ cm is concentric within a second coil of $N_2 = 2000$, $r_2 = 2.0$ cm and $l_2 = 50$ cm. Find the mutual inductance assuming free space conditions.

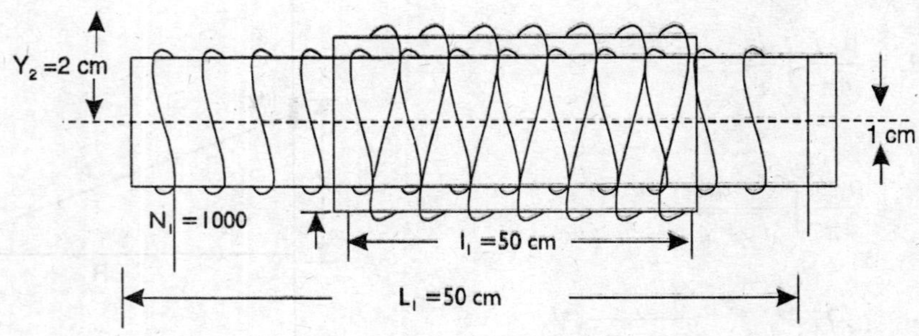

Fig. P. 3.15

Solution:

Let us find mutual inductance of solenoid 2 due to flux in solenoid 1.

$$M_{21} = \frac{N_2\phi_{21}}{I_1}\text{ H}$$

where ϕ_{21} = flux linking with solenoid 2 but produced by solenoid 1.

ϕ_1 = flux produced by solenoid 1 = $B_1 \times A_1$.

We know

$$B_1 = \mu_0 H_1 \text{ and } [\mu_r = 1]$$

$$H_1 = \frac{N_1 I_1}{l_1} = \frac{1000 \times I_1}{0.50}\text{ A/m}$$

$$\phi_1 = \frac{\mu_0 \times 1000 \times I_1}{0.5} \times \pi r_1^2$$

$$= \frac{4\pi \times 10^{-7} \times 1000 \times I_1}{0.5} \times \pi \times 1 \times 10^{-4}.$$

As all the flux will link with solenoid 2 (neglect fringing and leakage)

$$\phi_{21} = \phi_1$$

$$\therefore \quad M_{21} = \frac{2000 \times 4\pi \times 10^{-7} \times 1000 \times I_1 \times \pi \times 1 \times 10^{-4}}{0.5 I_1}$$

$$= 1.58 \text{ mH}$$

Problem 16

Calculate inductance of a ring-shaped coil having a mean diameter of 20 cm, wound on a wooden core of 2 cm diameter. The winding is uniformly distributed and contains 200 turns.

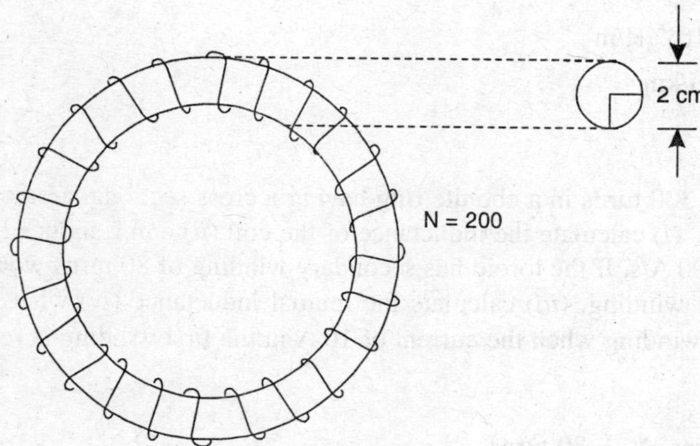

Fig. P. 3.16

Solution:

Inductance of toroid is

$$L = \frac{\mu_0 N^2 A}{l_m} \quad [\text{Let } \mu_r = 1]$$

$N = 200$

A = Area of cross-section $\pi r^2 = \pi \times 1 \times 10^{-4}\ \text{m}^2$

l_m = mean length = $\pi D = \pi \times 20 \times 10^{-2}$ m

$$L = \frac{4\pi \times 10^{-7} \times 200^2 \times \pi \times 1 \times 10^{-4}}{\pi \times 20 \times 10^{-2}}$$

$$= 25.13\ \mu\text{H}$$

Problem 17

Calculate the inductance per kilometer of two parallel round conductors spaced 80 cm apart and having conductor diameter of 1 cm.

Solution:

From eqn. (3.121), inductance per unit length

$$L = \frac{\mu}{\pi}\left[\frac{1}{4} + \ln\frac{d}{a}\right]$$

$d = 80$ cm, $a = 0.5$ cm

$$L = \frac{4\pi \times 10^{-7}}{\pi}\left[\frac{1}{4} + \ln\frac{80}{0.5}\right]$$

$= 2.13 \times 10^{-6}$ H/m

$= 2.13$ mH/km.

Problem 18

A toroid is wound with 300 turns in a ebonite ring having a cross sectional area of 4 cm^2 and a mean circumference of 35 cm. (*i*) calculate the inductance of the coil (*ii*) e.m.f. induced when the current is reduced at the rate of 200 A/s. If the toroid has secondary winding of 80 turns wound over the ebonite ring and inside the first winding, (*iii*) calculate the mutual inductance (*iv*) what will be the induced e.m.f. in the secondary winding when the current of 10 A in the first winding is reversed in 1 second?

Solution:

$N_1 = 300$ turns, $N_2 = 80$ turns

$A = 4$ cm^2, $L = 2\,\pi r = 35$ cm, $\mu_r = 1$.

$$\frac{dI_1}{dt} = 200\,\text{amp/s}$$

$$\frac{dI_2}{dt} = 10\,\text{amp/s.}$$

Inductance of the coil is given by the equation, $L = \dfrac{N\phi}{I}$,

where the flux is given by, Flux $= B_m \times$ A

$$= \frac{\mu NI}{2\pi r} \cdot \text{A}$$

$$\therefore \quad \text{Inductance} = L = \frac{N \cdot \dfrac{\mu NI}{2\pi r} \cdot \text{A}}{I}$$

$$= \frac{\mu N^2}{2\pi r} \times \text{A}$$

$$= \frac{4\pi \times 10^{-7} \times 300^2}{35 \times 10^{-2}} \times 4 \times 10^{-4}$$

$$\mathbf{L = 0.129\,mH}$$

(*ii*) Emf induced (self-induced e.m.f.) $E_1 = L \cdot \frac{dI_1}{dt}$

$$= 0.129 \times 10^{-3} \times 200$$

$$\mathbf{E_1 = 25.8\ mV}$$

(*iii*) Mutual inductance $= \frac{\mu \cdot N_2 \cdot N_1}{\text{length}} \times A$

$$= \frac{4\pi \times 10^{-7} \times 50 \times 300 \times 4 \times 10^{-4}}{35 \times 10^{-2}}$$

$$M = 0.0344 \text{ mH}$$

(*iv*) E.m.f. induced in second coil

$$= M \cdot \frac{dI_1}{dt}$$

$$= 0.0344 \times 10$$

$$= 0.0344 \text{ V}$$

$$E_2 = 344 \text{ mV}.$$

Problem 19

A straight 1 m long conductor is situated in a direction perpendicular to the uniform magnetic field of 0.8 tesla. If the conductor carries a steady current of 150 A find (*a*) the mechanical force on the conductor (*b*) the power required to move the conductor against the force at a uniform speed of 10 m/s. (*c*) The e.m.f. generated under this condition.

Solution:

$L = 1$ m; $\theta = 90°$; $I = 150$ A; $B = 0.8$ tesla; $v = 10$ m/s

We know that, (*a*) the force experienced by the conductor

$$F = B.I.L. \sin\theta$$

$$= 0.8 \times 150 \times 1 \times \sin 90$$

$$\mathbf{F = 120\ N.}$$

(*b*) Power required to move the conductor

$$P = \frac{\text{force} \times \text{distance}}{\text{time}}$$

$= \text{force} \times \text{velocity}$

$= 120 \times 10$

$P = 1200$ watt

(*c*) Dynamically induced e.m.f.

$V = Blv \sin\theta$

$= 0.8 \times 1 \times 10 \times \sin 90$

$V = 8$ volt

Problem 20

A solenoid has 2000 turns of copper wire wound on a former of length 1 m and diameter of 4 cm. It is placed co-axially within another solenoid with same length and number of turns but with a diameter of 7 cm. Determine the mutual inductance between two solenoids and also coefficient of coupling.

Solution:

$N_1 = N_2 = 2000.$

$l_1 = l_2 = 1$ m.

$d_1 = 4$ cm; $d_2 = 7$ cm.

From the Example 3.16

$$M_{21} = M = \frac{\mu \cdot N_1 N_2 \pi R_1^2}{l}$$

$$= 4\pi \times 10^{-7} (2000)^2 \times \pi \times \left(2 \times 10^{-2}\right)^2 / 1$$

$$= 6.31 \text{mH.}$$

Self-inductance of Coil 1

$$L_1 = \frac{\mu N_1^2 A_1}{l_1} = \frac{4\pi \times 10^{-7} \times 2000^2 \times \pi \times \left(2 \times 10^{-2}\right)^2}{1}$$

$$L_1 = 6.31 \text{mH.}$$

And self-inductance of Coil 2

$$L_2 = \frac{\mu N_2^2 A_2}{l_2} = \frac{4\pi \times 10^{-7} \times 2000^2 \times \pi \times \left(3.5 \times 10^{-2}\right)^2}{1}$$

$$L_2 = 0.019 \text{ mH}$$

We know that

$$M = k\sqrt{L_1 L_2}$$

$$k = \frac{M}{\sqrt{L_1 L_2}} = \frac{6.31 \times 10^{-3}}{\sqrt{6.31 \times 10^{-3} \times 0.019 \times 10^{-3}}}$$

$$= 0.576.$$

Problem 21

An iron ring of relative permeability 100 is wound uniformly with 2 coils of 100 and 400 turns of wire. The cross-section of ring is 4 cm^2 and the mean circumference is 50 cm. Calculate (*a*) self-inductance (*b*) mutual inductance (*c*) total inductance, when the coils are connected in series in the same sense and also in opposition.

Solution:

$\mu_r = 100$

$N_1 = 100$

$N_2 = 400$

Area = 4 cm^2

$2\pi r = 50$ cm.

$$L_1 = \frac{\mu N_1^2 A}{l} = \frac{\mu_0 \mu_r N_1^2 A}{l}$$

$$= \frac{4\pi \times 10^{-7} \times 100 \times 100^2 \times 4 \times 10^{-4}}{0.5}$$

$$L_1 = 1.005 \text{ mH}.$$

Self-inductance of coil 2.

$$L_2 = \frac{\mu N_2^2 A}{l}$$

$$= \frac{4\pi \times 10^{-7} \times 100 \times 400^2 \times 4 \times 10^{-4}}{0.5}$$

$$L_2 = 16.08 \text{ mH}.$$

(*b*) Mutual inductance between two coils

$$M = \frac{\mu N_1 N_2 \pi r^2}{l}$$

$$= \frac{4\pi \times 10^{-7} \times 100 \times 100 \times 400 \times 4 \times 10^{-4}}{0.5}$$

$$M = 4.02 \text{ mH}.$$

(*c*) Total inductance if the two coils are connected in series and flux of same sense

$$L = L_1 + L_2 + 2M.$$

$$= 1.005 + 16.08 + 4.02 \times 2$$

$$L_{eq} = 25.125 \text{ mH.}$$

Total inductance if the two coils are connected in series but with opposite senses

$$L_{eq} = L_1 + L_2 - 2M$$

$$= 1.005 + 16.08 - 4.02 \times 2$$

$$= 9.045 \text{ mH.}$$

Problem 22

Two coils are connected in series and their total self-inductance is 4.4 mH. When one coil is reversed, total self-inductance is 1.6 mH. All the flux in the first coil links with second coil. But only 40% of the of the second coil links with first coil. Find the self-inductance of each coil and the mutual inductance between the two.

Solution:

When current is not reversed, total self-inductance is

$$L_{eq} = L_1 + L_2 + 2M = 4.4 \text{ mH.} \qquad ...(i)$$

When current is reversed, total self-inductance is

$$L_{eq} = L_1 + L_2 - 2M = 1.6 \text{ mH} \qquad ...(ii)$$

$$k_1 = 1, \quad k_2 = 0.4, \quad k_1 k_2 = k^2 = 0.4.$$

Adding (*i*) and (*ii*)

$$2(L_1 + L_2) = 6 \text{ mH}$$

$$L_1 + L_2 = 3 \text{ mH.}$$

Subtracting (*i*) from (*ii*)

$$4M = 2.8 \text{ mH}$$

$$M = 0.7 \text{ mH.}$$

We know that

$$M = k\sqrt{L_1 L_2}\,.$$

$$M^2 = k^2(L_1 L_2).$$

$$(0.7)^2 = 0.4\, L_1(3 - L_1)$$

$$1.225 = 3L_1 - L_1^2$$

$$L_1^2 - 3L_1 + 1.225 = 0.$$

Solving

$$L_1 = 2.5 \text{ mH} \quad \text{(or)} \quad 0.5 \text{ mH}$$

$$L_2 = 0.5 \text{ mH} \quad \text{(or)} \quad 2.5 \text{ mH}.$$

Problem 23

Find the pull exerted on the plunger of an electromagnet when total flux uniformly distributed to 500 mwb and the diameter of plunger is 2.54 cm.

Solution:

$$\phi = 500 \times 10^{-6} \text{ wb}.$$

$$d = 2.54 \times 10^{-2} \text{ m}.$$

$$\phi = \frac{B_m}{A}$$

$$= \frac{500 \times 10^{-6}}{\pi\left(\frac{2.54 \times 10^{-2}}{2}\right)^2}$$

$$= 5.06 \times 10^{-4} \text{ m}^2$$

Pull (or) Force on the plunger

$$F = \frac{1}{2}\frac{B^2 A}{\mu}$$

$$= \frac{1}{2} \times \frac{5.06 \times 10^{-4} \times \pi \times \left(\frac{2.5 \times 10^{-2}}{2}\right)^{-2}}{4\pi \times 10^{-7}}$$

$$F = 196 \text{ N}.$$

Problem 24

In a receiver, the size of each of the pole is 1.2 cm × 0.2 cm and the flux between each pole and diaphram is 3μwb. Find the force with which the diaphram is attracted to the poles.

Solution:

Area of the pole $= 1.2 \times 0.2$

$= 2.4 \times 10^{-4} \text{m}^2.$

Flux of the pole $= 3 \times 10^{-6}$ wb

Number of poles = 2.

We know that the force between two magnets is given by

$$F = \frac{B^2 A}{2\mu_0} \text{ newton}$$

$$\phi = B \times \text{area}.$$

$$B = \frac{\phi}{A}.$$

Substituting for B,

$$F = \frac{\phi^2}{A^2} \cdot \frac{A}{2\mu_0} \text{ newton}$$

$$= \frac{\phi^2}{2\mu_0 A}.$$

Since there are two poles,

$$F = \frac{2 \cdot \phi^2}{2\mu_0 A} = \frac{\left(3 \times 10^{-6}\right)^2}{4\pi \times 10^{-7} \times \left(2.4 \times 10^{-4}\right)}$$

$$F = 0.3 \text{ N}.$$

Problem 25

Determine the mutual inductance between an infinitely long straight conductor along Y-axis and a rectangular single turn coil situated in XY-plane with its corner located at point $(a, 0)$, $(a + d, 0)$, (a, h), $(a, d + h)$.

Solution:

We can write, magnetic field produced by the straight conductor is

$$\vec{H} = \frac{I}{2\pi x} \vec{a}_z.$$

As $\quad \vec{B} = \mu \vec{H}$

$$\vec{B} = \frac{\mu I}{2\pi x} \cdot \vec{a}_z.$$

The flux linking with the rectangular coil is:

$$\text{Flux } \psi = \iint \vec{B} \cdot \vec{a}_z \, ds.$$

For the coil placed in the XY- plane, $ds = dx \cdot dy \cdot \vec{a}_z$. The flux produced by the conductor in Y-axis links with rectangular turn coil.

$$\therefore \quad d\lambda = d\psi = \vec{B} \cdot d\vec{s}$$

$$= \frac{\mu I}{2\pi x} \cdot dx \cdot dy.$$

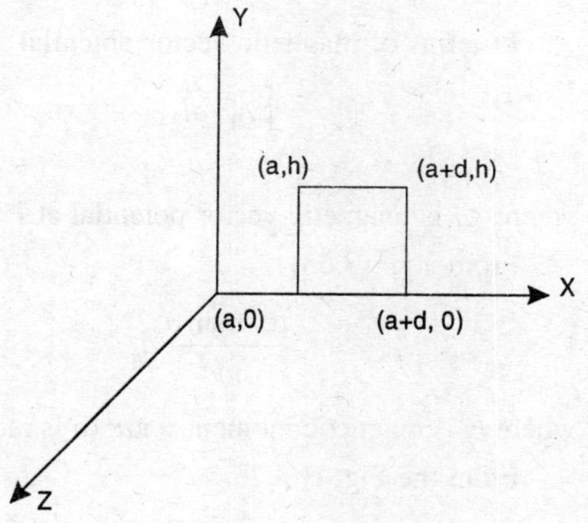

Fig. P. 3.25

Total flux linkage

$$\psi = \int_0^h \int_a^{a+d} B \cdot dx \cdot dy$$

$$= \int_{y=0}^{h} \int_{x=a}^{(a+d)} \frac{\mu I}{2\pi x} \cdot dx \cdot dy$$

$$= \frac{\mu I}{2\pi} \int_0^h (\ln x)_a^{(a+d)} \cdot dy$$

$$= \frac{\mu I}{2\pi} (\ln(a+d) - \ln a) \int_0^h dy$$

$$= \frac{\mu I h}{2\pi} \ln\left(\frac{a+d}{a}\right).$$

$$\therefore \quad \psi = \frac{\mu I h}{2\pi} \ln\left(\frac{a+d}{a}\right) \cdot \vec{a}_z$$

$$M = \frac{\psi}{I} = \frac{\mu h}{2\pi} \ln(a+d).$$

Problem 26

Determine the force between two co-axial circular coils of radii b_1 and b_2 separated by a distance d that is much larger than the radii ($d >> b_1, b_2$). The coil consists of N_1 and N_2 closely-wound turns and carry currents I_1 and I_2, respectively.

Initially the mutual inductance at a point on any of the coil is to be determined and then the force can be calculated using the equation.

Let us consider a point P on coil 2.

Mutual inductance on coil 2 due to coil 1 is

$$M_{21} = \frac{N_2 \psi_{21}}{I_1},$$

where ψ_{21} is the flux linking with coil 2 due to current in coil 1.

In terms of magnetic vector potential $\vec{A}$

$$\psi_{21} = \oint_{c_2} \vec{A}_1 \cdot d\vec{l}_2,$$

where A_1 is magnetic vector potential at P due to coil 1.

From eqn. (3.63)

$$\vec{A} = \frac{\mu_0 m \sin\theta}{4\pi r^2}\vec{a}_\phi,$$

where m is magnetic moment $= \pi a^2$ (a is radius of coil) and r is the distance of the point from the coil).

From the Fig. P. 3.25,

$$\vec{A}_1 = \frac{\mu_0 N_1 I_1 b_1^2}{4R^2}\sin\theta\,\vec{a}_\phi,$$

where b_1 is the radius of coil 1 with number of turns N_1 and R is the distance of the point P from the coil 1.

$$\vec{A}_1 = \frac{\mu_0 N_1 I_1 b_1^2}{4R^2}\left(\frac{b_2}{R}\right)\vec{a}_\phi$$

$$= \frac{\mu_0 N_1 I_1 b_1^2 b_2}{4\left(d^2 + b_2^2\right)^{3/2}}\vec{a}_\phi$$

$$\psi_{21} = \oint_{c_2} \vec{A}_1 \cdot d\vec{l}_2 = \frac{\mu_0 N_1 I_1 b_1^2 b_2}{4\left(d^2 + b_2^2\right)^{3/2}}(2\pi b_2).$$

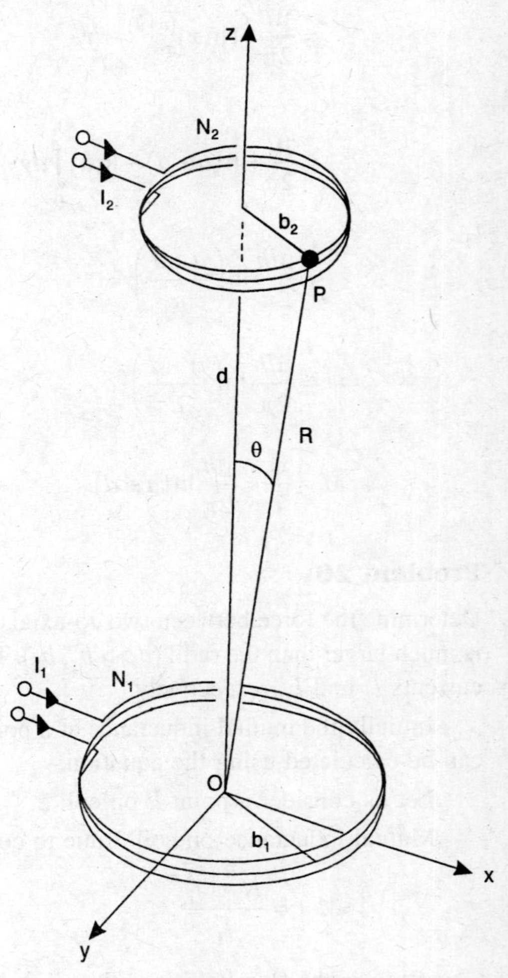

Fig. P. 3.26

Mutual inductance

$$M_{21} = \frac{N_2 \psi_{21}}{I_1}$$

$$M_{21} = \frac{\mu_0 N_2 N_1 b_1^2 b_2^2 \pi}{2\left(d^2 + b_2^2\right)^{3/2}}.$$

From eqns. (3.138) and (3.135)

Force $\vec{F} = \nabla W$ and

$$\nabla W = \frac{1}{2} I^2 dL \text{ for self-inductance.}$$

For mutual inductance, $W = M_{21} I_1 I_2$

$$\nabla W = I_1 I_2 \nabla M_{21}.$$

Hence, $\vec{F} = I_1 I_2 \nabla M_{21}.$

The force on coil 2 exerted by coil 1 due to change in distance d between the two coils is given by

$$\vec{F}_{21} = I_1 I_2 \left.\frac{dM_{21}}{dz}\right|_{z=d} \vec{a}_z$$

$$\vec{F}_{21} = I_1 I_2 \frac{d}{dz}\left[\frac{\mu_0 N_2 N_1 b_1^2 b_2^2 \pi}{2\left(z^2 + b_2^2\right)^{3/2}}\right]$$

$$= \frac{\mu_0 N_1 N_2 I_1 I_2 b_1^2 b_2^2 \pi}{2} \frac{d}{dz}\left[\frac{1}{\left(z^2 + b_2^2\right)^{3/2}}\right]$$

$$\frac{d}{dz}\left[\frac{1}{\left(z^2 + b_2^2\right)^{3/2}}\right]_{z=d} = \left.\frac{-3z}{\left(z^2 + b_2^2\right)^{5/2}}\right|_{z=d}$$

$$= \frac{-3d}{\left(d^2 + b_2^2\right)^{5/2}}$$

$$\vec{F}_{21} = \frac{3}{2} \frac{\mu_0 N_1 N_2 I_1 I_2 b_1^2 b_2^2 \pi d}{\left(d^2 + b_2^2\right)^{5/2}} - \vec{a}_z \text{ N}$$

$$= \frac{3}{2}\mu_0 \frac{m_1 m_2 d}{\left(d^2 + b_2^2\right)^{5/2}} (-\vec{a}_z) \text{ N},$$

where m_1 and m_2 are magnetic moment of coils 1 and 2 with number of turns N_1 and N_2 respectively.

$$m_1 = N_1 \pi b_1^2 I_1,\ m_2 = N_2 \pi b_2^2 I_2$$

As $d >> b_2$,

$$\vec{F}_{12} \approx -\frac{3\mu_0 m_1 m_2}{2\pi d^4} \text{ N}.$$

Negative sign indicates that $\vec{F}_{12}$ is a force of attraction for current flowing in the same direction.

3.18 MAGNETIC CIRCUITS

In this chapter, magnetic field problems are solved using magnetic circuit approach. Magnetic devices such as toroids, transformers, motors, generators and relays may be considered as magnetic circuits. The analysis can be easily done by using the analogy between electric circuit and magnetic circuit. Once the analogous electric circuit is drawn for the given magnetic circuit, then by applying the electric circuit concepts, a solution can be obtained directly.

The analogy between magnetic and electric circuits is summarised in the table below:

Electric circuit	Magnetic circuit
Conductivity σ	Permeability μ
Field intensity E	Field intensity H
Current $I = \int J \cdot ds$	Flux $\psi = \int B \cdot ds$
Current density $J = \frac{I}{S} = \sigma E$	Flux density $B = \frac{\Psi}{S} = \mu H$
e.m.f. V	MMF F
Resistance R	Reluctance $\Re$
Conductance $G = \frac{1}{R}$	Permeance $P = \frac{1}{\Re}$
Ohms law $R = \frac{V}{I} = \frac{l}{\sigma S}$	Ohms law $\Re = \frac{\rho}{\psi} = \frac{l}{\mu s}$
or $V = El = IR$	or $J = Hl = \psi\Re = NI$
Kirchoff's law	Kirchoff's law
$\Sigma I = 0$	$\Sigma\psi = 0$
$\Sigma V - \Sigma RI = 0$	$\Sigma F - \Sigma\Re\psi = 0$

Based on the analogy between electric and magnetic circuits, Kirchoff's current and voltage laws can be applied to nodes and loops of a given magnetic circuit just as in an electric circuit. The rules of adding voltages and for combining series and parallel resistance also hold for MMF's and reluctances. Thus for n magnetic elements in series,

$$\psi_1 = \psi_2 = \psi_3 = \ldots\ldots = \psi_n$$

and $F = F_1 + F_2 + \ldots + F_n$.

For n magnetic circuit in parallel,

$$\psi = \psi_1 + \psi_2 + \psi_3 + \ldots\ldots + \psi_n$$

$$F = F_1 = F_2 = \ldots = F_n.$$

There are also some differences between electric and magnetic circuits. Unlike an electric circuit where current I flows, magnetic flux does not flow. Also, conductivity σ is independent of current density J in an electric circuit, whereas permeability μ varies with flux density B in magnetic circuit. These differences notwithstanding, the magnetic circuit concept serves as an approximate analysis of practical magnetic devices.

Problem 27

The toroidal core shown in the figure below has $\rho_0 = 10$ cm and a circular cross-section of $a = 1.0$ cm. If the core is made of steel with $\mu = 1000\ \mu_0$ and has a coil with 200 turns, calculate the amount of current that will produce a flux of 0.5 mwb in the core.

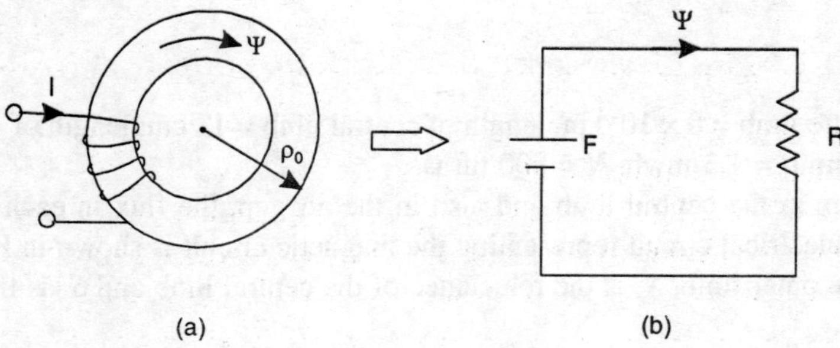

Fig. P. 3.27

Solution:

$\rho_0 = 10$ cm; $a = 1$ cm $\mu = 1000\ \mu_0$

N = 200 $\psi = 0.5$ mwb.

The problem can be solved using the electric analogy.

Mean length of toroid $= l = 2\ \pi\rho_0$

$= 2\ \pi \times 10 \times 10^{-2}$

$= 0.6283.$

Area of cross-section $= \pi a^2$

$= \pi \times (1 \times 10^{-2})^2$.

$= 3.14 \times 10^{-4}.$

$$\text{Reluctance } \Re = \frac{l}{\mu A} = \frac{l}{\mu_0 \mu_r a}$$

$$= \frac{0.6283}{4\pi \times 10^{-7} \times 1000 \times 3.14 \times 10^{-4}}$$

$= 1.592 \times 10^6.$

Magnetomotive force $F = NI = \psi\Re$

$$I = \frac{\psi\Re}{N} = \frac{0.5 \times 10^{-3} \times 1.592 \times 10^6}{200}$$

$= 3.98$ A.

Problem 28

A magnetic material is made of mild steel. The central limb is wound with 800 turns and has a cross-section of 8 cm^2. Each of the outer limb has a cross-section of 5 cm^2. The air gap length is 1mm. Calculate the current required to set up a flux of 1.3 mwb in central limb. Data for the *B-H* curve is given below.

B	1.1	1.3	1.35	1.5	1.625
H	500	850	1000	2000	3800

Solution:

$l_a = 10^{-3}$ m.

Area of the centre limb = 8×10^{-4} m; length of central limb = 12 cm; length of outer limb = 30 cm. Flux in the central limb = 1.3 mwb. N = 800 turns.

Let ψ is the flux in the central limb and also in the air gap, the flux in each of the side limb is $\psi/2$. An equivalent electrical circuit representing the magnetic circuit is shown in Fig P. 3.28 (*b*). S_0 is the reluctance of the outer limb, S_c is the reluctance of the central limb and S_a is the reluctance of the air gap.

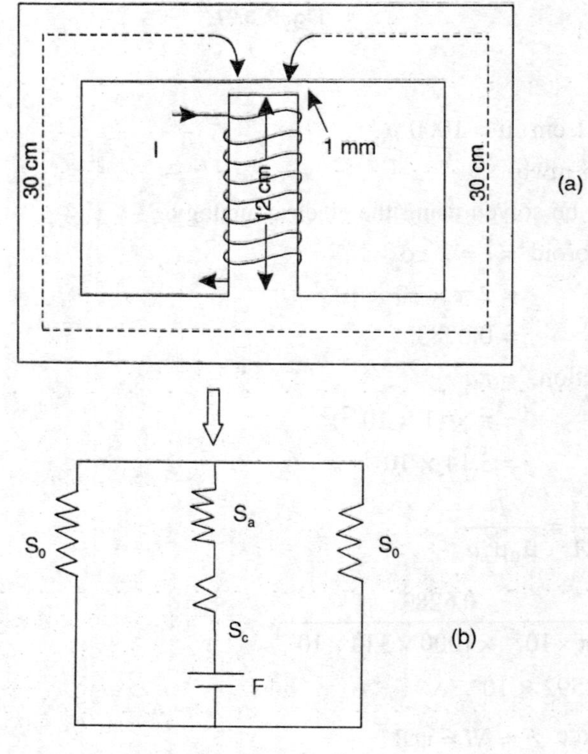

Fig. P. 3.28

MMF required for the air gap

$$B_A = \frac{\text{Flux}}{\text{Area}} = \frac{1.3 \times 10^{-3}}{8 \times 10^{-4}} = 1.625 \text{ tesla}$$

$$H_A = \frac{B}{\mu_0 \mu_r} = \frac{1.625}{4\pi \times 10^{-7}} \qquad [\mu_r = 1]$$

$$= 12.93 \times 10^5 \text{ AT/m}.$$

$\therefore$ MMF in the air gap $= \boldsymbol{H} \times \boldsymbol{l} = 12.93 \times 10^5 \times 10^{-3}$

$= 1293$ AT.

MMF required for the central limb

$$B_c = \frac{\phi}{A_c} = \frac{1.293 \times 10^{-3}}{8\pi \times 10^{-4}} 1.616 \text{ tesla}$$

$H_c = 3850$ AT/m [From the data]

MMF required for central limb $= H_c \times L_c$

$= 3800 \times 12 \times 10^{-2}$

$= 456$ AT.

MMF required for outer limb

Flux density $(B_0) = \frac{0.65 \times 10^{-3}}{5 \times 10^{-4}}$ 1.3 tesla

From data $H_0 = 850$ AT/m.

MMF required for outer limb $= H_0 \times l_0$.

$= 850 \times 30 \times w^{-2}$

$= 255$ AT.

$\therefore$ Total MMF $= 1293 + 456 + 255 = 2004$ AT.

$$I = \frac{\text{MMF}}{\text{Turns}} = \frac{2004}{500}$$

$I = 4.01$ A.

Problem 29

A mild steel ring having a cross-sectional area of 5 sq. cm and a mean circumference of 50 sq. cm has a coil of 250 turns wound uniformly around it. Calculate the current required to produce flux of 0.8 mwb in the ring. Assume $\mu = 750$.

Solution:

Area of cross-section = 5 sq. cm. = 5×10^{-4} m

Mean length = 50 cm = 50×10^{-2} m

Number of turns = 250.

Flux = 0.8 mwb.

$$\text{Reluctance} = \frac{l}{\mu_0 \mu_r A} \qquad \mu_r = 1.$$

$$= \frac{50 \times 10^{-2}}{4\pi \times 10^{-7} \times 750 \times 5 \times 10^{-4}}$$

$$= 106.1 \times 10^3$$

MMF required $= \phi \times$ Reluctance

$= 0.8 \times 10^{-3} \times 106.1 \times 10^3$

$= 84.88$ AT.

$$\text{MMF} = NI = \frac{84.88}{250}$$

$= 0.339$ A.

Problem 30

An iron ring with a cross-sectional area of 3 cm^2 and a mean circumference of 15 cm is wound with 250 turns wire carrying a current of 0.3A. The relative permeability of the ring is 1500. Calculate the flux established in the ring.

Solution:

Area of the ring = 3 cm^2 = 3×10^{-4} m.

Mean circumference = 15 cm = 15×10^{-2} m

Number of turns = 250

Current in the coil = 3.0 A

Permeability = 1500.

$$\text{Reluctance} = \frac{l}{\mu_0 \mu_r a}$$

$$= \frac{15 \times 10^{-2}}{4\pi \times 10^{-7} \times 1500 \times 3 \times 10^{-4}}$$

$= 26.525 \times 10^3$

MMF required = NI

$= 250 \times 0.3 = 75$

$$\therefore \quad \text{Flux} = \frac{NI}{\text{reluctance}} = \frac{75}{26 \cdot 52 \times 10^3}$$

= 2.82 mwb

SUMMARY

1. Magnetostatic field produced by moving charges or static current (d.c.).

2. Biot-Savart's law is similar to Coulomb's law. It states that the magnetic field intensity $d\vec{H}$ at a point P due to current element $Id\vec{l}$ is

$$d\vec{H} = \frac{Idl \sin\alpha}{4\pi R^2}\hat{a}_n \text{ A/m}.$$

where α is the angle between the differential current element and the line joining the point P to the element.

$$d\vec{H} = \frac{Id\vec{l} \times \vec{R}}{4\pi R^3}.$$

For surface or volume current distribution, $Id\vec{l}$ is replaced by $\vec{k}ds$ or $\vec{J}dv$ respectively.

$$Id\vec{l} \equiv \vec{k}ds \equiv \vec{J}dv$$

3. Magnetic field intensity due to finite length of conductor carrying current I amps is

$$\vec{H} = \frac{I}{4\pi h}(\cos\alpha_1 + \cos\alpha_2)\vec{a}_z.$$

4. Magnetic field intensity due to infinite length of conductor carrying current I amps is

$$\vec{H} = \frac{I}{2\pi h}\vec{a}_z \text{ (or) } \frac{I}{2\pi\rho}\vec{a}_\rho.$$

5. Magnetic field intensity due to a circular loop of conductor of radius 'a' at a point along the axis is

$$\vec{H} = \frac{Ia^2}{2\left(a^2 + h^2\right)^{3/2}}\vec{a}_z .$$

6. Magnetic field intensity due to a circular loop of radius 'a' m at its centre is

$$\vec{H} = \frac{I}{2a}\vec{a}_z.$$

7. Ampere's circuital law which is similar to Gauss's law states that the circulation of $\vec{H}$ around a closed path is equal to the current enclosed by the path

$$\oint_l \vec{H}\cdot d\vec{l} = I_{enc} = \int_s \vec{J}\cdot d\vec{s}$$

$$\nabla \times \vec{H} = \vec{J}.$$

8. The magnetic flux through a surface 'S' is given by

$$\psi = \int_s \vec{B}\cdot d\vec{s} \ (wb)$$

$$\vec{B} = \mu_0 \vec{H}.$$

9. As an isolated magnetic charge does not exist, the net magnetic flux through a closed surface is zero

$$\psi = \int_s \vec{B} \cdot d\vec{s} = 0$$

$$\nabla \cdot \vec{B} = 0.$$

Magnetostatic field is solenoidal.

10. Magnetic field intensity $\vec{H}$ at a point along the axis of a solenoid is

$$\vec{H} = \frac{NI}{l} \vec{a}_x \text{ A/m}.$$

11. The magnetic field intensity due to a toroid of mean radius ρ is

$$\vec{H} = \frac{NI}{2\pi\rho} \vec{a}_\phi = \frac{NI}{l_m} \vec{a}_\phi.$$

12. The magnetic scalar potential V_m (amp) is defined as

$$\vec{H} = -\nabla V_m \text{ if } \vec{J} = 0,$$

and the magnetic vector potential $\vec{A}$ (wb/m) as

$$\vec{B} = \nabla \times \vec{A}.$$

$$\vec{A} = \int_L \frac{\mu_0 I d\vec{l}}{4\pi R} \text{ for line current.}$$

$$= \int_S \frac{\mu_0 k d\vec{s}}{4\pi R} \text{ for surface current.}$$

$$= \int_S \frac{\mu_0 J d\vec{v}}{4\pi R} \text{ for volume current.}$$

13. Magnetic force due to moving change particle in a magnetic field $\vec{F} = Q\vec{u} \times \vec{B}$.

Magnetic force experienced by a current element in a magnetic field

$$\vec{F} = Id\vec{l} \times \vec{B} \text{ (or) } |F| = BIl \sin\theta.$$

Magnetic force between two current elements

$$\vec{F} = I_1 \vec{l} \times \vec{B}_2 \text{ (or) } \vec{F} = I_2 \vec{l} \times \vec{B}_1.$$

14. Lorentz force equation $\vec{F} = Q\left[\vec{E} + \vec{u} \times \vec{B}\right]$.

15. If two conductors carry current in the same direction, then the magnetic force will be attractive. The magnitude is given by

$$|F| = \frac{\mu_0 I_1 I_2}{2\pi D} \text{ N/m}.$$

If the direction of current flow is opposite, then the magnitude of force will be same but the direction is repulsive.

16. Ampere's law of force (or) ampere's work law

$$\vec{F}_{12} = \frac{\mu I_1 I_2}{4\pi} \oint_{c_1} \oint_{c_2} \frac{d\vec{l}_1 \times \left(d\vec{l}_2 \times \vec{a}_{12}\right)}{R^2}.$$

17. Magnetic dipole moment of a small current loop $\vec{m} = IA\vec{a}_n$, where A is area of current loop $\vec{a}_n$, is given by right hand rule.

18. The torque on a current loop with magnetic moment $\vec{m}$ in a uniform magnetic field $\vec{B}$ is

$$\vec{T} = \vec{m} \times \vec{B}$$

$$\vec{T} = \vec{r} \times \vec{F}$$

19. A magnetic dipole is a bar magnet or a small filamental current loop. It is so called due to the fact that its $\vec{B}$ field lines are similar to the $\vec{E}$ field lines of an electric dipole.

20. When a material is subjected to a magnetic field, it becomes magnetised. The magnetisation $\vec{M}$ is the magnetic dipole moment per unit volume of the material. For linear material

$$\vec{M} = \chi_m H,$$

where χ_m is the magnetic susceptability of the material.

21. In terms of their magnetic properties, materials are either linear (diamagnetic or paramagnetic) or non linear (ferromagnetic). For linear materials

$$\vec{B} = \mu\vec{H} = \mu_0\mu_l\vec{H} = \mu_0(1+\chi_m)\vec{H}$$

$$= \mu_0\left(\vec{H} + \vec{M}\right).$$

22. The boundary conditions that $\vec{H}$ or $\vec{B}$ must satisfy at the interface between two different media are

$$\vec{B}_{1n} = \vec{B}_{2n}$$

$$\vec{H}_{1t} - \vec{H}_{2t} = k \text{ (or) } \vec{H}_{1t} = \vec{H}_{2t} \text{ if } \vec{k} = 0.$$

$$\frac{\tan\theta_1}{\tan\theta_2} = \frac{\mu_1}{\mu_2}.$$

23. The inductance L of an inductor is defined as the ratio of the magnetic flux linkage to the current through the inductor.

$$L = \frac{\lambda}{I} = \frac{N\psi}{I}.$$

24. Self inductance of a solenoid is $L = \frac{\mu N^2 A}{l}$ H.

25. Self inductance of a toroid is $L = \frac{\mu N^2 A}{l_m}$ H.

26. Self inductance of a co-axial cable is

$$L = \frac{\mu}{2\pi}\left[\frac{1}{4} + \ln\frac{b}{a}\right] \text{H/m}.$$

27. Self inductance of a two-wire transmission line is

$$L = \frac{\mu}{\pi}\left[\frac{1}{4} + \ln\frac{d}{a}\right] \text{H/m}.$$

28. Mutual inductance of a co-axial solenoid of same length but different radius, in both the direction is same.

Mutual inductance of a co-axial solenoid of same radius but different length, in both the direction is same.

29. Self-inductance and mutual inductance are related to coefficient of coupling as $M = k\sqrt{L_1 L_2}$ if $k_1 = k_2 = k$ (or) $M = \sqrt{k_1 k_2 L_1 L_2}$ if $k_1 \neq k_2$.

30. Energy stored in magnetic field $w = \frac{1}{2} L I^2$ J.

31. The mutual inductance is defined as the ratio of the flux linkage in one circuit to the current flow in other circuit.

$$M_{12} = \frac{\lambda_{12}}{I_2} = \frac{N_1 \psi_{12}}{I_2} \text{H}.$$

32. Mutual inductance according to Neumann's formula

$$M_{12} = \frac{\mu N_1 N_2}{4\pi} \oint_{c_1} \oint_{c_2} \frac{\vec{dl_1} \cdot \vec{dl_2}}{r_{12}} \text{H}.$$

33. Energy density in magnetic field is

$$w = \frac{1}{2} BH \text{ J/m}^3$$

$$= \frac{1}{2} \mu H^2 \text{ J/m}^3$$

$$= \frac{1}{2}\frac{B^2}{\mu} \text{ J/m}^3$$

34. The magnetic force on a piece of magnetic material is

$$F = \frac{1}{2}\frac{B^2 a}{\mu_0 \mu_r} \text{N}.$$

35. Magnetic force in terms of energy stored $F = \nabla W_m$

In terms of self-inductance $F = \frac{1}{2} I^2 dL$

In terms of mutual inductance $F_{12} = I_1 I_2 \nabla M_{12}$.

36. Torque in terms of energy stored

$$T = \frac{dW_m}{d\phi}.$$

37. A magnetic circuit can be analysed in the same way as an electric circuit. The similarity between them is

$$F = NI = \oint_l H \cdot dl = \psi \cdot \Re \text{ and}$$

$$\text{e.m.f.} = V = IR$$

that is,

$$F \leftrightarrow V, \psi \leftrightarrow I \ R \leftrightarrow \Re.$$

EXERCISE

1. A long straight wire carries a current $I = 1 - \text{A}$. At what distance is the magnetic field $H = 1$ A/m? [1.59 m]

2. How much current must flow in a loop of radius 0.5 m to produce a magnetic field $H = |\text{mA}|\text{m}$? What is the magnetic field H at a distance of 2m from the loop along its axis? [1 mA, 14 μA/m]

3. A rectangular loop carrying 10A of current is placed on $Z = 0$ plane as shown in Fig. E.3.3. Evaluate at
(a) (2, 2, 0)
(b) (4, 2, 0)
(c) (4, 8, 0)
(d) (0, 0, 2).

[(*a*) $1.964\,\vec{a}_z$ A/m
(*b*) $1.78\,\vec{a}_z$ A/m
(*c*) $-0.1178\,\vec{a}_z$ A/m
(*d*) $-0.2457\,\vec{a}_x - 0.3165\,\vec{a}_y + 0.1798\,\vec{a}_z$ A/m]

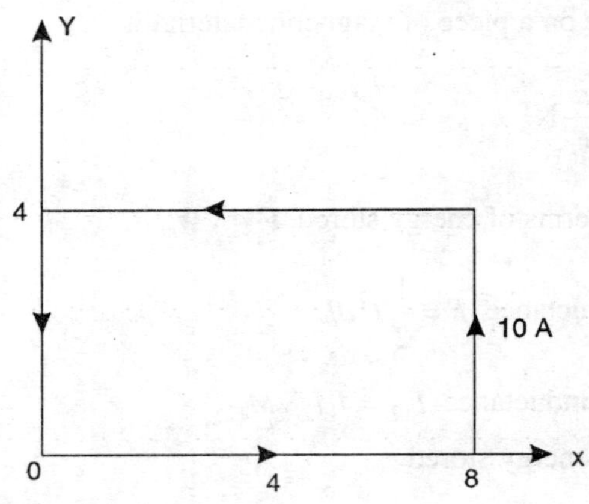

Fig. E. 3.3

4. Two identical current loops have their centres at (0, 0, 0) and (0, 0, 4) and their axes same as the Z-axis. If each loop has radius 2 m and carries current 5A in $\vec{a}_\phi$. Calculate $\vec{H}$ at

   (*a*) (0, 0, 0) (*b*) (0, 0, 2). [(*a*) $1.36\,\vec{a}_z$ A/m (*b*) $0.884\,\vec{a}_z$ A/m]

5. A 3 cm long solenoid carries a current of 400 mA. If the solenoid is to produce a magnetic flux density of 5 mwb/m^2, how may turns of wire are needed? [$N = 298$]

6. A solenoid of radius 4 mm and length 2 cm has 150 turns/m and carries current 500 mA. Find (*a*) $|\vec{H}|$ at the centre (*b*) $|\vec{H}|$ at the end of the solenoid.

   [(*a*) 69.63 A/m (*b*) 36.77 A/m]

7. An infinitely long solid conductor of radius *a* is placed along the *Z*-axis. If the conductor carries curent I in the +Z direction, show that $\vec{H} = \dfrac{I\rho}{2\pi a^2}\vec{a}_\phi$ within the conductor. Also find the corresponding density.

8. An infinitely long filamentary wire carries a current of 2A in the +*Z* direction. Calculate at $\vec{B}$ at (–3, 4, 7). [$80\,\vec{a}_\phi\; n\,\text{wb/m}^2$]

9. In a two-wire transmission line, each wire is of radius 2 cm and the wires are separated 10 cm. The wire centered at (0, 0) carries current 5 A while the other centered at (10cm, 0) carries the return current. Find $\vec{H}$ at (a) (5 cm, 0). [(*a*) $31.43\,\vec{a}_y$ A/m]

10. The magnetic vector potential of a current distribution in free space is given by

    $\vec{A} = 15e^{-\rho}\sin\phi\,\vec{a}_z$ wb/m. Find $\vec{H}$ at $(3, \pi/4, -10)$.

    $\left[\left(14\vec{a}_\rho + 42\vec{a}_\phi\right)\times 10^4 \text{ A/m}\right]$

11. Find the current density J if vector potential $\vec{A}$ is given as $\vec{A} = \dfrac{10}{\rho^2}\vec{a}_z$ wb/m in free space.

$$\left[-\frac{40}{\mu_0\rho^4}\vec{a}_z\ \text{A/m}^2\right]$$

12. A conductor 4 m long lies along the Y axis with a current of 10 A in the $\vec{a}_y$ direction. Find the force on the conductor if the field in the regions is $\vec{B} = 0.05\,\vec{a}_x$ tesla. $[-2\,\vec{a}_z\ \text{N}]$

13. Find the forces per unit length on two long straight parallel conductors if each carries a current of 10 A in the same direction and the separation distance is 0.20 m. $[10^{-4} - \vec{a}_x\ \text{N/m}]$

14. Find the maximum torque on an 85 turn, rectangle coil, 0.2 m by 0.3 m carrying a current of 2A in a field $\vec{B} = 6.5$ tesla.

[66.3 N-m]

15. A 200 turn, rectangular coil, 0.3 m by 0.15 m with a current of 5 A is in a uniform field $\vec{B} = 0.2$ tesla. Find the magnetic moment m and the maximum torque. [45 A/m^2, 9 N-m]

16. A solenoid with length 10 cm and radius 1 cm has 450 turns. Calculate its inductance.

[0.8 mH]

17. The core of a toroid is 12 cm^2 and is made of material with $\mu_r = 200$. If the mean radius of the toroid is 50 cm, calculate the number of turns needed to obtain an inductance of 2.5 H.

[5103 turns]

18. When two parallel identical wires are separated by 3m, the inductance per unit length is 2.5 μh/m. Calculate the diameter of each wire.

[14.87 mm]

19. In terms of the d.c. current I_1, how much magnetic energy is stored in the insulating medium of a 2 m long air-filled section of a co-axial transmission line, given that the radius of the inner conductor is 5 cm and the thinner radius of the outer conductor is 10 cm.

[139 I^2 nJ]

20. How many turns are required for a 30 cm-long solenoid to have an inductance of 10 mH? The solenoid diameter is 4 cm. The medium is air. [$N = 1380$]

21. A solenoid with $N_1 = 1000$, $r_1 = 1.0$ cm and $l_1 = 50$ cm is concentric within a second coil of $N_2 = 2000$, $r_2 = 2.0$ cm and $l_2 = 50$ cm. Find the mutual inductance assuming free space conditions.

[1.58 mH]

22. A magnetic track brake is energised by the car motors and each pole face has an area of 10 sq. cm. Calculate the pull per pole when the flux density is 1.0 tesla.

[400 N]

23. An iron ring with a cross-sectional area of 3 cm^2 and a mean circumference of 15 cm is wound with 250 turns of wire carrying a current of 0.3A. The relative permeability of the ring is 1500.

(*a*) Calculate the flux established in the ring.

(*b*) If a saw cut of width 2 mm is made in the above ring, find the new value of the flux in the circuit.

[0.283 mwb, 13.40 μwb]

24. Consider the magnetic circuit in Fig. E 3.9. Assuming that the core ($\mu = 1000\ \mu_0$) has a uniform cross-section of 4 cm^2, determine the flux density in the air gap.

[88.5 mwb/m^2]

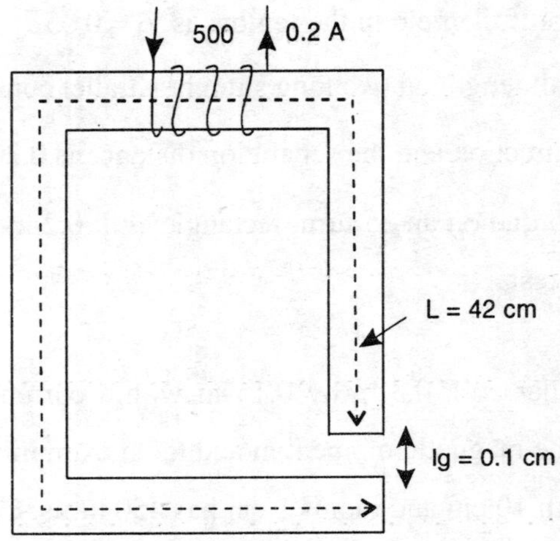

Fig. E. 3.9

25. The cast iron parallel circuit core in Fig. E 3.10 has a 500-turn coil and a uniform cross-section of 1.5 cm^2 throughout. The mean lengths are $l_1 = l_3 = 10$ cm and $l_2 = 4$ cm. Determine the coil current necessary to result in a flux density of 0.25 tesla in leg 3. [1.05A]

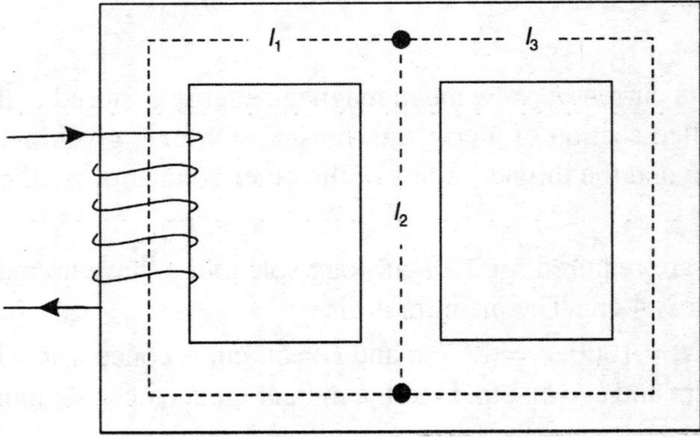

Fig. E. 3.10

4

Electromagnetic Induction and Maxwell's Equations

INTRODUCTION

So far the discussions are restricted to static or time-invariant. Electromagnetic fields, now it is extended to dynamic or time-varying electric and magnetic fields. In static electromagnetic fields, electric and magnetic fields are independent of each other whereas in dynamic electromagnetic fields, the two fields are interdependent. In other words, a time-varying electric field necessarily involves a corresponding time-varying magnetic field. It has been already mentioned that electrostatic fields are due to static electric charges whereas magnetostatic fields are produced due to motion of electric charges with uniform velocity (direct current) or static magnetic charges (magnetic poles). The time-varying fields or waves are usually due to accelerated charges or time-varying currents.

In summary, stationary charges (q) → electrostatic fields.

Steady currents (or) motion of charges $\left(\frac{dq}{dt} = I\right) \rightarrow$ Magnetostatic fields.

time-varying current or accelerated charges $\left(\frac{dI}{dt} = \frac{d^2q}{dt^2}\right) \rightarrow$ Electromagnetic fields (or) waves.

4.1 FARADAY'S LAW OF ELECTROMAGNETIC INDUCTION

According to Faraday's experiments, a static magnetic field produces no flow of current, but a time-varying field produces an induced voltage [called electromotive force or simply e.m.f.] in a closed circuit, which causes a flow of current. Faraday discovered that the induced e.m.f, $V_{\text{e.m.f}}$ (in volts), in any closed circuit is equal to time rate of change of magnetic flux linkage by the circuit. This is called Faraday's law and it can be expressed as

$$V_{emf} = -\frac{d\lambda}{dt} = -N\frac{d\psi}{dt} \quad ...(4.1)$$

where N is the number of turns in the circuit and ψ is the flux through each turn. The negative sign shows that the induced voltage acts in such a way as to oppose the flux producing it. This is known as

Lenz's law. It emphasises the fact that the direction of current flow in the circuit is such that the induced magnetic field produced by the induced current will oppose the original magnetic field.

The electric fields considered so far are caused by electric charges. However, there are other kinds of electric fields not directly caused by electric charges. These are e.m.f. produced fields. Sources of e.m.f. include electric generators, batteries, fuel cells etc., which all convert non-electrical energy into electrical energy. As electrostatic field E_e is conservative, it cannot maintain a steady current in a closed circuit, *i.e.*, $\oint_L E_e \cdot dl = 0 = IR$. But an e.m.f-produced field E_f is non-conservative so, it will maintain a steady current.

4.2 TRANSFORMER AND MOTIONAL EMFs

The relationship between electric and magnetic fields can be obtained using Faraday's law.

For a circuit with single turn ($N = 1$), eqn. (4.1) becomes

$$V_{emf} = -\frac{d\psi}{dt}. \quad \text{...(4.2)}$$

In terms of $\vec{E}$ and $\vec{B}$ eqn. (4.2) can be written as

$$V_{emf} = \oint_L \vec{E} \cdot d\vec{l} = -\frac{d}{dt}\int_s \vec{B} \cdot d\vec{s}, \quad \text{...(4.3)}$$

where ψ has been replaced by $\int_s \vec{B} \cdot d\vec{s}$ and s is the surface area of the circuit bounded by the closed path L. It is evident from eqn. (4.3) that in a time-varying situation, both electric and magnetic fields are present and inter-related.

The variation of flux with time in eqn. (4.3) may be caused in three ways.

1. by having a stationary loop in a time-varying B field.
2. by having a time-varying loop area in a static B field.
3. by having a time-varying loop area in a time-varying B field.

4.3 STATIONARY LOOP IN TIME-VARYING *B* FIELD (TRANSFORMER EMF)

When a stationary conducting loop is placed in a time-varying magnetic B field, then eqn. (4.3) becomes

$$V_{emf} = \oint_L \vec{E} \cdot d\vec{l} = -\int_s \frac{\partial \vec{B}}{\partial t} \cdot d\vec{s} \quad \text{...(4.4)}$$

This e.m.f. induced by the time-varying current (producing time-varying B field) in a stationary loop is often referred to as transformer e.m.f, since it is due to transformer action. By applying Stoke's theorem to the middle term in eqn. (4.4).

$$\int_s \left(\nabla \times \vec{E}\right) \cdot d\vec{s} = -\int_s \frac{\partial \vec{B}}{\partial t} \cdot d\vec{s}. \quad \text{...(4.5)}$$

So,

$$\nabla \times \vec{E} = -\frac{\partial \vec{B}}{\partial t}. \quad \text{...(4.6)}$$

Eqn (4.6) shows that the time-varying field is not conservative $\left(\nabla \times \vec{E} \neq 0\right)$.

4.4 MOVING LOOP IN STATIC B FIELD (MOTIONAL EMF)

When a conducting loop is moving in a static B field, an e.m.f. is induced in the loop. From eqn. (3.71), the force on a charge moving with uniform velocity u in a magnetic field $\vec{B}$ is

$$\vec{F}_m = Q\vec{u} \times \vec{B}.$$

The motional e.m.f E_m is defined as

$$\vec{E}_m = \frac{\vec{F}_m}{Q} = \vec{u} \times \vec{B}. \quad \text{...(4.7)}$$

If the conducting loop moving with uniform velocity u is considered to be consisting of a large number of free electrons, the e.m.f. induced in the loop is

$$V_{emf} = \oint_L \vec{E}_m \cdot d\vec{l} = \oint_L \left(\vec{u} \times \vec{B}\right) \cdot d\vec{l}. \quad \text{...(4.8)}$$

This type of e.m.f. is called motional e.m.f. or flux-cutting e.m.f. because it is due to motional action and it is found in electrical machines such as motors, generators and alternators.

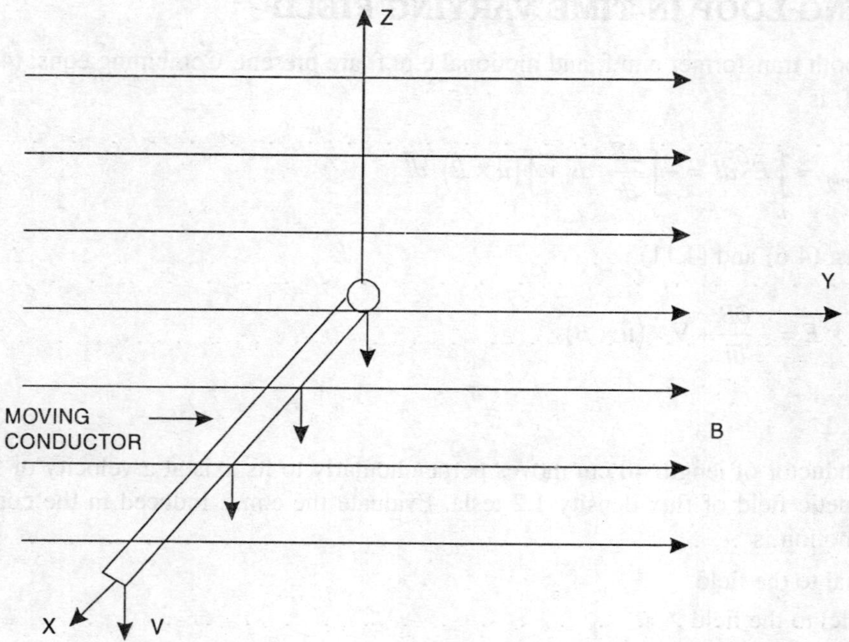

Fig. 4.1 Moving conductor in a static *B* field

For example, consider a straight conductor with its axis coincident with X-axis and moving downward (in $-Z$ direction) in a field directed along Y-axis as in Fig. 4.1.

$$\vec{u} = u(-\vec{a}_z)$$

$$\vec{B} = B(\vec{a}_y)$$

$$d\vec{l} = dl\,\vec{a}_x$$

$$\therefore \quad V_{emf} = \int \vec{E}_m \cdot d\vec{l} = \int u(-\vec{a}_z) \times B(\vec{a}_y) \cdot dl\,\vec{a}_x.$$

If the conductor is of length 'l' m

$$V_{emf} = uBl. \qquad ...(4.9)$$

This e.m.f. is maximum since the direction of motion is perpendicular to the magnetic field orientation.

If the angle between the direction of motion and field is θ, then

$$V_{emf} = uBl\sin\theta \qquad ...(4.10)$$

By applying Stoke's theorem to eqn. (4.8)

$$\int_s \left(\nabla \times \vec{E}_m\right) \cdot d\vec{s} = \int_s \nabla \times \left(\vec{u} \times \vec{B}\right) \cdot d\vec{s}$$

(or)

$$\nabla \times \vec{E}_m = \nabla \times \left(\vec{u} \times \vec{B}\right). \qquad ...(4.11)$$

4.5 MOVING LOOP IN TIME VARYING FIELD

In this case, both transformer e.m.f. and motional e.m.f. are present. Combining eqns. (4.4) and (4.8), the total e.m.f. is

$$V_{emf} = \oint_L \vec{E} \cdot d\vec{l} = -\int_s \frac{\partial \vec{B}}{dt} \cdot d\vec{s} + \oint_L \left(\vec{u} \times \vec{B}\right) \cdot d\vec{l} \qquad ...(4.12)$$

(or) from eqns. (4.6) and (4.11)

$$\nabla \times \vec{E} = -\frac{\partial \vec{B}}{dt} + \nabla \times \left(\vec{u} \times \vec{B}\right). \qquad ...(4.13)$$

Example 4.1

A straight conductor of length 40 cm moves perpendicularly to its axis at a velocity of 50 m/sec in a uniform magnetic field of flux density 1.2 tesla. Evaluate the e.m.f. induced in the conductor if the direction of motion is

(*a*) normal to the field

(*b*) parallel to the field

(*c*) at an angle 60° to the orientation of the field.

From eqn. (4.10)

$V = uBl \sin\theta$

(*a*) $\theta = 90°$

$V = uBl$

$= 50 \times 1.2 \times .4$

$= 24$ volt

(*b*) $\theta = 0$

$V = 0$

As there is no flux linkage, no e.m.f. is induced.

(*c*) $\theta = 60°$

$V = uBl \sin 60°$

$= 50 \times 1.2 \times .4 \times .866$

$= 20.8$ volt.

Example 4.2

A square coil with loop area 0.01 sq.m and 50 turns is rotated about its axis, at right angles to a uniform magnetic field $\vec{B} = 1.0$ tesla. Calculate the instantaneous value of the e.m.f. induced in the coil when its plane is (*a*) at right angles to the field (*b*) at 45° to the field (*c*) in the plane of the field. Speed of rotation = 1000 rpm.

(*a*) If the plane of coil is at right angles to the field, then its motion is parallel to the field.

So, $\theta = 0°$, from eqn. (4.10)

$V = Bl\,u \sin\theta$

$\therefore V = 0.$

(*b*) If the plane is parallel with field, its motion is perpendicular to the field and

$\theta = 90°$

So, $V = Bl\,u$

Velocity $u = \omega \times r$

where $\omega \rightarrow$ angular velocity rad/s

$$\omega = \frac{2\pi N}{60}.$$

r is the axis of rotation

$$r = \frac{l}{2}$$

l is length of coil in m

$$\omega = \frac{2\pi \times 1000}{60} \text{ rad/s.}$$

$$V = Br\left(\frac{2\pi N}{60}\right)\left(\frac{l}{2}\right) \text{ volt}$$

The above equation gives the value of e.m.f. induced in one conductor.

e.m.f./turn = 2 × e.m.f./conductor

No. of turns given = 50

∴ Total e.m.f. induced

$$= 2 \times Bl\left(\frac{2\pi N}{60}\right)\left(\frac{l}{2}\right) \times 50$$

$$= 2\frac{Bl^2}{2}\left(\frac{2\pi N}{60}\right) \times 50$$

$B = 1.0$ tesla

$l = 0.1$ m (as area = .01 sq. m)

$N = 1000$ rpm

$V = 52.4$ volt.

Example 4.3

Figure 4.2 shows an arrangement for Faraday disc generator which consists of a circular disc rotating at a constant angular velocity ω rad/s with the axis horizontal and cutting a uniform magnetic field of flux density B wb/m^2 along Z-axis. Two brushes make contact with the disc, one at the centre and the other at the rim of the disc.

Find the e.m.f. induced between the brushes.

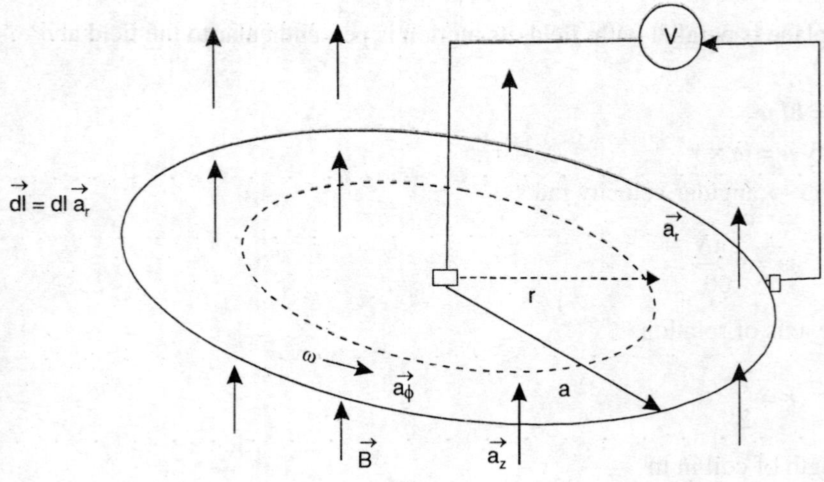

Fig. 4.2 Faraday's disc generator (or) homopolar generator

Consider one radial element at a distance of r m from the centre of the disc. Any point on this radial element has velocity $\vec{u} = u\vec{a}_\phi$, where $u = \omega \times r$ m/s.

$$V_{emf} = \int \vec{u} \times \vec{B} \cdot d\vec{l},$$

where $d\vec{l}$ is the differential length of the circular disc

$$d\vec{l} = dr\,\vec{a}_r$$

$$V_{emf} = \int \left(\omega r\,\vec{a}_\phi \times B\vec{a}_z\right) \cdot dr\,\vec{a}_r$$

$$= \int_{r=0}^{r=a} (\omega r B)\,dr \qquad \begin{bmatrix} \vec{a}_\phi \times \vec{a}_z = \vec{a}_r \\ \vec{a}_r \cdot \vec{a}_r = 1 \end{bmatrix}$$

where 'a' is the radius of the disc.

$$V_{emf} = \left.\frac{\omega B r^2}{2}\right]_{r=0}^{r=a}$$

$$V_{emf} = \frac{1}{2}\omega B a^2 \text{ volt.} \qquad \text{...(4.14)}$$

Example 4.4

A copper disc, 0.5 m in diameter, is rotated at a constant speed of 2000 rpm on a horizontal axis perpendicular to and through the centre of the disc, the axis lying in the magnetic meridian. Two brushes make contact with the disc, one at the edge and other at the centre. If the horizontal component of the earth's field B is 0.2 gauss (0.02 milli tesla), calculate the e.m.f. induced between the brushes.

From the Example 4.3

$$V_{emf} = \frac{1}{2}\omega B a^2 \text{ volt}$$

$$B = 0.02 \text{ milli tesla}$$

$$= 0.02 \times 10^{-3} \text{ wb/m}^2$$

$$\omega = \frac{2\pi N}{60} \text{ rad/s}$$

$$a = \frac{0.5}{2} = 0.25 \text{ m}$$

$$\therefore \quad V_{emf} = 0.131 \text{ mV}.$$

Example 4.5

A circuit has 2000 turns enclosing a magnetic circuit 30 sq. cm in section. A current of 5 A in the circuit produces a field of flux density 1 (tesla) and when the current is doubled, flux density increases only by 50%. Determine the mean value of inductance of the circuit between 5 A and 10 A and also the induced e.m.f. when the current increases uniformly from 5 A to 10 A in 0.1 second.

Induced e.m.f. V for 'N' number of turns is given by

$$V = -N\frac{d\phi}{dt} \text{ volt}$$

$$= -N\left(\frac{d\phi}{di}\right)\left(\frac{di}{dt}\right), \quad \text{...(4.15)}$$

where inductance $L = N\frac{d\phi}{dt}$ henry.

$$\therefore \quad V = -L\frac{di}{dt} \text{ volt.} \quad \text{...(4.16)}$$

Mean value of inductance

$$L = N\left[\frac{\phi_2 - \phi_1}{i_2 - i_1}\right]$$

when $i_1 = 5\text{ A}$, $B_1 = 1 \text{ wb/m}^2$

$\phi_1 = B_1 \times A$

$A \rightarrow$ Area of section = 30 sq. cm

$\phi_1 = 1 \times 30 \times 10^{-4} = 30 \times 10^{-4}$ wb

and when $i_2 = 10$ A

$\phi_2 = 150\%$ of ϕ_1

$= 1.5 \times 30 \times 10^{-4}$

$= 4.5 \times 10^{-4}$ wb.

Substitute all the values in eqn. (4.15)

$$L = 2000\left[\frac{45 \times 10^{-4} - 30 \times 10^{-4}}{10 - 5}\right]$$

$$= 2000 \times \frac{15 \times 10^{-4}}{5}$$

$$= 0.6 \text{ henry.}$$

The current increases from $i_1 = 5$ A to $i_2 = 10$ A in 0.1 s.

Substitute $dt = 0.1$ s in eqn. (4.16)

$$V_{emf} = -0.6 \times \frac{5}{0.1}$$

$$= -30 \text{ volt.}$$

4.6 DISPLACEMENT CURRENT

In the previous section, the equations for electrostatic fields are modified for time-varying situations and also to satisfy Faraday's law. Again, the equation for magnetostatic fields is to be modified for time-varying condition.

For static electromagnetic fields, Ampere circuital law states that

$$\nabla \times \vec{H} = \vec{J}. \qquad \text{...(4.17)}$$

From the vector identity, the divergence of the curl of any vector field is zero.
Hence,

$$\nabla \cdot \left(\nabla \times \vec{H}\right) = \nabla \cdot J = 0. \qquad \text{...(4.18)}$$

But the equation of continuity in eqn. (2.125) states that

$$\nabla \cdot \vec{J} = -\frac{\partial \rho_v}{\partial t} \neq 0. \qquad \text{...(4.19)}$$

Thus eqns. (4.18) and (4.19) are incompatible for time-varying conditions. So, eqn. (4.17) must be modified to agree with eqn. (4.19). For this, a term is added to eqn. (4.17) and so it becomes

$$\nabla \times \vec{H} = \vec{J} + \vec{J}_d. \qquad \text{...(4.20)}$$

where $\vec{J}_d$ is to be determined and defined.

Again the divergence of the curl of any vector is zero. Hence,

$$\nabla \cdot \left(\nabla \times \vec{H}\right) = 0 = \nabla \cdot \vec{J} + \nabla \cdot \vec{J}_d \qquad \text{...(4.21)}$$

$$\nabla \cdot \vec{J}_d = -\nabla \cdot \vec{J} = \frac{\partial \rho_v}{\partial t}$$

$$= \frac{\partial}{\partial t}\left(\nabla \cdot \vec{D}\right) \qquad \text{(since from Gauss's law in eqn. } \nabla \cdot \vec{D} = \rho_v\text{)}$$

$$\nabla \cdot \vec{J}_d = \nabla \cdot \frac{\partial \vec{D}}{\partial t}$$

$$\text{So } \vec{J}_d = \frac{\partial \vec{D}}{\partial t}. \qquad \text{...(4.22)}$$

Substituting eqn. (4.22) in eqn. (4.20)

$$\nabla \times \vec{H} = \vec{J} + \frac{\partial \vec{D}}{\partial t}. \qquad \text{...(4.23)}$$

The above eqn. (4.23) is modified form of Ampere's circuital law for time-varying field.

The term $\vec{J}_d = \dfrac{\partial \vec{D}}{\partial t}$ is known as displacement current density and $\vec{J}$ is the conduction current density.

Without the term $\vec{J}_d$, the electromagnetic wave propagation is impossible (for eg. TV or radio waves). At low frequencies, $\vec{J}_d$ is neglected compared with $\vec{J}$, but at radio frequencies the two terms are comparable.

Based on the displacement current density, the displacement current is defined as

$$I_d = \int \vec{J}_d \cdot d\vec{s} = \int \frac{\partial \vec{D}}{\partial t} \cdot d\vec{s}. \qquad ...(4.24)$$

This displacement current is a result of time-varying electric field.

A typical example of such current is the current through a capacitor when an alternating voltage source is applied to its plates.

To illustrate the need for the displacement current, let us apply the unmodified form of Ampere's circuital law to a closed path L shown in Fig. 4.3.

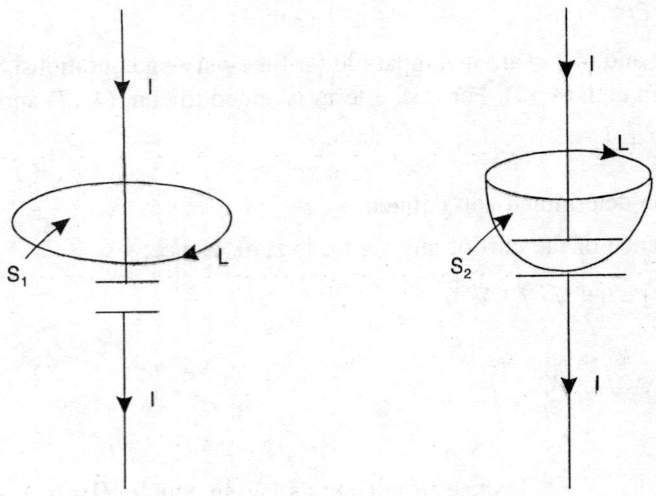

Fig. 4.3 Need for J_d in Ampere's circuital law

From Ampere's circuital law

$$\oint_L \vec{H} \cdot d\vec{l} = \int_{S_1} \vec{J} \cdot d\vec{s} = I_{enc} = I. \qquad ...(4.25)$$

where I is the current through the conductor and S_1 is the flat surface bounded by L.

If a balloon shaped surface S_2 is used and it is passing between the capacitor plates as in Fig. 4.3,

then
$$\oint_L \vec{H} \cdot d\vec{l} = \int_{S_2} \vec{J} \cdot d\vec{s} = I_{enc} = 0. \qquad ...(4.26)$$

Because no conduction current ($J = 0$) flows through S_2, the above statement (or) result is contradictory, since the same closed path L is used. To solve this conflict, the displacement current density is added in eqn. (4.25).

$$\oint_l \vec{H} \cdot d\vec{l} = \int_s \left(\vec{J} + \vec{J}_d\right) \cdot ds.$$

For the surface S_1, $\vec{J}_d = 0$. So eqn. (4.25) remains valid and for the surface S_2, $\vec{J} = 0$ so that

$$\oint \vec{H} \cdot d\vec{l} = \int_{S_2} \vec{J}_d \cdot d\vec{s}$$

$$= \frac{d}{dt} \int_{S_2} \vec{D} \cdot d\vec{s}$$

$$= \frac{d}{dt}[Q] \qquad \text{(from Gauss's law)}$$

$$= \frac{dQ}{dt} = I.$$

So, the same current is obtained for either surface, though it is conduction current in S_1 and displacement current in S_2.

Example 4.6

Show that the displacement current in the dielectric of a parallel-plate capacitor is equal to the conduction current in the leads.

The capacitance of the parallel plate capacitor is

$$C = \frac{\varepsilon A}{d}, \qquad \text{...(4.27)}$$

where A is the plate area and d is separation. The conduction current is

$$i_c = C\frac{dV}{dt}. \qquad \text{...(4.28)}$$

Substitute eqn. (4.27) in eqn. (4.28)

$$i_c = \frac{\varepsilon A}{d}\left(\frac{dV}{dt}\right). \qquad \text{...(4.29)}$$

On the other hand, the electric field in the dielectric in $E = V/d$. Hence,

$$D = \varepsilon E = \varepsilon \cdot \frac{V}{d}.$$

So
$$\frac{\partial D}{\partial t} = \frac{\varepsilon}{d}\frac{dV}{dt} = J_d.$$

The displacement current $i_d = \int_s \vec{J}_d \cdot d\vec{s}$

$$i_D = \int_s \frac{\partial \vec{D}}{\partial t} \cdot d\vec{s} = \int_S \frac{\varepsilon}{d}\frac{dV}{dt} ds$$

$$= \frac{\varepsilon}{d}\frac{dV}{dt}\int_s ds$$

$$= \frac{\varepsilon}{d}\frac{dV}{dt}\cdot A$$

$$i_D = \frac{\varepsilon A}{d}\frac{dV}{dt}. \qquad \text{...(4.30)}$$

From eqn. (4.29) and eqn (4.30), it is proved that conduction and displacement currents are same.

4.7 MAXWELL'S EQUATIONS

Maxwell's equations are derived on the basis of Ampere's circuital law, Faraday's law and Gauss's law both in point (differential) and integral form. These equations which are compiled by Maxwell led to the discovery of electromagnetic waves. For a field to be qualified as an electromagnetic field, it must satisfy all four of Maxwell's equations. The integral form of Maxwell's equations depicts the underlying physical laws, whereas the differential form is used more frequently in solving problems.

Generalised form of Maxwell's equations.

Table 4.1

Differential form	*Integral form*	*Remarks*
$\nabla\cdot\vec{D} = \rho_v$	$\oint_s \vec{D}\cdot d\vec{s} = \int_v \rho_v\, dv$	Gauss's law for electrostatics.
$\nabla\cdot\vec{B} = 0$	$\oint_s \vec{B}\cdot d\vec{s} = 0$	Gauss's law for magnetostatics
$\nabla\times\vec{E} = -\frac{\partial\vec{B}}{\partial t}$	$\oint_L \vec{E}\cdot d\vec{l} = -\frac{\partial}{\partial t}\int_s \vec{B}\cdot d\vec{s}$	Faraday's law
$\nabla\times\vec{H} = \vec{J} + \frac{\partial\vec{D}}{\partial t}$	$\oint_L \vec{H}\cdot d\vec{l} = \int_s \left(\vec{J} + \frac{\partial\vec{D}}{\partial t}\right)\cdot d\vec{s}$	Ampere's circuital law.

4.8 TIME HARMONIC FIELDS

Let us assume that the electromagnetic fields are time-harmonic, *i.e.*, they vary periodically or sinusoidally with time. So, the phasor form is applied to electromagnetic fields.

A phasor is a complex number which can be written as $re^{j\phi}$

$$\phi = \omega t + \theta,$$

where θ may be function of time or space co-ordinates or constant.

So, any vector field $\vec{A}$ which is sinusoidally time-varying can be represented as

$$\vec{A}(t) = \vec{A}_0\, e^{j(\omega t+\theta)}. \qquad \text{...(4.31)}$$

The real part of $\vec{A}(t) = \vec{A}_0 \cos(\omega t + \theta)$ and imaginary part of $\vec{A}(t) = \vec{A}_0 \sin(\omega t + \theta)$.

While performing mathematical expressions, either the real part or the imaginary part of a quantity should be used but not both at same time.

By dropping the time factor $e^{j\omega t}$ in $\vec{A}(t)$, the resulting complex term $\vec{A}_0 e^{j\theta}$ is called the phasor form of vector field and is denoted by $\vec{A}_s$ *i.e.*

$$\vec{A}_s = \vec{A}_0\, e^{j\theta}, \qquad \text{...(4.32)}$$

where subscript s denotes the phasor form of $\vec{A}(t)$.

Thus $\vec{A}(t) = \vec{A}_0 \cos(\omega t + \theta)$ is the instantaneous form and it can be expressed as

$$\vec{A}(t) = Re\left(\vec{A}_s\, e^{j\omega t}\right) \qquad \text{...(4.33)}$$

$$\frac{\partial \vec{A}(t)}{\partial t} = \frac{\partial}{\partial t}\left(Re\,\vec{A}_s e^{j\omega t}\right)$$

$$= Re\left(j\omega\, \vec{A}_s e^{j\omega t}\right)$$

$$\frac{\partial \vec{A}(t)}{\partial t} = j\,\omega\, \vec{A}_s. \qquad \text{...(4.34)}$$

The above eqn. (4.34) shows that the time derivative of instantaneous quantity is equivalent to multiplying its phasor form by $j\omega$.
Similarly,

$$\int \vec{A}\,\partial t = \frac{\vec{A}_s}{j\omega}. \qquad \text{...(4.35)}$$

So, the Maxwell's equations for time-varying fields are modified and given in Table 4.2.

All the field quantities $\vec{E}, \vec{D}, \vec{H}, \vec{B}, \vec{J}$ and ρ_v and their derivatives can be expressed in phasor form using eqn. (4.33) and eqn. (4.34).

Table 4.2

Point form	*Integral form*
$\nabla \cdot \vec{D}_s = \rho_{vs}$	$\oint \vec{D}_s \cdot d\vec{s} = \int \rho_{vs} dv$
$\nabla \cdot \vec{B}_s = 0$	$\oint \vec{B}_s \cdot d\vec{s} = 0$
$\nabla \times \vec{E}_s = -j\omega \vec{B}_s$	$\oint \vec{E}_s \cdot d\vec{l} = -j\omega \int \vec{B}_s \cdot d\vec{s}$
$\nabla \times \vec{H}_s = \vec{J}_s + j\omega \vec{D}_s$	$\oint \vec{H}_s \cdot d\vec{l} = \int \left(\vec{J}_s + j\omega \vec{D}_s\right) \cdot d\vec{s}$

4.9 ELECTRO MAGNETIC POTENTIAL

In Faraday's and Ampere's laws, there exist interconnection between electric and magnetic fields, when fields are time-varying. Similarly for time-varying fields, we can deduce a relationship between electric scalar potential V and vector magnetic potential $\vec{A}$.

From Maxwell's third equation (Faraday's law) for time-varying fields

$$\nabla \times \vec{E} = -\frac{\partial \vec{B}}{\partial t}.$$

As
$$\vec{B} = \nabla \times \vec{A}$$

$$\nabla \times \vec{E} = -\frac{\partial}{\partial t}\left(\nabla \times \vec{A}\right)$$

$$\nabla \times \left(\vec{E} + \frac{\partial \vec{A}}{\partial t}\right) = 0. \qquad \text{...(4.36)}$$

From vector identity eqn. (1.67)

$$\nabla \times \nabla \phi = 0$$

In eqn. (4.36), assume scalar ϕ as potential V,

$$\nabla \phi = \vec{E} + \frac{\partial \vec{A}}{\partial t} = -\nabla V.$$

Hence
$$\vec{E} = -\nabla V - \frac{\partial \vec{A}}{\partial t}. \qquad \text{...(4.37)}$$

The above equation is potential gradient for time-varying fields.

4.10 RELATION BETWEEN CIRCUIT THEORY AND FIELD THEORY

Analysis of any electrical network comprising R, L, C would be extremely complicated from the approach of field theory, because the spatial distribution of the elements and the orientation of the connecting conducting wires are geometrically highly irregular. But large number of such networks can be solved by circuit theory concepts.

Circuit theory is an approximate theory and yet the accuracy of its results, in a very large number of cases, has been so high that no better solution is required.

Also, circuit theory is essentially an one-dimensional simplification of the field theory which deals with two- and three-dimensional spatial distribution. Circuit theory cannot consider the effects of space boundary conditions as in field theory problems.

In this section, some of the circuit theory relations like Kirchoff's current law, voltage law are derived on the basis of field theory concepts.

Kirchoff's Voltage Law:

Consider a resistor, which is constructed from a resistive material of conductivity σ, length l and cross sectional area A, connected to a battery of V volts.

Let I be the current at all points in the circuit and E be the total field at the same points. From eqn. (2.106), Ohm's law at a point

$$\frac{\vec{J}}{\sigma} = \vec{E} \qquad \text{...(4.38)}$$

where $\vec{E} = \vec{E}_c + \vec{E}_e$

$\vec{E}_c$ is the static electric field due to charges and $\vec{E}_e$ is the field generated by other causes, for example by a battery (e.m.f. producing field)

$$\vec{E} = \frac{\vec{J}}{\sigma} = \frac{\vec{I}}{A} \cdot \frac{RA}{l} \qquad \left[\because R = \frac{l}{\sigma A}\right]$$

$$\vec{E} = \vec{I} \cdot \frac{R}{l}.$$

$$\vec{E}_c + \vec{E}_e = \vec{I} \cdot \frac{R}{l}. \qquad \text{...(4.39)}$$

Integrate over the closed path on both sides of the eqn. (4.39)

$$\oint \vec{E}_c \cdot d\vec{l} + \oint \vec{E}_e \cdot d\vec{l} = \oint \vec{I} \cdot R/l \cdot d\vec{l}. \qquad \text{...(4.40)}$$

As the electric field by static charges E_c over the closed path is zero, $\oint \vec{E}_c \cdot d\vec{l} = 0$.

For e.m.f. producing field (E_e), electric field over the closed path is equal to battery potential V. So eqn. (4.40) becomes

$$V = IR. \qquad \text{...(4.41)}$$

Eqn. (4.41) gives Kirchoff's voltage law which states that, around a loop, sum of voltage rise is equal to sum of voltage drop.

Kirchoff's Current Law:

In Section 3.3 we have seen that magnetic flux always exists in closed path and mathematically it can be written as

$$\oiint \vec{B} \cdot d\vec{s} = 0$$

which indicates that magnetic flux coming out from a closed surface or volume is zero.
Similar to magnetic flux, current will exist in a closed path (circuit). Hence it can be written in terms of current density as

$$\oiint_S \vec{J} \cdot d\vec{s} = 0. \qquad \text{...(4.42)}$$

So, it is inferred that

- Current flow is continuous and solenoidal.

- It has no source or sink.
- As much current must flow into the volume as leaves it.

Differential form of eqn. (4.42) is

$$\nabla \cdot \vec{J} = 0, \quad ...(4.43)$$

which indicates that current distribution will not diverge or converge with respect to a point, it is continuous.

In other words, total current at any one junction is zero ($\sum I = 0$).

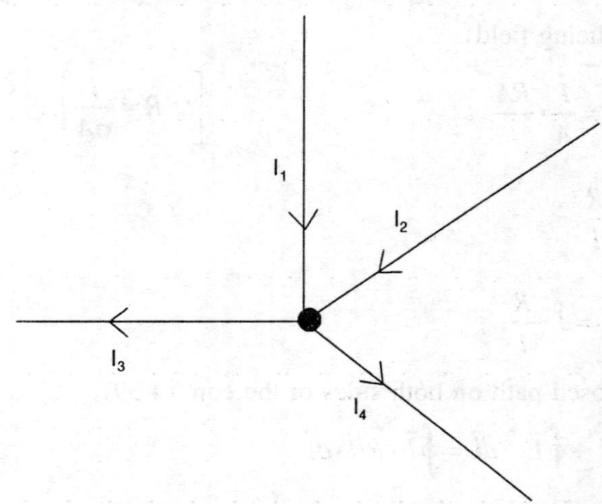

Fig. 4.4 Kirchoff's current law $I_1 + I_2 = I_3 + I_4$

Equations (4.42) and (4.43) are mathematical form of Kirchoff's current law.

RLC Series Circuit

In this section, voltage equation for a *RLC* series circuit is derived using Maxwell's field equations.

Consider a *RLC* series circuit energised by an a.c. generator of V volts.

As already mentioned, the total electric field in the circuit

$$\vec{E}_{\text{tot}} = \vec{E}_e + \vec{E}_c.$$

From eqn. (4.38)

$$\vec{E}_{\text{tot}} = \frac{\vec{J}}{\sigma}$$

$$\vec{E}_e = \vec{E}_{\text{tot}} - \vec{E}_c. \quad ...(4.44)$$

From eqn. (4.37),

$$\vec{E}_c = -\nabla V - \frac{\partial \vec{A}}{\partial t}.$$

So,

$$\vec{E}_e = \frac{\vec{J}}{\sigma} + \nabla V + \frac{\partial \vec{A}}{\partial t}. \quad ...(4.45)$$

Integrate around the closed path on both sides of the equation.

$$\oint \vec{E}_e \cdot d\vec{l} = \oint \frac{\vec{J}}{\sigma} \cdot d\vec{l} + \oint \nabla V \cdot d\vec{l} + \oint \frac{\partial \vec{A}}{\partial t} \cdot d\vec{l} \qquad \text{...(4.46)}$$

$$\oint \frac{\vec{J}}{\sigma} \cdot d\vec{l} = \frac{Jl}{\sigma}$$

$$= \frac{I}{a} \frac{l}{\sigma}$$

$$= I \cdot \frac{l}{\sigma a} = IR \qquad \text{...(4.47)}$$

$$\oint \nabla V \cdot d\vec{l} = \oint \vec{E} \cdot d\vec{l} = Ed,$$

where d is the distance between the plates of the capacitor as in Fig. 4.5.

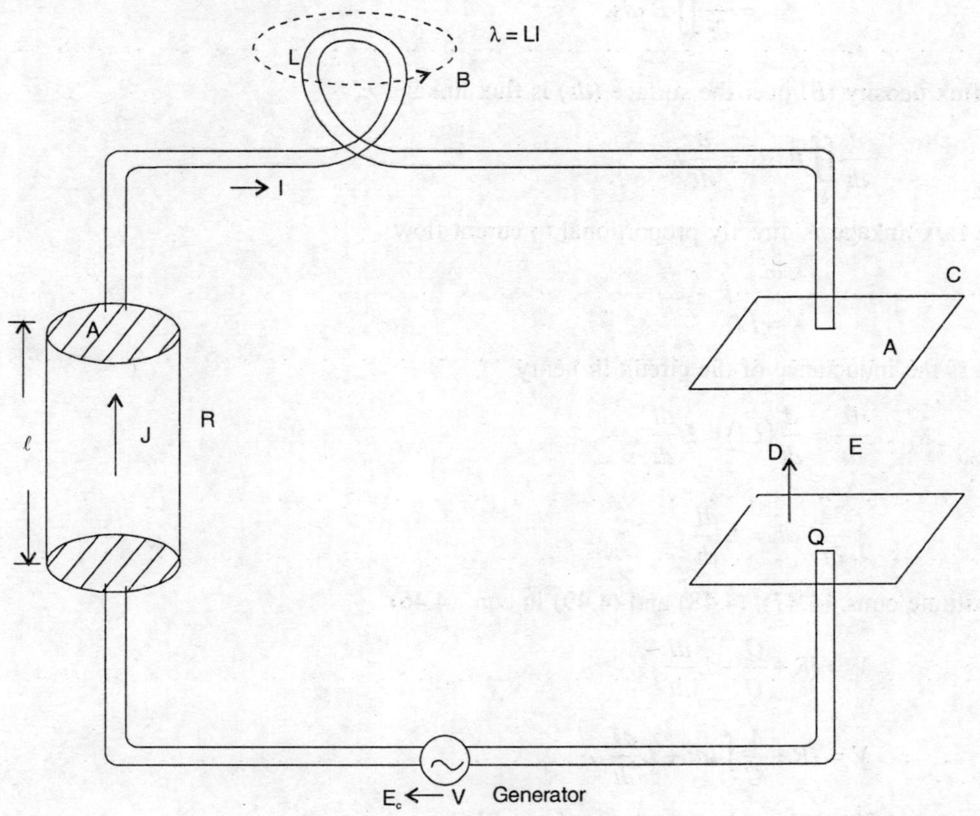

Fig. 4.5 *RLC* series circuit

$$Ed = \frac{D}{\varepsilon}\cdot d.$$

$$= \frac{Q}{A\varepsilon}d = \frac{Q}{\varepsilon A/d} = \frac{Q}{c}.$$

Hence, $$\oint \nabla V \cdot d\vec{l} = \frac{Q}{c}, \qquad ...(4.48)$$

where A and ε are area of the plate and permittivity of the dielectric medium respectively.

$$\oint \frac{\partial \vec{A}}{\partial t}\cdot d\vec{l} = \frac{d}{dt}\oint \vec{A}\cdot d\vec{l}.$$

According to Stoke's theorem,

$$\frac{d}{dt}\oint \vec{A}\cdot d\vec{l} = \frac{d}{dt}\iint_S \left(\nabla \times \vec{A}\right)\cdot d\vec{s}$$

$$= \frac{d}{dt}\iint_S \vec{B}\cdot d\vec{s}.$$

As flux density (B) over the surface (ds) is flux linkage λ,

$$\frac{d}{dt}\iint_S \vec{B}\cdot d\vec{s} = \frac{d}{dt}\lambda.$$

The flux linkage is directly proportional to curent flow.

$$\lambda \propto I$$

$$\lambda = LI,$$

where L is the inductance of the circuit in henry.

$$\frac{d\lambda}{dt} = \frac{d}{dt}(LI) = L\frac{dI}{dt}.$$

So, $$\oint \frac{\partial \vec{A}}{\partial t}\cdot d\vec{l} = L\frac{dI}{dt}. \qquad ...(4.49)$$

Substitute eqns. (4.47), (4.48) and (4.49) in eqn. (4.46)

$$V = IR + \frac{Q}{C} + L\frac{dI}{dt}$$

$$V = IR + \frac{1}{C}\int Idt + L\frac{dI}{dt}. \qquad ...(4.50)$$

The eqn. (4.50) is the voltage equation for a *RLC* series circuit.

In deriving the above equation under field theory concepts, the following assumptions are made:

1. At any instant, the current is same at all parts of the circuit *i.e.* disturbance is propagated around the circuit instantaneously.

2. Circuit length is small.
3. Radiation is ignored which is important at high frequencies.

SOLVED PROBLEMS

Problem 1

A conductor of length 100 cm moves at right angles to a uniform field of strength 10,000 lines per cm^2 with a velocity of 50 m/s. Calculate the e.m.f. induced in it. Find also the value of induced e.m.f. when the conductor moves at an angle of 30° to the direction of the field.

(1 $wb/m^2 = 10^4$ lines/m^2 or gauss)

Solution:

length of the conductor = 100×10^{-2} m.

velocity $v = 50$ m/s.

$B = 10{,}000$ lines/m^2 = 1wb/m^2

$\theta = 30°$

Case I

$e = Blv \sin\theta$

$\theta = 90\,\theta$

$e = 1 \times 100 \times 10^{-2} \times 50$

$e = 50$ V

Case II

$\theta = 30°$

$e = 1 \times 100 \times 10^{-2} \times 50 \times \sin 30°$

$= 25$ V.

Problem 2

A single-turn rectangular loop of enclosed area 1 sq.m is situated in air with its plane normal to a magnetic field which varies at the rate of 1 wb/m^2 sec. Estimate e.m.f. induced in the loop.

Solution:

Here the magnetic field is changing, so, as per Faraday's law of transformer e.m.f.,

The induced e.m.f. is $V = -\iint \frac{\partial B}{\partial t} \cdot ds$.

Taking the magnitude

$$V = \iint \frac{\partial B}{\partial t} \cdot ds .$$

Area is given as 1 m^2

$$\frac{\partial B}{\partial t} = 1\,\text{wb}/\text{m}^2\ \text{s}.$$

$V = 1 \times 1$

$= 1$ volt.

Problem 3

A conducting cylinder of radius 5 cm and height 20 cm rotates at 600 rev/s in a radial field $B = 0.50$ tesla. Sliding contacts at the top and bottom connect to voltmeter as shown in Fig. P. 4.3. Find the induced voltage.

Solution:

Here the magnetic field is fixed, but conductor area exposed to field is varying. From Faraday's law of motional induced e.m.f. voltage induced is $V = \int \vec{E} \cdot d\vec{l}$.

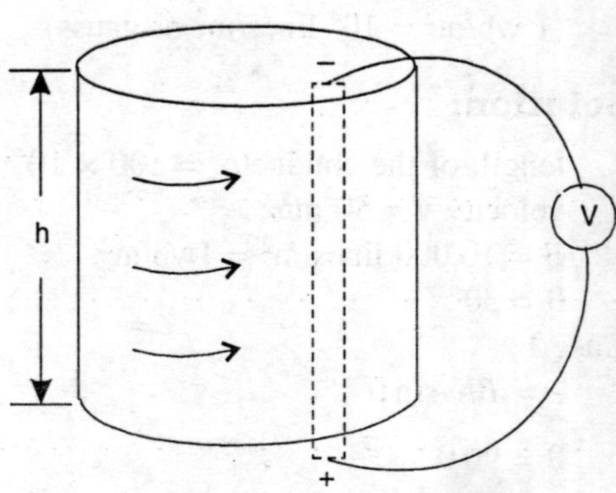

Fig. P. 4.3

$$\vec{E} = \vec{U} \times \vec{B}$$

Velocity $u = \omega \times r$

$\omega = 2\pi N$ (N is in rps)

$\therefore \quad u = 2\pi \times 600 \times 0.05\vec{a}_\phi$

[$\because$ the cylinder is revolving about the axis]

$B = 0.50\vec{a}_\rho$ [Field is in radial direction]

$$\vec{E} = 2\pi \times 600 \times 0.05\vec{a}_\phi \times 0.50\,\vec{a}_\rho$$

$$= 94.25(-\vec{a}_z)\,\text{V/m}.$$

As length is varying along Z axis,

$$d\vec{l} = dz\,\vec{a}_z$$

$$V = \int_{z=0}^{0.20\,m} E(-\vec{a}_z) \cdot dz\,\vec{a}_z$$

$$V = \int_{z=0}^{0.20\,m} = 94.25(-\vec{a}_z) \cdot dz\,\vec{a}_z = -94.25 \times 0.2$$

$$V = -18.85\,\text{V}.$$

Problem 4

A circular cross-section conductor of radius 2 mm carries a current $i_c = 2.5 \sin (5 \times 10^8 t)$ μA. What is the amplitude of the displacement current density if $\sigma = 35$ Ms/m and $\varepsilon_r = 1$?

Solution:

The ratio of conduction to displacement current density is

$$J_c = \sigma E$$

$$J_D = \varepsilon \frac{\partial E_s}{\partial t}$$

if $E_s = E \sin \omega t$

$$\frac{\partial E_s}{\partial t} = E\omega \cos \omega t.$$

Taking only the amplitude,

$$\varepsilon \frac{\partial E_s}{\partial t} = \omega \varepsilon E$$

$$\therefore \quad \frac{J_c}{J_D} = \frac{\sigma E}{\omega \varepsilon E} = \frac{\sigma}{\omega \varepsilon} = \frac{\sigma}{\omega \varepsilon_0 \varepsilon_r}$$

$$= \frac{35 \times 10^6}{(5 \times 10^8) \cdot 8.85 \times 10^{-12}} \quad [\because \ \varepsilon_r = 1]$$

$$= 79.09 \times 10^8.$$

We know $J_c = I_c / \text{Area}$

$$\text{Area} = \pi a^2 = \pi \times (2 \times 10^{-3})^2$$

$$J_c = \frac{2.5 \times 10^{-6}}{\pi \times (2 \times 10^{-3})^2} = 0.198$$

$$J_D = \frac{J_c}{79.09 \times 10^8}$$

$$= \frac{0.198}{79.09 \times 10^8}$$

$$= 2.515 \times 10^{-11} \text{ A/m}^2.$$

Problem 5

The conduction current flowing through a wire with conductivity $\sigma = 3 \times 10^7$ s/m and relative permittivity $\varepsilon_r = 1$ is given by $I_c = 3 \sin \omega t$ (mA). If $\omega = 10^8$ rad/s, find the displacement current.

Solution:

The displacement current density $J_D = \varepsilon \dfrac{\partial E}{\partial t}$.

Since I_c is given, 'E' is found from conduction current density as

$$J_c = \sigma E.$$

$$I_c = \sigma E \times A$$

$$E = \frac{I_c}{\sigma A} = \frac{3\times 10^{-3}\sin\omega t}{3\times 10^{7}\times A}\,\text{V/m}$$

$$= \frac{1\times 10^{-10}}{A}\sin\omega t.$$

$$\therefore \quad \frac{\partial E}{\partial t} = \left(\frac{1\times 10^{-10}}{A}\right)\omega\cos\omega t$$

$$J_d = \varepsilon\omega\left(\frac{1\times 10^{-10}}{A}\right)\cos\omega t$$

$$= 8.84\times 10^{-12}\times 10^{8}\left(\frac{1\times 10^{-10}}{A}\right)\cos\omega t$$

$$= \frac{8.85\times 10^{-14}}{A}\cos\omega t$$

$$I_d = J_d \cdot A$$

$$= 8.85 \times 10^{-14} \cos \omega t \text{ amp.}$$

Problem 6

A poor conductor is characterized by a conductivity σ = 100 (s/m) and permittivity $\varepsilon = 4\varepsilon_0$. At what angular frequency ω is the amplitude of the conduction current density J equal to the amplitude of the displacement density J_d?

$$J_c = \sigma E$$

$$J_D = \varepsilon\omega E \text{ (only amplitude)}$$

$$\sigma E = \varepsilon\omega E$$

$$\sigma = \omega\varepsilon$$

$$100 = \omega \times 4\,\varepsilon_0.$$

$$\omega = \frac{100}{4\times 8.85\times 10^{-12}}$$

$$= 2.82 \times 10^{12} \text{ rad/s.}$$

Problem 7

In a material for which σ = 5 s/m and ε_r = 1, the electric field intensity is $E = 250 \sin 10^{10}t$ V/m. Find the conduction and displacement current densities and the frequency at which they have equal magnitudes.

Solution:

Conduction current density $J_c = \sigma E$.

$$J_c = 5 \times 250 \sin 10^{10} t$$

$$= 1250 \sin 10^{10} t \text{ A/m}^2.$$

Displacement current density $J_d = \varepsilon \dfrac{\partial E}{\partial t}$

$$= \varepsilon \frac{\partial}{\partial t} 250 \sin 10^{10} t$$

$$= \varepsilon_0 \left[250 \times 10^{10} \cos 10^{10} t\right]$$

$$= 22.13 \cos 10^{10} t \text{ A/m}^2.$$

For $J_C = J_D$.

$\sigma E = \omega \varepsilon E$ [Take the amplitude only]

$$\sigma = \omega \varepsilon$$

$$\omega = \frac{\sigma}{\varepsilon_0} = \frac{5}{8.85 \times 10^{-12}} = 5.65 \times 10^{11} \text{ rad/s.}$$

$$\omega = 2\pi f$$

$$f = \frac{5.65 \times 10^{11}}{2 \times \pi} = 8.99 \times 10^{10} \text{Hz}$$

$$f = 8.99 \times 10^{10} \text{Hz}.$$

SUMMARY

1. Time-varying current (or) accelerated charge produces electromagnetic field.
2. Faraday's law states that the induced e.m.f. is given by ($N = 1$)

$$V_{emf} = -\frac{\partial \psi}{\partial t}$$

3. For transformer e.m.f., $V_{emf} = -\int_S \dfrac{\partial \vec{B}}{\partial t} \cdot d\vec{S}$ for motional e.m.f.,

$$V_{emf} = \int_l \left(\vec{u} \times \vec{B}\right) \cdot d\vec{l}.$$

4. The displacement current $I_d = \int_S \vec{J}_d \cdot d\vec{s}$, where $\vec{J}_d = \frac{\partial \vec{D}}{\partial t} = \varepsilon \frac{\partial \vec{E}}{\partial t}$.

5. Modified form of Ampere's circuital law is

$$\oint_l \vec{H} \cdot d\vec{l} = I_c + I_d = \iint_S \left(\vec{J}_c + \vec{J}_d\right) \cdot d\vec{s}$$

$$\nabla \times \vec{H} = \vec{J}_c + \vec{J}_d = \sigma \vec{E} + \varepsilon \frac{\partial \vec{E}}{\partial t}.$$

EXERCISE

1. A coil consists of 100 turns of wire wrapped around a square frame of sides 0.25 m. The coil is centered at the origin with each of its sides parallel to the *X* or *Y*-axis. Find the induced e.m.f. across the open-circuited ends of the coil if the magnetic field is given by

   (*a*) $\vec{B} = 10e^{-2t}\,\vec{a}_z$ tesla [$125\,e^{-2t}$ V]

   (*b*) $\vec{B} = 10\cos x \cos 10^3 t\,\vec{a}_z$ tesla [$62.3 \sin 10^3 t$ V]

   (*c*) $B = 10 \cos x \sin 2y \cos 10^3 t\,\vec{a}_z$ tesla. [0 V]

2. A 30 cm × 40 cm rectangular loop rotates at 130 rad/s in a magnetic field 0.06 wb/m^2 normal to the axis of rotation. If the loop has 50 turns, determine the induced voltage in the loop. [–54 V]

3. An area of 0.65 m^2 in the $Z = 0$ plane is enclosed by a filamentary conductor. Find the induced voltage, given that

   $\vec{B} = 0.05 \cos 10^3 t \left(\frac{\vec{a}_y + a_z}{\sqrt{2}}\right)$ tesla. [$23.0 \sin 10^3 t$ V]

4. What voltage is developed by a Faraday disk generator with the meter connections at $r_1 = 1$ mm and $r_2 = 100$ mm when the disk turns at 500 rev/min in a flux density of 0.80 T? [0.209 V]

5. A parallel plate capacitor of separation 0.6 mm and with a dielectric of $\varepsilon_r = 15.3$ has an applied rms voltage of 25 V at a frequency of 15 Ghz. Find the rms displacement current density. [5.32×10^5 A/m^2]

6. A coaxial capacitor of length $l = 6$ cm uses an insulating dielectric material $\varepsilon_r = 9$. The radii of the cylindrical conductors are 0.5 cm and 1 cm. If the voltage applied across the capacitor is $V(t) = 100 \sin(120\,\pi t)$, what is the displacement current? [1.63 cos (120 πt) μA]

7. Moist soil has a conductivity of 10^{-3} s/m and $\varepsilon_r = 2.5$. Find J_c and J_0 where

   $\vec{E} = 6 \times 10^{-6} \sin 9 \times 10^9 t$ V/m. [$1.2 \times 10^{-6} \cos 9 \times 10^9 t$ A/m^2]

5

Electromagnetic Wave Propagation

INTRODUCTION

The existence of electromagnetic waves was predicted by Maxwell's equations and it was investigated by Heinrich Hertz. Hertz succeeded in generating and detecting radio waves which are sometimes called Hertzian waves.

In general, waves are means of transporting energy or information. Typical examples of electromagnetic waves include radiowaves, TV signals, radar beams and light rays. All forms of electromagnetic energy have three fundamental characteristics:

- They all travel at high velocity
- While travelling, they assume the properties of waves
- They radiate outward from a source

If a physical phenomenon that occurs at one place at a given time is reproduced at other places at later times, the time delay being proportional to the space separation from the first location, then the group of phenomena constitute a wave.

5.1 WAVE EQUATION

Let us consider a perfect dielectric region free of charges ($\rho_v = 0$) and for which μ and ε are constants. Maxwell's equations for such a medium are modified as below.

$$\nabla \cdot \vec{D} = \rho_v = 0 \qquad ...(5.1)$$

$$\nabla \cdot \vec{B} = 0 \text{ (or) } \nabla \cdot \vec{H} = 0 \qquad \left[\because \vec{B} = \mu\vec{H}\right] \qquad ...(5.2)$$

$$\nabla \times \vec{E} = \frac{-\partial \vec{B}}{\partial t}$$

$$\nabla \times \vec{E} = -\mu \frac{\partial \vec{D}}{\partial t} \qquad \left[\because \vec{B} = \mu\vec{H}\right] \qquad ...(5.3)$$

$$\nabla \times \vec{H} = \vec{J} + \frac{\partial \vec{D}}{\partial t}$$

$\vec{J} = 0$ in dielectric medium,

So,

$$\nabla \times \vec{H} = \frac{\partial \vec{D}}{\partial t}$$

$$\nabla \times \vec{H} = \varepsilon \frac{\partial \vec{E}}{\partial t} \qquad \left(\because \vec{D} = \varepsilon \vec{E}\right). \qquad \text{...(5.4)}$$

Taking the curl of both sides of eqn. (5.04)

$$\nabla \times \left(\nabla \times \vec{H}\right) = \varepsilon\left(\nabla \times \frac{\partial \vec{E}}{\partial t}\right). \qquad \text{...(5.5)}$$

Interchanging $\partial/\partial t$ and curl operation on the right side of eqn. (5.5) and applying the vector identity on left side

$$\nabla \times \left(\nabla \times \vec{H}\right) = \nabla\left(\nabla \cdot \vec{H}\right) - \nabla^2 H$$

$$\nabla\left(\nabla \cdot \vec{H}\right) - \nabla^2 \vec{H} = \varepsilon \frac{\partial}{\partial t}\left(\nabla \times \vec{E}\right). \qquad \text{...(5.6)}$$

From eqn. (5.2) $\nabla \cdot \vec{H} = 0$ and from eqn. (5.3) $\nabla \times \vec{E} = -\mu \frac{\partial \vec{H}}{\partial t}$

So, eqn. (5.6) becomes

$$-\nabla^2 \vec{H} = -\mu\varepsilon \frac{\partial}{\partial t}\left(\frac{\partial \vec{H}}{\partial t}\right)$$

$$-\nabla^2 \vec{H} = -\mu\varepsilon \frac{\partial^2 \vec{H}}{\partial t^2}$$

$$\nabla^2 \vec{H} = \mu\varepsilon \frac{\partial^2 \vec{H}}{\partial t^2}. \qquad \text{...(5.7)}$$

Similarly, the equation relating $\vec{E}$ is obtained by taking curl on both sides of eqn. (5.3) and substituting from other equations.

$$\nabla^2 \vec{E} = \mu\varepsilon \frac{\partial^2 \vec{E}}{\partial t^2}. \qquad \text{...(5.8)}$$

The above two eqns. (5.7) and (5.8) are said to be the wave equations.

For free space

$$\mu = \mu_0 = 4\pi \times 10^{-7} \text{ H/m}.$$

$$\varepsilon = \varepsilon_0 = \frac{1}{36\pi} \times 10^{-9} \text{ F/m}.$$

Eqn. (5.7) can be written as

$$\frac{\partial^2 \vec{H}}{\partial t^2} = \frac{1}{\mu\varepsilon} \nabla^2 \vec{H}.$$

Let the velocity of wave $u = \frac{1}{\sqrt{\mu\varepsilon}}$.

Then, $$\frac{\partial^2 \vec{H}}{\partial t^2} = u^2 \nabla^2 \vec{H}. \quad \text{...(5.9)}$$

Similarly, eqn. (5.8) can be modified as

$$\frac{\partial^2 \vec{E}}{\partial t^2} = u^2 \nabla^2 \vec{E}. \quad \text{...(5.10)}$$

Let us calculate the wave velocity u for free space:

$$u = \frac{1}{\sqrt{\mu_0 \varepsilon_0}} = \frac{1}{\sqrt{4\pi \times 10^{-7} \times \frac{10^{-9}}{36\pi}}}$$

$$= 3 \times 10^8 \text{ m/s}.$$

The value of velocity is same as velocity of light.

The eqns. (5.9) and (5.10) are three-dimensional wave equations. Consider the one-dimensional wave equations by assuming that the wave is propagating in Z direction only.

The wave equations in one-dimensional form are

$$\frac{\partial^2 \vec{E}}{\partial t^2} = u^2 \frac{\partial^2 \vec{E}}{\partial z^2} \quad \text{...(5.11)}$$

$$\frac{\partial^2 \vec{H}}{\partial t^2} = u^2 \frac{\partial^2 \vec{H}}{\partial z^2}. \quad \text{...(5.12)}$$

Assume that field vectors $\vec{E}$ and $\vec{H}$ are harmonic (or sinusoidal) time-dependent.

The above two equations can be modified by substituting $\vec{E} = \vec{E}_s e^{j\omega t}$ as given in eqn. (4.33).

So, eqn. (5.11) becomes

$$\frac{\partial^2}{\partial t^2}\left(\vec{E}_s e^{j\omega t}\right) = u^2 \frac{\partial^2}{\partial z^2}\left(\vec{E}_s e^{j\omega t}\right)$$

$$\vec{E}_s \frac{\partial^2}{\partial t^2}\left(e^{j\omega t}\right) = u^2\, e^{j\omega t}\left(\frac{\partial^2 \vec{E}_s}{\partial z^2}\right)$$

$$\vec{E}_s\left(-\omega^2\, e^{j\omega t}\right) = u^2\, e^{j\omega t}\left(\frac{\partial^2 \vec{E}_s}{\partial z^2}\right)$$

$$\vec{E}_s = -\frac{u^2}{\omega^2}\frac{\partial^2 \vec{E}_s}{\partial z^2}$$

$$\frac{\partial^2 \vec{E}_s}{\partial z^2} = \frac{-\omega^2}{u^2}\vec{E}_s.$$

Let $\beta = \dfrac{\omega}{u}.$

So,

$$\frac{\partial^2 \vec{E}_s}{\partial z^2} = -\beta^2 \vec{E}_s. \qquad \text{...(5.13)}$$

Similarly,

$$\frac{\partial^2 \vec{H}_s}{\partial z^2} = -\beta^2 \vec{H}_s. \qquad \text{...(5.14)}$$

Solving for $\vec{E}$, considering eqn. (5.13),

$$\frac{\partial^2 \vec{E}_s}{\partial z^2} + \beta^2 \vec{E}_s = 0$$

$$\left(\frac{\partial^2}{\partial z^2} + \beta^2\right)\vec{E}_s = 0.$$

The solution of this differential equation is

$$\vec{E}_s = A\,e^{-j\beta z} + B\,e^{j\beta z}. \qquad \text{...(5.15)}$$

Substituting $\vec{E} = \vec{E}_s\, e^{j\omega t}$ in eqn. (5.15)

$$\vec{E} = \left(Ae^{-j\beta z} + Be^{+j\beta z}\right)e^{j\omega t}$$

$$\vec{E} = Ae^{j(\omega t-\beta z)} + Be^{j(\omega t+\beta z)}. \qquad \text{...(5.16)}$$

By using the same procedure, $\vec{H}$ is given as

$$\vec{H} = Ce^{j(\omega t-\beta z)} + De^{j(\omega t+\beta z)}. \qquad \text{...(5.17)}$$

In eqn (5.16),

let $$E^+ = Ae^{j(\omega t-\beta z)} \qquad \text{...(5.18)}$$

$$E^- = Be^{j(\omega t+\beta z)}. \qquad \text{...(5.19)}$$

Consider the eqn (5.18) and take the imaginary part of it.

$$\vec{E}^+ = A\sin(\omega t - \beta z). \qquad \text{...(5.20)}$$

In the above equation

- A is called amplitude of the wave and has the same units as $\vec{E}$
- $(\omega t - \beta z)$ is the phase in radians of the wave: it depends on time t and space variable z.
- ω is the angular frequency in (radians/second); β is the phase constant or wave number (radians/meter).

Since E is varying with both time t and space variable z, $\vec{E}$ can be plotted as a function of t by keeping z constant and vice-versa. The plots of E are shown in Fig. (5.1)

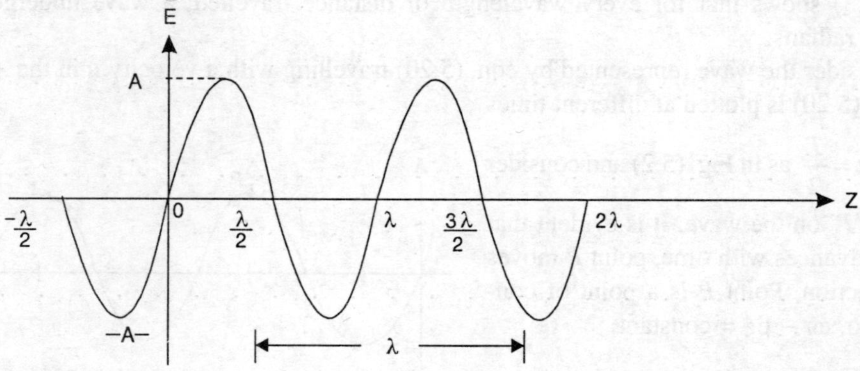

Fig. 5.1(a) $\vec{E}(z,t) = A\sin(\omega t - \beta z)$ **with constant** z

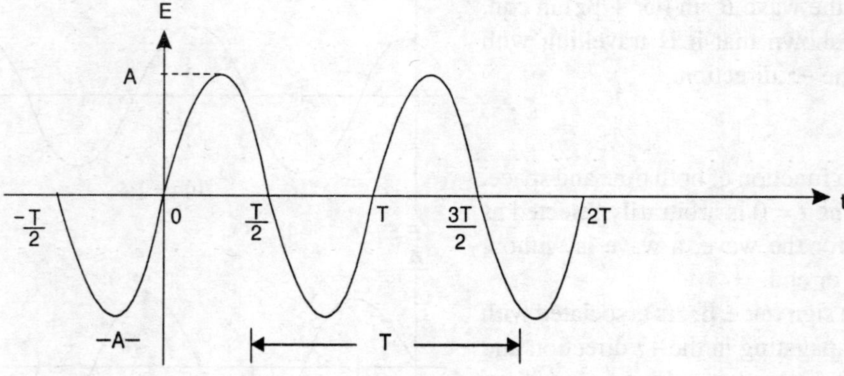

Fig. 5.1(b) $\vec{E}(z,t)$ **with constant** t

From the Fig. (5.1a), it is observed that the wave takes distance λ to repeat itself and hence λ is called wavelength (in meters). From Fig. (5.1b), the wave takes time T to repeat itself and T is known as the period (in seconds). The wave travels the distance λ m in T s at a speed of u m/s,

$$\lambda = uT \qquad ...(5.21)$$

But $T = 1/f$ where f is the frequency of wave in Hertz. Hence,

$$u = f\lambda \qquad ...(5.22)$$

Since $\omega = 2\pi f$, β can be represented as

$$\beta = \frac{\omega}{u} = 2\pi\frac{f}{u}$$

$$\beta = 2\pi\left(\frac{f}{u}\right).$$

From eqn. (5.22)

$$\beta = \frac{2\pi}{\lambda}. \qquad ...(5.23)$$

Eqn. (5.23) shows that for every wavelength of distance travelled, a wave undergoes a phase change of 2π radians.

Now, consider the wave represented by eqn. (5.20) travelling with a velocity u in the $+z$ direction.

The eqn. (5.20) is plotted at different times $t = 0$, $t = \frac{T}{4}$, $t = \frac{T}{2}$ as in Fig. (5.2) and consider a fixed point 'P' on the wave. It is evident that as the wave advances with time, point P moves along $+z$ direction. Point P is a point of constant phase, so, $\omega t - \beta z$ = constant.

$$\text{(or)} \quad \frac{dz}{dt} = \frac{\omega}{\beta} = u. \qquad ...(5.24)$$

Similarly, the wave $B \sin(\omega t + \beta z)$ in eqn. (5.19) can be shown that it is travelling with velocity u in the $-z$ direction.

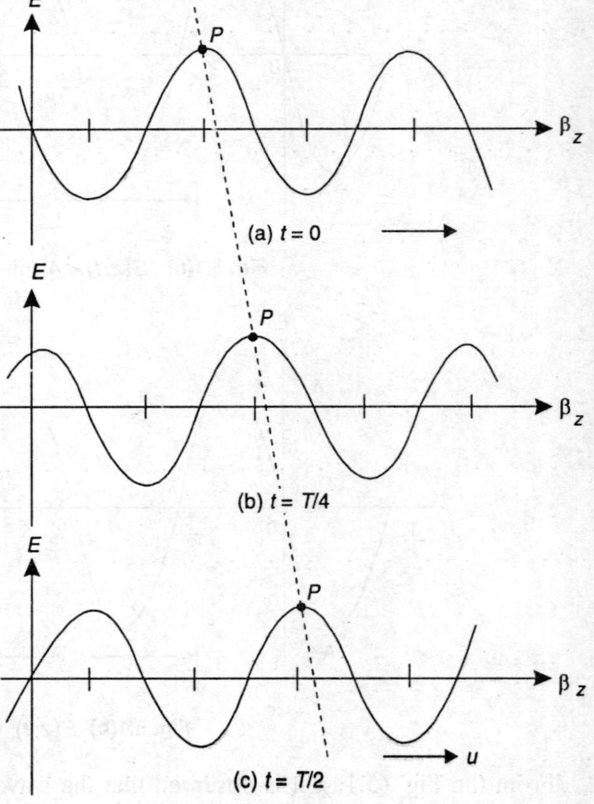

Fig. 5.2 $\bar{E}(z, t) = A\sin(\omega t - \beta z)$ at time $t = 0$, $t = T/4$, $t = T/2$

Note:

- A wave is a function of both time and space.
- Though time $t = 0$ is arbitrarily selected as reference for the wave, a wave is without beginning or end.
- A negative sign $(\omega t \pm \beta z)$ is associated with a wave propagating in the $+z$ direction and it is called forward travelling or positive-going wave, whereas a positive sign indicates that the wave is travelling in the $-z$ direction and is called as backward travelling or negative-going wave.

P moves along $+z$ direction with velocity u.

5.2 WAVE PROPAGATION IN LOSSY DIELECTRICS

A lossy dielectric medium is a medium in which an electromagnetic wave loses power as it propagates due to poor conduction. In other words, a lossy dielectric is a partially-conducting medium (imperfect dielectric or imperfect conductor) with $\sigma \neq 0$, whereas a lossless dielectric is a perfect or good dielectric in which $\sigma = 0$.

Consider a linear, isotropic, homogeneous, lossy dielectric medium that is charge free ($\rho_v = 0$). Take the Maxwell's equation from Table 4.2 and modify as below:

(*i*) $\nabla \cdot \vec{D}_s = \rho_{vs} = 0$

$$\nabla \cdot \varepsilon \vec{E}_s = 0$$

So, $\nabla \cdot \vec{E}_s = 0$ ...(5.25)

(*ii*) $\nabla \cdot \vec{B}_s = 0$

$$\nabla \cdot \mu \vec{H}_s = 0$$

So, $\nabla \cdot \vec{H}_s = 0$ ...(5.26)

(*iii*) $\nabla \times \vec{E}_s = -j\omega \vec{B}_s$

$$\nabla \times \vec{E}_s = -j\omega \mu \vec{H}_s \qquad ...(5.27)$$

(*iv*) $\nabla \times \vec{H}_s = \vec{J}_s + j\omega \vec{D}_s$

$$\nabla \times \vec{H}_s = \sigma \vec{E}_s + j\omega\varepsilon \vec{E}_s$$

$$\nabla \times H_s = (\sigma + j\omega\varepsilon)\vec{E}_s. \qquad ...(5.28)$$

Taking the curl of both sides of eqn. (5.27)

$$\nabla \times \nabla \times \vec{E}_s = -j\omega\mu \nabla \times \vec{H}_s. \qquad ...(5.29)$$

Applying the vector identity

$$\nabla \times \nabla \times \vec{A} = \nabla\left(\nabla \cdot \vec{A}\right) - \nabla^2 A$$

to the left hand side of eqn. (5.29) and substituting eqn. (5.28) to the right-hand side.

$$\nabla\left(\nabla \cdot \vec{E}_s\right) - \nabla^2 \vec{E}_s = -j\omega\mu(\sigma + j\omega\varepsilon)\vec{E}_s.$$

From eqn. (5.25), $\nabla \cdot \vec{E}_s = 0.$

Hence,

$$\nabla \cdot \vec{E}_s = j\omega\mu(\sigma + j\omega\varepsilon)\vec{E}_s. \qquad ...(5.30)$$

Let $\gamma^2 = j\omega\mu(\sigma + j\omega\varepsilon).$ ...(5.31)

Eqn. (5.30) becomes

$$\nabla^2 \vec{E}_s = \gamma^2 \vec{E}_s, \quad \text{...(5.32)}$$

where γ is the propagation constant of the medium.

By similar procedure, it can be shown for $\vec{H}$ field as

$$\nabla^2 \vec{H}_s = \gamma^2 \vec{H}_s. \quad \text{...(5.33)}$$

The two eqns. (5.31) and (5.32) are called Helmholtz's equations or vector wave equations.

Since γ is a complex quantity, let

$$\gamma = \alpha + j\beta \quad \text{...(5.34)}$$

α and β can be obtained as below.

From eqn. (5.31)

$$\gamma^2 = j\omega\mu(\sigma + j\omega\varepsilon).$$

From eqn. (5.34)

$$\gamma^2 = (\alpha + j\beta)^2 = \alpha^2 - \beta^2 + j2\alpha\beta \quad \text{...(5.35)}$$

$$\alpha^2 - \beta^2 + j2\alpha\beta = j\omega\mu\sigma - \omega^2\mu\varepsilon. \quad \text{...(5.36)}$$

Equate the real and imaginary part

$$\alpha^2 - \beta^2 = -\omega^2\mu\varepsilon \text{ (or)}$$

$$\beta^2 - \alpha^2 = \omega^2\mu\varepsilon \quad \text{...(5.37)}$$

$$2\alpha\beta = \omega\mu\sigma. \quad \text{...(5.38)}$$

Also, in eqn. (5.35)

$$|\gamma^2| = \sqrt{(\alpha^2 - \beta^2) + (2\alpha\beta)^2}$$

$$|\gamma^2| = \alpha^2 + \beta^2. \quad \text{...(5.39)}$$

Similarly, from eqn. (5.31)

$$|\gamma^2| = \omega\mu\sqrt{\sigma^2 + \omega^2\varepsilon^2}. \quad \text{...(5.40)}$$

Equate eqn. (5.39) and (5.40)

$$\alpha^2 + \beta^2 = \omega\mu\sqrt{\sigma^2 + \omega^2\varepsilon^2}. \quad \text{...(5.41)}$$

Adding eqns. (5.37) and (5.41)

$$2\beta^2 = \omega^2\mu\varepsilon + \omega\mu\sqrt{\sigma^2 + \omega^2\varepsilon^2}$$

$$2\beta^2 = \omega^2\mu\varepsilon\left[1 + \frac{1}{\omega\varepsilon}\sqrt{\sigma^2 + \omega^2\varepsilon^2}\right]$$

$$\beta^2 = \frac{\omega^2\mu\varepsilon}{2}\left[1 + \frac{\omega\varepsilon}{\omega\varepsilon}\sqrt{\frac{\sigma^2}{\omega^2\varepsilon^2} + 1}\right]$$

$$\beta^2 = \frac{\omega^2\mu\varepsilon}{2}\left[1+\sqrt{\frac{\sigma^2}{\omega^2\varepsilon^2}+1}\right]$$

$$\beta = \omega\left[\sqrt{\frac{\mu\varepsilon}{2}\left[1+\sqrt{\frac{\sigma^2}{\omega^2\varepsilon^2}+1}\right]}\right]. \quad ...(5.42)$$

Now subtract eqn. (5.37) from eqn. (5.41)

$$\omega\mu\sqrt{\sigma^2+\omega^2\varepsilon^2} - \omega^2\mu\varepsilon = 2\alpha^2$$

$$\alpha^2 = \frac{\omega^2\mu\varepsilon}{2}\left[\frac{1}{\omega\varepsilon}\sqrt{\sigma^2+\omega^2\varepsilon^2}-1\right]$$

$$= \frac{\omega^2\mu\varepsilon}{2}\left[\frac{\omega\varepsilon}{\omega\varepsilon}\sqrt{\frac{\sigma^2}{\omega^2\varepsilon^2}+1}-1\right]$$

$$= \frac{\omega^2\mu\varepsilon}{2}\left[\sqrt{\frac{\sigma^2}{\omega^2\varepsilon^2}+1}-1\right]$$

$$\alpha = \omega\left[\sqrt{\frac{\mu\varepsilon}{2}\left[\sqrt{\left(\frac{\sigma^2}{\omega^2\varepsilon^2}+1\right)}-1\right]}\right]. \quad ...(5.43)$$

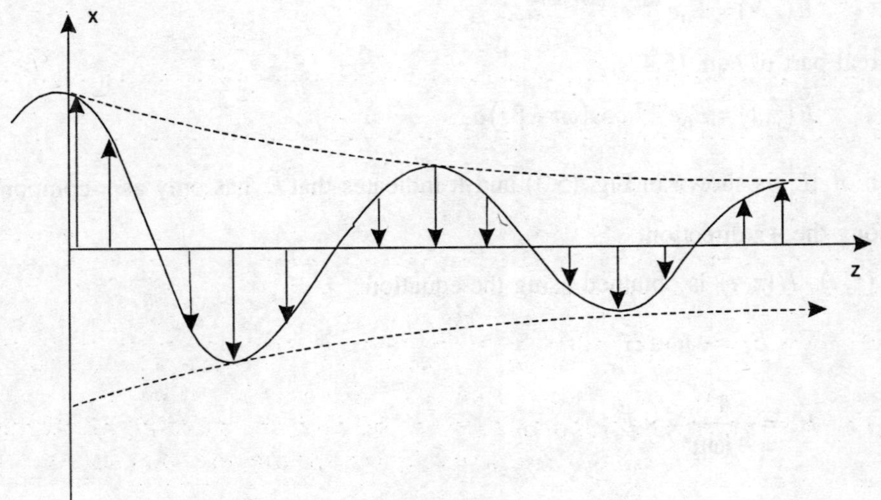

Fig 5.3 Field with *x* component travelling along *z* direction

Now, assume the wave propagates along $+\vec{a}_z$, and $\vec{E}_s$ has only x-component, then

$$\vec{E}_s = E_{xs}(z)\vec{a}_x. \qquad ...(5.44)$$

Substituting this in eqn. (5.32)

$$(\nabla^2 - \gamma^2)E_{xs}(z) = 0$$

$$\nabla^2 E_{xs}(z) - \gamma^2 E_{xs}(z) = 0$$

in cartesian co-ordinates.

$$\frac{\partial^2 E_{xs}(z)}{\partial x^2} + \frac{\partial^2 E_{xs}(z)}{\partial y^2} + \frac{\partial^2 E_{xs}(z)}{\partial z^2} - \gamma^2 E_{xs}(z) = 0.$$

Since the wave is propagating only in $+z$ direction, the above equation becomes

$$\frac{\partial^2 E_{xs}(z)}{\partial z^2} - -\gamma^2 E_{xs}(z) = 0. \qquad ...(5.45)$$

Solution for the above eqn. (5.45) is

$$E_{xs}(z) = E_0 e^{-\gamma z} + E_0' e^{\gamma z}.$$

Taking only the positive-going wave, *i.e.*, wave travelling along $+z$ direction.

$$E_{xs}(z) = E_0 e^{-\gamma z} \qquad ...(5.46)$$

$$E(z, t) = E_{xs}(z)e^{j\omega t}$$

$$\vec{E}(z, t) = E_0 e^{-\gamma z} e^{j\omega t}\,\vec{a}_x$$

$$\vec{E}(z, t) = E_0 e^{-\alpha z} e^{j(\omega t - \beta z)}\vec{a}_x. \qquad ...(5.47)$$

Taking real part of eqn. (5.47),

$$\vec{E}(z, t) = E_0 e^{-\alpha z}\cos(\omega t - \beta z)\vec{a}_x. \qquad ...(5.48)$$

A sketch of $|\vec{E}|$ is shown in Fig. (5.3) and it indicates that $\vec{E}$ has only as x-component and it is travelling along the $+z$ direction.

From $\vec{E}(z, t)$, $\vec{H}(z, t)$ is obtained using the equation

$$\nabla \times \vec{E}_s = -j\omega\mu\,\vec{H}_s$$

$$\vec{H}_s = \frac{1}{-j\omega\mu}\nabla \times \vec{E}_s. \qquad ...(5.49)$$

As $\vec{E}$ has only x-component and the wave is travelling along z direction, eqn. (5.49) becomes

$$\nabla \times \vec{E}_s = \begin{vmatrix} \vec{a}_x & \vec{a}_y & \vec{a}_z \\ \frac{\partial}{\partial x} & \frac{\partial}{\partial y} & \frac{\partial}{\partial z} \\ E_{sx} & \underset{0}{\underset{\Downarrow}{E_{sy}}} & \underset{0}{\underset{\Downarrow}{E_{sz}}} \end{vmatrix}$$

$$= -\vec{a}_y \left(-\frac{\partial E_{sx}}{\partial z} \right)$$

$$\vec{H}_s = \frac{1}{-j\omega\mu} \left(\frac{\partial E_{sx}}{\partial z} \right) \vec{a}_y.$$

As $\vec{E}_{sx} = E_0 e^{-\gamma z}$

$$\frac{\partial}{\partial z}\left(\vec{E}_{sx}\right) = -\gamma\, E_0\, e^{-\gamma z}.$$

$$\therefore \qquad \vec{H}_s = \frac{-1}{j\omega\mu}(-\gamma)\, E_0\, e^{-\gamma z}\, \vec{a}_y$$

$$\vec{H}_s = \frac{\gamma}{j\omega\mu} E_0\, e^{-\gamma z}\, \vec{a}_y.$$

As $\vec{H} = H_s e^{j\omega t}$

$$\vec{H} = \frac{\gamma}{j\omega\mu} E_0\, e^{-\gamma z} e^{j\omega t}\, \vec{a}_y$$

Substituting for γ from eqn. (5.31)

$$\vec{H}(z, t) = \sqrt{\frac{j\omega\mu(\sigma + j\omega\varepsilon)}{(j\omega\mu)^2}}\, E_0 e^{-\gamma z} e^{j\omega t}\, \vec{a}_y$$

$$\vec{H}(z, t) = \sqrt{\frac{\sigma + j\omega\varepsilon}{j\omega\mu}}\, E_0 e^{-\alpha z} e^{j(\omega t - \beta z)}\, \vec{a}_y.$$

Let us take $\eta = \sqrt{\frac{j\omega\mu}{\sigma + j\omega\varepsilon}}$ and $H_0 = \frac{E_0}{\eta}$; then, ...(5.50)

$$\vec{H}(z, t) = H_0\, e^{-\alpha z}\, e^{j(\omega t - \beta z)}\, \vec{a}_y,$$

where η is a complex quantity known as intrinsic impedance (in ohms) of the medium. Since η is a complex quantity

$$\eta = |\eta| \underline{|\theta_\eta} = |\eta| e^{j\theta_\eta}$$

$$\eta = \sqrt{\frac{j\omega\mu}{\sigma + j\omega\varepsilon}} = \sqrt{\frac{j\omega\mu}{j\omega\varepsilon\left(\frac{\sigma}{j\omega\varepsilon} + 1\right)}}$$

$$\eta = \sqrt{\frac{\frac{\mu}{\varepsilon}}{1 + \frac{\sigma}{j\omega\varepsilon}}} = \sqrt{\frac{\frac{\mu}{\varepsilon}}{1 - j\frac{\sigma}{\omega\varepsilon}}}$$

$$|\eta| = \frac{\sqrt{\frac{\mu}{\varepsilon}}}{\left[1 + \left(\frac{\sigma}{\omega\varepsilon}\right)^2\right]^{1/4}} \quad \text{...(5.51)}$$

$$\theta_\eta = 0 - \frac{\tan^{-1}\left(\frac{-\sigma}{\omega\varepsilon}\right)}{2}$$

$$2\theta_\eta = \tan^{-1}\left(\frac{\sigma}{\omega\varepsilon}\right), \quad \text{...(5.52)}$$

where $0 \le \theta_\eta \le 45°$.

So,

$$\vec{H} = \text{Re}\left\{\frac{E_0}{|\eta| e^{j\theta_\eta}} e^{-\alpha z} e^{j(\omega t - \beta z)}\right\}\vec{a}_y$$

$$\vec{H} = \frac{E_0}{|\eta|} e^{-\alpha z} \cos\left(\omega t - \beta z - \theta_\eta\right)\vec{a}_y. \quad \text{...(5.53)}$$

From eqns. (5.48) and (5.53), it is known that as the wave propagates along $\vec{a}_z$, it decreases or attenuates in amplitude by a factor $e^{-\alpha z}$, and hence α is known as the attenuation constant or attenuation factor of the medium. It is a measure of the spatial rate of decay of the wave in the medium, measured in nepers per meter (Np/m) or is decibels per meter (dB/m).

$$N_p = 20 \log_{10} e = 8.686 \, \text{dB} \quad \text{...(5.54)}$$

From eqn. (5.53), it is noted that if $\sigma = 0$, $\alpha = 0$ and the wave is not attenuated as it propagates. The quantity β is a measure of the phase shift per length and is called phase constant or wave number.

Also from eqns. (5.48) and (5.53) that $\vec{E}$ and $\vec{H}$ are out of phase by θ_η at any instant of time due to complex intrinsic impedance of the medium. At any time, $\vec{E}$ leads $\vec{H}$ by θ_η.

The ratio of the magnitude of the conduction current density $\vec{J}$ to displacement current density $\vec{J}_d$ in a lossy medium is

$$\frac{|\vec{J}_s|}{|\vec{J}_{ds}|} = \frac{|\sigma \vec{E}_s|}{|j\omega\varepsilon \vec{E}_s|} = \frac{\sigma}{\omega\varepsilon} = \tan\theta. \quad ...(5.55)$$

Here tan θ is known as loss tangent and θ is the loss angle of the medium.

A medium is said to be a good (loss less or perfect) dielectric if tan θ is very small ($\sigma << \omega\varepsilon$) or a good conductor if tan θ is very large ($\sigma >> \omega\varepsilon$). Also a medium is regarded as a good conductor at low frequencies and as a good dielectric at high frequencies.

5.3 PLANE WAVES IN LOSSLESS DIELECTRICS

In a lossless dielectric, $\sigma << \omega\,\varepsilon$.

Let

$$\sigma = 0,\ \varepsilon = \varepsilon_0\varepsilon_r,\ \mu = \mu_0\mu_r. \quad ...(5.56)$$

Substituting these in eqns. (5.42) and (5.43) gives

$$\alpha = 0,\ \beta = \omega\sqrt{\mu\varepsilon} \quad ...(5.57)$$

$$u = \frac{\omega}{\beta} = \frac{1}{\sqrt{\mu\varepsilon}},\ \lambda = \frac{2\pi}{\beta}. \quad ...(5.58)$$

Also,

$$\eta = \sqrt{\frac{\mu}{\varepsilon}}\,\angle 0°.$$

Thus $\vec{E}$ and $\vec{H}$ are in time phase with each other.

$$\vec{E} = \text{Re}\left[E_0\, e^{j(\omega t - \beta z)}\right]\vec{a}_x$$

$$\vec{E} = E_0\cos(\omega t - \beta z)\,\vec{a}_x \quad ...(5.59)$$

$$\vec{H} = H_0\cos(\omega t - \beta z)\,\vec{a}_y, \quad ...(5.60)$$

where $$H_0 = \frac{E_0}{\eta}$$

5.4 PLANE WAVES IN FREE SPACE

In this case,

$$\sigma = 0, \varepsilon = \varepsilon_0, \mu = \mu_0$$

Substituting the above quantities in eqns. (5.43) and (5.42) then,

$$\alpha = 0, \ \beta = \omega\sqrt{\mu_0\varepsilon_0} = \frac{\omega}{c} \quad ...(5.61)$$

$$\mu = \frac{\omega}{\beta} = \frac{1}{\sqrt{\mu_0\varepsilon_0}} = c, \ \lambda = \frac{2\pi}{\beta}, \quad ...(5.62)$$

where $c = 3 \times 10^8$ m/s, the speed of light in vacuum. Light is characteristically electromagnetic.

Also, $\theta_\eta = 0$ and $\eta = \eta_0$,

where η_0 is called the intrinsic impedance of free space and is given by

$$\eta_0 = \sqrt{\frac{\mu_0}{\varepsilon_0}} = 120\,\pi = 377\,\Omega \quad ...(5.63)$$

$$\vec{E} = E_0 \cos(\omega t - \beta z)\vec{a}_x \quad ...(5.64)$$

$$\vec{H} = H_0 \cos(\omega t - \beta z)\vec{a}_y. \quad ...(5.65)$$

The plots of $\vec{E}$ and $\vec{H}$ are shown in Fig. (5.4). In general, if $\vec{a}_E$, $\vec{a}_H$ and $\vec{a}_k$ are unit vectors along $\vec{E}$ field, $\vec{H}$ field and the direction of wave propagation, then

$$\vec{a}_k \times \vec{a}_E = \vec{a}_H \quad \text{(or)}$$

$$\vec{a}_k \times \vec{a}_H = -\vec{a}_E \quad \text{(or)}$$

$$\vec{a}_E \times \vec{a}_H = -\vec{a}_k \quad ...(5.66)$$

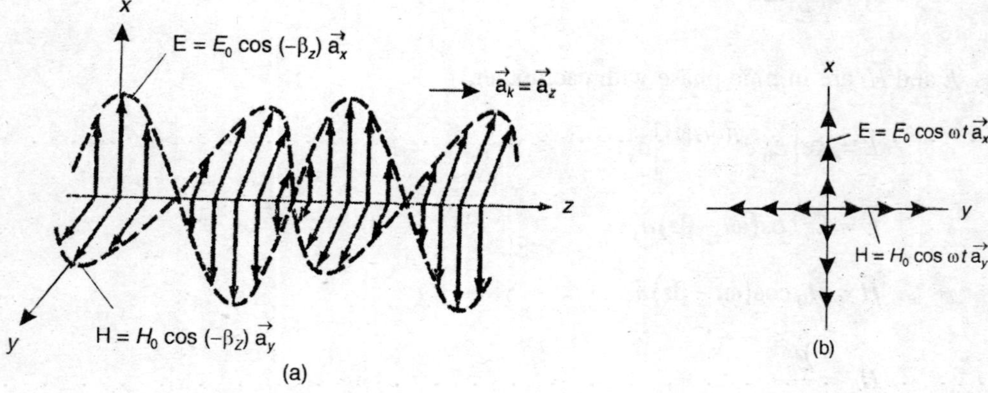

Fig. 5.4 Plot of $\vec{E}$ and $\vec{H}$ as functions of Z at $t = 0$

Both $\vec{E}$ and $\vec{H}$ fields (or electromagnetic waves) are everywhere normal to the direction of wave propagation and these fields lie in a plane that is transverse or orthogonal to the direction of wave propagation. The electromagnetic wave has no electric or magnetic field components along the direction of propagation. Such a wave is called a transverse electromagnetic wave (TEM) wave.

Each $\vec{E}$ and $\vec{H}$ is called a uniform plane wave because $\vec{E}$ or $\vec{H}$ has the same magnitude throughout any transverse plane, defined by Z = constant. The direction of the electric field gives the polarization of the TEM wave. For example, the wave in eqn (5.64) is polarized in x-direction.

5.5 PLANE WAVES IN GOOD CONDUCTORS

A perfect or good conductor has $\sigma >> \omega\varepsilon$. Hence

$$\sigma \approx \infty, \ \varepsilon = \varepsilon_0 \varepsilon_r, \ \mu = \mu_0 \mu_r \qquad ...(5.67)$$

Substituting these in eqns. (5.42) and (5.43)

$$\alpha = \beta = \sqrt{\frac{\omega\mu\sigma}{2}} = \sqrt{2\pi f\mu\sigma/2}$$

$$\alpha = \beta = \sqrt{\pi f\mu\sigma} \qquad ...(5.68)$$

$$u = \frac{\omega}{\beta} = \frac{\omega}{\sqrt{\pi f\mu\sigma}} = \frac{2\pi f}{\sqrt{\pi f\mu\sigma}}$$

$$u = \sqrt{\frac{(2\pi f)^2}{\pi f\mu\sigma}} = \sqrt{\frac{2\omega}{\mu\sigma}} \qquad ...(5.69)$$

$$\lambda = \frac{2\pi}{\beta}. \qquad ...(5.70)$$

Also, as $\sigma >> \omega\varepsilon$ neglect $j\omega\varepsilon$ in eqn. (5.30),

η becomes

$$\eta = \sqrt{\frac{j\omega\mu}{\sigma}}$$

$$|\eta| = \sqrt{\frac{\omega\mu}{\sigma}} \qquad ...(5.71)$$

$$\tan 2\theta_\eta = \frac{\sigma}{\omega u}.$$

Since $\sigma \approx \infty$

$$\tan 2\theta_\eta = \infty$$

$$2\theta_\eta = \infty$$

$$\theta_\eta = 45°. \qquad ...(5.72)$$

So,

$$\eta = \sqrt{\frac{\omega\mu}{\sigma}} \angle 45°. \qquad ...(5.73)$$

That is, $\vec{E}$ leads $\vec{H}$ by 45°.

If $\vec{E} = E_0 e^{-\alpha z} \cos(\omega t - \beta z)\vec{a}_x$, ...(5.74)

then,

$$\vec{H} = \frac{E_0}{\eta} e^{-\alpha z} \cos(\omega t - \beta z - 45°)\vec{a}_y \qquad ...(5.75)$$

$$H_0 = \frac{E_0}{\sqrt{\frac{\omega\mu}{\sigma}}}$$

$$\vec{H} = H_0 e^{-\alpha z} \cos(\omega t - \beta z - 45°)\vec{a}_y. \qquad ...(5.76)$$

From eqns. (5.74) and (5.76), it is seen that as $\vec{E}$ and $\vec{H}$ waves travel in a conducting medium, its amplitude is attenuated by the factor $e^{-\alpha Z}$. The distance δ, through which the wave amplitude decreases by a factor e^{-1} (about 37%) is called skin depth or penetration depth of the medium.

That is

$$E_0 e^{-\alpha\delta} = E_0 e^{-1} \qquad ...(5.77)$$

$$\delta = \frac{1}{\alpha}. \qquad ...(5.78)$$

Substituting α from eqn. (5.68)

$$\delta = \frac{1}{\sqrt{\pi f \mu \sigma}}. \qquad ...(5.79)$$

The skin depth is a measure of the depth to which an electromagnetic wave can penetrate the medium.

The phenomenon whereby the field intensity in a conductor rapidly decreases is known as skin effect. The fields and associated currents are confined to a very thin layer of the conductor surface.

Surface or skin resistance Rs (in Ω/m^2) is defined as the real part of intrinsic impedance (η) for a good conductor.

$$\eta = \sqrt{\frac{\omega\mu}{\sigma}} \angle 45°$$

$$= \sqrt{\frac{(2\pi f\mu)}{\sigma}} \angle 45^\circ$$

$$= \sqrt{\frac{2\pi f\mu\sigma}{\sigma^2}} \angle 45^\circ.$$

As $\alpha = \sqrt{\pi f\mu\sigma}$

$$\eta = \sqrt{2}\,\frac{\alpha}{\sigma} \angle 45^\circ$$

$$= \frac{\sqrt{2}}{\sigma\delta} \angle 45^\circ \left(\because \delta = \frac{1}{\alpha}\right)$$

$$\eta = \frac{\sqrt{2}}{\sigma\delta} (\cos 45^\circ + j \sin 45^\circ)$$

$$\eta = \frac{\sqrt{2}}{\sigma\delta} \left(\frac{1}{\sqrt{2}} + \frac{j}{\sqrt{2}}\right)$$

$$\eta = \frac{1+j}{\sigma\delta}. \qquad ...(5.80)$$

So, the surface resistance $R_s = \frac{1}{\sigma\delta}$ ...(5.81)

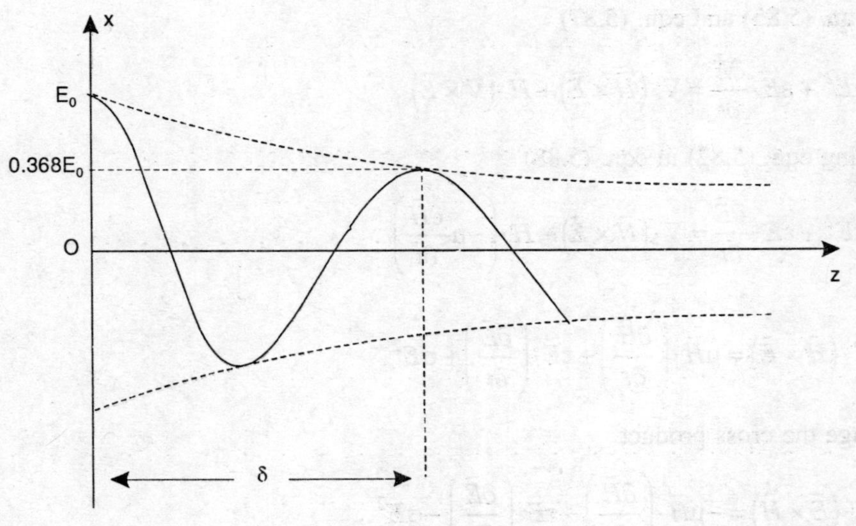

Fig. 5.5 Skin depth

5.6 POWER AND POYNTING VECTOR

The rate of energy transported from one point (where a transmitter is located) to another point (where a receiver is located) by means of electromagnetic waves is obtained by using the Maxwell's equations in Table 4.1.

$$\nabla \times \vec{E} = -\mu \frac{\partial \vec{H}}{\partial t} \qquad ...(5.82)$$

$$\nabla \times \vec{H} = \sigma \vec{E} + \varepsilon \frac{\partial \vec{E}}{\partial t}. \qquad ...(5.83)$$

Dotting both sides of eqn. (5.83) with $\vec{E}$

$$\vec{E} \cdot \left(\nabla \times \vec{H}\right) = \sigma \vec{E} \cdot \vec{E} + \varepsilon \vec{E} \cdot \frac{\partial E}{\partial t} \qquad ...(5.84)$$

$$= \sigma E^2 + \varepsilon \vec{E} \cdot \frac{\partial \vec{E}}{\partial t}. \qquad ...(5.85)$$

From vector identity,

$$\nabla \cdot \left(\vec{A} \times \vec{B}\right) = \vec{B} \cdot \left(\nabla \times \vec{A}\right) - \vec{A} \cdot \left(\nabla \times \vec{B}\right).$$

Let $\vec{A} = \vec{H}$ and $\vec{B} = \vec{E}$

$$\nabla \cdot \left(\vec{H} \times \vec{E}\right) = \vec{E} \cdot \left(\nabla \times \vec{H}\right) - \vec{H} \cdot \left(\nabla \times \vec{E}\right) \qquad ...(5.86)$$

$$\vec{E} \cdot \left(\nabla \times \vec{H}\right) = \nabla \cdot \left(\vec{H} \times \vec{E}\right) + \vec{H} \cdot \left(\nabla \times \vec{E}\right). \qquad ...(5.87)$$

Equate eqn. (5.85) and eqn. (5.87)

$$\sigma E^2 + \varepsilon \vec{E} \cdot \frac{\partial \vec{E}}{\partial t} = \nabla \cdot \left(\vec{H} \times \vec{E}\right) + \vec{H} \cdot \left(\nabla \times \vec{E}\right) \qquad ...(5.88)$$

Substituting eqn. (5.82) in eqn. (5.88)

$$\sigma E^2 + \varepsilon \vec{E} \cdot \frac{\partial \vec{E}}{\partial t} = \nabla \cdot \left(\vec{H} \times \vec{E}\right) + \vec{H} \cdot \left(-\mu \frac{\partial \vec{H}}{\partial t}\right)$$

$$\nabla \cdot \left(\vec{H} \times \vec{E}\right) = \mu \vec{H} \cdot \left(\frac{\partial \vec{H}}{\partial t}\right) + \varepsilon \vec{E} \cdot \left(\frac{\partial \vec{E}}{\partial t}\right) + \sigma E^2.$$

Interchange the cross product

$$\nabla \cdot \left(\vec{E} \times \vec{H}\right) = -\mu \vec{H} \cdot \left(\frac{\partial \vec{H}}{\partial t}\right) - \varepsilon \vec{E} \cdot \left(\frac{\partial \vec{E}}{\partial t}\right) - \sigma E^2.$$

Substitute the vector identities

$$-\mu\vec{H}\cdot\left(\frac{\partial\vec{H}}{\partial t}\right)=-\frac{1}{2}\mu\frac{\partial H^2}{\partial t}$$

$$-\varepsilon\vec{E}\cdot\left(\frac{\partial E}{\partial t}\right)=-\frac{1}{2}\varepsilon\frac{\partial E^2}{\partial t}$$

$$\nabla\cdot\left(\vec{E}\times\vec{H}\right)=-\frac{1}{2}\mu\frac{\partial\vec{H}^2}{\partial t}-\frac{1}{2}\cdot\varepsilon\frac{\partial E^2}{\partial t}-\sigma E^2. \quad ...(5.89)$$

Taking volume integral on both sides and interchange the order of differentiation and integration.

$$\int_v \nabla\cdot\left(\vec{E}\times\vec{H}\right)dv=-\frac{\partial}{\partial t}\int_v\left(\frac{1}{2}\mu H^2+\frac{1}{2}\varepsilon E^2\right)dv-\int_v \sigma E^2\,dv. \quad ...(5.90)$$

Applying divergence theorem to left hand side of eqn. (5.91)

$$\oint_s\left(\vec{E}\times\vec{H}\right)\cdot d\vec{s}=-\frac{\partial}{\partial t}\int_v\left(\frac{1}{2}\varepsilon E^2+\frac{1}{2}\mu H^2\right)dv-\int\sigma E^2\,dv. \quad ...(5.91)$$

The above equation is known as Poynting's theorem.

$\oint_s\left(\vec{E}\times\vec{H}\right)\cdot d\vec{s}$ represents the total power leaving the volume.

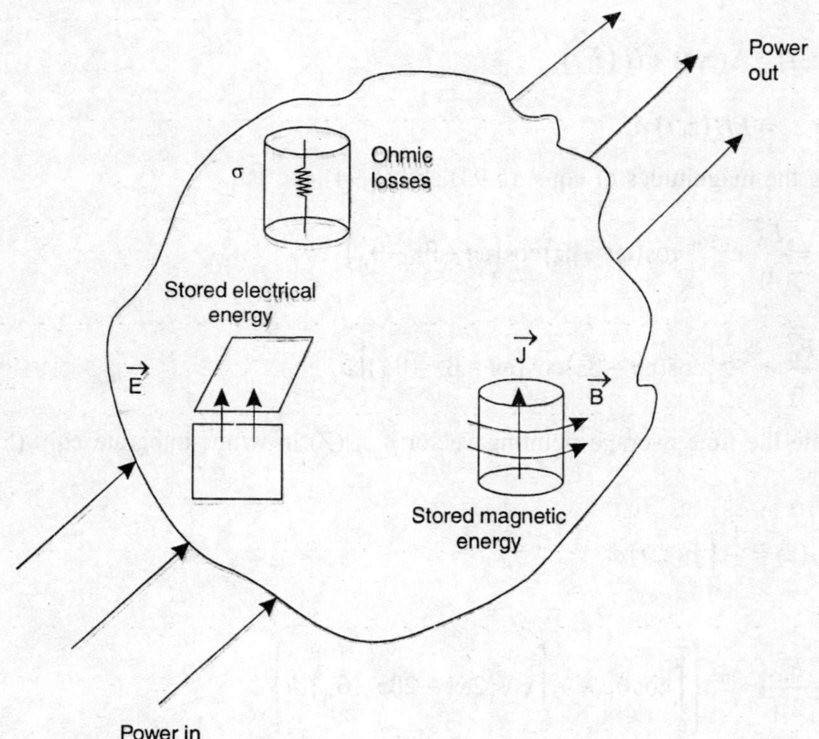

Fig. 5.6 Power balance for *EM* fields

The first term in the right hand side represents rate of decrease of energy stored in electric and magnetic fields. The second term represents the ohmic power dissipated.

Poynting theorem states that the net power flowing out of a given volume v is equal to the time rate of decrease in the energy stored within the volume v minus the conduction losses. The theorem is illustrated in Fig. 5.6.

Moreover, the quantity $\vec{E} \times \vec{H}$ on left hand side is known as Poynting vector '$\vec{p}$' in watt per square meter *i.e.*,

$$\vec{p} = \vec{E} \times \vec{H}. \quad \text{...(5.92)}$$

It represents the instantaneous power density vector associated with the electromagnetic field at a given point. The integration of the Poynting vector over any closed surface gives the net power flowing out of that surface.

It should be noted that $\vec{p}$ is normal to both $\vec{E}$ and $\vec{H}$ and is therefore along the direction of wave propagation $\vec{a}_k$ for uniform plane waves.

Thus, $\quad \vec{a}_k = \vec{a}_E \times \vec{a}_H.$

Let

$$\vec{E}(z, t) = E_0 e^{-\alpha z} \cos(\omega t - \beta z)\vec{a}_x \quad \text{...(5.93)}$$

$$\vec{H}(z, t) = \frac{E_0}{\eta} e^{-\alpha z} \cos\left(\omega t - \beta z - \theta_\eta\right)\vec{a}_y \quad \text{...(5.94)}$$

$$\vec{p}\,(z, t) = \vec{E}(z, t) \times \vec{H}\,(z, t)$$

$$= EH(z, t)\,\vec{a}_z.$$

Multiplying the magnitudes of eqns. (5.93) and (5.94),

$$EH = \frac{E_0^2}{\eta} e^{-2\alpha z} \cos(\omega t - \beta z)\cos\left(\omega t - \beta z - \theta_\eta\right)$$

$$\vec{p}(z, t) = \frac{E_0^2}{\eta} e^{-2\alpha z}\left[\cos(\omega t - \beta z)\cos\left(\omega t - \beta z - \theta_\eta\right)\right]\vec{a}_z \quad \text{...(5.95)}$$

To determine the time average pointing vector p_{ave} (Z) in w/m^2, integrate eqn. (5.105) over the period $T = 2\pi/\omega$.

$$\vec{p}_{ave}(z) = \frac{1}{T}\int_0^T \vec{p}(z, t)\,dt \quad \text{...(5.96)}$$

$$= \frac{1}{T}\frac{E_0^2}{2\eta} e^{-2\alpha z}\left\{\int_0^T \cos\theta_\eta dt + \int_0^T \cos\left(2\omega t - 2\beta z - \theta_\eta\right)dt\right\}$$

$$= \frac{1}{2T}\frac{E_0^2}{|\eta|}e^{-2\alpha z}\left\{\cos\theta_\eta T + \frac{\sin\left(2\omega t - 2\beta z - \theta_\eta\right)}{2\omega}\right\}_0^{T=2\frac{\pi}{\omega}}$$

$$= \frac{1}{2T}\frac{E_0^2}{|\eta|}e^{-2\alpha z}\left\{\cos\theta_\eta T + \frac{\sin\left(2\omega \times t - 2\beta z - \theta_\eta\right)}{2\omega}\right\}_0^{T=2\frac{\pi}{\omega}}$$

$$= \frac{1}{2T}\frac{E_0^2}{|\eta|}e^{-2\alpha z}\left\{\cos\theta_\eta T + \frac{\sin\left(4\pi - 2\beta z - \theta_\eta\right)}{2\omega} - \frac{\sin\left(0 - 2\beta z - \theta_\eta\right)}{2\omega}\right\}$$

$$= \frac{1}{2T}\frac{E_0^2}{|\eta|}e^{-2\alpha z}\cos\theta_\eta T \cdot \vec{a}_z$$

$$\vec{p}_{ave} = \frac{E_0^2}{2|\eta|}e^{-2\alpha z}\cos\theta_\eta \cdot \vec{a}_z \text{ watt/m}^2. \qquad ...(5.97)$$

It can be shown that the instantaneous Poynting vector in eqn. (5.96) can also be written as

$$\vec{p}_{ave}(z) = \frac{1}{2}\text{Re}\left(\vec{E}_s \times \vec{H}_s{*}\right). \qquad ...(5.98)$$

In exponential focus, $\vec{E}_s$ and $\vec{H}_s$ can be expressed as

$$\vec{E}_s = E_0\, e^{-\alpha z} e^{j(\omega t - \beta z)}\vec{a}_x$$

$$\vec{H}_s = \frac{E_0}{\eta}e^{-\alpha z}e^{j(\omega t - \beta z - \theta\eta)}\vec{a}_y$$

$$\vec{E}_s \times \vec{H}_s{}^* = \frac{E_0^2}{\eta}e^{-2\alpha z}e^{j(\omega t - \beta z)}e^{-j(\omega t - \beta z - \theta\eta)}\vec{a}_z$$

$$\vec{p}_{ave}(z) = \frac{1}{2}\text{Re}\left(\frac{E_0^2}{\eta}e^{-2\alpha z}e^{j\theta\eta}\right)\vec{a}_z$$

$$p_{ave}(z) = \frac{1}{2}\frac{E_0^2}{\eta}e^{-2\alpha z}\cos\theta_\eta\, \vec{a}_z$$

which is same as eqn. (5.97).

The average power crossing a given surface 'S' is

$$p_{ave} = \int_s \vec{p}_{ave} \cdot d\vec{s} \text{ watt} \qquad ...(5.99)$$

5.7 REFLECTION OF A PLANE WAVE AT NORMAL INCIDENCE

When a plane wave from one medium meets a different medium, it is partly reflected and partly transmitted. The proportion of the incident wave that is refracted or transmitted depends on the parameters (ε, μ, σ) of the two media involved. Here, assume that the incident wave plane is normal to the boundary between the media.

5.7.1 General Case: Lossy Dielectric-dielectric Boundary

Assume that a plane wave propagating along the +Z direction is incident normally on the boundary $Z = 0$ between medium 1 ($Z < 0$) characterised by σ_1, ε_1, μ_1 and medium 2 ($Z > 0$) characterised by σ_2, ε_2, μ_2 as in Fig. 5.7. In this figure, subscripts *i*, *r*, and *t* denote incident, reflected and transmitted waves respectively. Incident wave $\left(\vec{E}_i, \vec{H}_i\right)$ is travelling along $+\vec{a}_z$ in medium 1.

In phasor form, $\vec{E}$ is represented as in eqn. (5.46)

$$\vec{E}_{is}(z) = E_{i0} e^{-\gamma_1 z} \vec{a}_x.$$

Then

$$\vec{H}_{is}(z) = H_{i0} e^{-\gamma_1 z} \vec{a}_y$$

$$= \frac{E_{i0}}{\eta_1} e^{-\gamma_1 z} \vec{a}_y \qquad \text{...(5.100)}$$

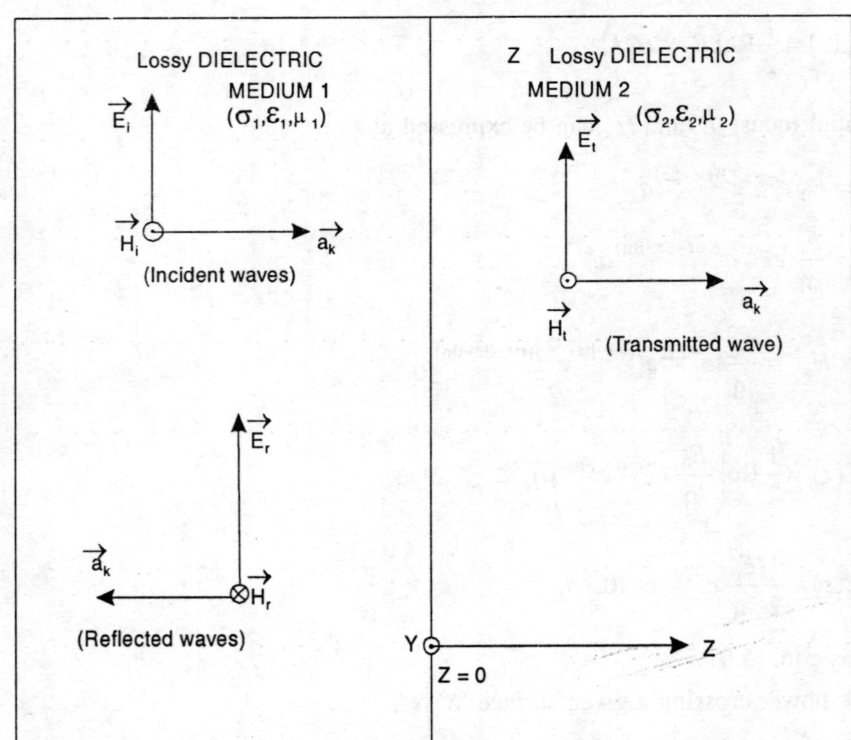

Fig. 5.7 A plane wave incident normally on an interface between two media

Reflected wave: $\left(\vec{E}_r, \vec{H}_r\right)$ is travelling along $-\vec{a}_z$ in medium 1. If

$$\vec{E}_{rs}(z) = E_{r0}e^{\gamma_1 z}\,\vec{a}_x, \quad \text{...(5.101)}$$

then

$$\vec{H}_{ts}(z) = H_{r0}e^{\gamma_1 z}\left(-\vec{a}_y\right)$$

$$= \frac{-E_{r0}}{\eta_1}e^{\gamma_1 z}\,\vec{a}_y. \quad \text{...(5.102)}$$

Assume that for normal incident $\vec{E}_i$, $\vec{E}_r$ and $\vec{E}_t$ have the same polarisation, *i.e.* their directions are same along $+x$ direction $(\vec{a}_x)$.

Transmitted wave:

$\left(\vec{E}_t, \vec{H}_t\right)$ is travelling along $+\vec{a}_z$ in medium 2. If

$$\vec{E}_{ts}(z) = E_{t0}\,e^{-\gamma_2 z}\,\vec{a}_x, \quad \text{...(5.103)}$$

then

$$\vec{H}_{ts}(z) = H_{t0}\,e^{-\gamma_2 z}\,\vec{a}_y$$

$$= \frac{E_{t0}}{\eta_2}e^{-\gamma_2 z}\,\vec{a}_y. \quad \text{...(5.104).}$$

E_{io}, E_{ro} and E_{to} are respectively the magnitudes of the incident, reflected and transmitted electric fields at $Z = 0$.

From the Fig. 5.7, the total field in medium 1 is having both incident and reflected fields, whereas in medium 2, only the transmitted field exists. That is,

$$\vec{E}_1 = \vec{E}_i + \vec{E}_r,\ \vec{H}_1 = \vec{H}_i + \vec{H}_r$$

$$\vec{E}_2 = \vec{E}_t, \quad \vec{H}_2 = H_t. \quad \text{...(5.105)}$$

At the interface $Z = 0$, the boundary conditions require that the tangential components of $\vec{E}$ and $\vec{H}$ fields must be continuous, since the waves are transverse, and fields are entirely tangential to the interface. Hence at $Z = 0$

$$\vec{E}_{1\tan} = \vec{E}_{2\tan} \text{ and } \vec{H}_{1\tan} = \vec{H}_{2\tan}$$

That is,

$$\vec{E}_i(0) + \vec{E}_r(0) = \vec{E}_t(0)$$

$$E_{i0} + E_{r0} = E_{t0} \quad \text{...(5.106)}$$

$$\vec{H}_i(0) + \vec{H}_r(0) = \vec{H}_t(0) \quad \text{(or)}$$

$$H_{i0} + H_{r0} = H_{t0} \quad \text{...(5.107)}$$

$$\frac{1}{\eta_1}(E_{i0} - E_{r0}) = \frac{E_{t0}}{\eta_2}. \quad ...(5.108)$$

Adding eqn. (5.106) and eqn. (5.108), we get

$$\boxed{\begin{aligned} E_{i0} &= E_{t0}\left[\frac{\eta_2 + \eta_1}{2\eta_2}\right] \\ E_{t0} &= \left[\frac{2\eta_2}{\eta_2 + \eta_1}\right]E_{i0} \end{aligned}} \quad ...(5.109)$$

Eqn. (5.108) can be rearranged as

$$\frac{\eta_2}{\eta_1}[E_{i0} - E_{r0}] = E_{t0}$$

$$\frac{\eta_2}{\eta_1}E_{i0} - \frac{\eta_2}{\eta_1}E_{r0} = E_{t0}. \quad ...(5.110)$$

Subtracting eqn. (5.110) from eqn. (5.106)

$$E_{i0}\left[1 - \frac{\eta_2}{\eta_1}\right] + E_{r0}\left[1 + \frac{\eta_2}{\eta_1}\right] = 0$$

$$E_{r0} = E_{i0}\frac{\left[\frac{\eta_2}{\eta_1} - 1\right]}{\left[1 + \frac{\eta_2}{\eta_1}\right]}$$

$$E_{r0} = E_{i0}\left[\frac{\eta_2 - \eta_1}{\eta_2 + \eta_1}\right] \quad ...(5.111)$$

$$\frac{E_{r0}}{E_{i0}} = \mathcal{T} = \text{Reflection coefficient}$$

$$\frac{E_{t0}}{E_{i0}} = \tau = \text{Transmission coefficient.}$$

$$\mathcal{T} = \frac{\eta_2 - \eta_1}{\eta_2 + \eta_1} \quad ...(5.112)$$

$$\tau = \frac{2\eta_2}{\eta_2 + \eta_1} \quad ...(5.113)$$

Note that

- $1 + T = \tau$
- Both T and τ are dimensionless and may be complex.
- $0 \le |T| \le 1$

5.7.2 Special Case Perfect Dielectric-conductor Boundary

Consider medium 1 is a perfect dielectric $(\text{lossless } \sigma_1 = 0)$ and medium 2 is a perfect conductor $(\sigma_2 \approx \infty)$. In this case, $\eta_2 = 0$, since η_2 for a conducting medium is $\sqrt{\frac{\omega\mu}{\sigma}}$ $\sigma \approx \infty$. So, $T = -1$ and $\tau = 0$ which indicates that the wave is totally reflected. This is because fields in a perfect conductor must vanish. There can be no transmitted wave $\left(\vec{E}_2 = 0\right)$. The totally reflected wave combines with the incident wave to form a standing wave. A standing wave just 'stands' and does not travel and consists of two travelling waves $\left(\vec{E}_i \text{ and } \vec{E}_r\right)$ of equal amplitudes but in opposite direction.

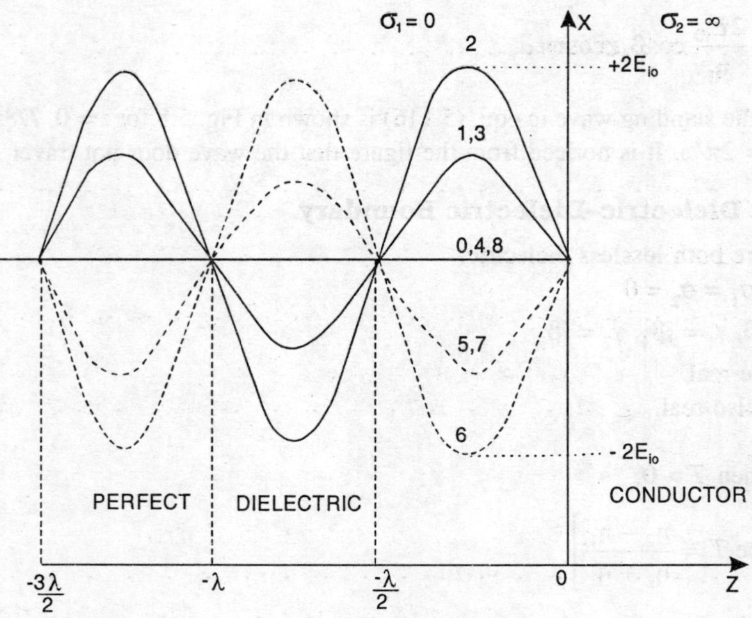

Fig. 5.8 Standing waves curves 0, 1, 2, 3, 4 are respectively at times $t = 0, T/8, T/4, 3T/8, T/2, \ldots\ldots; \lambda = 2\pi/\beta,$

The standing wave in medium 1 can be obtained by combining eqns. (5.99) and (5.103)

$$\vec{E}_{1s} = \vec{E}_{is} + \vec{E}_{rs}$$

$$\vec{E}_{1s} = \left(E_{i0}e^{-\gamma_1 z} + E_{r0}e^{\gamma_1 z}\right)\vec{a}_x.$$

But $$\mathcal{T} = \frac{E_{r0}}{E_{i0}} = -1, \qquad ...(5.114)$$

$$\sigma_1 = 0, \alpha_1 = 0, \gamma_1 = j\beta_1.$$
(for a dielectric)

Hence,

$$\vec{E}_{1s} = -E_{i0}\left[e^{j\beta_1 z} - e^{-j\beta_1 z}\right]\vec{a}_x$$

$$\vec{E}_{1s} = -2jE_{i0}\sin\beta_1 z\,\vec{a}_x \qquad ...(5.115)$$

$$\vec{E}_1 = \text{Re}\left(E_{1s}e^{j\omega t}\right)$$

(or) $$\vec{E}_1 = -2jE_{i0}\sin\beta_1 z(\cos\omega t + j\sin\omega t)\vec{a}_x$$

$$\vec{E}_1 = 2E_{i0}\sin\beta_1 z\,\sin\omega t\,\vec{a}_x. \qquad ...(5.116)$$

By taking similar steps, it can be shown that magnetic field component of the wave is

$$\vec{H}_1 = \frac{2E_{i0}}{\eta_1}\cos\beta_1 z\cos\omega t\,\vec{a}_y. \qquad ...(5.117)$$

A sketch of the standing wave in eqn. (5.116) is shown in Fig. 5.8 for $t = 0, T/8, T/4, 3T/8, T/2$ and so on, where $T = 2\pi/\omega$. It is noticed from the figure that the wave does not travel, but oscillates.

5.7.3 Perfect Dielectric–Dielectric Boundary

Media 1 and 2 are both lossless dielectric.

In this case $\sigma_1 = \sigma_2 = 0$

$\alpha_1 = \alpha_2 = 0, \gamma_1 = j\beta_1, \gamma_2 = j\beta_2$

η_1 and η_2 are real

$\mathcal{T}$ and τ are also real.

Case I:

If $\eta_2 > \eta_1$, then $\mathcal{T} > 0$.

$$\left[\text{Since } \mathcal{T} = \frac{\eta_2 - \eta_1}{\eta_2 + \eta_1}\right].$$

Again, there is also a transmitted wave in medium 2 and a standing wave in medium 1. However, the incident and reflected waves have amplitudes that are not equal in magnitude.

$$\vec{E}_{1s} = \vec{E}_{is} + \vec{E}_{rs}$$

$$\vec{E}_{is} = \vec{E}_{i0}\,e^{-\gamma_1 z}\,\vec{a}_x,\ \vec{E}_{rs} = E_{r0}\,e^{\gamma_1 z}\,\vec{a}_x$$

$$\frac{E_{r0}}{E_{i0}} = \mathcal{T}.$$

$$\therefore \qquad \vec{E}_{1s} = E_{io}\left[e^{-\gamma 1z} + \mathcal{T}\, e^{\gamma_1 z}\right]\vec{a}_x$$

$$= E_{i0}\left[e^{-j\beta_1 z} + \mathcal{T}\, e^{j\beta_1 z}\right]\vec{a}_x$$

$$= E_{i0}\left[(\cos\beta_1 z - j\sin\beta_1 z) + \mathcal{T}(\cos\beta_1 z + j\sin\beta_1 z)\right]\vec{a}_x$$

$$= E_{i0}\left\{(1+\mathcal{T})\cos\beta_1 z - j\sin\beta_1 z(1-\mathcal{T})\right\}\vec{a}_x$$

$$\left|\vec{E}_{1s}\right| = E_{io}\sqrt{\cos^2\beta_1 z(1+\mathcal{T})^2 + \sin^2\beta_1 z(1-\mathcal{T})^2}.$$

Maximum values of $\left|\vec{E}_{1S}\right|$ occur at $-\beta_1 z_{\max} = n\pi$

The $-ve$ sign indicates that standing wave is in $-z$ direction

$$z_{\max} = \frac{-n\pi}{\beta_1}$$

$$= \frac{-n\lambda_1}{2}\left(\because \beta_1 = \frac{2\pi}{\lambda_1}\right)$$

$n = 0, 1, 2, \ldots..$

Minimum values of $\left|\vec{E}_{1S}\right|$ occur at

$$-\beta_1 z_{\min} = (2n+1)\frac{\pi}{2}$$

$$z_{\min} = \frac{-(2n+1)\pi}{2\beta_1}$$

$$= \frac{-(2n+1)\lambda_1}{4},$$

where $n = 0, 1, 2, \ldots..$

So, $$\left|\vec{E}_1\right|_{\max} = E_{i0}(1+\mathcal{T}) \qquad \ldots(5.118)$$

$$\left|\vec{E}_1\right|_{\min} = E_{i0}(1-\mathcal{T}). \qquad \ldots(5.119)$$

Case II:

If $\eta_2 < \eta_1$, $\mathcal{T} < 0$. For this case, the locations of $\left|\vec{E}_1\right|$ maximum $\left|\vec{E}_1\right|$ and minimum are interchanged.

Note:

1. $\left|\vec{H}_1\right|_{min}$ occurs whenever there is $\left|\vec{E}_1\right|_{max}$ and vice versa.
2. The transmitted wave in medium 2 is a purely travelling wave and consequently, there are no maxima or minima in this region.

The ratio of $|E_1|_{max}$ to $|E_1|_{min}$ $\left(\text{or } |H_1|_{max} \text{ to } |H_1|_{min}\right)$ is called the standing wave ratio 'S'; that is,

$$S = \frac{\left|\vec{E}_1\right|_{max}}{\left|\vec{E}_1\right|_{min}} = \frac{\left|\vec{H}_1\right|_{max}}{\left|\vec{H}_1\right|_{min}} = \frac{1+|\mathcal{T}|}{1-|\mathcal{T}|} \qquad \text{...(5.120)}$$

(or) $$|\mathcal{T}| = \frac{S-1}{S+1}. \qquad \text{...(5.121)}$$

Standing wave due to reflection at normal incidence between two lossless media is given in Fig. 5.9.

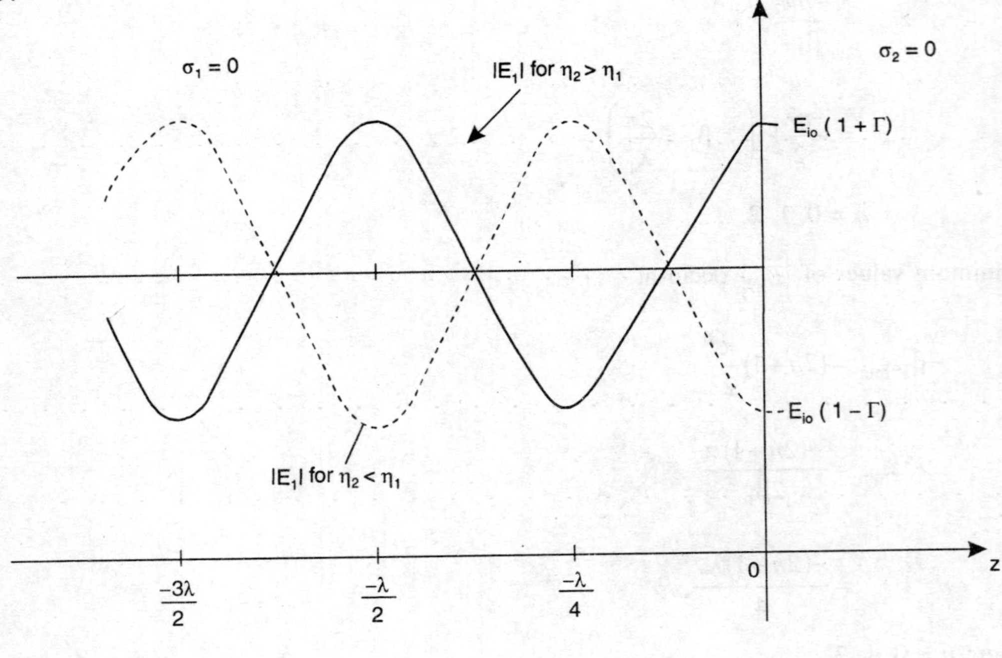

Fig. 5.9

5.8 REFLECTION OF A WAVE AT OBLIQUE INCIDENCE

Consider a linearly polarised plane wave obliquely incident on a boundary between two lossless dielectric media as in Fig. 5.10 (*a*).

The plane defined by the propagation vector $\vec{k}$ and unit normal vector $\vec{a}_n$ to the boundary is called the plane of incidence. The angle θ_1 between $\vec{k}$ and $\vec{a}_n$ is the angle of incidence.

From the Fig. 5.10 (*b*), x' and z' are orthogonal with z' is oriented along wave propagation direction.

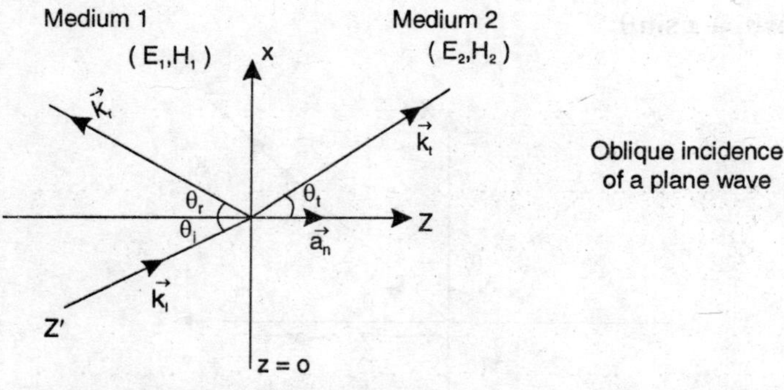

Direction of propagation of wave

Fig. 5.10 (a)

$$Z' = z\cos\theta_i + x\sin\theta_i \quad ...(5.122)$$

Fig. 5.10 (b) Incident wave

$$-Z' = z\cos\theta_r + x\sin\theta_r \quad ...(5.123)$$

Fig. 5.10 (c) Reflected wave

$$-Z' = z\cos\theta_r + x\sin\theta_t \qquad ...(5.124)$$

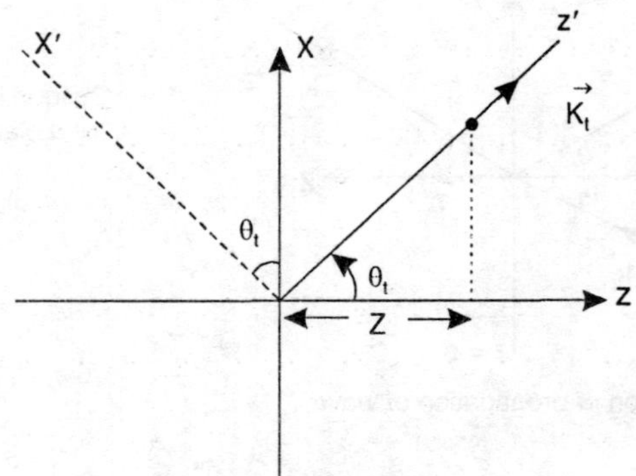

Fig. 5.10 (d) Transmitted wave

Consider two cases given below:

(*i*) The electric field $\vec{E}$ is perpendicular to the plane of incidence known as perpendicular or vertical polarisation.

(*ii*) The electric field $\vec{E}$ is parallel to the plane of incidence known as parallel or horizontal polarisation.

5.8.1 Perpendicular Polarisation

In this case, $\vec{E}$ field is perpendicular to the plane of incidence (the *XZ* plane) *i.e.* it is along *Y*-axis. Here $\vec{H}$ field is parallel to plane of incidence (along *X'*).

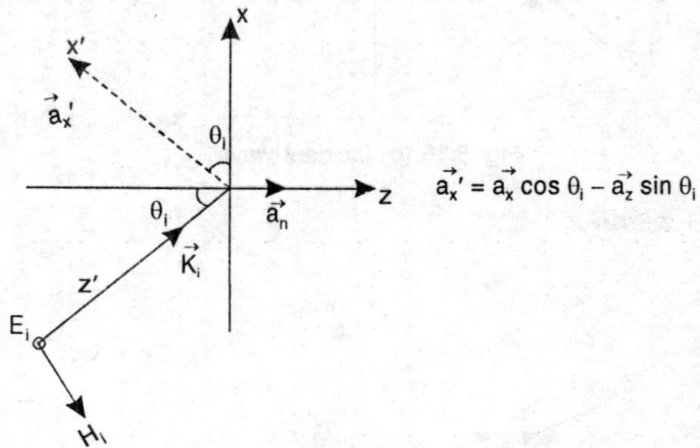

Fig. 5.11 (a)

We know that

$$\vec{E}_{is} = E_{i0}\, e^{-\gamma_1 z'}\, \vec{a}_y$$

$$\gamma_1 = \alpha_1 + j\beta_1$$

$$\alpha_1 = 0 \;\; (\because \text{ lossless media}).$$

Substitute Z' from eqn. (5.122)

$$\vec{E}_{is} = E_{i0}\, e^{-j\beta(x\sin\theta_i + z\cos\theta_i)}\, \vec{a}_y$$

$$\vec{H}_{is} = \frac{E_{i0}}{\eta_1} e^{-j\beta_1(x\sin\theta_i + z\cos\theta_i)} (-a_{x'})$$

$$\left(\because\ \vec{k} = \vec{E} \times \vec{H}\right)\ a_{z'} = \vec{a}_y \times (-\vec{a}_{x'}) \qquad \text{...(5.125)}$$

$$\vec{H}_{is} = \frac{E_{i0}}{\eta_1} e^{-j\beta_1}\,(x\sin\theta_i + z\cos\theta_i)(-\cos\theta_i \vec{a}_x + \sin\theta_i\, \vec{a}_z). \qquad \text{...(5.126)}$$

REFLECTED WAVE

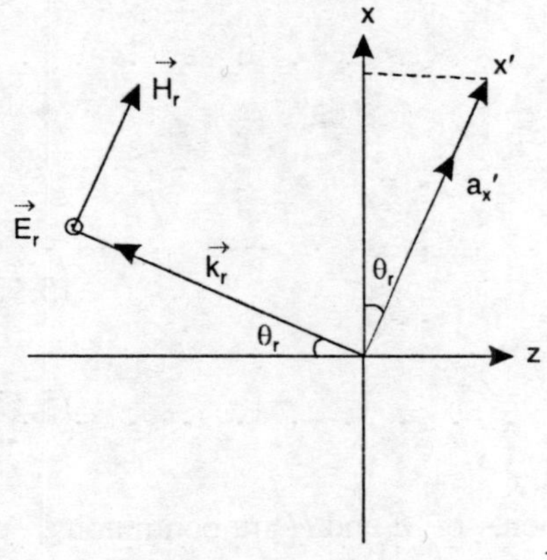

Fig. 5.11 (b)

$$\vec{k} = \vec{E} \times \vec{H}$$

$$-\vec{a}_{z'} = \vec{a}_y \times \vec{a}_{x'}$$

$$\vec{a}_{x'} = \vec{a}_x \cos\theta_r + \vec{a}\sin\theta_r$$

The reflected wave is backward travelling wave $(\vec{a}_k = -\vec{a}_{z'})$.

From the Fig. 5.11 (*b*),

$$\vec{E}_{rs} = \vec{E}_{r0}\, e^{-j\beta_1(-z')}\, \vec{a}_y$$

$$\vec{H}_{rs} = \frac{\vec{E}_{r0}}{\eta_1} e^{-j\beta_1(-z')}\, \vec{a}_x.$$

Again substitute Z' from the eqn. (5.123)

$$\vec{E}_{rs} = E_{r0}\, e^{-j\beta_1(x\sin\theta_r - z\cos\theta_r)}\, \vec{a}_y$$

$$\vec{H}_{rs} = \frac{E_{r0}}{\eta_1} e^{-j\beta_1(x\sin\theta_r - z\cos\theta_r)} (\vec{a}_x \cos\theta_r + \vec{a}_z \sin\theta_r).$$

TRANSMITTED WAVE

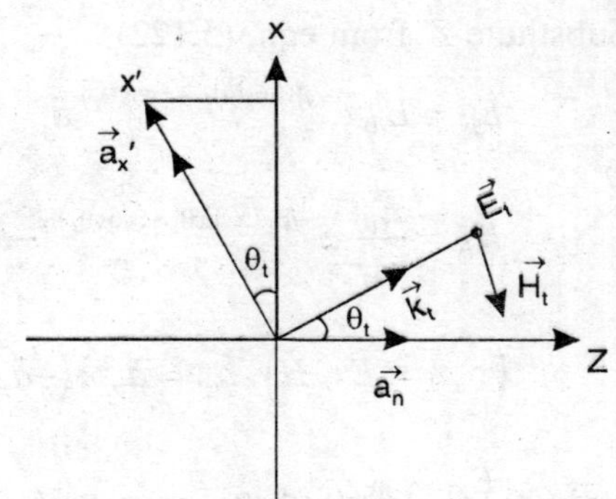

Fig. 5.11 (c)

$$\vec{k} = \vec{E} \times \vec{H}$$

$$\vec{a}_{z'} = \vec{a}_y \times (-\vec{a}_{x'})$$

$$\vec{a}_{x'} = -\vec{a}_z \sin\theta_t + \vec{a}_x \cos\theta_t$$

The transmitted wave is propagating in the medium 2 in forward direction $(\vec{a}_{z'})$

$$\vec{E}_{ts} = E_{t0} e^{-j\beta_2 z'} \vec{a}_y$$

$$\vec{H}_{ts} = \frac{E_{t0}}{\eta_2} e^{-j\beta_2 z'} (-a_{x'}).$$

Substitute Z' from eqn. (5.124)

$$\vec{E}_{ts} = E_{t0}\, e^{-j\beta_2 (z\cos\theta_t + x\sin\theta_t)}\, \vec{a}_y \quad ...(5.127)$$

$$\vec{H}_{ts} = \frac{E_{t0}}{\eta_2} e^{-j\beta_2 (z\cos\theta_t + x\sin\theta_t)} (\vec{a}_z \sin\theta_t - \vec{a}_x \cos\theta_t). \quad ...(5.128)$$

From boundary conditions, at $Z = 0$, the tangential components of $\vec{E}$ and $\vec{H}$ are continuous.

$$E_{i0}\, e^{-j\beta_1 x \sin\theta_i} + E_{r0}\, e^{-j\beta_1 x\sin\theta_r} = E_{t0}\, e^{-j\beta_2 x \sin\theta_t}$$

$$\mathcal{T} = \frac{E_{r0}}{E_{i0}} \text{ and } \tau = \frac{E_{t0}}{E_{i0}}$$

$$E_{i0}\, e^{-j\beta_1 x\sin\theta_i} + \mathcal{T} E_{i0}\, e^{-j\beta_1 x \sin\theta_r} = \tau E_{i0}\, e^{-j\beta_2 x\sin\theta_t}$$

$$e^{-\beta_1 \sin\theta_i} + \mathcal{T} e^{-\beta_1 \sin\theta_r} = \tau e^{-\beta_2 \sin\theta_t}.$$

We know that

$$1 + \mathcal{T} = \tau.$$

So,

$$e^{\beta_1 \sin\theta_i} = e^{\beta_1 \sin\theta_r} = e^{\beta_2 \sin\theta_t}$$

$$\beta_1 \sin\theta_i = \beta_1 \sin\theta_r = \beta_2 \sin\theta_t.$$

From the first equality

$$\theta_i = \theta_r \quad ...(5.129)$$

Also from second equalities,

$$\beta_1 \sin\theta_i = \beta_2 \sin\theta_t$$

$$\frac{\sin\theta_i}{\sin\theta_t} = \frac{\beta_2}{\beta_1},$$

where $\beta = \dfrac{w}{u} = \dfrac{w}{\dfrac{1}{\sqrt{u\varepsilon}}}$

$$\beta_1 = \omega\sqrt{\mu_1\varepsilon_1} \qquad ...(i)$$

$$\beta_2 = \omega\sqrt{\mu_2\varepsilon_2} \qquad ...(ii)$$

$$\therefore \quad \frac{\sin\theta_i}{\sin\theta_t} = \frac{\sqrt{\mu_2\varepsilon_2}}{\sqrt{\mu_1\varepsilon_1}}. \qquad ...(5.130)$$

The above eqn. (*i*) is Snell's law of reflection and eqn. (*ii*) is Snell's law of refraction. Similarly,

$$\vec{H}_{is} + \vec{H}_{rs} = \vec{H}_{ts}$$

$$\frac{E_{i0}}{\eta_1} e^{-j\beta_1 x \sin\theta_i}(-\cos\theta_i) + \mathcal{T}\frac{E_{i0}}{\eta_1} e^{-j\beta_1 x \sin\theta_r \cos\theta_r}$$

$$= \tau\frac{E_{i0}}{\eta_2} e^{-j\beta_2 x \sin\theta_t}(-\cos\theta_t).$$

Substitute $\theta_i = \theta_r$ and $\beta_1 \sin\theta_i \sin\theta_t$

$$\frac{e^{-j\beta_1 \sin\theta_i}}{\eta_1}(-\cos\theta_i) + \frac{\tau}{\eta_1} e^{-j\beta_1 \sin\theta_i}(\cos\theta_r) = \frac{\tau}{\eta_2} e^{-j\beta_1 \sin\theta_i}(-\cos\theta_t)$$

$$\frac{\cos\theta_i}{\eta_1} - \mathcal{T}\frac{\cos\theta_r}{\eta_1} = \frac{\tau\cos\theta_t}{\eta_2}$$

$$\cos\theta_i - \mathcal{T}\cos\theta_i = \frac{\eta_1}{\eta_2}\tau\cos\theta_t.$$

We know

$$1 + \mathcal{T}_\perp = \tau_\perp$$

$$\cos\theta_i - \mathcal{T}\cos\theta_i = \frac{\eta_1}{\eta_2}(1+\mathcal{T})\cos\theta_t$$

$$\cos\theta_i(1-\mathcal{T}) = \frac{\eta_1}{\eta_2}(1+\mathcal{T})\cos\theta_t$$

$$\cos\theta_i - \frac{\eta_1}{\eta_2}\cos\theta_t = \left(\frac{\eta_1}{\eta_2}\cos\theta_t + \cos\theta_i\right)\mathcal{T}$$

$$\frac{\eta_2\cos\theta_i - \eta_1\cos\theta_t}{\eta_1\cos\theta_t + \eta_2\cos\theta_i} = \mathcal{T}_\perp \text{ and} \qquad ...(5.131)$$

$$\frac{2\eta_2\cos\theta_i}{\eta_1\cos\theta_t + \eta_2\cos\theta_i} = \tau_\perp. \qquad ...(5.132)$$

Assume dielectric media are not only lossless but non-magnetic also. So, $\mu_1 = \mu_2 = \mu_0$. Snell's law of refraction becomes

$$\frac{\sin\theta_i}{\sin\theta_t} = \sqrt{\frac{\varepsilon_2}{\varepsilon_1}}. \qquad ...(5.133)$$

Eqn. (5.142) can be written as

$$\mathcal{T}_\perp = \frac{\cos\theta_i - \frac{\eta_1}{\eta_2}\cos\theta_t}{\cos\theta_i + \frac{\eta_1}{\eta_2}\cos\theta_t} \qquad ...(5.134)$$

$$\mathcal{T}_\perp = \frac{\cos\theta_i - \frac{\eta_1}{\eta_2}\sqrt{1-\sin^2\theta_t}}{\cos\theta_i + \frac{\eta_1}{\eta_2}\sqrt{1-\sin^2\theta_t}} \qquad ...(5.135)$$

$$\frac{\eta_1}{\eta_2} = \sqrt{\frac{\mu_0}{\varepsilon_1}} \times \sqrt{\frac{\varepsilon_2}{\mu_0}} = \sqrt{\frac{\varepsilon_2}{\varepsilon_1}}. \qquad ...(5.136)$$

Also, $$\sin^2\theta_t = \frac{\varepsilon_1}{\varepsilon_2}\sin^2\theta_i. \qquad ...(5.137)$$

Substitute above two eqns. (5.136) and (5.137) in eqn. (5.135) and simplify, we get

$$\mathcal{T}_\perp = \frac{\cos\theta_i - \sqrt{\left(\frac{\varepsilon_2}{\varepsilon_1}\right) - \sin^2\theta_i}}{\cos\theta_i + \sqrt{\left(\frac{\varepsilon_2}{\varepsilon_1}\right) - \sin^2\theta_i}}. \qquad ...(5.138)$$

If the medium 2, in a more dense dielectric than medium 1 ($\varepsilon_2 > \varepsilon_1$), the quantity in the square root will be positive and $\mathcal{T}$ will be real. If $\varepsilon_1 > \varepsilon_2$ and $\sin^2\theta_i \geq \left(\frac{\varepsilon_2}{\varepsilon_1}\right)$, then $\mathcal{T}$ will be complex but $|\mathcal{T}| = 1$.

Under these conditions, the incident wave is totally reflected back into more dense medium. The incident angle for which $\mathcal{T} = 1\angle 0°$ is called critical angle θ_{ic}.

$$\theta_{ic} = \sin^{-1}\sqrt{\frac{\varepsilon_2}{\varepsilon_1}}.$$

When a wave incident from the more dense medium onto the less dense medium at an equal to (or) exceeding the critical angle, the wave will be totally internally reflected.

From the eqn. (5.131), it is possible that $\mathcal{T}_\perp = 0$. Under this condition, there is no reflection $E_{r0} = 0$, and the incident angle at which this takes place is called Brewster angle $\theta_{B\perp}$.

$$\theta_i = \theta_{B\perp}, \text{ when } \mathcal{T}_\perp = 0$$

$$\eta_2^2 \cos^2\theta_{B\perp} = \eta_1^2 \cos^2\theta_t.$$

From Snell's law of refraction,

$$\sin^2\theta_t = \frac{\mu_1\varepsilon_1}{\mu_2\varepsilon_2}\sin^2\theta_{B\perp}.$$

So, eqn. (5.151) becomes

$$1 - \sin^2\theta_{B\perp} = \frac{\eta_1^2}{\eta_2^2}\left(1 - \frac{\mu_1\varepsilon_1}{\mu_2\varepsilon_2}\sin^2\theta_{B\perp}\right)$$

$$\eta_1 = \sqrt{\frac{\mu_1}{\varepsilon_1}},\ \eta_2 = \sqrt{\frac{\mu_2}{\varepsilon_2}}$$

$$\frac{\eta_1^2}{\eta_2^2} = \frac{\mu_1\varepsilon_2}{\mu_2\varepsilon_1}$$

$$1 - \sin^2\theta_{B\perp} = \frac{\mu_1\varepsilon_2}{\mu_2\varepsilon_1}\left(1 - \frac{\mu_1\varepsilon_1}{\mu_2\varepsilon_2}\sin^2\theta_{B\perp}\right)$$

$$\sin^2\theta_{B\perp}\left(1 - \frac{\mu_1^2}{\mu_2^2}\right) = 1 - \frac{\mu_1\varepsilon_2}{\mu_2\varepsilon_1}$$

$$\sin^2\theta_{B\perp}\frac{1 - \dfrac{\mu_1\varepsilon_2}{\mu_2\varepsilon_1}}{1 - \left(\dfrac{\mu_1}{\mu_2}\right)^2}.$$

For non-magnetic media $\mu_1 = \mu_2 = \mu_0$, which implies $\sin^2\theta_{B\perp} \to \infty$. So, $\theta_{B\perp}$ does not exist because sine of angle is never greater than unity. If $\mu_1 \neq \mu_2$ and $\varepsilon_1 - \varepsilon_2$, then

$$\sin\theta_{B\perp} = \sqrt{\frac{\mu_2}{\mu_1 + \mu_2}}. \qquad ...(5.139)$$

5.9 PARALLEL POLARISATION

In this case, $\vec{E}$ field is parallel to the plane of incidence (the XZ plane) *i.e.* it is along x'. The $\vec{H}$ field is perpendicular to the plane of incidence (along Y axis)

Incident wave

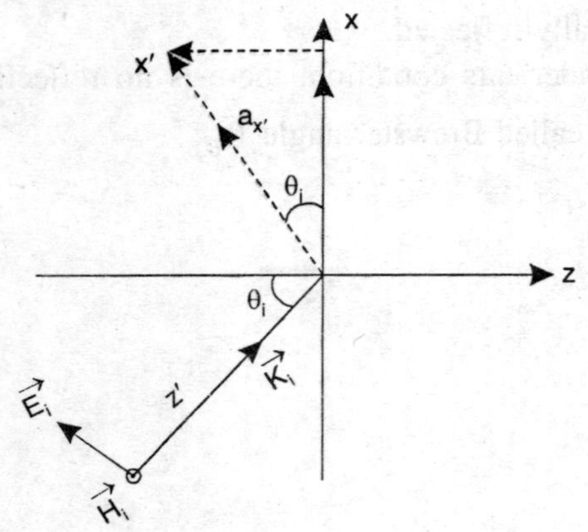

$$\vec{k} = \vec{E} \times \vec{H}$$

$$\vec{a}_{z'} = \vec{a}_{x'} \times \vec{a}_y$$

$$\vec{a}_{x'} = \vec{a}_x \cos\theta_i - \vec{a}_z \sin\theta_i$$

Fig. 5.12 (a)

As in previous case.

$$\vec{E}_{is} = E_{i0}\, e^{-j\beta_1 z'}\, \vec{a}_{x'}.$$

Substitute z' from eqn. (5.122)

$$\vec{E}_{is} = E_{i0}\, e^{-j\beta_1(x\sin\theta_i + z\cos\theta_i)} \left(\vec{a}_x \cos\theta_i - \vec{a}_z \sin\theta_i\right) \quad ...(5.140)$$

$$\vec{H}_{is} = \frac{E_{i0}}{\eta_1} e^{-j\beta_1(x\sin\theta_i + z\cos\theta_i)}\, \vec{a}_y. \quad ...(5.141)$$

Reflected Wave

$$\vec{k} = \vec{E} \times \vec{H}$$

$$-\vec{a}_{z'} = \vec{a}_{x'} \times \left(-\vec{a}_y\right)$$

$$\vec{a}_{x'} = \vec{a}_x \cos\theta_r + \vec{a}_z \sin\theta_r$$

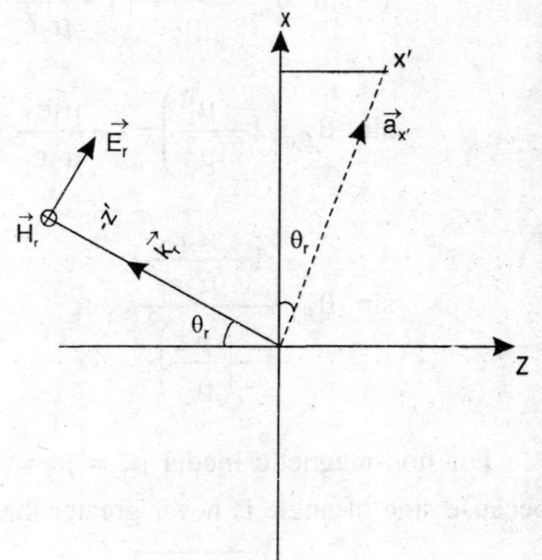

Fig. 5.12 (b)

$$\vec{E}_{rs} = E_{r0}\, e^{-j\beta_1(-z')}\, \vec{a}_{x'}\,.$$

Substitute $(-z')$ from eqn. (5.123)

$$\vec{E}_{rs} = E_{r0}\, e^{-j\beta_1(x\sin\theta_r - z\cos\theta_r)}(\vec{a}_x \cos\theta_r + \vec{a}_z \sin\theta_r)$$

$$\vec{H}_{rs} = H_{r0}\, e^{-j\beta_1(-z')}(-\vec{a}_y) \quad \text{...(5.142)}$$

$$= \frac{E_{r0}}{\eta_1} e^{-j\beta_1(x\sin\theta_r - z\cos\theta_r)}(-\vec{a}_y) \quad \text{...(5.143)}$$

Transmitted Wave

$$\vec{k}_t = \vec{E}_t \times \vec{H}_t$$

$$\vec{a}_{z'} = a_{x'} \times \vec{a}_y$$

$$a_{x'} = -\vec{a}_z \cos\theta_t + \vec{a}_x \sin\theta_t$$

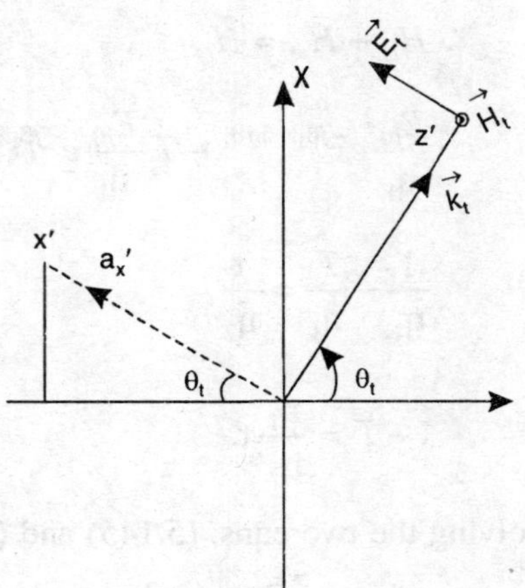

Fig. 5.12 (c)

$$\vec{E}_{ts} = E_{t0}\, e^{-j\beta_2 z'}\, \vec{a}_{x'}.$$

Substituting z' from eqn. (5.124)

$$\vec{E}_{ts} = E_{t0}\, e^{-j\beta_2(x\sin\theta_t + z\cos\theta_t)}(\vec{a}_x \cos\theta_i - \vec{a}_z \sin\theta_t) \quad \text{...(5.144)}$$

$$\vec{H}_{ts} = \frac{E_{t0}}{\eta_2} e^{-j\beta_2(x\sin\theta_t + z\cos\theta_t)}\, \vec{a}_y. \quad \text{...(5.145)}$$

According to boundary conditions at $z = 0$, the tangential components of $\vec{E}$ and $\vec{H}$ are continuous:

$$\vec{E}_{is} + \vec{E}_{rs} = \vec{E}_{ts} \qquad \text{(only tangential components)}$$

$$E_{i0} e^{-j\beta_1 x\sin\theta_i} \cos\theta_i + E_{r0}\, e^{-j\beta_1 \times \sin\theta_r} \cos\theta_r = E_{t0} e^{-j\beta_2 \sin\theta_t} \cos\theta_t$$

$$\frac{E_{r0}}{E_{i0}} = \mathcal{T},\ \frac{E_{t0}}{E_{i0}} = \tau.$$

Also substitute Snell's law of reflection and refraction:

$$\theta_i = \theta_r \text{ and } \beta_1 \sin\theta_i = \beta_2 \sin\theta_t$$

$$E_{i0}e^{-j\beta_1 x \sin\theta_i}\cos\theta_i + \mathcal{T} E_{i0}e^{-j\beta_1 x \sin\theta_i}\cos\theta_t$$

$$= \tau E_{i0}e^{-j\beta_1 x \sin\theta_i}\cos\theta_t$$

$$\cos\theta_i + \mathcal{T}\cos\theta_i = \tau\cos\theta_t$$

$$1 + \mathcal{T} = \tau\frac{\cos\theta_t}{\cos\theta_i}.$$

Similarly,

$$\vec{H}_{is} + \vec{H}_{rs} = \vec{H}_{ts} \qquad \text{(only tangential components)}$$

$$\frac{E_{i0}}{\eta_1}e^{-j\beta_1 x \sin\theta_i} - \mathcal{T}\frac{E_{i0}}{\eta_1}e^{-j\beta_1 x \sin\theta_i} = \tau\frac{E_{i0}}{\eta_2}e^{-j\beta_1 x \sin\theta_i}$$

$$\frac{1}{\eta_1} - \frac{\mathcal{T}}{\eta_1} = \frac{\tau}{\eta_2}$$

$$1 - \mathcal{T} = \frac{\eta_1}{\eta_2}\tau. \qquad \text{...(5.146)}$$

Solving the two eqns. (5.145) and (5.146)

$$\tau_{\parallel} = \frac{2\eta_2\cos\theta_i}{\eta_1\cos\theta_i + \eta_2\cos\theta_t} \qquad \text{...(5.147)}$$

$$\mathcal{T}_{\parallel} = \frac{\eta_2\cos\theta_t - \eta_1\cos\theta_i}{\eta_1\cos\theta_t + \eta_2\cos\theta_i}. \qquad \text{...(5.148)}$$

For no reflection, $\mathcal{T}_{\parallel} = 0$.
By replacing θ_i with Brewster angle, we obtain

$$\eta_2\cos\theta_t = \eta_1\cos\theta_{B_\parallel}.$$

By similar procedure as in perpendicular polarisation, we get

$$\sin^2\theta_{B_{11}} = \frac{1 - \frac{\mu_2\varepsilon_1}{\mu_1\varepsilon_2}}{1 - \left(\frac{\varepsilon_1}{\varepsilon_2}\right)^2}.$$

Assume the media is non-magnetic, $\mu_1 = \mu_2 = \mu_0$.

$$\sin^2\theta_{B_{11}} = \frac{1}{1 + \frac{\varepsilon_1}{\varepsilon_2}}$$

$$\sin\theta_{B_{11}} = \sqrt{\frac{\varepsilon_2}{\varepsilon_1 + \varepsilon_2}}$$

$$\tan\theta_{B_{11}} = \sqrt{\frac{\varepsilon_2}{\varepsilon_1}} = \frac{\eta_2}{\eta_1}.$$

SOLVED PROBLEMS

Problem 1

In free space $E(z, t) = 10^3 \sin(\omega t - \beta z)\vec{a}_y$ V/m. Obtain *H*(*z*, *t*) and find the propagation constant γ for the given frequency *f* = 95.5 MHz.

Solution:

$$E(z, t) = 10^3 \sin(\omega t - \beta z)\vec{a}_y$$

The wave is propagating along +*z* direction. Electric field is in *y* direction.

$\vec{E} \times \vec{H}$ must be in the +*z* direction.

So, $\vec{H}$ should be along $(-\vec{a}_x)$ direction.

$$\vec{H}(z, t) = \frac{E_0}{\eta}\sin(\omega t - \beta z)(-\vec{a}_x) \qquad \left[\because\ \vec{a}_y \times (-\vec{a}_x) = \vec{a}_z\right]$$

$$E_0 = 10^3$$

$$\eta = \sqrt{\frac{\mu_0}{\varepsilon_0}}$$

$$\eta = \sqrt{\frac{4\pi \times 10^{-7}}{\frac{1}{36\pi} \times 10^{-9}}} = 120\pi$$

$$H(z, t) = \frac{10^3}{120\pi}\sin(\omega t - \beta z)(-\vec{a}_x)$$

$$= -2.65 \sin(\omega t - \beta z)\vec{a}_x \text{ A/m}.$$

Propagation constant γ = α + *j*β

In free space, the attenuation factor α is zero. γ = *j*β where β is phase constant in rad/m.

$$\beta = \frac{\omega}{u} = \frac{2\pi f}{u}$$

$u \to$ velocity of wave.

For free space
$u = 3 \times 10^8$ m/s.

$$\beta = \frac{2\pi \times 95.5 \times 10^6}{3 \times 10^8} = 2 \text{ rad/m}.$$

Problem 2

The electric field of a 1MHz plane wave travelling in the $+z$ direction in air point along the x-direction. If the peak value of E is 1.2π (mV/m), obtain the expressions for $E(z, t)$ and $H(z, t)$.

Solution:

$f = 1$ MHz.
wave length $\lambda = u/f$;
u is velocity of the wave in free air $= 3 \times 10^8$ m/s.

$$\lambda = \frac{3 \times 10^8}{1 \times 10^6} = 300 \text{ m}.$$

$$\text{Phase constant } \beta = \frac{\omega}{u} = \frac{2\pi f}{u} = \frac{2\pi}{300}.$$

$$\therefore \quad \vec{E}(z, t) = E_0 \sin(\omega t - \beta z)\vec{a}_x$$

$$= 1.2\pi \sin\left(2\pi \times 1 \times 10^6 - \frac{2\pi}{300} z\right)\vec{a}_x \text{ mV/m}$$

$$= 1.2\pi \sin\left(2\pi \times 10^6 - 0.021z\right)\vec{a}_x \text{ mV/m}$$

$$\vec{H}(z, t) = \frac{E_0}{\eta} \sin(\omega t - \beta z)\vec{a}_y \qquad \left[\because \ \vec{E} \times \vec{H} = \vec{a}_z\right]$$

$$\eta = 120\pi$$

$$\vec{H}(z, 6) = \frac{1.2\pi}{120\pi} \sin\left(2\pi \times 10^6 - 0.021z\right)\vec{a}_y \text{ mA/m}$$

$$= .01 \sin\left(2\pi \times 10^6 - 0.021z\right)\vec{a}_y \text{ mA/m}$$

$$= 10 \sin\left(2\pi \times 10^6 - 0.021z\right)\vec{a}_y \ \mu\text{A/m}.$$

Problem 3

The electric field of a uniform plane wave travelling in a lossless medium with an intrinsic impedance 188.5 Ω is given by $\vec{E} = 10e^{-j4\pi y}$ mV/m. Determine (*a*) the associated magnetic field and write $\vec{E}$ in instantaneous form. If the medium is non magnetic ($\mu = \mu_0$).

Intrinsic impedance for a lossless medium

$$\eta = \sqrt{\frac{\mu}{\varepsilon}} = 188.5$$

$$\eta = \sqrt{\frac{\mu_0\mu_r}{\varepsilon_0\varepsilon_r}} = \sqrt{\frac{4\pi \times 10^7}{\frac{1}{36\pi}\times 10^9 \times \varepsilon_r}} = 188.5$$

$$188.5 = \frac{120\pi}{\sqrt{\varepsilon_r}}$$

$$\varepsilon_r = \left(\frac{120\pi}{188.5}\right)^2 = 4$$

$$\varepsilon_r = 4.$$

Given

$$\vec{E} = 10e^{-j4\pi y}\,\vec{a}_z.$$

Comparing the eqn. (5.15), it is known that the wave is travelling in $\vec{a}_y$ direction. The electric field is along $\vec{a}_z$ direction. Phase constant $\beta = 4\,\pi$.

$$\beta = \frac{\omega}{u} = \frac{2\pi f}{u}$$

$u \to$ velocity of the wave in $\vec{a}_y$ the lossless medium.

$$u = \frac{1}{\sqrt{\mu\varepsilon}} = \frac{1}{\sqrt{\mu_0\varepsilon_0\varepsilon_r}} = \frac{3\times 10^8}{\sqrt{\varepsilon_r}} = \frac{3\times 10^8}{2}$$

$$= 1.5 \times 10^8.$$

$$\omega = u \times \beta$$

$$= 1.5 \times 10^8 \times 4\pi$$

$$\omega = 6\pi \times 10^8 \text{ rad/s}.$$

Instantaneous form of $\vec{E} = E_0 \sin(\omega t - \beta y)\vec{a}_z$

$$= 10\sin\left(6\pi \times 10^8 t - 4\pi y\right)\vec{a}_z \text{ mV/m}$$

$$H_0 = \frac{E_0}{\eta} = \frac{10}{188.5} = 0.053 \text{ AT/m}$$

$\vec{E} \times \vec{H}$ should be along $\vec{a}_y$ (wave directions if $\vec{E}$ is along $\vec{a}_z$, $\vec{H}$ should be along $\vec{a}_x$).

$$\vec{H} = 0.053 e^{-j4\pi y} \vec{a}_x \text{ (mA/m)}.$$

Instantaneous form of $\vec{H} = 0.053 \sin\left(6\pi \times 10^8 t - 4\pi y\right) \text{mA/m}$.

Problem 4

A plane wave propagating through a medium with $\varepsilon_r = 8$, $\mu_r = 2$ has $\vec{E} = 0.5 e^{-z/3} \sin\left(10^8 t - \beta z\right) \vec{a}_x$ V/m.

Determine (*a*) β (*b*) the loss tangent (*c*) wave impedance (*d*) wave velocity (*e*) $\vec{H}$ field.

$\vec{E} = 0.5 e^{-z/3} \sin\left(10^8 t - \beta z\right) \vec{a}_x$. From eqn. (8.48), $\alpha = \dfrac{1}{3}$, $\omega = 10^8$.

(*a*) Velocity of wave travelling in the medium is

$$u = \frac{1}{\sqrt{\mu\varepsilon}} = \frac{1}{\sqrt{\mu_0 \mu_r \varepsilon_0 \varepsilon_r}} = \frac{3 \times 10^8}{\sqrt{8 \times 2}} = \frac{3}{4} \times 10^8$$

$$= 0.75 \times 10^8 \text{ m/s}.$$

$$\beta = \frac{\omega}{u}$$

$$\omega = 10^8$$

$$\beta = \frac{10^8}{0.75 \times 10^8} = 1.33 \text{ rad/m}.$$

(*b*) Loss tangent:

$$\alpha = \omega \left\{ \sqrt{\frac{\mu\varepsilon}{2} \left[\sqrt{\left(\frac{\sigma}{\omega\varepsilon}\right)^2 + 1} - 1 \right]} \right\}$$

$$\alpha^2 = \omega^2 \left[\frac{\mu\varepsilon}{2} \left[\sqrt{\frac{\sigma^2}{(\omega\varepsilon)^2} + 1} - 1 \right] \right]$$

$$\left(\frac{1}{3}\right)^2 = \left(10^8\right)^2 \cdot \left(\frac{4}{3}\right)^2 \times \frac{10^{-16}}{2} \left[\sqrt{1 + \left(\frac{\sigma}{\omega\varepsilon}\right)^2} - 1 \right]$$

$$\frac{1}{9} = \frac{16}{18} \left[\sqrt{1 + \left(\frac{\sigma}{\omega\varepsilon}\right)^2} - 1 \right]$$

$$\frac{18}{16}\times\frac{1}{9}=\left[\sqrt{1+\left(\frac{\sigma}{\omega\varepsilon}\right)^2}-1\right]$$

$$(1.125)^2=1+\left(\frac{\sigma}{\omega\varepsilon}\right)^2$$

$$1.2656-1=\left(\frac{\sigma}{\omega\varepsilon}\right)^2$$

$$\left(\frac{\sigma}{\omega\varepsilon}\right)^2=0.2656$$

$$\frac{\sigma}{\omega\varepsilon}=0.5153$$

Loss tangent = 0.5153.

(*c*) Wave impedance or intrinsic impedance: From eqns. (5.51) and (5.52),

$$|\eta|=\frac{\sqrt{\frac{\mu}{\varepsilon}}}{\left[1+(\sigma/\omega\varepsilon)^2\right]^{1/4}}$$

$$=\frac{\sqrt{\frac{\mu_0\mu_r}{\varepsilon_0\varepsilon_r}}}{\left(1+\left(\frac{\sigma}{\omega\varepsilon}\right)^2\right)^{1/4}}=\frac{120\pi\sqrt{\frac{2}{8}}}{1.0606}=177.72\,\Omega$$

$$\theta_\eta=+\frac{\tan^{-1}\left(\frac{\sigma}{\omega\varepsilon}\right)}{2}=\frac{\tan^{-1}(0.5153)}{2}$$

$$\eta=177.72\angle 13.63°$$

(*d*) Wave velocity $u = 0.75 \times 10^8$ m/s.

(*e*) $\vec{H}=\frac{E_0}{\eta}e^{-\alpha z}\sin\left(\omega r-\beta z-\theta_\eta\right)\vec{a}_y$ A/m

$$\vec{H}=\frac{0.5}{177.72}e^{-z/3}\sin\left(10^8-\beta_z-13.63°\right)\vec{a}_y\ \text{A/m}$$

$$\vec{H}=2.817\,e^{-z/3}\sin\left(10^8-\beta_z-13.63°\right)\vec{a}_y\ \text{A/m}.$$

Problem 5

In a non-magnetic medium $\vec{E} = 4\sin\left(2\pi \times 10^7 t - 0.8x\right)\vec{a}_z$ V/m. Find (*a*) ε_r, η (*b*) the time average power carried by the wave.

Solution:

$$\vec{E} = 4\sin\left(2\pi \times 10^7 t - 0.8x\right)\vec{a}_z \text{ V/m}$$

$$\alpha = 0,\ \theta_\eta = 0.$$

The wave is travelling along $\vec{a}_x$ direction. Since $\vec{E}$ is along $\vec{a}_z$ direction, H is along $\left(-\vec{a}_y\right)$

$$\omega = 2\pi \times 10^7 t \text{ rad/s } E_0 = 4 \text{ V/m},\ \beta = 0.8 \text{ rad/m}$$

$$\beta = \frac{\omega}{\mu}$$

$$u = \frac{2\pi \times 10^7}{0.8} = 7.85 \times 10^7 \text{ m/s.}$$

$$u = \frac{1}{\sqrt{\mu_0 \mu_r \varepsilon_0 \varepsilon_r}} = \frac{3 \times 10^8}{\sqrt{\mu_r \varepsilon_r}}.$$

For non-magnetic medium $\mu_r = 1$

$$\varepsilon_r = \left(\frac{3 \times 10^8}{7.85 \times 10^7}\right) = 14.6$$

$$\varepsilon_r = 14.6.$$

$$\text{Intrinsic impedance } \eta = \sqrt{\frac{\mu}{\varepsilon}}$$

$$\eta = \sqrt{\frac{\mu_0 \mu_r}{\varepsilon_0 \varepsilon_r}} = 120\pi \sqrt{\frac{1}{14.6}} = 98.7\,\Omega.$$

From eqn. (5.97),

$$\text{Average power } \vec{P}_{\text{ave}} = \frac{E_0^2}{2|\eta|} e^{-2\alpha x} \cos\theta_\eta\, \vec{a}_x.$$

As $\alpha = \theta_\eta = 0$,

$$\vec{P}_{\text{ave}} = \frac{E_0^2}{2\eta}\vec{a}_x = \frac{4^2}{2 \times 98.7} = 81\vec{a}_x \text{ mW/m}^2$$

Problem 6

Consider a straight wire of radius ρ_0 oriented along *Z*-axis carrying steady state current *I*. Determine the total power entering a unit length of wire using Poynting vector concept.

Solution:

The magnetic field intensity due to a straight wire of radius ρ_0 at the surface of the conductor is

$$\vec{H} = \frac{I}{2\pi\rho_0}\vec{a}_\phi,$$

where *I* is the current flow through the wire.

For the conductor

$$\vec{J} = \sigma\vec{E}.$$

So, electric field due to the conductor is $\vec{E} = \dfrac{J\vec{a}_z}{\sigma} = \dfrac{I}{(\pi\rho_0^2)\sigma}\vec{a}_z.$

The Poynting vector $\vec{S} = \vec{E} \times \vec{H}$

$$= \frac{I}{(\pi\rho_0^2)\sigma}\vec{a}_z \times \frac{I}{2\pi\rho_0}\vec{a}_\phi$$

$$\vec{S} = \frac{I^2}{2\pi^2\rho_0^3\sigma}(-\vec{a}_\rho).$$

Total power $= \oiint \vec{S}\cdot d\vec{s}$

$$= \oiint \frac{I^2}{2\pi^2\rho_0^3\sigma}(-\vec{a}_\rho)\cdot \rho\, d\phi\, dz\, \vec{a}_\rho$$

$$= \int_{z=0}^{l}\int_{\phi=0}^{2\pi} \frac{I^2}{2\pi^2\rho_0^2\sigma}\, d\phi\, dz$$ [Assume length of wire = '*l*' m]

$$= \frac{I^2}{2\pi^2\rho_0^2\sigma} \times 2\pi \times l$$

$$= \frac{I^2 l}{\pi\rho_0^2\sigma}.$$

Power entering/unit length $= \dfrac{I^2}{A\sigma} = I^2 R$ watt.

The result indicates that the field supplies the energy to balance out the I^2R heating losses in the wire. Contribution to the surface integral is only from the cylindrical part.

Problem 7

A co-axial cable is operating at 1000 V (d.c.) The radius of core is 2 cm inner and radius of sheath is 8 cm. Relative permittivity of insulating medium is 5. Length of the cable is 2 km. The cable carries a current of 10 A. Compute power carried by the cable by applying concept of Poynting vector.

Solution:

From eqn. (2.178), electric field intensity $\vec{E}$ due to a cable is

$$\vec{E} = \frac{V}{\rho \ln \frac{b}{a}} \vec{a}_\rho$$

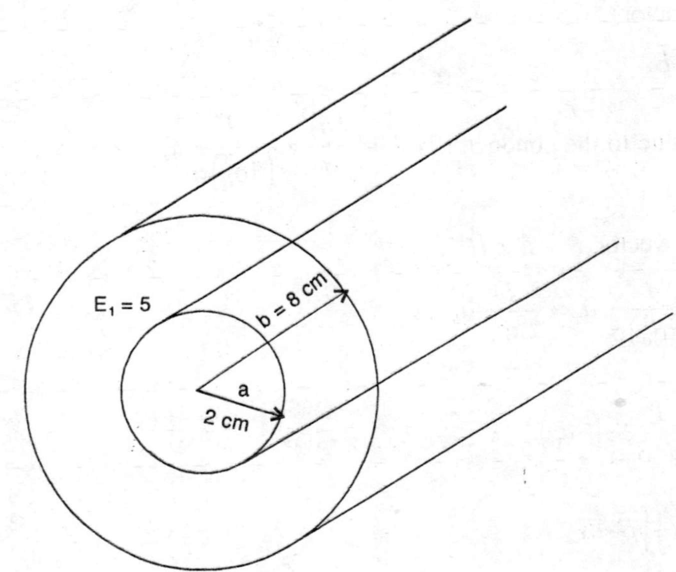

V is the potential of the cable
b is the outer radius in m
a is the inner radius in m
ρ is the distance of a point in the dielectric.

Magnetic field intensity $\vec{H}$ due to the cable is

$$\vec{H} = \frac{I}{2\pi\rho} \vec{a}_\phi .$$

Poynting vector $\vec{S} = \vec{E} \times \vec{H}$

$$= \frac{V}{\rho \ln \frac{b}{a}} \vec{a}_\rho \; \frac{I}{2\pi\rho} \cdot \vec{a}_\phi$$

$$= \frac{VI}{2\pi\rho^2 \ln b/a} \vec{a}_z .$$

Average power $P = \oiint \vec{S} \cdot d\vec{s}$

$$= \int_{\phi=0}^{2\pi} \int_{\rho=a}^{b} \frac{VI}{2\pi\rho^2 \ln\frac{b}{a}} \vec{a}_z \cdot \rho d\rho \, d\phi \, \vec{a}_z$$

$$P = \frac{VI}{2\pi \ln\frac{b}{a}} \int_{\phi=0}^{2\pi} \int_{\rho=0}^{b} \frac{d\rho}{\rho} d\rho \, d\phi$$

$$= \frac{VI}{2\pi \ln\frac{b}{a}} \times 2\pi \ln\frac{b}{a}$$

$= VI$ watt.

$P = 1000 \times 10$

$= 10$ kW

Problem 8

A perpendicular polarised wave propagates from region 1 ($\varepsilon_{r_1} = 8.5$, $\mu_r = 1$, $\sigma_1 = 0$) to region 2, in free space, with an angle of incidence 15°. Given $E_{i0} = 1.0\mu$ V/m. Find E_{r0}, E_{t0}, H_{r0} and H_{t0}.

Solution:

In Region 1, $\varepsilon_{r1} = 8.5$, $\mu_{r_1} = 1$, $\sigma_1 = 0$

In Region 2, $\varepsilon_{r_2} = 1$, $\mu_{r_2} = 1$

Angle of incidence $\theta_i = 15°$

$E_{i0} = 1\mu$ V/m.

From eqn. (5.131)

$$T_{\perp} = \frac{E_{r0}}{E_{i0}} = \frac{\eta_2 \cos\theta_i - \eta_1 \cos\theta_t}{\eta_1 \cos\theta_i + \eta_2 \cos\theta_t}$$

$$\eta_2 = \sqrt{\frac{\mu_2}{\varepsilon_2}} = \sqrt{\frac{\mu_0 \mu_{r_2}}{\varepsilon_0 \varepsilon_{r_2}}} = 120\pi\,\Omega$$

$$\eta_1 = \sqrt{\frac{\mu_1}{\varepsilon_1}} = \sqrt{\frac{\mu_0 \mu_{r_1}}{\varepsilon_0 \varepsilon_{r_1}}} = \sqrt{\frac{\mu_0 \times 1}{\varepsilon_0 \times 8.5}} = 129\pi\,\Omega$$

From eqn. (5.133)

$$\frac{\sin\theta_i}{\sin\theta_t} = \sqrt{\frac{\varepsilon_2}{\varepsilon_1}}$$

$$\frac{\sin\theta_i}{\sin\theta_t} = \sqrt{\frac{\varepsilon_0}{\varepsilon_0 \cdot 8.5}}$$

$$\sin\theta_t = \sin 15° \times 8.5$$

$$\theta_t = 48.98°$$

$$\frac{E_{r0}}{E_{i0}} = \frac{120\pi\cos 15° - 129\cos 48.98}{129\cos 48.98 + 120\pi\cos 15°}$$

$$\frac{E_{r0}}{E_{i0}} = 0.623\,.$$

Given $E_{i0} = 1\mu\ \text{V/m}$

$$E_{r0} = 0.625\mu\ \text{V/m}\,.$$

From eqn. (5.132)

$$\frac{E_{t0}}{E_{i0}} = \tau_{\perp} = \frac{2\eta_2\cos\theta_i}{\eta_1\cos\theta_t + \eta_2\cos\theta_i}$$

$$\frac{E_{t0}}{E_{i0}} = \frac{2 \times 120\pi \times \cos 15°}{129\cos 48.98 + 120\pi\cos 15}$$

$$E_{t0} = 1.623\mu\ \text{V/m}\,.$$

$$H_{i0} = \frac{E_{i0}}{\eta_1} = \frac{1\mu\ \text{V/m}}{129} = 7.75n\ \text{A/m}\,.$$

$$H_{r0} = \frac{E_{r0}}{\eta_1} = \frac{0.625\ \mu\ \text{V/m}}{129} = 4.84n\ \text{A/m}\,.$$

$$H_{t0} = \frac{E_{t0}}{\eta_2} = \frac{0.625\ \mu\ \text{V/m}}{120\pi} = 4.31n\,\text{A/m}\,.$$

Problem 9

The amplitude of E_i in free space (region 1) at the interface with region 2 is 1 V/m. If $H_{r0} = +1.41 \times 10^{-3}$ A/m, $\varepsilon_{r2} = 18.5$ and $\sigma_2 = 0$. Find μ_{r2}.

Solution:

Given $E_{i0} = 1$ V/m.

$$H_{r0} = +1.41 \times 10^{-3} \text{ A/m}$$

$$\frac{E_{r0}}{\eta_1} = H_{r0}$$

$$\eta_1 = \sqrt{\frac{\mu_0 \mu_{r1}}{\varepsilon_0 \varepsilon_{r1}}} = 120\pi$$

$$E_{r0} = +1.41 \times 10^{-3} \times 120\pi = +531.5 \times 10^{-3} \text{ V/m}.$$

From eqn. (5.131)

$$\frac{E_{r0}}{E_{i0}} = \frac{\eta_2 \cos\theta_i - \eta_1 \cos\theta_t}{\eta_1 \cos\theta_i + \eta_2 \cos\theta_i}$$

$$\theta_i = \theta_t = 0$$

$$\frac{E_{r0}}{E_{i0}} = \frac{\eta_2 - \eta_1}{\eta_1 + \eta_2} = +531.5 \times 10^{-3}$$

$$\eta_1 = 120\pi\Omega = 377\Omega$$

$$\frac{\eta_2 - 377}{\eta_2 + 377} = +531.5 \times 10^{-3}$$

$$\eta_2 - 377 = +531.5 \times 10^{-3} \eta_2 + 200.4$$

$$0.469\,\eta_2 = 577.4$$

$$\eta_2 = 1231\Omega$$

$$\eta_2 = \sqrt{\frac{\mu_2}{\varepsilon_2}} = \sqrt{\frac{\mu_0 \mu_{r2}}{\varepsilon_0 \varepsilon_{r2}}} = 1231$$

$$\mu_{r2} = 197.2$$

Problem 10

A free space–silver interface has E_{i0} = 100 V/m on the free space side. The frequency is 15 MHz and the silver constants are $\varepsilon_r = \mu_r = 1$, $\sigma = 61.7\,\text{Ms/m}$. Determine E_{r0} and E_{t0} at the interface.

Solution:

From eqn. (5.114), for a free space–conductor interface

$$\mathcal{T} = -1$$

$$\frac{E_{i0}}{E_{r0}} = -1$$

$$\therefore \quad E_{r0} = -100 \text{ V/m}.$$

Also from eqn. (5.113)

$$\tau = \frac{E_{t0}}{E_{i0}} = \frac{2\eta_2}{\eta_1 + \eta_2}.$$

η_2 at the silver medium is from eqn. (5.73)

$$\eta_2 = \sqrt{\frac{\omega\mu}{\sigma}} \angle 45°$$

$$= \sqrt{\frac{2\pi \times 15 \times 10^6 \times 4\pi \times 10^{-7} \times 1}{61.7 \times 10^{-6}}}$$

$$= 1.38 \times 10^{-3}\, \Omega$$

η_1 at the free space is 120π

$$\tau = \frac{E_{t0}}{E_{i0}} = \frac{2 \times 1.38 \times 10^{-3}}{377 + 1.38 \times 10^{-3}}$$

$$E_{t0} = 7.35 \times 10^{-4} \angle 45° \text{ V/m}.$$

5.10 WAVES IN TRANSMISSION LINES

A transmission line can be modelled as a simple circuit which consists of an generator at the sending end and a number of series elemental inductors and shunt capacitors. An electrical field is set up across the transmission line when a potential difference is established between the conductors of transmission line and the flow of current creates a magnetic field.

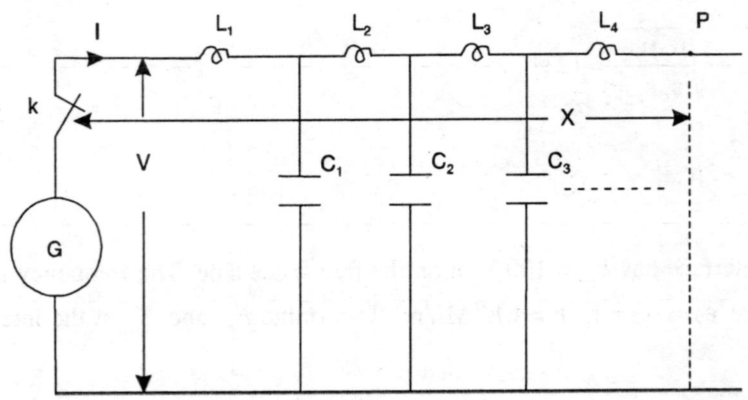

Fig. 5.13

When the switch k is closed at a instant, current I will flow through the line and potential drop across the line is V.

From the definition of wave, in case there is any disturbance in any part of the circuit at a time, it will take time for the remaining part of the line to experience the consequent change.

Hence, the electric field and magnetic field set up at the sending end creates a electric and magnetic wave and they travel towards the receiving end along the direction of power flow.

Let us determine the velocity of wave created by the transmission line.

Let L_1, C_1 be the inductance in henry per km and capacitance in farad per km.

Consider any point P, x km distant from the sending end of the line.

[As inductance per km is L_l, total inductance at $P = L_l x$.]

Magnetic flux linkage at $P = L_l I x$ $\quad [\because \lambda = LI]$

As per Faraday's law, potential difference at P is the rate of change of flux linkage.

$$V = L_1 I \frac{dx}{dt} = L_1 I v \qquad ...(5.148)$$

where v is the velocity of electromagnetic wave.

Similarly, C_l is the capacitor per km and at 'P', the capacitance is $C_l x$.

Charge accumulated at $P(Q) = C_l V x$

As flow of current is rate of flow of charge, $\quad [\because c = Q/v]$

$$I = \frac{dQ}{dt} = C_1 V \frac{dx}{dt}. \qquad ...(5.149)$$

Multiply (1) and (2)

$$VI = C_1 L_1 I V v^2$$

$$v^2 = \frac{1}{L_1 C_1}$$

$$v = \frac{1}{\sqrt{L_1 C_1}}.$$

Substitute the values of inductance and capacitance of tranmission line from eqns. (2.165) and (3.103).

$$L_1 = \frac{\mu_0}{2\pi}\left[\frac{1}{4} + \ln\frac{d}{a}\right], \; C_1 = \frac{2\pi\varepsilon_0}{\ln\frac{d}{a}}$$

$$\text{neglect } \frac{1}{4}, \text{ as } \frac{1}{4} << \ln\frac{d}{a}.$$

$$L_1 = \frac{\mu_0}{2\pi}\ln\frac{d}{a}, \; C_1 = \frac{2\pi\varepsilon_0}{\ln\frac{d}{a}}$$

$$v = \frac{1}{\sqrt{L_1 C_1} = \frac{1}{\sqrt{\mu_0 \varepsilon_0}}} = 3 \times 10^8 \text{ m/s},$$

which is same as before for the wave travelling in free space.
Consider dividing (4) and (2)

$$\frac{V}{I} = \frac{I}{V} \cdot \frac{L_1}{C_1}$$

$$\frac{V^2}{I^2} = \frac{L_1}{C_1}$$

$$\frac{V}{I} = \sqrt{\frac{L_1}{C_1}} = Z,$$

where Z is known as natural impedance of the transmission line.

SUMMARY

1. The wave equation is of the form

$$\frac{\partial^2 \vec{H}}{\partial t^2} = u^2 \nabla^2 \vec{H}, \frac{\partial^2 \vec{E}}{\partial t^2} = u^2 \nabla^2 \vec{E}.$$

2. One-dimensional wave equation, for a wave travelling in Z direction is

$$\frac{\partial^2 \vec{H}}{\partial t^2} = u^2 \frac{\partial^2 H}{\partial z^2}, \frac{\partial^2 \vec{E}}{\partial t^2} = u^2 \frac{\partial^2 E}{\partial z^2}, u = \frac{1}{\sqrt{\mu\varepsilon}},$$

where u is the velocity of the wave. In free space it is 3×10^8 m/s.

3. The solution of wave equation in general is

$$\vec{E} \text{ or } \vec{H} = A \sin(\omega t - \beta z) + B \sin(\omega t + \beta z),$$

where ω is angular frequency in rad/s; β is phase constant or wave number is rad/m; z represents the direction of wave propagation and A and B represents amplitude.

$A \sin(\omega t - \beta z)$ – positive going wave or forward travelling wave.

$A \sin(\omega t + \beta z)$ – negative going wave or backward travelling wave.

4. In a lossy dielectric, charge-free medium, the wave equation is of the form

$$\nabla^2 \vec{A}_s - \gamma^2 \vec{A}_s = 0,$$

where $\vec{A}_s$ is either $\vec{E}_s$ or $\vec{H}_s$ and γ is propagation constant $\nu = \alpha + j\beta$.

If wave is assumed in z direction and $\vec{E}$ is assumed in x direction, then

$$\vec{E}(z,t) = E_0 e^{-\alpha z} \cos(\omega t - \beta z)\vec{a}_x$$

$$\vec{H}(z,t) = H_0 e^{-\alpha z} \cos(\omega t - \beta z - \theta_\eta)\vec{a}_y,$$

where α is attenuation constant, β is phase constant, and η is intrinsic impedance of the medium.

$$\eta = |\eta| \underline{|\theta_\eta}$$

$$|\eta| = \frac{\sqrt{\mu/\varepsilon}}{\left[1 + (\sigma/\omega\varepsilon)^2\right]^{1/4}} \quad \underline{|\theta_n} = \frac{\tan^{-1}\sigma/\omega\varepsilon}{2}.$$

5. $\tan\theta$ is loss tangent and θ is the loss angle of the medium.

$$\tan\theta = \frac{|\vec{J}_s|}{|\vec{J}_{ds}|} = \frac{\sigma}{\omega\varepsilon}.$$

6. If $\tan\theta$ is small and at high frequencies, a medium is said to the good dielectric, and if $\tan\theta$ is very large and at low frequencies, a medium is said to be a good conductor.
7. Wave propagation in other media can be derived from that for lossy media as special cases.

   For free space, let $\sigma = 0,\ \varepsilon = \varepsilon_0,\ \mu = \mu_0$

   For lossless dielectric media, let $\sigma = 0,\ \varepsilon = \varepsilon_0\varepsilon_r,\ \mu = \mu_0\mu_r$

   For good conductor, let $\sigma = \infty,\ \varepsilon = \varepsilon_0,\ \mu = \mu_0$
8. If the electromagnetic wave has no electric or magnetic field components along the direction of propagation, then the wave is called transverse electromagnetic wave (TEM).

   If $\vec{E}$ is along x direction and the wave along z direction, then $\vec{H}$ is along y direction.
9. The direction of the electric field gives the polarisation of the wave.
10. In good conducting medium,

$$|\eta| = \sqrt{\frac{\omega\mu}{\sigma}} \quad \theta_\eta = 45°$$

$$\vec{E} = E_0 e^{-\alpha z} \cos(\omega t - \beta z)\vec{a}_x$$

$$\vec{H} = \frac{H_0}{\eta}{}^{-\alpha z} \cos(\omega t - \beta z - 45°)\vec{a}_y.$$

11. In a good conductor, the field tends to concentrate within the initial distance δ from the conductor surface. This phenomenon is called skin effect.

    Surface resistance (d.c.) or skin resistance R_s in Ω/m^2 is

$$R_s = \frac{1}{\sigma\delta}\ \Omega/\text{m}^2.$$

12. The distance δ through which the wave amplitude decreases by a factor e^{-1} (37%) is called skin depth or penetration depth of the medium.

$$\delta = \frac{1}{\alpha} = \frac{1}{\sqrt{\pi f \mu \sigma}}.$$

13. For free space and lossless dielectric medium

$$\vec{E} = E_0 \cos(\omega t - \beta z)\vec{a}_x$$

$$\vec{H} = \frac{E_0}{\eta} \cos(\omega t - \beta z)\vec{a}_y.$$

For free space $\eta = \sqrt{\frac{\mu_0}{\varepsilon_0}} = 377\,\Omega.$

For lossless dielectric $\eta = \sqrt{\frac{\mu}{\varepsilon}}.$

14. The Poynting vector $\vec{P}$ is the power flow vector in watt/m^2 whose direction is same as the direction of wave propagation

$$\vec{P} = \vec{E} \times \vec{H},\ \vec{P}_{\text{ave}} = \frac{1}{2}\text{Re}\left(\vec{E}_s \times \vec{H}_s^*\right)\ \text{W/m}^2.$$

Total power is $P = \iint \vec{P} \cdot d\vec{s}.$

Also $\vec{P}_{\text{ave}} = \frac{E_0^2}{2|\eta|} e^{-2\alpha z} \cos\theta_\eta\, \vec{a}_z.$

15. If a plane wave is incident normally from medium 1 to medium 2, the reflection coefficient $\mathcal{T}$ and transmission coefficient τ are given by

$$\mathcal{T} = \frac{E_{r0}}{E_{i0}} = \frac{\eta_2 - \eta_1}{\eta_2 + \eta_1}, \quad \tau = \frac{E_{t0}}{E_{i0}} = 1 + \mathcal{T}$$

$$\tau = \frac{2\eta_2}{\eta_2 + \eta_1}.$$

16. If the medium 1 is a perfect dielectric and medium 2 is a perfect conductor, then there is no transmission $\mathcal{T} = -1$, $\tau = 0$. All the wave will be reflected back known as total internal reflection and produces standing wave.
Standing wave ratio is

$$S = \frac{1 + |\mathcal{T}|}{1 - |\mathcal{T}|}.$$

17. For oblique incidence from lossless medium 1 to lossless medium 2:

For parallel polarisation,

$$T_{\|} = \frac{\eta_2 \cos\theta_t - \eta_1 \cos\theta_i}{\eta_2 \cos\theta_t + \eta_1 \cos\theta_i}$$

$$\tau_{\|} = \frac{2\eta_2 \cos\theta_i}{\eta_2 \cos\theta_t + \eta_1 \cos\theta_i}.$$

For perpendicular polarisation,

$$T_{\perp} = \frac{\eta_2 \cos\theta_i - \eta_1 \cos\theta_t}{\eta_2 \cos\theta_i + \eta_1 \cos\theta_t}$$

$$\tau_{\perp} = \frac{2\eta_2 \cos\theta_i}{\eta_2 \cos\theta_i + \eta_1 \cos\theta_t}.$$

18. $\theta_r = \theta_i$

Snell's law $\dfrac{\sin\theta_t}{\sin\theta_i} = \dfrac{\beta_1}{\beta_2} = \sqrt{\dfrac{\mu_1\varepsilon_1}{\mu_2\varepsilon_2}}.$

19. Total transmission or no reflection ($T = 0$) occurs when the angle incidence θ_i is equal to Brewster angle (polarising angle).

$$\sin^2\theta_{B\|} \frac{1 - \mu_2\varepsilon_1/\mu_1\varepsilon_2}{1 - (\varepsilon_1/\varepsilon_2)^2}$$

$$\sin^2\theta_{B\perp} \frac{1 - \mu_1\varepsilon_2/\mu_2\varepsilon_1}{1 - (\mu_1/\mu_2)^2}.$$

If dielectric media is lossless and non-magnetic

$$\sin\theta_{B\|} = \sqrt{\frac{\varepsilon_2}{\varepsilon_1 + \varepsilon_2}}.$$

If $\mu_1 \neq \mu_2$ and $\varepsilon_1 = \varepsilon_2$

$$\sin\theta_{B\perp} = \sqrt{\frac{\mu_2}{\mu_1 + \mu_2}}.$$

20. If $\vec{E}$ field lies in the place of incidence, then it is called parallel polarisation.

If $\vec{E}$ field lies perpendicular to the place of incidence, then it is called perpendicular polarisation.

21. The nature impedance of transmission line is $Z = \sqrt{\dfrac{L_1}{C_1}}$.

EXERCISE

1. The electric field of a travelling electromagnetic wave is given by

$$\vec{E}(z,t) = 10\cos\left(\pi \times 10^7 t + \pi z/15\right) \text{V/m}.$$

Determine (*a*) direction of wave propagation (*b*) The wave frequency (*c*) its wavelength λ (*d*) its phase velocity.

[(*a*) $-z$ direction; (*b*) $f = 5$ MHz; (*c*) $\lambda = 30$m; (*d*) $u = 1.5 \times 10^8$ m/s]

2. An electromagnetic wave is propagatory in the Z-direction in a lossy medium with attenuation constant $\alpha = 0.5$ Np/m. If the wave's electric field amplitude is 100 V/m at $z = 0$, how far can the wave travel before its amplitude will have been reduced to (*a*) 10 V/m (*b*) 1 V/m (*c*) 1μV/m.

[(*a*) 4.6 m; (*b*) 9.2 m; (*c*) 37 m]

3. Determine the propagation constant γ for a material having $\mu_r = 1$, $\varepsilon_r = 8$ and $\sigma = 0.25$ ps/m, if the wave frequency is 1.6 MHz. [$\gamma = j9.48 \times 10^{-2}\ \text{m}^{-1}$]

4. Find the skin depth d at a frequency of 1.6 MHz in aluminium, where s = 38.2 Ms/m and $m_r = 1$. Also find g and the wave velocity *u*.

[$\delta = 6.44 \times 10^{-5}\ \text{m}\ ;\ \gamma = 2.2 \times 10^4\ \angle 45° \text{m}^{-1};\ u = 647\ \text{m/s}$]

5. In free space $\vec{E}(z,t) = 50\cos(\omega t - \beta z)\vec{a}_x$ V/m. Find the average power crossing a circular area of radius 2.5 m in the plane Z = constant (use eqn. 5.98). [65.1 W]

6. A wave in air is incident upon the flat boundary of a soil medium with $\varepsilon_r = 4$ and $\mu_r = 1$ at $\theta_i = 50°$. Determine $\mathcal{T}_\perp$, $\tau_\perp$, $\mathcal{T}_\parallel$ and $\tau_\parallel$ and Brewster angle q_B.

[$\mathcal{T}_\perp = -0.48,\ \tau_\perp = 0.52,\ \mathcal{T}_\parallel = -0.16,\ \tau_\parallel = 0.58,\ \theta_B = 63.4°$]

7. A free space-conductor interface has $H_{i0} = 1$ A/m on the free space side. The frequency is 31.8 MHz and the conductor constants are $\varepsilon_r = \mu_r = 1$, $\sigma = 1.26$ Ms/m. Determine H_{r0} and H_{t0} and the depth of penetration of H_t. [$H_{r0} = 1$ A/m, $H_{t0} = 2$ A/m, $\delta = 80$ μm]

8. A wave propagates from a dielectric medium to the interface with free space. If the angle of incidence is the critical angle of 20°, find the relative permittivity. [8.55]

9. A parallel polarised wave propagates from air into a dielectric at Brewster angle of 75°. Find ε_r.

[13.93]

6

Numerical Methods

INTRODUCTION

With the advent of fast digital computers, numerical methods have become a more attractive technique for solving EM problems. We have discussed several analytic techniques for solving EM problems in the preceding chapters. These analytic solutions are obtained under certain assumed conditions. For example, in calculating the capacitance of a parallel plate capacitor, the assumptions made are, negligible fringing effect and the separation distance is very small compared to the width and length of the plates. Moreover, the analytic solutions are suitable only for simple configurations.

The three most commonly used simple numerical techniques in EM are (1) moment method (2) finite difference method and (3) finite element method. As most of the EM problems involve either partial differential equations or integral equations, partial differential equations are solved using finite difference or finite element method and the integral equations are solved using moment method. In this chapter, we will discuss about finite difference method and moment method only.

6.1 THE FINITE DIFFERENCE METHOD

Any problem is uniquely defined by the following three things:

1. A partial differential equation such as Laplace's or Poisson's equations.
2. A solution region.
3. Boundary and/or initial conditions.

In order to solve using finite difference method, initially the solution region is divided into a grid of nodes. Then the differential equations and the boundary conditions are approximated into a set of linear algebraic equations called difference equations. Finally, the algebraic equations are to be solved.

Now we will discuss the basic principle behind the numerical method *i.e.* approximating the difference equations. Let us assume that we know the potentials V_1, V_2, V_3 and V_4 at four points equidistant from a point $P(0, 0, 0)$ and lying on mutually perpendicular axes as shown in Fig. 6.1 and that we wish to find the potential V_0 at P in terms of V_1, V_2, V_3 and V_4. Assuming no variations of V in the Z direction, the problem reduces to a two-dimensional one and simplifies to

$$\frac{\partial^2 V}{\partial x^2} + \frac{\partial^2 V}{\partial y^2} = 0. \qquad ...(6.1)$$

Now $\frac{V_2 - V_0}{\Delta x}$ is the slope of V between points 2 and P. This is approximately equal to $\frac{\partial V}{\partial x}$ [as $\Delta x \to 0$]. Also, $\frac{V_0 - V_1}{\Delta x}$ is the slope of V between points P and 1.

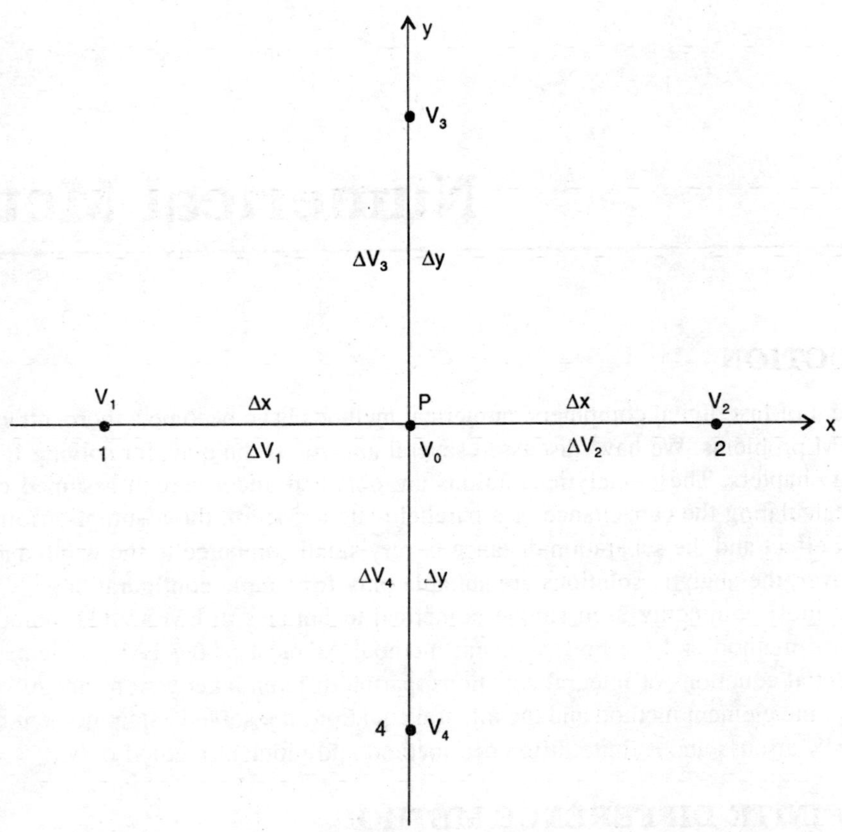

Fig 6.1 Finite difference solution pattern

The difference of these slopes (per distance increment Δx) is approximately equal to $\partial^2 V/\partial x^2$. Thus from eqn. (6.1)

$$\frac{\partial(\partial V/\partial x)}{\partial x} = -\frac{\partial(\partial V/\partial y)}{\partial y} \quad \text{...(6.2)}$$

$$\frac{\frac{V_2 - V_0}{\Delta x} - \frac{V_0 - V_1}{\Delta x}}{\Delta x} = \frac{-\left[\frac{V_3 - V_0}{\Delta y} - \frac{V_0 - V_4}{\Delta y}\right]}{\Delta y} \quad \text{...(6.3)}$$

Assuming $\Delta x = \Delta y$,

$$V_1 + V_2 + V_3 + V_4 - 4V_0 = 0. \quad \text{...(6.4)}$$

$$V_0 = \frac{1}{4}\left[V_1 + V_2 + V_3 + V_4\right]. \qquad \text{...(6.5)}$$

If we know the potential at points 1, 2, 3 and 4, then according to Laplace's equation the potential at the point P is as given by eqn. (6.5). Therefore, the potential at a point must be the average of the potential at four surrounding points.

To apply the eqn. (6.5) to a given problem, iterative method is commonly used. In the iteration method, the initial values of the potentials at free nodes are set equal to zero. The potentials at fixed nodes are kept unchanged at all times. Equation (6.5) is applied to every free node in turn until the potentials at all free nodes are calculated. As the potentials obtained at the end of this first iteration are not accurate, we repeat the calculation at every node using old values to determine new ones. This iteration is continued until a prescribed degree of accuracy is achieved or the old and new values at each node are satisfactorily close.

Example 6.1

An infinitely long square trough of sheet metal is shown in cross-section in Fig. 6.2a. The sides and bottom of the trough are at zero potential. The lid, separated by small gaps from the trough, is at a potential of 40 V. Obtain the potential distribution.

The potential at the centre of the trough is from eqn. (6.5).

$$\frac{40+0+0+0}{4} = 10\,\text{V}.$$

To find the potential at the centre of the four quadrants of the troughs as in Fig. 6.2(*b*), we rotate the *XY*-axes 45° and take the potential at the cover-trough gap as 20 V $\left[\frac{40+0}{2}\right]$.

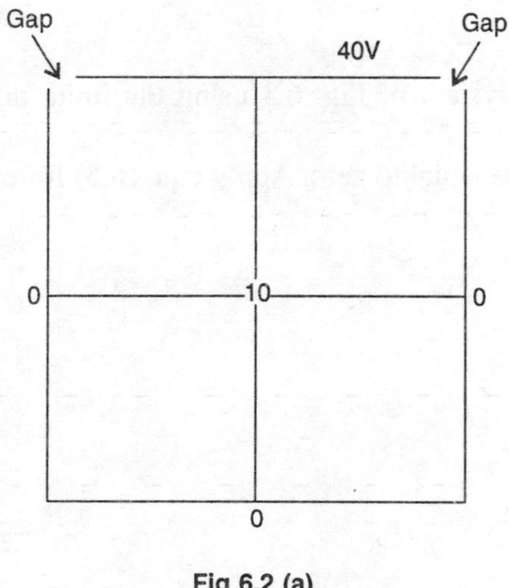

Fig 6.2 (a)

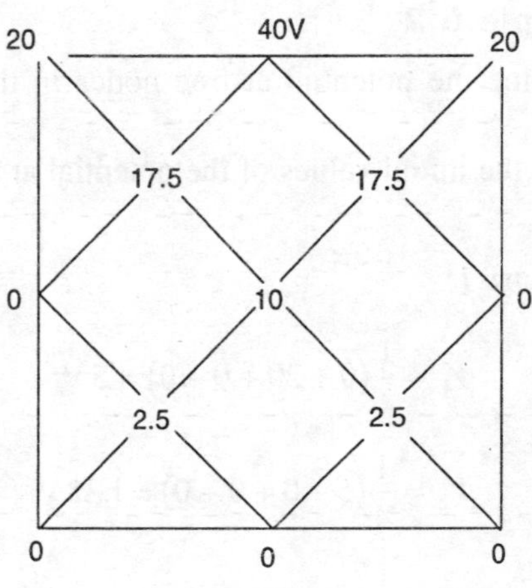

Fig 6.2 (b)

Use the eqn. (5.6) for remaining free nodes inside the trough as $\frac{20+40+0+10}{4} = 17.5$ V in upper left and right quadrants. For lower left and right quadrants, $\frac{0+10+0+0}{4} = 2.5$ V.

Next, we find V at four more points with voltages as shown in Fig. 6.2(*c*) (*XY* axes in usual orientation).

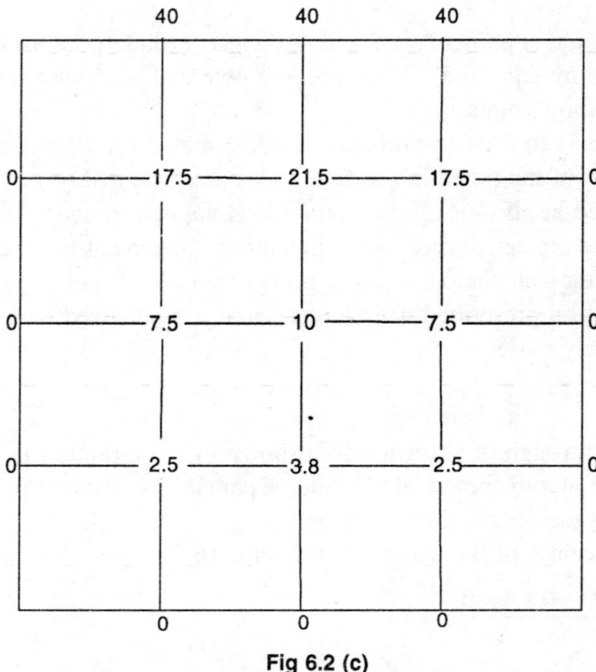

Fig 6.2 (c)

We do several iterations until the values no longer change. This method is well-adapted for very accurate calculations using digital computers. A still more accurate solution can be obtained by using a larger number of points (several grids).

Example 6.2

Determine the potential at free nodes in the potential system of Fig. 6.3 using the finite difference method.

Set the initial values of the potential at the free nodes equal to zero. Apply eqn. (6.5) for each free node.

Iteration 1:

$$V_1 = \frac{1}{4}(0 + 20 + 0 + 0) = 5\,\text{V}$$

$$V_2 = \frac{1}{4}(5 + 0 + 0 + 0) = 1.25\,\text{V}$$

$$V_3 = \frac{1}{4}(5 + 20 + 0 + 0) = 6.25\,\text{V}$$

$$V_4 = \frac{1}{4}(1.25 + 6.25 + 0 + 0) = 1.875\,\text{V}$$

$$V_5 = \frac{1}{4}(6.25 + 20 + 30 + 0) = 14.06\,\text{V}$$

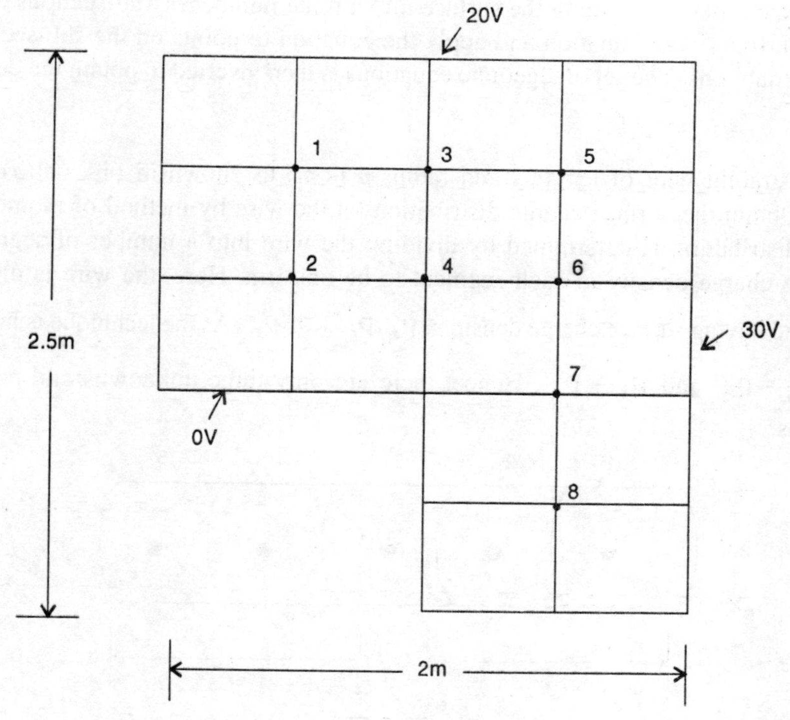

Fig. 6.3

$$V_6 = \frac{1}{4}(1.875 + 14.06 + 0 + 30) = 11.48\,\text{V}$$

$$V_7 = \frac{1}{4}(0 + 11.48 + 30 + 0) = 10.37\,\text{V}$$

$$V_8 = \frac{1}{4}(10.37 + 0 + 30 + 0) = 10\,\text{V}$$

Continue the iterations with the new values of V_1 to V_8. After 10 iterations, we obtain

$V_1 = 10.04$ V $\quad V_2 = 4.956$ V $\quad V_3 = 15.22$ V $\quad V_4 = 9.786$ V

$V_5 = 21.05$ V $\quad V_6 = 18.97$ V . $V_7 = 15.06$ V $\quad V_8 = 11.26$ V.

6.2 THE MOMENT METHOD

While the finite difference method is used in solving equations, the moment method is commonly used in solving integral equations. The technique consists of finding the charge distribution on the conductors held at known constant potentials.

Consider a surface charge distribution ρ_s on a given surface. The potential due to the charge distribution can be expressed as

$$V = \frac{1}{4\pi\varepsilon_0}\iint_s \frac{\rho_s\, ds}{R}. \qquad \text{...(6.6)}$$

The procedure consists of dividing the surface into a finite number of sub-sections to approximate the integral in eqn. (6.6) by summation and apply the equation to points on the sub-sections to obtain a set of algebraic equations. The set of algebraic equations is then inverted to obtain the desired solution.

Example 6.3

Consider a thin, straight wire of length l and radius a ($<<l$) as shown in Fig. 6.4(a) and held at a potential of 1V. Obtain the surface charge distribution on the wire by method of moments.

The charge distribution is determined by dividing the wire into a number of segments and also assuming that the charge density in each segment to be uniform. Here, the wire is divided into five equal segments and having surface charge densities $\rho_{s_1}, \rho_{s_2} \ldots\ldots \rho_{s_5}$. As the technique is moment method, by symmetry $\rho_{s_4} = \rho_{s_2}$ and $\rho_{s_5} = \rho_{s_1}$. Hence, there are only three unknowns and need three independent equations.

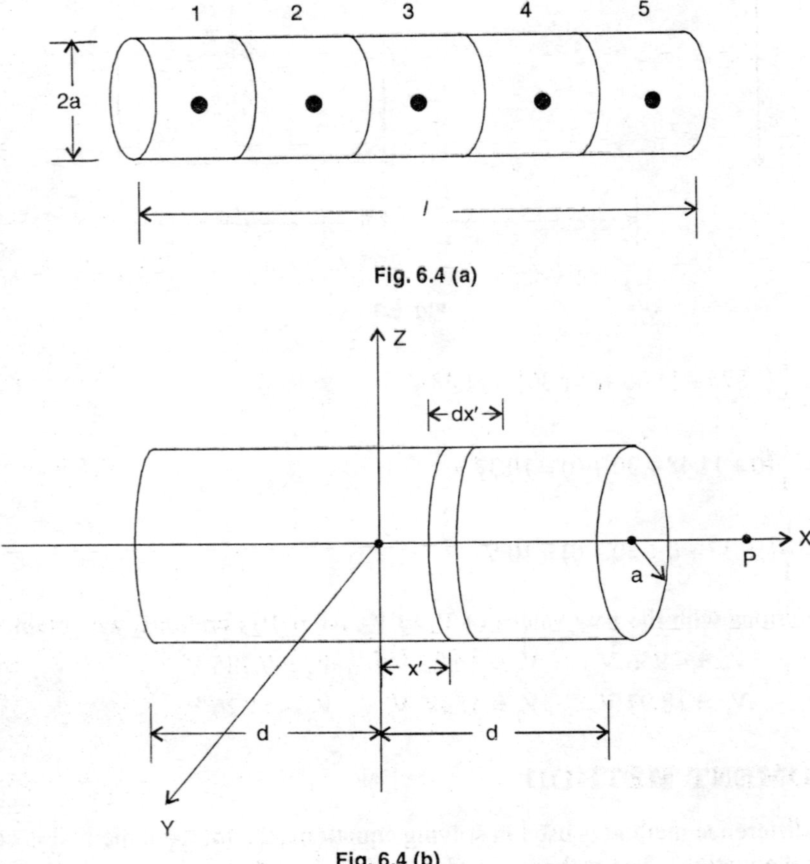

Fig. 6.4 (a)

Fig. 6.4 (b)

To derive an equation, consider one segment at a time and write the potential at its centre due to the charges in the five segments.

Consider a cylindrical surface charge of uniform density ρ_s coaxial with X-axis and located symmetrically about the origin as in 6.4(b) and find the potential due to the charge (1) at origin and (2) at a point (0, 0, x) where $x > d$ using the approximation ($a<<d$). In finding the potential at origin, we will

compute the potential due to the charge in a given segment at its centre point and in the second case of finding at a point (0, 0, x), we will compute the potential due to the charge in a given segment at the centre point of another segment.

Divide the cylindrical surface charge in Fig. 6.4(a) into a number of ring charges. One segment is shown in Fig. 6.4(b).

From Fig. 6.3(b), the potential at centre is

$$V_0 = \int_{x'=-d}^{d} \int_{\phi=0}^{2\pi} \frac{\rho_s \, a \, d\phi \, dx'}{4\pi\varepsilon_0 \sqrt{a^2 + (x')^2}} \qquad \text{...(6.7)}$$

$$= \int_{x'=-d}^{d} \left[\frac{\rho_s \, a \phi}{4\pi\varepsilon_0 \sqrt{a^2 + (x')^2}} \right]_{\phi=0}^{2\pi} dx' \qquad \text{...(6.8)}$$

$$= \frac{\rho_s a}{2\,\varepsilon_0} \int_{x'=-d}^{d} \frac{dx'}{\sqrt{a^2 + (x')^2}} \qquad \text{...(6.9)}$$

$$= \frac{\rho_s \, a}{2\varepsilon_0} \ln \left[x' + \sqrt{a^2 + (x')^2} \right]_{x'=-d}^{d} \qquad \text{...(6.10)}$$

$$= \frac{\rho_s \, a}{2\varepsilon_0} \ln \frac{d + \sqrt{a^2 + d^2}}{-d + \sqrt{a^2 + d^2}} \qquad \text{[As } a << d\text{]} \qquad \text{...(6.11)}$$

$$= \frac{\rho_s \, a}{2\varepsilon_0} \ln \frac{2d}{-d + \left(a^2 + d^2\right)^{1/2}}$$

$$= \frac{\rho_s \, a}{2\varepsilon_0} \ln \frac{2d}{-d + d\left(1 + \dfrac{a^2}{d^2}\right)^{1/2}}, \qquad \text{...(6.12)}$$

using the formula

$$\left(1 + \frac{a^2}{d^2}\right)^{1/2} = 1 + \frac{a^2}{2d^2} + \left(\frac{a^2}{3d^2}\right)^2 + \text{.........}.$$

By neglecting the higher order terms eqn. (6.12) becomes

$$V_0 = \frac{\rho_s \, a}{2\varepsilon_0} \ln \frac{2d}{-d + d\left(1 + \dfrac{a^2}{2d^2}\right)} \qquad \text{...(6.13)}$$

$$V_0 = \frac{\rho_s a}{2\varepsilon_0} \ln \frac{2d}{a^2/2d}$$

$$= \frac{\rho_s a}{2\varepsilon_0} \ln \left(\frac{2d}{a}\right)^2$$

$$= \frac{\rho_s a}{2\varepsilon_0} \ln \frac{2d}{a}. \qquad ...(6.14)$$

To find the potential (for the second case) at P (0, 0, x), where $x > d$,

$$V = \int_{x=-d}^{d} \int_{\phi=0}^{2\pi} \frac{\rho_s a\, d\phi\, dx'}{4\pi\varepsilon_0 (x - x')} \qquad ...(6.15)$$

$$= \int_{x=-d}^{d} \frac{\rho_s a\, dx'}{2\varepsilon_0 (x - x')}$$

$$= \frac{\rho_s a}{2\varepsilon_0} \int_{x=-d}^{d} \frac{dx'}{x - x'}$$

$$= \frac{\rho_s a}{2\varepsilon_0} \left[-\ln(x - x')\right]_{x=-d}^{d}$$

$$= \frac{\rho_s a}{2\varepsilon_0} \ln \frac{x+d}{x-d}. \qquad ...(6.16)$$

Applying eqn. (6.12) and eqn. (6.16) in writing the equation for potential at the centre of the segment 1.

(1) At the centre of segment 1, due to the same segment the potential is obtained by eqn. (6.12) as

$$\frac{\rho_{s_1} a}{\varepsilon_0} \ln \frac{l}{5a}, \quad \text{since} \quad 2d = \frac{l}{5}. \qquad ...(6.17)$$

(2) At the centre of segment s, due to segment 2, the potential is obtained by eqn. (6.16) as

$$\frac{\rho_{s_2} a}{2\varepsilon_0} \ln \left[\frac{l/5 + l/10}{l/5 - l/10}\right], \quad \text{since} \quad \begin{array}{l} x = l/5 \\ d = l/10 \end{array}$$

$$\frac{\rho_{s_2} a}{2\varepsilon_0} \ln 3. \qquad ...(6.18)$$

Similarly, the potential at the centre of segment 1 due to other segments (3, 4 and 5) is obtained using the eqn. (6.16) as

$$\frac{\rho_{s_3} a}{2\varepsilon_0} \ln \left[\frac{\frac{2l}{5} + \frac{l}{10}}{\frac{2l}{5} - \frac{l}{10}} \right]$$

$$= \frac{\rho_{s_3} a}{2\varepsilon_0} \ln \frac{5}{3} \qquad \text{...(6.19)}$$

$$\frac{\rho_{s_4} a}{2\varepsilon_0} \ln \left[\frac{3l/5 + l/10}{3l/5 - l/10} \right]$$

$$= \frac{\rho_{s_4} a}{2\varepsilon_0} \ln \frac{7}{5} \qquad \text{...(6.20)}$$

$$\frac{\rho_{s_5} a}{2\varepsilon_0} \ln \left[\frac{4l/5 + l/10}{4l/5 - l/10} \right]$$

$$= \frac{\rho_{s_5} a}{2\varepsilon_0} \ln \frac{9}{7}. \qquad \text{...(6.21)}$$

Adding eqn. (6.17) to eqn. (6.21) and equating the potential to 1V,

$$\frac{\rho_{s_1} a}{\varepsilon_0} \ln \frac{l}{5a} + \frac{\rho_{s_2} a}{2\varepsilon_0} \ln 3 + \frac{\rho_{s_3} a}{2\varepsilon_0} \ln \frac{5}{3} + \frac{\rho_{s_4} a}{2\varepsilon_0} \ln \frac{7}{5} + \frac{\rho_{s_5} a}{2\varepsilon_0} \ln \frac{9}{7} = 1. \qquad \text{...(6.22)}$$

Using the symmetry $\rho_{s_5} = \rho_{s_1}$ and $\rho_{s_4} = \rho_{s_2}$

$$\rho_{s_5} \left[2 \ln \frac{l}{5a} + \ln \frac{9}{7} \right] + \rho_{s_2} (\ln 3 + \ln 1.4) + \rho_{s_3} \ln \frac{5}{3} = \frac{2\varepsilon_0}{a}. \qquad \text{...(6.23)}$$

Similarly, writing the equations for the potentials at the centre points of segments 2 and 3 and arranging the three equations in matrix form, we get

$$\begin{bmatrix} 2 \ln \frac{l}{5a} + \ln \frac{9}{7} & \ln 3 + \ln 1.4 & \frac{5}{3} \\ \ln 3 + \ln 1.4 & 2 \ln \frac{l}{5a} + \ln \frac{5}{3} & \ln 3 \\ 2 \ln \frac{5}{3} & 2 \ln 3 & 2 \ln \frac{l}{5a} \end{bmatrix} \cdot \begin{bmatrix} \rho_{s_1} \\ \rho_{s_2} \\ \rho_{s_3} \end{bmatrix} = \begin{bmatrix} \frac{2\varepsilon_0}{a} \\ \frac{2\varepsilon_0}{a} \\ \frac{2\varepsilon_0}{a} \end{bmatrix}. \qquad \text{...(6.24)}$$

Using eqn. (6.24), the solutions for ρ_{s_1}, ρ_{s_2} and ρ_{s_3} can be obtained.

Example 6.4

Use the moment method to find the capacitance of the parallel plate capacitor of Fig. 6.5. Assume the potential in upper plate is at 1V and the lower plate is at the potential of –1V.

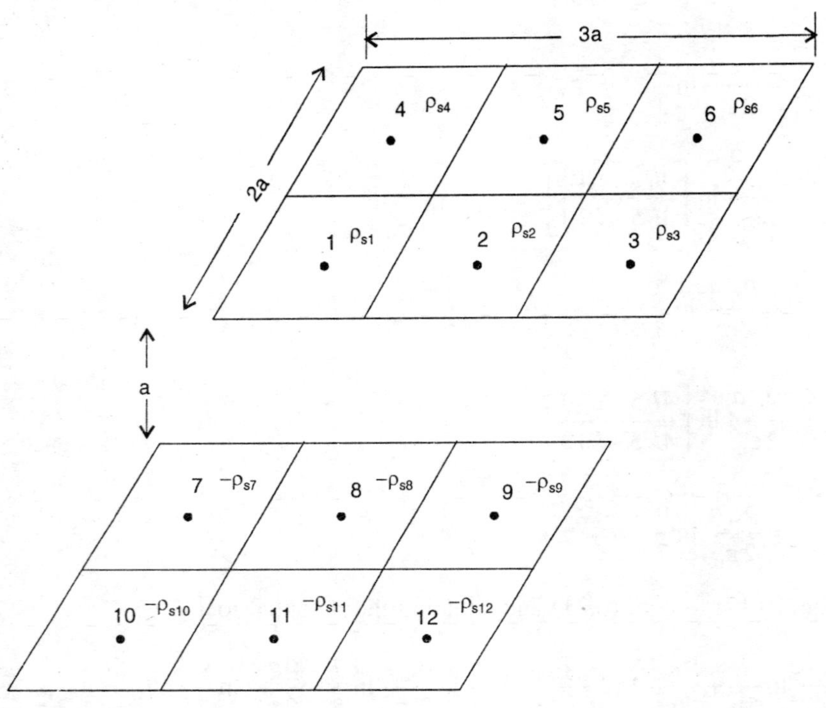

Fig. 6.5

Divide each plate into 2×3 set of squares as in Fig 6.5 and assume that each square consists of uniformly-distributed surface charge density. Since moment method is used, by symmetry, it is enough to find only two unknown charge densities ρ_{s_1} and ρ_{s_2}. So it is sufficient to write two independent equations.

To find ρ_{s_1} and ρ_{s_2}, consider the squares 1 and 2 and equate the potentials at the centre points of these squares to 1V.

We know the potential at a point due to a charge Q is $V = \dfrac{Q}{4\pi\varepsilon_0 r}$, where r is the distance between the point and the charge.

So, the sum of potentials at a point 1 due to the charge densities ρ_{s_1} in squares 3, 4, 6 are

$$\frac{\rho_{s_1} a^2}{4\pi\varepsilon_0 2a} + \frac{\rho_{s_1} a^2}{4\pi\varepsilon_0 a} + \frac{\rho_{s_1} a^2}{4\pi\varepsilon_0 \sqrt{a^2 + (2a)^2}}. \qquad ...(6.25)$$

$$\left[Q = \rho_{s_1} a^2\right]$$

Similarly, the sum of potentials at the point 1 due to the charge densities $-\rho_{s_1}$ in squares 7, 9, 10 and 12 are

$$\frac{-\rho_{s_1} a^2}{4\pi\varepsilon_0 a} - \frac{\rho_{s_1} a^2}{4\pi\varepsilon_0\sqrt{(2a)^2 + a^2}} - \frac{\rho_{s_1} a^2}{4\pi\varepsilon_0\sqrt{a^2 + a^2}} - \frac{\rho_{s_1} a^2}{4\pi\varepsilon_0\left[\left(\sqrt{2a}\right)^2 + (2a)^2\right]}. \quad ...(6.26)$$

Now by considering the squares with charge densities ρ_{s_2}, the sum of potential at 1 is

$$\frac{\rho_{s_2} a^2}{4\pi\varepsilon_0}\left[\frac{1}{a} + \frac{1}{\sqrt{2}a} - \frac{1}{\sqrt{2}a} - \frac{1}{\sqrt{3}a}\right]. \quad ...(6.27)$$

Finally, the potential at 1 due to the same square 1 is calculated as follows.

Divide the square into many circular rings of various radii (ρ). Consider a single elemental surface ds $(\rho d\rho d\phi)$ with charge of $dQ = \rho_{s_1} \cdot \rho\, d\rho\, d\phi$. The potential at 1 due to this charge dQ from the Fig. 6.6 is

$$dV = \frac{dQ}{4\pi\varepsilon_0 r} = \frac{\rho_{s_1} \cdot \rho d\rho d\phi}{4\pi\varepsilon_0 \cdot \rho}.$$

The total potential at 1 is

$$\int_{\rho=0}^{a/2}\int_{\phi=0}^{2\pi} \frac{\rho_{s_1}\, \rho\, d\rho d\phi}{4\pi\varepsilon_0\, \rho} = \frac{\rho_{s_1} a}{4\varepsilon_0}. \quad ...(6.28)$$

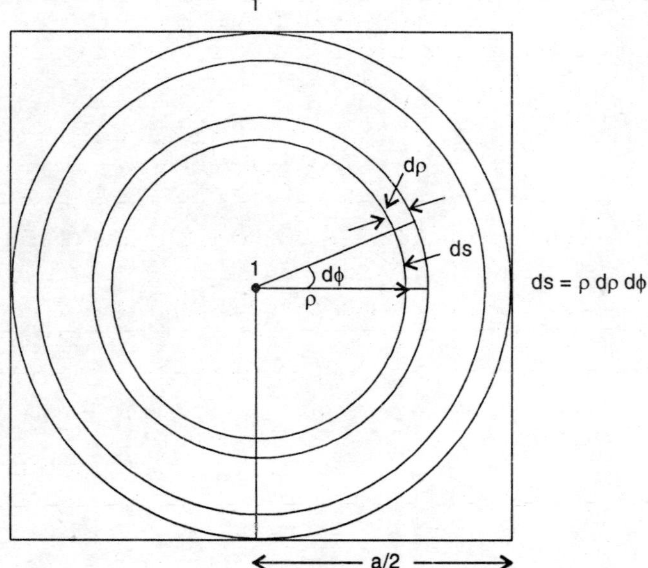

Fig. 6.6

The total potential at point 1 is summing all eqns. (6.26) to (6.28) which is equal to 1V.

$$\frac{\rho_{s_1} a^2}{4\pi\varepsilon_0}\left[\frac{1}{2a}+\frac{1}{a}+\frac{1}{\sqrt{5}a}-\frac{1}{a}-\frac{1}{\sqrt{5}a}-\frac{1}{\sqrt{2}a}-\frac{1}{\sqrt{6}a}\right]$$

$$+\frac{\rho_{s_2} a^2}{4\pi\varepsilon_0}\left[\frac{1}{a}+\frac{1}{\sqrt{2}a}-\frac{1}{\sqrt{2}a}-\frac{1}{\sqrt{3}a}\right]+\frac{\rho_{s_1} a}{4\varepsilon_0}=1. \quad ...(6.29)$$

Similarly, by finding the total potential at point 2 and equating it to 1V, the sum of potentials due to all the charge densities in various squares is

$$\frac{\rho_{s_1} a^2}{4\pi\varepsilon_0}\left[\frac{2}{a}+\frac{2}{\sqrt{2}\,a}-\frac{2}{\sqrt{2}\,a}-\frac{2}{\sqrt{3}\,a}\right]+\frac{\rho_{s_2} a^2}{4\pi\varepsilon_0}\left[\frac{1}{a}-\frac{1}{a}-\frac{1}{\sqrt{2}\,a}\right]+\frac{\rho_{s_r} a}{4\varepsilon_0}=1. \quad ...(6.30)$$

By simplifying eqns. (6.29) and (6.30),

$$0.2010\,\rho_{s_1}+0.336\,\rho_{s_2}=\frac{\varepsilon_0}{a} \quad ...(6.31)$$

$$0.0672\,\rho_{s_1}+0.19\,\rho_{s_2}=\frac{\varepsilon_0}{a}. \quad ...(6.32)$$

After solving for ρ_{s_1} and ρ_{s_2}, the charge on either plate is $(4a^2 \cdot \rho_{s_1} + 2a^2\rho_{s_2})$. As the potential difference between the plates is 2V, the capacitance is

$$C=Q/V \quad C=\left(4a^2\,\rho_{s_1}+2a^2\rho_{s_2}\right)/2\,. \quad ....(6.33)$$

By dividing the each plate into a larger number of squares, the result will be more accurate.

7

Miscellaneous Solved Problems

1. Determine the divergence of the given vector field at specified points.

$$\overline{B} = \rho Z \sin\phi \, \vec{a}_\rho + 3\rho Z^2 \cos\phi \, \vec{a}_\phi \text{ at } \left(5, \frac{\pi}{2}, 1\right)$$

$$\nabla \cdot \overline{B} = \frac{1/\rho}{\partial/\partial\rho}(B_\rho) + \frac{1/\rho}{\partial/\partial\phi}(B_\phi) + \frac{\partial}{\partial z}(B_z)$$

$$= \frac{1/\rho}{\partial/\partial\rho}(\rho \cdot \rho_z \sin\phi) + \frac{1/\rho}{\partial/\partial\phi}\left(3\rho z^2 \cos\phi\right)$$

$$= \frac{1}{\rho}(2\rho z \sin\phi) + \frac{1}{\rho}\left(-3\rho z^2 \sin\phi\right)$$

$$= 2z \sin\phi - 3z^2 \sin\phi$$

$$= 2\sin\frac{\pi}{2} - 3\sin\frac{\pi}{2} = -1.$$

2. Determine the curl of the vector field at specified points.

$$\vec{C} = r^2 \cot\theta \, \vec{a}_r + 2\sin\theta\cos\phi \, \vec{a}_\phi \text{ at } \left(2, \frac{\pi}{6}, \frac{\pi}{4}\right)$$

$$\nabla \times \vec{C} = \frac{1}{r^2 \sin\theta} \begin{vmatrix} \vec{a}_r & r\vec{a}_\theta & \vec{a}_\phi \\ \dfrac{\partial}{\partial r} & \dfrac{\partial}{\partial \theta} & \dfrac{\partial}{\partial \phi} \\ r^2 \cot\theta & 0 & 2\sin\theta\cos\phi \end{vmatrix}$$

$$\left[\vec{a}_r \frac{\partial}{\partial\theta}(2\sin\theta\cos\phi) - r\vec{a}_\theta\left(\frac{\partial}{\partial r}2\sin\theta\cos\phi - \frac{\partial}{\partial\phi} r\cot\theta\right) + \vec{a}_\phi\left(\frac{-\partial}{\partial\theta} r^2 \cot\theta\right)\right]\frac{1}{r^2 \sin\theta}$$

$$= \frac{1}{r^2 \sin\theta}(2\cos\theta\cos\phi)\vec{a}_r - \frac{1}{r^2 \sin\theta}(0)\vec{a}_\phi + \frac{1}{r^2 \sin\theta} r^2 \text{cosec}^2\theta\, \vec{a}_\phi.$$

$$= \frac{2}{r^2}\cot\theta\cos\phi\,\vec{a}_r + \text{cosec}^3\theta\,\vec{a}_\phi$$

$$= \frac{2}{4}\cot\frac{\pi}{6}\cos\frac{\pi}{4}\vec{a}_r + \text{cosec}^3\frac{\pi}{6}\vec{a}_\phi$$

$$= \frac{2}{4}\sqrt{3}\frac{1}{\sqrt{2}}\vec{a}_r + \left(\frac{2}{\sqrt{3}}\right)^3 \vec{a}_\phi$$

$$= \frac{\sqrt{3}}{2\sqrt{2}}\vec{a}_r + \frac{8}{3\sqrt{3}}\vec{a}_\phi.$$

3. Determine the flux $\vec{D} = \rho^2\cos^2\phi\,\vec{a}_\rho + z\sin\phi\,\vec{a}_\phi$ over the closed surface if the cylinder $0 \le z \le 1$, $\rho = 4$. Also verify the divergence theorem.

$$\oint\!\!\oint_S \vec{D}\cdot d\vec{s} = \iiint_V \nabla\cdot\vec{D}\,dv$$

$$\int_{z=0}^{1}\int_{\phi}^{2\pi}\left(\rho^2\cos^2\phi\,\vec{a}_\rho + z\sin\phi\,\vec{a}_\phi\right)\cdot\rho\,d\phi\,dz\,\vec{a}_\rho$$

$$\int_{z=0}^{1}\int_{\phi=0}^{2\pi}\rho^2\cdot\cos^2\phi\,\rho\,d\phi\,dz$$

$$= \int_{z=0}^{1}\int_{\phi=0}^{2\pi}\rho^2\cdot\cos^2\phi\,d\phi\,dz$$

$$= \int_{z=0}^{1}\int_{\phi=0}^{2\pi}\frac{(1+\cos 2\phi)}{2}d\phi\,dz$$

$$= \int_{z=0}^{z=1}\left[\int_{\phi=0}^{2\pi}32\,d\phi + \int_{\phi=0}^{2\pi}32\cos 2\phi\,d\phi\right]dz$$

$$= \int_{z=0}^{z=1}\left(32\times 2\pi + 32\frac{\sin 2\phi}{2}\right)_{\phi=0}^{2\pi}dz$$

$$= 64\pi$$

$$\nabla \cdot \vec{D} = \frac{1}{\rho}\frac{\partial}{\partial \rho}(\rho D\rho) + \frac{1/\rho}{\partial/\partial\phi}(D_\phi)$$

$$= \frac{1}{\rho}\frac{\partial}{\partial \rho}\left(\rho \cdot \rho^2 \cos^2\phi\right) + \frac{1}{\rho}\frac{\partial}{\partial \phi}(z \sin\phi)$$

$$= \frac{1}{\rho}\left(3\rho^2 \cos^2\phi\right) + \frac{1}{\rho}(z\cos\phi)$$

$$\iiint_V \nabla \cdot \vec{D}\, dv = \int_{z=0}^{1}\int_{\phi=0}^{2\pi}\int_{\rho=0}^{4} 3\rho^2 \cos^2\phi \; d\rho\, d\phi\, dz + \rho \cdot \rho^2 \cos^2\phi + \int_{z=0}^{1}\int_{\phi=0}^{2\pi}\int_{\rho=0}^{4} z\rho \cos\phi \; d\rho\, d\phi\, dz$$

$$= \int_{z=0}^{1}\int_{\phi=0}^{2\pi}\left(\frac{3\rho^3}{3}\right)_0^4 \cos^2\phi\, d\phi\, dz$$

$$= \int_{z=0}^{1}\int_{\phi=0}^{2\pi} 64\left(\frac{1+\cos 2\phi}{2}\right) d\phi\, dz$$

$$= \int_{z=0}^{1}\int_{\phi=0}^{2\pi} 32\, d\phi\, dz + \int_{z=0}^{1}\int_{\phi=0}^{2\pi} 64\frac{\cos 2\phi}{2}\, d\phi\, dz$$

$$= 64\pi$$

Flux = 64π wb.

4. Verify the divergence theorem for the following case. $A = xy^2\,\vec{a}_x + y^3\vec{a}_y + y^2 z\,\vec{a}_z$ and the surface is a cuboid defined by $0 < x < 1$, $0 < y < 1$ and $0 < z < 1$.

$$\oiint \vec{A} \cdot d\vec{s} = \iiint_V \nabla \cdot \vec{A}\, dv$$

$$\iint \left(xy^2\vec{a}_x + y^3\vec{a}_y + y^2 z\vec{a}_z\right) \cdot \left(dy\,dz\,\vec{a}_x + dx\,dz\,\vec{a}_y + dx\,dy\,\vec{d}_z\right)$$

$$= \int_{z=0}^{1}\int_{y=0}^{1} xy^2 dy\,dz \int_{y=0}^{z=1}\int_{x=0}^{=1} y^3 dx\,dz + \int_{y=0}^{1}\int_{x=0}^{1} y^2 z\, dx\,dy$$

$$= \left(\frac{xy^3}{3}z + xy^3 z + \frac{xy^3}{3}z\right)_0^1$$

$$= \frac{5}{3}.$$

$$\nabla \cdot \vec{A} = \frac{\partial}{\partial x}\vec{A}_x + \frac{\partial}{\partial y}\vec{A}_y + \frac{\partial}{\partial z}\vec{A}_z$$

$$= y^2 + 3y^2 + y^2 = 5y^2$$

$$= \int_{z=0}^{1}\int_{y=0}^{1}\int_{x=0}^{1} 5y^2\, dx\, dy\, dz = \left.\frac{5xy^3}{3}z\right)_0^1 = \frac{5}{3}.$$

Hence proved.

5. Given that $\vec{F} = x^2 y\vec{a}_x - \vec{a}_y$. Find

(*a*) $\oint_L \vec{F} \cdot d\vec{l}$ where L is shown in Fig. M.5.

(*b*) $\iint_S \left(\nabla \times \vec{F}\right) \cdot d\vec{s}$ where s is the area bounded by L.

(*c*) Whether Stoke's theorem is satisfied.

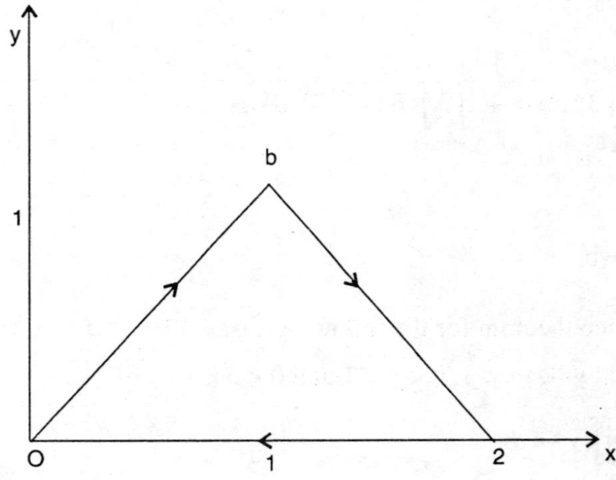

Fig. M.5 (a)

$$\oint_L \vec{F} \cdot d\vec{l} = \int_a^b + \int_b^c + \int_c^a$$

$$\int_a^b \vec{F} \cdot d\vec{l} = \int_{\substack{x=0\\y=0}}^{\substack{x=1\\y=1}} \left(x^2 y\vec{a}_x - y\vec{a}_y\right) \cdot \left(dx\,\vec{a}_x + dy\,\vec{a}_y\right)$$

$$= \int_{x=0}^{1} x^2 y\, dx - \int_{y=0}^{1} y\, dy.$$

The straight line $ab \to x = y$

$$= \int_{x=0}^{1} x^3 dx - \int_{y=0}^{1} y\, dy = \frac{-1}{4}.$$

$$= \int_{b}^{c} \vec{F} \cdot d\vec{l} = \int_{\substack{x=1\\y=1}}^{\substack{y=0\\x=2}} \left(x^2 y \vec{a}_x - y \vec{a}_y\right) \cdot \left(dx\, \vec{a}_x + dy\, \vec{a}_y\right)$$

$$= \int_{x=1}^{x=2} x^2 y\, dx - \int_{y=1}^{y=0} y\, dy.$$

The straight line equation for $bc \to x + y = 2$

$$= \int_{x=1}^{2} x^2 (2 - x)\, dx - \int_{y=1}^{0} y\, dy$$

$$= \frac{14}{3} - \frac{15}{4} + \frac{1}{2}$$

$$= \frac{56 - 45 + 6}{12} = \frac{17}{12}.$$

$$\int_{c}^{a} \vec{F} \cdot d\vec{l} = \int_{x=2}^{x=0} \left(x^2 y \vec{a}_x - y \vec{a}_y\right) \cdot dx\, \vec{a}_x$$

$$= \int_{x=2}^{x=0} x^2 y\, dx$$

$$= 0, \qquad [\because y = 0].$$

$$\int_{l} \vec{F} \cdot d\vec{l} = \frac{-1}{4} - \frac{17}{12} = \frac{-3 + 17}{12} = \frac{14}{12} = \frac{7}{6}.$$

(*b*) $\iint_{s} \left(\nabla \times \vec{F}\right) \cdot ds$

$$\nabla \times \vec{F} = \begin{vmatrix} \vec{a}_x & \vec{a}_y & \vec{a}_z \\ \dfrac{\partial}{\partial x} & \dfrac{\partial}{\partial y} & \dfrac{\partial}{\partial z} \\ x^2 y & -y & 0 \end{vmatrix} \quad -\vec{a}_z (x)^2$$

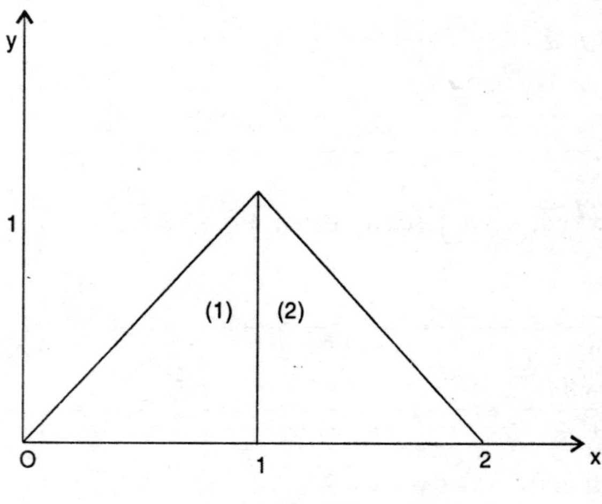

Fig. M-5 (b)

As the direction of area covered in clockwise, the surface is $ds = dx\,dy(-\vec{a}_z)$.

$$\iint_S \nabla \times \vec{F} \cdot ds = \iint x^2 (-\vec{a}_z) \cdot dx\,dy(-\vec{a}_z)$$

$$= \int_{x=0}^{1} \int_{y=0}^{x} x^2\,dy\,dx + \int_{x=1}^{2} \int_{y=2-x}^{0} x^2\,dy\,dx$$

$$= \int_{x=0}^{1} x^2 (y)_0^x\,dx + \int_{1}^{2} x^2 (y)_{2-x}^{0}\,dy\,dx$$

$$= \left(\frac{x^4}{4}\right)_0^1 + \left(2\frac{x^3}{3} - \frac{x^4}{4}\right)^2$$

$$= \frac{1}{4} + \frac{16}{3} - \frac{2}{3} - \frac{16}{4} + \frac{1}{4}$$

$$= \frac{7}{6}.$$

(*c*) Yes, Stokes theorem is satisfied.

6. Two concentric perfectly conducting spheres of radius a and b contains charges $+Q$ and $-Q$ respectively. The region between the spheres is a field with dielectric of permittivity ε. Find the total energy contained in the system.

As per Gauss's law

Electric field intensity

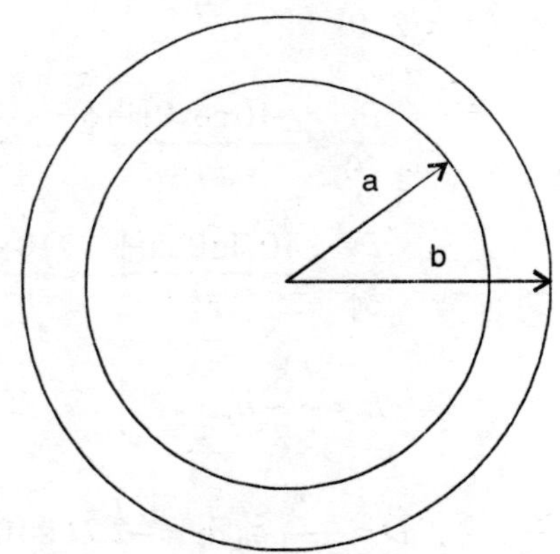

Fig. M.6

$$\vec{E} = \begin{cases} 0 & r < a \\ \dfrac{Q}{4\pi \in_0 r^2}\, \vec{a}_r & a < r < b \\ 0 & r < b \end{cases}$$

Energy in the system $W = \dfrac{1}{2}\dfrac{1}{2}\int_V \varepsilon_0 E^2\, dv$

$$= \frac{1}{2}\int_{\phi=0}^{2\pi}\int_{\theta=0}^{\pi}\int_{r=a}^{b} \in_0 \left(\frac{Q}{4\pi \in_0 r^2}\right) \cdot r^2 \sin\theta\, dr\, d\theta\, d\phi$$

$$= \frac{1}{2}\iiint \frac{Q^2}{(4\pi)^2 \in_0 r^2} \sin\theta\, dr\, d\theta\, d\phi$$

$$= \frac{Q^2}{8\pi^2 \in_0}\iint\int_{r=a}^{b} \frac{dr}{r^2} \sin\theta\, d\theta\, d\phi$$

$$= \frac{Q^2}{8\pi^2 \in_0}\iint\left[\frac{-1}{r}\right]_a^b \sin\theta\, d\theta\, d\phi$$

$$= \frac{Q^2}{8\pi^2 \in_0}\cdot\left[\frac{1}{a} - \frac{1}{6}\right]\int_{b=\phi=0}^{\pi}\int_{\theta=0}^{\pi} \sin\theta\, d\theta\, d\phi$$

$$= \frac{Q^2}{2\pi \in_0}\left(\frac{b-a}{ab}\right) \text{ joule.}$$

7. Given the potential $v = \dfrac{10\sin\theta\cos\phi}{r^2}$, find the (*a*) Electric flux density D at $\left(2, \dfrac{\pi}{2}, 0\right)$.

$\vec{E} = -\nabla V.$

∇V in spherical co-ordinate is

$$\frac{\partial v}{\partial r}\vec{a}_r + \frac{1}{r}\frac{\partial v}{\partial \theta}\vec{a}_\theta + \frac{1}{r\sin\theta}\frac{\partial v}{\partial \phi}\vec{a}_\phi$$

$$\left.\frac{\partial v}{\partial r}\right|_{(2,\,\pi/2,\,0)} = \frac{-20\sin\theta\cos\phi}{r^3} \rightarrow \frac{-20\sin\pi/2\cos\theta}{2^3} = \frac{-20}{8}$$

$$\frac{\partial v}{\partial \theta} = \frac{-10\cos\theta\sin\phi}{r^2} \rightarrow \frac{-10\cos\pi/2}{2^2} = 0$$

$$\frac{\partial v}{\partial \phi} = \frac{10\sin\theta\sin\phi}{r^2} \rightarrow \frac{10\sin\pi/2\sin\theta}{2^2} = 0$$

$$\vec{E} = -\frac{5}{2}\,\vec{a}_r$$

$$\vec{D} = -\frac{5}{2}\,\epsilon_0\,\vec{a}_r = -2.21\times10^{-11}\ \text{wb/m}^2.$$

8. Calculate the work done in moving a 10 μC (*b*) charge from point *A* (1, 30°, 120°) to (4, 90°, 60°) use spherical (r, θ, ϕ) co-ordinate system.
   As V is known.

$$W = -Q\int_A^B \vec{E}\cdot d\vec{l} = QV_{AB}$$

$$= Q(V_B - V_A)$$

$$V_A = \left.\frac{10\sin\theta\cos\phi}{r^2}\right|(1,\,30°,\,120°) = 10\cdot\frac{1}{2}\left(\frac{-1}{2}\right) = \frac{-5}{2}$$

$$V_B = \left.\frac{10\sin\theta\cos\phi}{r^2}\right|(4,\,90°,\,60°) = \frac{10\cdot(1)\cdot\left(\frac{1}{2}\right)}{16} = \frac{5}{16}$$

$$W = 10\times10^{-6}\left(\frac{5}{16}+\frac{5}{2}\right)$$

$$= 28.125\,\mu\text{J}.$$

9. Two electrodes at potentials $V = V_0$ and $V = 0$ volt are separated by a distance d along Z-axis. The region between the electrodes contains a uniform charge density ρ_0, which is generated by electrode at V_0 and collected at electrode at zero volts. Calculate the electric field at any point z. What is the direction of the field?
   Since $\rho_v \neq 0$, apply Poisson's equation

$$\nabla^2 V = \frac{-\rho_0}{\varepsilon}.$$

The boundary conditions $V(z=0)=V_0$ and $V(z=d)=0$. V depends only on z.

$$\frac{\partial^2 V}{\partial z^2}=-\frac{\rho_0}{\varepsilon}.$$

Integrating once,

$$\frac{\partial V}{\partial z}=\frac{-\rho_0 z}{\varepsilon}+A.$$

Integrating again,

$$V=\frac{-\rho_0 z^2}{2\varepsilon}+Az+B. \qquad \text{...(1)}$$

Applying boundary conditions in (1)

when $z=0$, $V=V_0$

$$\boxed{V_0=B}$$

when $z=d$, $V=O$

$$O=\frac{-\rho_0 d^2}{2\varepsilon}+Ad+V_0.$$

$$\boxed{A=\frac{\rho_0 d}{2\varepsilon}-\frac{V_0}{d}}.$$

Electric potential

$$V=\frac{-\rho_0 Z^2}{2\varepsilon}+\frac{\rho_0 dZ}{2\varepsilon}-\frac{V_0 Z}{d}+V_0.$$

Electric field intensity E is

$$\vec{E}=-\nabla V=-\frac{dV}{dz}\vec{a}_z$$

$$\vec{E}=\left[\frac{V_0}{d}+\frac{\rho_0}{\varepsilon}\left(z-\frac{d}{2}\right)\right]\vec{a}_z.$$

Direction of the field along Z-axis.

To find net force

$$\vec{F}=\iiint \rho_v E\,dv \left[\because \; F=Eq \text{ and } q=\iiint_v \rho_v dv\right]$$

$$=\rho_0\iint ds\int_{z=0}^{d} E\,dz$$

$$= \rho_0 \rho \left[\frac{V_0 z}{d} + \frac{\rho_0}{2 \in} \left(z^2 - d \cdot z \right) \right]_{z=0}^{d} d\vec{a}_z$$

$$= \rho_0 \rho \left[V_0 + \frac{\rho_0}{2 \in} \left(d^2 - d^2 \right) \right] \vec{a}_z$$

$$\vec{F} = \rho_0 S V_0 \vec{a}_z .$$

The pressure in force per unit area

$$P = \frac{F}{S} = \rho_0 V_0 \vec{a}_z .$$

10. Two infinitely long, concentric conducting cylinders of radix r_a and r_b are located on the Z-axis. If ε_r in between two cylinders, $V = V_0$ at r_a and $V = 0$ at r_b. Formulate the problem and calculate the voltage and E, C per length using direct integration method.

As V depends on radius (ρ), in cylindrical coordinate, the Laplace's equation

$$\nabla^2 v = \frac{1}{\rho} \frac{d}{d\rho} \left(\rho \frac{\partial v}{\partial y} \right) = 0$$

$$\frac{d^2 v}{d\rho^2} = \frac{1}{\rho} \frac{d}{d\rho} \left(\rho \frac{dv}{d\rho} \right) = 0$$

$$\frac{d}{d\rho} \left(\rho \frac{dv}{d\rho} \right) = 0$$

$$\rho \frac{dv}{d\rho} = A$$

$$\frac{dv}{d\rho} = \frac{A}{\rho}$$

$$\boxed{V = A \ln \rho + B} \cdot \quad \text{...(1)}$$

Applying boundary conditions in (1),

$$V = V_0 \qquad \rho = r_a$$
$$V = O \qquad \rho = r_b$$

$$V_0 = A \ln r_a + B$$

$$O = A \ln r_b + B.$$

Solving

$$A = \frac{V_0}{\ln \frac{r_a}{r_b}} \quad B = \frac{-V_0}{\ln\left(\frac{r_a}{r_b}\right)} \ln(r_b).$$

So,

$$V = \frac{V_0 \ln \rho}{\ln \frac{r_a}{r_b}} - \frac{V_0}{\ln \frac{r_a}{r_b}} \cdot \ln r_b$$

$$V = \frac{V_0 \cdot \ln \frac{\rho}{r_b}}{\ln \frac{r_a}{r_b}}.$$

To find $\vec{E}$

$$\vec{E} = -\nabla V = \frac{-dv}{d\rho} \cdot \vec{a}_\rho$$

$$\vec{E} = \frac{-V_0}{\ln \frac{r_a}{r_b}} \left(\frac{1}{\rho/r_b} \cdot \frac{1}{r_b} \right) \vec{a}_\rho$$

$$\boxed{\vec{E} = \frac{-V_0}{\rho \ln \frac{r_a}{r_b}} \vec{a}_\rho}.$$

To find Q

$$Q = \iint \varepsilon \vec{E} \cdot d\vec{s}$$

$$= \int_{z=0}^{\ell} \int_{\phi=0}^{2\pi} \frac{-v_0}{\rho \ln \frac{r_a}{r_b}} \vec{a}_\rho \cdot \rho \, d\phi \, dz \left(-\vec{a}_\rho\right)$$ [Assume length of cylinder as 'l' m]

$$Q = \frac{v_0 \, 2\pi\varepsilon l}{\ln \frac{r_a}{r_b}} \text{ coulomb}$$

$$C = \frac{Q}{v_0} = \frac{2\pi\varepsilon l}{\ln \frac{r_a}{r_b}} \text{ F.}$$

$$\text{C/m. length} = \frac{2\pi\varepsilon}{\ln\frac{r_a}{r_b}} \text{ F/m.}$$

11. A 500 MHz electromagnetic wave is propagating through a perfect non-magnetic dielectric with relative permittivity $\in_r = 6$.
   (1) Calculate the wavelength and phase constant.
   (2) With the wave travelling in $+z$ direction, the sinusoidal electric field at $z = 80$ cm in delayed relative to the field at $z = 65$ cm. Calculate the time delay in nano second and the phase delay in degree.
   (3) Calculate the average power density in the wave if the peak value of magnetic field $H = 0.5$ A/m.

(1) Given $f = 500$ MHz

$$\in_r = 6,\ \mu_r = 1.$$

$$\text{Velocity of wave} = \frac{1}{\sqrt{\mu \in}} = \frac{3\times10^8}{\sqrt{\mu_r \in_r}} \text{ m/s}$$

$$\mu = \frac{3\times10^8}{\sqrt{6}} = 1.22\times10^8 \text{ m/s.}$$

$$\text{Phase constant } \beta = \frac{\omega}{\mu} = \frac{2\times\pi\times500\times10^6}{1.22\times10^8}$$

$$= 25.75 \text{ rad/m.}$$

$$\text{Wavelength } \lambda = \frac{2\pi}{\beta} = \frac{u}{f}$$

$$= 0.244 \text{ m.}$$

(2) $|\vec{E}| = E_0 \sin(\omega t + \beta z)$

$$\sin(\omega t + \beta z) = \sin(314\times10^7 t + 25.75\, z).$$

When $z = 80$ cm $= 0.8$ m

$$\sin(314\times10^7 t + 20.6 \text{ rad})$$

$$20.6 \text{ rad} \Rightarrow 1180°.$$

When $z = 65$ cm $= 0.65$ m

$$\sin(314\times10^7 t + 16.74 \text{ rad})$$

$$16.74 \text{ rad} \Rightarrow 959°$$

Phase delay in degree = 1180° – 959° = 221°.

Time t at $z = 0.8$ m or $\lambda = 0.8$ m

$$t_1 = \frac{0.8}{1.22 \times 10^8}\left(\frac{\lambda}{u} = t\right)$$

t at $z = 0.65$ m ($\lambda = 0.65$ m)

$$t_2 = \frac{0.65}{1.22 \times 10^8}$$

$= 5.32$ ns.

Time delay = 1.18 ns.

(3) Average power density $|P_{ave}| = EH$

$$E = \eta H$$

$$P_{ave} = \eta H^2$$

$$\eta = \sqrt{\frac{\mu}{\in}} = \sqrt{\frac{\mu_0 \in_r}{\in_0 \in_r}} = 120\pi\sqrt{\frac{1}{6}}$$

$$\eta = 154.$$

$$P_{ave} = 154 \times 0.5^2 = 38.5 \text{ W/m}^2.$$

12. A plane electromagnetic wave having a frequency of 10 MHz has an average pointing vector of 1 W/m^2. If the medium is lossless with $\in_r = 3$ and $\mu_r = 2$, find the velocity of propagation, wave length, impedance of the medium and RMS value of electric field.

Given $f = 10$ MHz

$$P = 1 \text{ W/m}^2.\ \in_r = 3,\ \mu = 2.$$

Velocity of wave $\mu = \dfrac{1}{\sqrt{\mu \in}}$

$$= \frac{1}{\sqrt{\mu_0 \in_0 \mu_r \in_r}}$$

$$\mu = \frac{3 \times 10^8}{\sqrt{6}} = 1.22 \times 10^8 \text{ m/s.}$$

Wavelength $\lambda = \dfrac{u}{f}$

$$= \frac{1.22 \times 10^8}{10 \times 10^6}$$

$= 12.2$ m

Impedance of the medium

$$\eta = \sqrt{\frac{\mu}{\in}} = \sqrt{\frac{\mu_0 \mu_r}{\in_0 \in_r}}$$

$$= 120\pi \sqrt{\frac{2}{3}}$$

$$= 307.8\ \Omega.$$

RMS value of E

$$P = E_m H_m = \frac{E_m^2}{\eta}$$

$$1 = \frac{E_m^2}{307.8}$$

$$E_m^2 = 307.8$$

$$E_m = 17.55\ \frac{\text{V}}{\text{m}}$$

$$E_{\text{RMS}} = \frac{E_m}{\sqrt{2}} = 12.4\ \frac{\text{V}}{\text{m}}$$

13. A wire 6 mm in diameter and 5 km in length is suspended at a constant height 20 m above sea-water. Calculate the capacitance between conductor and earth.

Capacitance per unit length is

$$C = \frac{\pi \in_0}{\ln \frac{2H}{r}}\ \text{F/m}$$

$$H = 10\ \text{m}$$

$$r = 3\ \text{mm}.$$

Capacitance for 5 km length

$$C = \frac{\pi \in_0 \times 5 \times 10^3}{\ln \frac{2 \times 10}{3 \times 10^{-3}}}$$

$$= 1.58 \times 10^{-8}\ \text{F}$$

$$= 15.8\ \text{nF}.$$

14. A linearly polarised plane wave propagates through free space at an angle θ with respect to a plane circular surface on XY plane. If the peak amplitude of $\vec{E} = 10$ m V/m, calculate the average power that passes through 1 m^2 surface.

Average power $P = \dfrac{E_0^2}{2\eta}\ \text{W/m}^2$

$$\eta \text{ in free space} = \sqrt{\frac{\mu_0}{\varepsilon_0}}$$

$$= 120\ \pi.$$

$E_0 = 10$ m V/m.

$$P = \frac{(10 \times 10^{-3})^2}{2 \times 120\pi}$$

$$P = 1.32 \times 10^{-7}$$

$$= 0.132\ \mu\text{W/m}^2$$

For a surface of 1 m^2

$$P = 0.132\ \mu\text{W}.$$

15. A travelling $\vec{E}$ field in free space strikes a partially conducting medium with $\epsilon_r = 55$, $\mu_r = 1$, $\sigma = 1$ Ms/m. Given a frequency of 500 MHz and $E_i = 100$ V/m, determine E_i and H_i.

As σ is very large, it can be taken as conducting medium.

$$\alpha = \beta = \sqrt{\frac{\omega\mu\sigma}{2}} = 44.43 \times 10^3\ \text{m}^{-1}.$$

$$\therefore \qquad E_i = E_0\, e^{-\alpha Z} \sin(\omega t - \beta Z)$$

$$E_i = 100\, e^{-44.43 \times 10^3 z} \sin\left(1000\,\pi t - 44.43 \times 10^3 Z\right) \frac{\text{V}}{\text{m}}$$

For a good conductor

$$\eta = \sqrt{\frac{\omega\mu}{\sigma}}\ \underline{|45^\circ}$$

$$= 0.063\ \underline{|45^\circ}$$

$$H_{max} = \frac{E_{max}}{\eta}$$

$$= 1587.3\ \frac{\text{A}}{\text{m}}.$$

$$H_i = 1587.3 e^{-44.43 \times 10^3 z} \sin\left(1000\pi t - 44.43 \times 10^3 Z - 45^\circ\right) \frac{\text{A}}{\text{m}}.$$

16. If distilled water has $\mu_r = 1$, $\epsilon_r = 81$, $\tan\delta = 0.05$ at 1.0 GHz, calculate

(1) The 1/e depth of penetration at 1 GHz.

(2) The 1% depth of penetration.

(3) The depth when the field is almost zero.

(1) $E_0\, e^{-\alpha\delta} = E_0\, e^{-1}$

$$\delta = \frac{1}{\alpha} = \frac{1}{\sqrt{\pi f \mu \sigma}}$$

Given $\mu r = 1$, $\epsilon_r = 81$, $\omega = 1\,\text{GHz}$

$$\tan\theta = \tan\delta = \frac{\sigma}{\omega\,\epsilon} = 0.05$$

$$\sigma = 0.225 \text{ sm}^{-1}.$$

Depth of penetration is

$$\delta = 0.0336 \text{ m}.$$

(2) $E_0\, e^{-\alpha\delta} = 0.01\, E_0$

$$e^{-\alpha\delta} = 0.01$$

$$-\delta\alpha = \ln 0.01$$

$$\delta = \frac{-\ln 0.1}{\alpha} = 0.077 \text{ m}.$$

(3) $E_0\, e^{-\alpha\delta} = 0$

$$e^{-\alpha\delta} = e^{-\infty}$$

$$\delta = \infty.$$

17. A uniform plane wave of 200 MHz travelling in free space impinges normally on a large block of material having $\epsilon_r = 4$, $\mu_r = 9$ and $\sigma = 0$.

Calculate transmission and reflection coefficient at the interface.

For the medium 1 (free space)

$$\eta_1 = \sqrt{\frac{\mu_0}{\epsilon_0}} = 377\,\Omega.$$

For medium 2, $\mu_{r_2} = 9$, $\epsilon_{r_2} = 4$.

$$\eta_2 = \sqrt{\frac{\mu_0 \mu_{r_2}}{\epsilon_0 \epsilon_{r_2}}} = 565.23\,\Omega.$$

Reflection coefficient

$$T = \frac{\eta_2 - \eta_1}{\eta_2 + \eta_1}$$

$$= 0.2.$$

Transmission coefficient

$$\tau = \frac{2\eta_2}{\eta_1 + \eta_2}$$

$$= 1.2.$$

18. A 300 MHz uniform plane wave travelling in free space strikes a large block of copper [$\mu_r = 1$, $\epsilon_r = 1$ and $\sigma = 5.8 \times 10^7$ s/m] normal to the surface. If the surface of copper lies in the *YZ* plane and the wave is propagating in the *x* direction, write complete time domain expressions for incident reflected and transmitted waves.

Medium 1 (free space)

$$\eta_1 = 377\ \Omega.$$

Medium 2

$$\eta_2 = \sqrt{\frac{\mu_2}{\epsilon_2}} = 377\ \Omega.$$

Transmission coefficient

$$\tau = \frac{E_{to}}{E_{io}} = \frac{2\eta_2}{\eta_1 + \eta_2}$$

$$= 1$$

$$\boxed{E_{to} = E_{io}}.$$

Reflection coefficient

$$T = \frac{E_{ro}}{E_{io}} = \frac{\eta_2 - \eta_1}{\eta_1 + \eta_2}$$

$$= 0\ [\because \eta_1 = \eta_2]$$

$$H_i = \frac{E_i}{\eta_1}$$

$$H_i = 2.65 \times 10^{-3}\ E_i$$

$$H_r = 0.$$

$\therefore \qquad E_i = E_0 \cos(\omega t - \beta z)\vec{a}_z$

$$H_i = 2.65 \times 10^{-3} E_0 \cos(\omega t - \beta z)\vec{a}_y$$

$$E_t = E_i, \; H_t = H_i$$

$$E_r = H_r = 0.$$

19. For silver, the conductivity $\sigma = 3 \times 10^6$ Mho/m. At what frequency will the depth of penetration be 1 mm? (Given $\delta = 1 \times 10^{-3}$ m)

$$\delta = \sqrt{\frac{2}{\omega\mu\sigma}}$$

$$\delta^2 = \frac{2}{2\pi f \mu\sigma}$$

$$f = \frac{1}{\pi\delta^2 \mu_0\mu_r\sigma} = \frac{1}{\pi\left(1\times10^{-3}\right)\times\mu_0 \times 1\times 3\times 10^6}$$

$$= 84.4 \text{ KHz.}$$

20. After what frequency can the earth be considered a perfect dielectric? Assume $\frac{\sigma}{\omega \in} = \frac{1}{100}$. (Given $\sigma = 5 \times 10^{-3}$ s/m, $\mu_r = 10$, $\in_r = 8$).

For a perfect dielectric $\frac{\sigma}{\omega \in} << 1$

$$\frac{\sigma}{\omega \in} = \frac{1}{100}$$

$$\omega = \frac{100 \times \sigma}{\in}$$

$$\omega = \frac{100 \times 5 \times 10^{-3}}{8 \in_0}$$

$$f = \frac{\omega}{2\pi} = \frac{100 \times 5 \times 10 - 3}{2\pi \times 8 \in_0}$$

$$f = 1.123 \text{ GHz.}$$

21. Given a ferrite material, operating in linear mode, with $B = 0.05\ T$, $\mu_r = 50$, calculate magnetic susceptibility X_m, M and H.

$$\mu_r = 1 + X_m \qquad \text{...(1)}$$

$$\beta = \mu_0 \mu_r H \quad \text{...(2)}$$

$$M = X_m H. \quad \text{...(3)}$$

From (1),

$$X_m = 50 - 1 = 49.$$

From (2),

$$H = \frac{B}{\mu_0 \mu_r} = \frac{0.05}{4\pi \times 10^{-7} \times 50} = 795.77 \text{ H/m}.$$

From (3),

$$M = 49 \times 795.77$$

$$= 38992.96 \text{ A/m}.$$

22. Find the force/length between two long, straight parallel conductors carrying a current of 10 A in the same direction. A distance of 0.2 m separates the conductors. Repeat the above problem when the conductors carry currents in opposite direction.

If the conductors are assumed to be placed along $\vec{a}_x$ direction and the current flow is assumed to be in $-\vec{a}_x$ (inside the conductor), then

$$\text{Force } \vec{F}_1 = \frac{\mu_0 I_1 I_2}{2\pi D} - \vec{a}_y \frac{\text{N}}{\text{m}}.$$

$$\text{On 1st conductor} = \frac{4\pi \times 10^{-7} \times 100}{2\pi \times 0.2}$$

$$= 1 \times 10^{-4} \text{ N/m } (-\vec{a}_y).$$

$$\text{Force on 2nd conductor } \vec{F}_2 = 1 \times 10^{-4} \text{ N/m } - \vec{a}_y.$$

If the direction of the current flow is opposite in two conductors

$$\vec{F}_1 = 1 \times 10^{-4} (\vec{a}_y) \text{ N/m}$$

$$\vec{F}_2 = 1 \times 10^{-4} (-\vec{a}_y) \text{ N/m}.$$

23. The interface between two different regions in normal to Z-axis. If

$\vec{B}_1 = \mu_0 (43.5\,\vec{a}_x + 24.0\,\vec{a}_z)$ and

$\vec{B}_2 = \mu_0 (22\,\vec{a}_x + 24\,\vec{a}_z)$, what is the ratio of relative permeability?

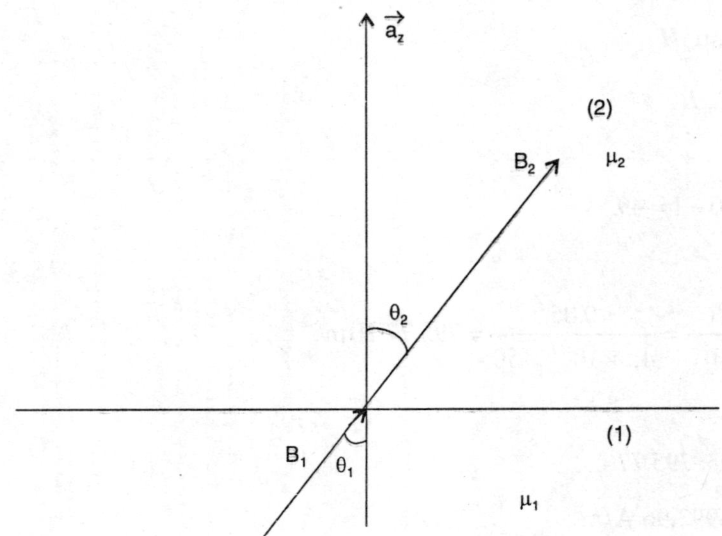

Fig. M-23

From Fig. M-23,

$$\cos\theta_1 = \frac{\vec{B}_1 \cdot \vec{a}_z}{|B_1|}$$

$$\cos\theta_2 = \frac{\vec{B}_2 \cdot \vec{a}_z}{|B_2|}$$

$$\cos\theta_1 = \frac{(43.5\,\vec{a}_x + 24\,\vec{a}_z) \cdot \mu_0 \vec{a}_z}{\sqrt{(43.5)^2 + 24^2}}$$

$$= 0.483\,\mu_0$$

$$\tan\theta_1 = 1.8125\,\mu_0.$$

Similarly,

$$\tan\theta_2 = 0.917\,\mu_0.$$

From the boundary conditions,

$$\frac{\tan\theta_1}{\tan\theta_2} = \frac{\mu r_1}{\mu r_2}$$

$$= \frac{1.8125}{0.917}$$

$$\frac{\mu r_1}{\mu r_2} = 1.976.$$

24. A flat, perfectly-conducting surface in XY plane is situated in a magnetic field $\vec{H} = 3\cos x\,\vec{a}_x + z\cos x\,\vec{a}_y$ A/m for $Z \geq 0$ and $\vec{H} = 0$ for $Z < 0$. Find the current density on the conductor surface.

We know, as per Ampere's circuital law in differential form

$$\nabla \times \vec{H} = \vec{J}$$

$$\begin{vmatrix} \vec{a}_x & \vec{a}_y & \vec{a}_z \\ \dfrac{\partial}{\partial x} & \dfrac{\partial}{\partial y} & \dfrac{\partial}{\partial z} \\ 3\cos x & z\cos x & 0 \end{vmatrix} = \nabla \times \vec{H}$$

$$\vec{a}_x(-\cos x) - \vec{a}_y(0) + \vec{a}_z(-Z\sin x)$$

$$\nabla \times \vec{H} = \cos x(-\vec{a}_x) + Z\sin x(-\vec{a}_z)$$

$$\vec{J} = \cos x(-\vec{a}_x) + Z\sin x(-\vec{a}_z)\,Z \geq 0$$

$$= 0 \qquad Z < 0.$$

25. Two conductors in Fig. M-25 are filamentary and the one located on the Z-axis is infinite in length. Find the vector force exerted by the wire on each of the four sides of the rectangular loop and then give the total vector force on the loop.

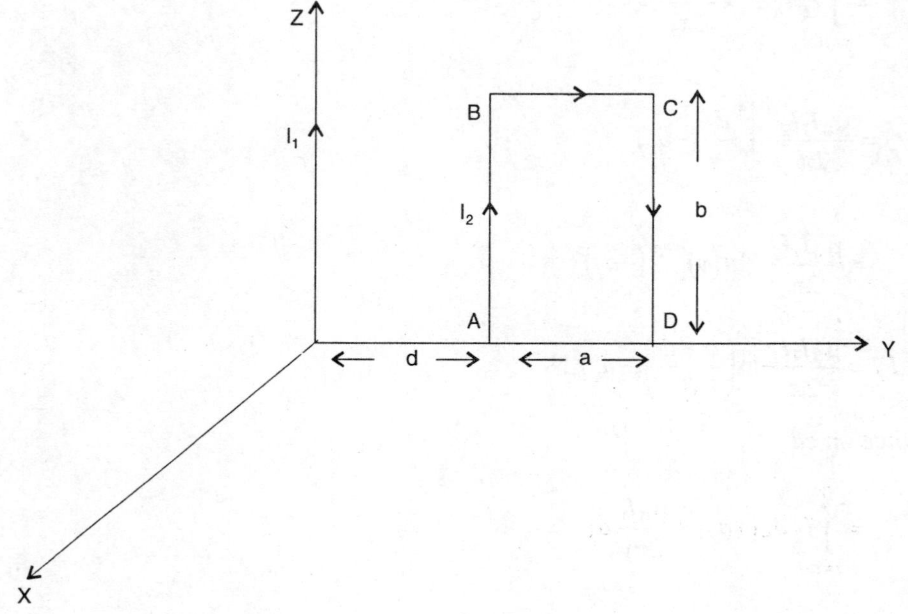

Fig. M-25

Magnetic field intensity due to filament carrying current I_1 at any point (x) is

$$\vec{H} = \frac{I_1}{2\pi x}\vec{a}_x$$

$$\vec{B} = \frac{\mu_0 I_1}{2\pi y}\vec{a}_x.$$

Force on the loop is

$$\vec{F} = \oint I\,d\vec{l} \times \vec{B}.$$

The force on AB is $= \int_{Z=0}^{b} I_2 dZ(\vec{a}_z) \times \frac{\mu_0 I_1}{2\pi y}\vec{a}_x$

$$= \int_{Z=0}^{b} \frac{I_2 \mu_0 I_1}{2\pi y} dy\,\vec{a}_x \qquad [\because\ y = d]$$

$$\vec{F}_1 = \frac{\mu_0 I_1 I_2}{2\pi d} \times b\,\vec{a}_y \text{ N}$$

Now, the force on bc

$$= \int_{y=d}^{d+a} I_2 dy\,\vec{a}_y \times \frac{\mu_0 I_1}{2\pi y}\vec{a}_x$$

$$= \frac{\mu_0 I_1 I_2}{2\pi} \int_{y=d}^{d+a} \frac{dy}{y}(-\vec{a}_z)$$

$$= \frac{\mu_0 I_1 I_2}{2\pi} \ \text{In}(y)_d^{a+d}(-\vec{a}_z)$$

$$\vec{F}_2 = \frac{\mu_0 I_1 I_2}{2\pi}\text{In}\left(\frac{a+d}{d}\right)(-\vec{a}_z) \text{ N.}$$

The force on cd

$$= \int_{z=b}^{0} I_2\,dz(+\vec{a}_z) \times \frac{\mu_0 I_1}{2\pi y}\vec{a}_x$$

$$= \frac{\mu_0 I_1 I_2}{2\pi(a+d)} \int_b^0 dz(+\vec{a}_y)$$

$$\vec{F}_3 = \frac{-\mu_0 I_1 I_2 b}{2\pi(a+d)} \vec{a}_y.$$

The force on *da*

$$= \int_{y=d+a}^{d} I_2\, dy(+\vec{a}_y) \times \frac{\mu_0 I_1}{2\pi y} \vec{a}_x$$

$$= \frac{\mu_0 I_1 I_2}{2\pi} \int_{d+a}^{d} \frac{dy}{y}(-\vec{a}_z).$$

$$\vec{F}_4 = \frac{\mu_0 I_1 I_2 b}{2\pi} \ln \frac{d}{d+a}(-\vec{a}_z)$$

$$\vec{F} = \vec{F}_1 + \vec{F}_2 + \vec{F}_3 + \vec{F}_4$$

$$= \frac{\mu_0 I_1 I_2 b}{2\pi d} \vec{a}_y + \frac{\mu_0 I_1 I_2 b}{2\pi} \ln \frac{d+a}{d}(-\vec{a}_z) + \frac{\mu_0 I_1 I_2 b}{2\pi(a+d)}(-\vec{a}_y) + \frac{\mu_0 I_1 I_2}{2\pi} \ln \frac{d+a}{d}(+\vec{a}_z).$$

$$\vec{F} = \frac{\mu_0 I_1 I_2 b}{2\pi}\left(\frac{1}{d} - \frac{1}{a+d}\right)\vec{a}_y$$

$$\vec{F} = \frac{\mu_0 I_1 I_2 b}{2\pi}\left(\frac{1}{d} + \frac{1}{a+d}\right) \text{N}$$

26. Find $\vec{H}$ at (–3, 4, 0) due to current filament shown in Fig. M. 26(*a*)

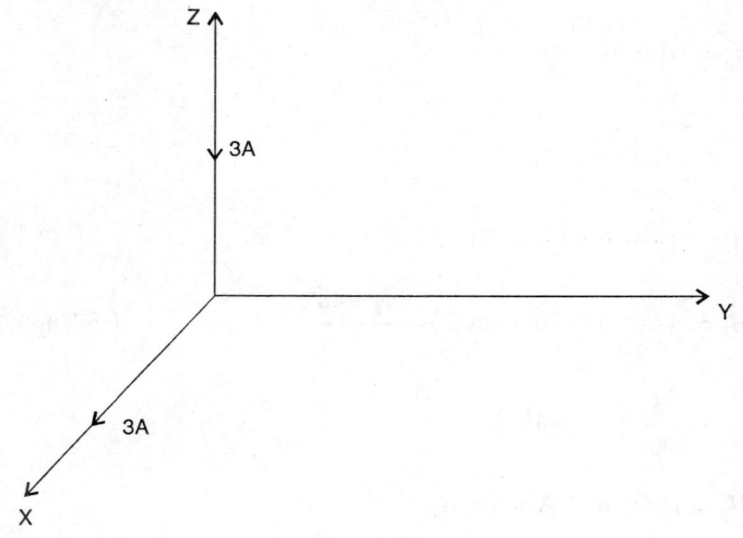

Fig. M. 26 (a)

Consider due to current through 'Z' axis

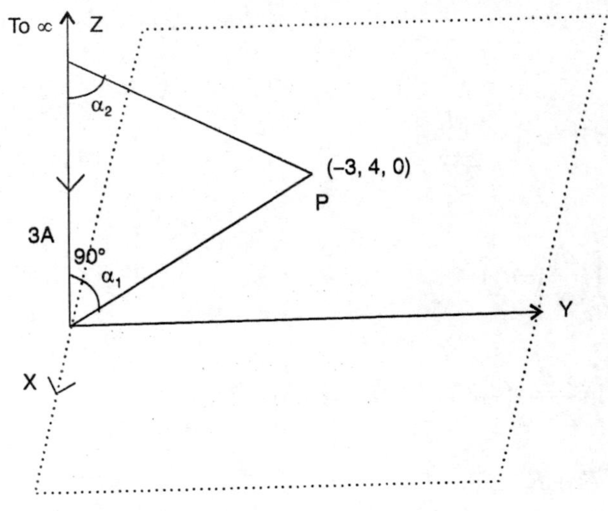

Fig. M. 26 (b)

$$\vec{H}_1 = \text{at } p \text{ due to } 3\text{A along } (-\vec{a}_z)$$

$$\vec{H}_1 = \frac{1}{4\pi\rho}(\cos\alpha_1 + \cos\alpha_2)\,\vec{a}_\rho.$$

Direction of magnetic field intensity $\vec{H}\,(\vec{a}_\rho)$ = Direction of current $(-\vec{a}_z)$ × position vector of P with respect to current element.

Position vector of P = [(–3, 4, 0) – (0, 0, Z)]

$$\vec{a}_\rho = -\vec{a}_z \times \frac{(-3\vec{a}_x + 4\vec{a}_y)}{5}$$

$$= \frac{3}{5}\vec{a}_y + \frac{4}{5}\vec{a}_x$$

$$\rho = 5 \text{ (distance of point)}$$

$$\vec{H}_1 = \frac{3}{4\pi \times 5}(\cos 90° + \cos 0°) \cdot \frac{3\vec{a}_y + 4\vec{a}_x}{5} \qquad (\because \text{ length is } \alpha_2 = 0 \text{ infinite})$$

$$= \frac{3}{100\pi}(3\vec{a}_y + 4\vec{a}_x).$$

$\vec{H}_2$ at P due to 3 A along $\vec{a}_x$

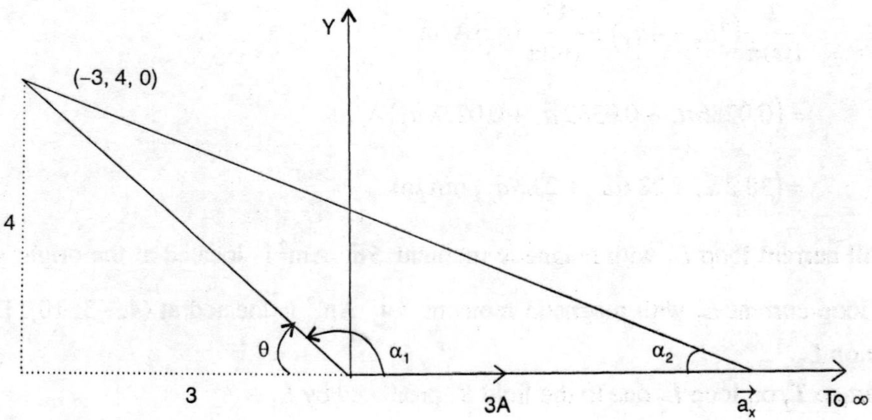

Fig. M.26 (c)

$$\alpha_1 = 180° - \theta$$

$$\tan\theta = \frac{4}{3} = 1.33$$

$$\theta = 53.13°$$

$$\cos\alpha_1 = -0.6$$

$$\alpha_2 = 0, \cos\alpha_2 = 1$$

$$\vec{H}_2 = \frac{I}{4\pi\rho}(\cos\alpha_1 + \cos\alpha_2)\vec{a}_\rho$$

$\vec{a}_\rho = \vec{a}_x \times$ (direction of distance vector of P from X-axis)

$$(-3, 4, 0) \rightarrow (x, 0, 0)$$

$$= \frac{\vec{a}_x \times 4\vec{a}_y}{4}$$

$$= \vec{a}_z$$

$$\rho = 4$$

$$\vec{H}_2 = \frac{3}{4\pi \times 4}(1 - 0.6)\vec{a}_z.$$

$$= \frac{3}{16\pi}0.4 = \frac{12}{160\pi}\vec{a}_z$$

Total $\vec{H} = \vec{H}_1 + \vec{H}_2$

$$\frac{3}{100\pi}\left(3\vec{a}_y + 4\vec{a}_x\right) + \frac{12}{160\pi}\left(\vec{a}_z\right) \text{A/m}.$$

$$= \left(0.0286\,\vec{a}_y + 0.0382\,\vec{a}_x + 0.0238\,\vec{a}_z\right) \text{A/m}.$$

$$= \left(38.2\,\vec{a}_x + 28.6\vec{a}_y + 23.8\,\vec{a}_z\right) \text{mA/m}.$$

27. A small current loop L_1 with magnetic moment $5\vec{a}_z$ Am² is located at the origin while another small loop current L_2 with magnetic moment $3\vec{a}_y$ Am² is located at (4, –3, 10). Determine the torque on L_2.

The torque T_2 on loop L_2 due to the field B_1 produced by L_1 is

$$\vec{T}_2 = \vec{M}_2 \times \vec{B}_1,$$

where $\vec{M}_2$ is the magnetic moment on loop L_2

$$\vec{M}_2 = 3\,\vec{a}_y.$$

Convert into spherical co-ordinates

$$\vec{M}_2 = 3\vec{a}_y = 3\left(\sin\theta\sin\phi\,\vec{a}_r + \cos\theta\sin\phi\,\vec{a}_\theta + \cos\phi\,\vec{a}_\phi\right)$$

$\vec{M}_2$ at $(4, -3, 10)$

$$r = \sqrt{4^2 + (-3)^2 + 10^2} = 5\sqrt{5}$$

$$\theta = \tan^{-1}\frac{\sqrt{x^2 + y^2}}{Z} = \tan^{-1}\frac{\sqrt{16+9}}{10}$$

$$\boxed{\theta = 26.56}.$$

$$\phi = \tan^{-1}\frac{y}{x} = \tan^{-1}\frac{-3}{4}$$

$$\boxed{\theta = 36.86^\circ}$$

$$\vec{M}_2 = 3\left(-0.268\,\vec{a}_r - 0.536\,\vec{a}_\theta + 0.8\,\vec{a}_\phi\right).$$

Magnetic flux density $\vec{B}_1$ due to L_1 is

$$\vec{B}_1 = \frac{\mu_0 M_1}{4\pi r^3}(2\cos\theta\,\vec{a}_r + \sin\theta\,\vec{a}_\theta).$$

$$= \frac{4\pi \times 10^{-7} \times 5}{4\pi\left(5\sqrt{5}\right)^3}\left(1.788\,\vec{a}_r + 0.447\,\vec{a}_\theta\right).$$

$$\vec{B}_1 = 6.4\,\vec{a}_r + 1.6\,\vec{a}_\theta$$

$$T_2 = \vec{M}_2 \times \vec{B}_1$$

$$= \begin{vmatrix} \vec{a}_r & \vec{a}_\theta & \vec{a}_\phi \\ -0.804 & -1.608 & 2.4 \\ 6.4 & 1.6 & 0 \end{vmatrix}$$

$$T_2 = \left(-0.804\,\vec{a}_r - 1.608\,\vec{a}_\theta + 2.4\,\vec{a}_\phi\right)3\left(6.4\,\vec{a}_r + 1.6\,\vec{a}_\theta\right)$$

$$= -3.84\,\vec{a}_r + 15.36\,\vec{a}_\theta + 9.01\,\vec{a}_\phi.$$

28. A conductor of length 2.5 m located at $Z = 0$, $x = 4$ m carries a current of 12.0 A in the $-\vec{a}_y$ direction. Find the uniform $\vec{B}$ in the region if the force on the conductor in 1.20×10^{-2} N in the direction $(-\vec{a}_x + \vec{a}_z)|\sqrt{2}$.

$$\vec{F} = I\vec{l} \times \vec{B}$$

$$\vec{F} = 1.20 \times 10^{-2}\left(\frac{-\vec{a}_x + \vec{a}_z}{\sqrt{2}}\right)$$

$$I\vec{l} = 12.0 \times 2.5\left(-\vec{a}_y\right)$$

$$\vec{B} = B_x\vec{a}_x + B_y\vec{a}_y + B_z\vec{a}_z.$$

$$I\vec{l} \times \vec{B} = \begin{vmatrix} \vec{a}_x & \vec{a}_y & \vec{a}_z \\ 0 & -30 & 0 \\ B_x & B_y & B_z \end{vmatrix}$$

$$= \vec{a}_z\left(-30\,B_z\right) + \vec{a}_z\left(30\,B_x\right)$$

$$= \frac{1.2 \times 10^{-2}}{\sqrt{2}}\left(-\vec{a}_x\right) + \frac{1.2 \times 10^{-2}}{\sqrt{2}}\,\vec{a}_z$$

$$= +30\,B_z\left(-\vec{a}_x\right) + 30\,B_x\left(\vec{a}_z\right)$$

$$B_x = B_z = 0.028 \times 10^{-2}$$

$$= 2.8 \times 10^{-4} \text{ T.}$$

Y component of B has any value.

29. Find the flux crossing the portion of the plane $\phi = \pi/4$ defined by $0.01 < r < 0.05$ m and $0 < z < 2$. A current filament of 2.5 A along the Z-axis is in the $\vec{a}_z$ direction.

As the current element is in $\vec{a}_z$

$$\vec{H} = \frac{1}{2\pi\rho}\vec{a}_\phi$$

$$\vec{B} = \frac{\mu_0 I}{2\pi\rho}\vec{a}_\phi.$$

To find flux

$$\phi = \iint B \cdot ds$$

$$\phi = \iint \frac{\mu_0 I}{2\pi\rho}\vec{a}_\phi \cdot d\rho\, dz\, \vec{a}_\phi$$

$$= \int_{z=-0}^{2}\int_{\rho=0.01}^{0.05} \frac{\mu_0 I}{2\pi\rho} d\rho\, dz$$

$$= \frac{\mu_0 I}{2\pi}(\ln\rho)_{0.01}^{0.05}\int_{z=-0}^{2} dz$$

$$= \frac{\mu_0 I}{2\pi}\ln\frac{0.05}{0.01} \times 2$$

$$= 1.6\ \mu\text{ wb}.$$

30. Given magnetic vector potential $\vec{A} = \frac{-\rho^2}{4}\vec{a}_z \frac{\text{wb}}{\text{m}}$, calculate the total magnetic flux crossing the surface $\phi = \frac{\pi}{2}$ $(1 \le \rho \le 2 \text{ m}, \quad 0 \le z \le 5 \text{ m})$.

We know

$$\vec{B} = \nabla \times \vec{A}$$

$$= \frac{1}{\rho}\begin{vmatrix} \vec{a}_\rho & \rho\vec{a}_\phi & \vec{a}_z \\ \frac{\partial}{\partial_\rho} & \frac{\partial}{\partial_\phi} & \frac{\partial}{\partial_z} \\ A_\rho & \rho A_\phi & A_z \end{vmatrix}$$

$$= \frac{1}{\rho}\begin{vmatrix} \vec{a}_\rho & \rho\vec{a}_\phi & \vec{a}_z \\ \frac{\partial}{\partial_\rho} & \frac{\partial}{\partial_\phi} & \frac{\partial}{\partial_z} \\ 0 & 0 & -\frac{\rho^2}{4} \end{vmatrix}$$

$$= \frac{-1}{\rho}\rho\vec{a}_\phi\left(\frac{-2\rho}{4}\right)$$

$$\vec{B} = \frac{\rho}{2}\vec{a}_\phi.$$

Hence,

$$\phi \iint \vec{B}\cdot d\vec{s} = \int_{z=-0}^{5}\int_{\rho=1}^{2} \frac{\rho}{2}\vec{a}_\phi \cdot d\rho\, dz$$

$$= \int_0^5 \left(\frac{3}{4}\right)_1^2 dz$$

$$= \int_0^5 \frac{3}{4} dz$$

$$\phi = 3.75 \text{ wb.}$$

31. An area of 0.65 m^2 in the $Z = 0$ plane is enclosed by a filamentary conductor. Find the induced voltage, given that

$$\vec{B} = 0.05\cos 10^3 t\left(\frac{(\vec{a}_y + \vec{a}_z)}{\sqrt{2}}\right) \text{ T}$$

$$V = -\int_S \frac{\partial B}{\partial t}\cdot ds\,\vec{a}_z$$

$$= +\int_S 50\sin 10^3 t\left(\frac{\vec{a}_y + \vec{a}_z}{\sqrt{2}}\right)\cdot 0.65\,\vec{a}_z$$

$$= 23\sin 10^3 t \text{ volt.}$$

32. Use energy integral to find the internal inductance per unit length of a cylindrical conductor of radius a.

Equivalent to electrical stored energy

$$W = \frac{1}{2}\int_V \vec{D}\cdot\vec{E}\,dv.$$

Magnetic stored energy is

$$W = \frac{1}{2}\int_V \vec{B}\cdot\vec{H}\,dv = \frac{1}{2}LI^2 \text{ J}$$

$$L = \int_V \frac{B\cdot H}{I^2}\,dv.$$

We know that $\vec{H}$ due to cylindrical conductor at any point inside is

$$\vec{H} = \frac{I\rho}{2\pi a^2}\vec{a}_\phi$$

$$\vec{B} = \frac{\mu_0 I\rho}{2\pi a^2}\vec{a}_\phi$$

$$\vec{B}\cdot\vec{H} = \frac{\mu_0 I^2\rho^2}{4\pi^2 a^4}$$

$$L = \int_V \frac{\vec{B}\cdot\vec{H}}{I^2}\,dv = \int_{z=0}^{l}\int_{\phi=0}^{2\pi}\int_{\rho=0}^{a} \frac{\mu_0 I^2\rho^2}{4\pi^2 a^4 I^2}\rho\,d\rho\,d\phi\,dz$$

$$= \frac{\mu_0}{4\pi^2 a^4}\int_{\rho=0}^{a}\rho^3\,d\rho\int_{\phi=0}^{2\pi}d\phi\int_{z=0}^{l}dz$$

$$= \frac{\mu_0}{4\pi^2 a^4}\times\frac{a^4}{4}\times 2\pi\times l$$

$$= \frac{\mu_0 l}{8\pi}$$

Inductance per unit length $\dfrac{L}{l} = \dfrac{\mu_0}{8\pi}$ H/m.

33. A circular loop of N turns of conducting wire lie in the X–Y plane with its center at the origin of a magnetic field specified by $\vec{B} = B_0\cos(\pi t/2b)\vec{a}_z$, when b is the radius of loop and ω is the angular frequency. Find the e.m.f. induced in the loop.

$$\text{e.m.f. } (v) = -\int_s \frac{\partial B}{\partial t} \cdot d\vec{s}, \qquad \text{where area } s = \pi b^2$$

$$= \int_\sigma \left[B_0 \sin\left(\frac{\pi t}{2b}\right) \cdot \frac{\pi}{2b} \vec{a}_z \cdot \pi b^2 \vec{a}_z \right]$$

$$= \frac{B_0 \times \pi^2 b}{2} \sin\left(\frac{\pi t}{2b}\right) \text{ volt.}$$

34. A long solenoid S has 220 turns/cm and carries a current $I = 1.5$ A. Its diameter is $d = 3.2$ cm. At its center we place a 130 turn closely packed coil C of diameter $d_c = 2.1$ cm. Current in the solenoid is increased from 0 – 1.5 A in a steady rate over a period of 0.16 s. What is the absolute value of induced e.m.f. that appears in the central coil, while the current in the solenoid is being changed?

H_1 due to solenoid with $\frac{N}{l} = 220$ turns/cm.

$$H_1 = \frac{NI}{l} = 220 \times 1.5 = 330 \text{ AT/m} = 33 \times 10^3 \text{ AT/m}$$

$$B_1 = \frac{\mu_0 NI}{l} = 0.04146 \text{ wb/m}^2.$$

Flux linking with central coil of diameter 2.1 cm

$$\phi_{21} = B \times A = B \times \pi r^2$$

$$\phi_{21} = 0.04146 \times \pi \times \left(\frac{2.1}{2}\right)^2 \times 10^{-4}$$

$$= 1.436 \times 10^{-4} = 14.36 \text{ mwb}$$

$$M = \frac{N_2 \phi_{21}}{I_1} = \frac{130 \times 14.36 \times 10^{-3}}{1.5}$$

$$= 1.24 \text{ H.}$$

e.m.f. induced in central coil is

$$V = M\frac{di}{dt}$$

$$1.24 \times \frac{1.5}{0.16} = 11.62 \text{ V.}$$

35. Calculate the inductance of a ring-shaped coil having a mean diameter of 20 cm. Wound on a wooden core of 2 cm diameter, the winding is uniformly distributed and contains 200 turns.

$$L = \frac{\mu_0 N^2 A}{lm}$$

$$= \frac{4\pi \times 10^{-7} \times 200^2 \times \pi \times 1 \times 10^{-4}}{\pi \times 20 \times 10^{-2}}$$

$$= 25.13\ \mu\text{H}.$$

36. A coil has a self inductance of 1 henry and a resistance of 4 hours. If it is connected to a 40 V d.c. supply, estimate the energy stored in the magnetic field when the current has attained its final steady value. **(Anna Univ. May '04 ECE).**

Steady current is $\frac{V}{R} = \frac{40}{4} = 10\text{A}$

$\therefore$ Energy stored is $\frac{1}{2} L I^2$

$$= \frac{1}{2} \times 1 \times 10^2 = 50 \text{ J}.$$

37. Use differential form of Faraday's law to determine $\vec{B}$ if $\vec{E} = xy^2 e^{-t}\,\vec{k}$ and $\vec{B} = 0$ at $t = 0$.

$$\nabla \times \vec{E} = \frac{-\partial B}{\partial t}$$

$$\nabla \times \vec{E} = \begin{vmatrix} \vec{i} & \vec{j} & \vec{k} \\ \frac{\partial}{\partial x} & \frac{\partial}{\partial y} & \frac{\partial}{\partial z} \\ 0 & 0 & xy^2 e^{-t} \end{vmatrix}$$

$$= \vec{i}\left(2xye^{-t}\right) - \vec{j}\left(y^2 e^{-t}\right)$$

$$B = \int \left(\nabla \times \vec{E}\right) dt$$

$$= \int 2xye^{-t} dt\,\vec{i} - \int y^2 e^{-t} \vec{k}$$

$$\vec{B} = -2xye^{-t}\,\vec{i} + y^2 e^{-t}\,\vec{k} + c.$$

At $t = 0$, $B = 0$.

$$C = -2xy\vec{i} + y^2\,\vec{k}$$

$$\vec{B} = -2xye^{-t}\,\vec{i} + y^2 e^{-t} k^2 - 2xy\vec{i} + y^2\vec{k}$$

$$\vec{B} = -2xy\left(1 + e^{-t}\right)\vec{i} + y^2\left(1 + e^{-t}\right)\vec{k}.$$

38. The loop is inside a uniform magnetic field $\vec{B} = 50\,a\vec{x}$ mwb/m^2. If side *DC* of the loop cuts the flux lines at the frequency of 50 Hz and the loop lies in the *Y–Z* plane at time $t = 0$ find, (*a*) The induced emf at $t = 1$ ms, (*b*) The induced current at $t = 3$ ms.

The motional induced e.m.f, V_{emf}

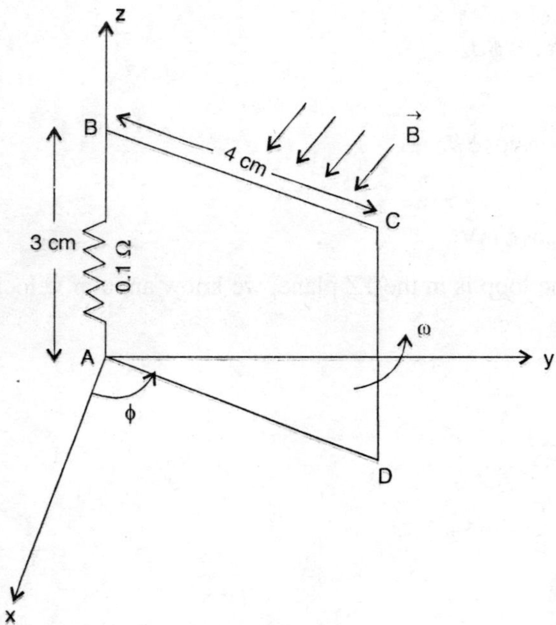

Fig. M-38

$$V_{emf} = \int (\vec{u} \times \vec{B}) \cdot d\vec{l}$$

$d\vec{l}$ for side $CD = dz\,\vec{a}_z$

$$\vec{u} = r\vec{\omega} = 4 \cdot 2\pi f\,\vec{a}_\phi \qquad [r = AD]$$

$$= 400\,\pi\,\vec{a}_\phi.$$

Transform $\vec{B}$ in to cylindrical coordinates

$$\vec{B} = B_0\,\vec{a}_x = B_0 \left(\cos\phi\,\vec{a}_\rho - \sin\phi\,\vec{a}_\phi\right)$$

$$= 0.05\left(\cos\phi\,\vec{a}_\rho - \sin\phi\,\vec{a}_\phi\right)$$

$$\vec{u} \times \vec{B} = \begin{vmatrix} \vec{a}_\rho & \vec{a}_\phi & \vec{a}_z \\ 0 & 400\pi & 0 \\ 0.05\cos\phi & -0.05\sin\phi & 0 \end{vmatrix}$$

$$= -\vec{a}_z\left(400\pi \cdot 0.05 \cos \vec{a}_\phi\right)$$

$$= -20\pi \cos\phi\, \vec{a}_z$$

$$\left(\vec{u} \times \vec{B}\right) \cdot d\vec{l} = -20\pi \cos\phi\, \vec{a}_z \cdot dz\, \vec{a}_z$$

$$= -20\,\pi \cos\phi\, dz$$

$$V_{emf} = \int_{z=0}^{0.03} -20\pi \cos\phi\, dz$$

$$= -6\,\pi \cos\phi\, \text{mV}.$$

Given that at $t = 0$, the loop is in the YZ plane, we know angular velocity is the rate of change of angular deflection

$$\omega = \frac{d\phi}{dt}$$

$$\phi = \omega t + C_0.$$

At $\quad t = 0,\ \phi = \frac{\pi}{2},\ \text{so},\ C_0 = \frac{\pi}{2}$

$$\phi = \omega t + \frac{\pi}{2}.$$

$$\therefore \qquad V_{emf} = -6\pi \cos\left(\omega t + \frac{\pi}{2}\right) \cdot \text{mV}$$

$$= 6\pi \sin \omega t\, \text{mV}$$

$$= 6\pi \sin 100\pi t\, \text{mV}.$$

At $t = 1$ ms

$$V_{emf} = 6\pi \sin\left(100\pi \times 1 \text{ ms}\right) = 5.825 \text{ mV}.$$

Induced current is

$$i = \frac{V_{emf}}{R} = 60\pi\ \sin 100\pi t\, \text{mA}.$$

At $t = 3$ ms

$$i = 60\pi \sin\left(100\pi \times 3 \text{ ms}\right)$$

$$i = 0.1525 \text{ A}.$$

39. A cylindrical capacitor has radii $a = 1$ cm and $b = 2.5$ cm. If the space between the plates is filled with an inhomogeneous dielectric with $Er = (10 + \rho)/\rho$, where ρ is in centimeters, find the capacitance per meter of the capacitor.

Electric field intensity in the dielectric medium $\vec{E} = \frac{\lambda}{2\pi \in_0 \in_r \rho} \vec{a}_\rho$

$$\text{Potential } V = -\int_{\rho=b}^{a} \vec{E} \cdot d\vec{l}$$

$$= -\int_{b}^{a} E\vec{a}_\rho \cdot d\rho \vec{a}_\rho$$

$$= -\int_{b}^{a} \vec{E} = \frac{\lambda}{2\pi \in_0 \in_r \rho} \vec{a}_\rho \cdot d\rho \bar{a}_\rho$$

$$V = -\frac{\lambda}{2\pi \in_0} \int_{\rho=b}^{a^{\in R}} \frac{d\rho}{\in_r \rho}$$

$$= \frac{-\lambda}{2\pi \in_0} \int_{\rho=b}^{a} \frac{d\rho}{\frac{(10+\rho)}{\rho} \cdot \rho}$$

$$= \frac{-\lambda}{2\pi \in_0} \int_{b}^{a} \frac{d\rho}{\rho + 10}$$

$$V = \frac{\lambda}{2\pi \in_0} \ln(10+b) \Big| (10+a)$$

$$c \text{ is f/m} = \frac{\lambda}{V} = \frac{2\pi \in_0}{\ln \frac{10+b}{10+a}}$$

$$= \frac{2\pi \times 8.85 \times 10^{-2}}{\ln \frac{12.5}{11}}$$

$$c = 434.6 \text{ pf/m}.$$

40. The region $x > 0$ contains a dielectric material for which $\in_{r_1} = 3$ and region $x < 0 \left(\in_{r_2} = 4\right)$. If in region 2, $\vec{E}_2 = 10\vec{i} + 15\vec{j} - 20\vec{k}$, find $\vec{E}_1$ in region 1.

$$\vec{E}_2 = 10\vec{a}_x + 15\vec{a}_y - 20\vec{a}_z$$

$$\vec{D}_2 = \epsilon_0 \cdot \epsilon_{r_2} \vec{E}_2$$

$$\vec{D}_2 = 4\epsilon_0 \left(10\vec{a}_x + 15\vec{a}_y - 20\vec{a}_z\right).$$

The normal components of D are continuous and the normal component is $\vec{a}_x$

$$\vec{D}_1 = 40\epsilon_0 \vec{a}_x + D_{y_1}\vec{a}_y + D_{z_1}\vec{a}_z.$$

The tangential components of E are continuous

$$\vec{E}_1 = E_{x_1}\vec{a}_x + 15\vec{a}_y - 20\vec{a}_z.$$

We know $D = \epsilon\vec{E} = \epsilon_0 \epsilon_{r\phi} \vec{E}$

$$\vec{D}_1 = \epsilon_0 \epsilon_{r_1} \vec{E}_1$$

$$40\epsilon_0 \vec{a}_x + D_{y_1}\vec{a}_y + D_{z_1}\vec{a}_z = 3\epsilon_0 \left(E_{x_1}\vec{a}_x + 15\vec{a}_y - 20\vec{a}_z\right)$$

$$40\epsilon_0 = 3\epsilon_0 E_{x_1} = \frac{40}{3},\ D_{y_1} = 45\epsilon_0$$

$$E_{y_1} = \frac{D_{y_1}}{\epsilon_0 \epsilon_r} = \frac{45}{3} = 15$$

$$D_{z_1} = -60\epsilon_0$$

$$E_{z_1} = \frac{D_{z_1}}{\epsilon_0 \epsilon_{r_1}}$$

$$= \frac{-60}{3} = -20$$

$$\vec{E}_1 = \frac{40}{3}\vec{a}_x + 15\vec{a}_y - 20\vec{a}_z.$$

41. Two electrically insulated plates of area 1.0 m^2 are 3 mm apart and have equal and opposite charges of 10 μC neglecting any fringing effects of the field at the edges of the plate, find the force between the plates due to their electrical charge when the space between the plates is filled with air.

From eqn. (2.190)

$$\vec{F} = \frac{Q^2}{2\varepsilon A}\vec{a}_z$$

$A = 1.0\ m^2$, $Q = 10\ \mu C$, $\varepsilon_r = 1$

$$\vec{F} = \frac{(10 \times 10^{-6})^2}{2 \times \varepsilon_0 \times 1 \times 1} \vec{a}_z$$

$$= 5.65 \text{ N.}$$

42. Eight identical spherical drops of mercury charged to 12 V above the earth potential are made to coalesce into a single spherical drop. What is the new potential?

In general work done by a charge Q in electric field is

$$W = Q\ V \text{ Joules} \quad \text{...}(i)$$

As charges are made to coalesce, work done in collecting all charges together is given by eqn. (2.145)

$$W = \frac{1}{2} \Sigma QV$$

$$W = \frac{1}{2} \times 8 \times Q \times 12 = 48\ Q \quad \text{...}(ii)$$

Comparing (*i*) and (*ii*), new potential is 48 V.

Bibliography

1. *Elements of Electromagnetics*, 2nd edition, Mathew N.O, Sadiku, Saunders College Publishing.
2. *Electromagnetics with Applications*, 5th edition, Kraus Fleisch, McGraw-Hill International Edition.
3. *Field Theory*, 4th edition, K.A. Gangadhar, Khanna Publishers.
4. *Fundamentals of Electromagnetic Theory*, P.M. Ramanathan, Khanna Publishers.
5. *Basic Electromagnetic Fields*, B.K. Bhat, CBS Publishers & Distributors.
6. *Basic Electromagnetic Fields*, Herbert P. Nefer Jr., Harper & Row Publishers, New York.
7. *Electromagnetics*, 3rd edition, John D. Kraus, McGraw-Hill International Publishers.
8. *Electromagnetic Theory and Applications*, 4th edition, Umesh Sinha, Tech India Publications.
9. *Electromagnetic Fields & Waves*, V.V. Sarvate, New Age International (P) Ltd.
10. *Applied Electromagnetics*, Martin A. Ploms, McGraw-Hill Kogakusha Ltd.
11. *Introductory Course in Electromagnetic Fields*, 3rd edition, P.V. Gupta, Dhanpat Rai & Sons.
12. *Field and Wave Electromagnetics*, 2nd edition, David K. Cheng, Pearson Education, Asia.
13. *Fundamentals of Applied Electromagnetics*, 2001 Media edition, Fawaaz T. Ulaby, Prentice-Hall of India Ltd.
14. *Electromagnetic Theory*, 3rd revised edition, Dr. K.K. Chopra and G.C. Agarwall, K. Nath & Co., Meerut.
15. *Theory and Problems of Electromagnetics*, 2nd edition, Schaums Online Series, Joseph Administrator.
16. *Electromagnetic Fields and Waves*, 2nd Edition, Paul Lorrain & Dale R Corson, CBS Publishers & Distributors.
17. *Fields and Waves in Communication Electronics*, 3rd edition, Simon Ramo, John R. Whinnery, Theodore Van Duzer, John Wiley & Sons.
18. *Engineering Electromagnetics*, 6th edition, William H. Hayt, Jr., John A. Buck, Tata McGraw Hill Edition.
19. *Introduction to Electromagnetic Fields*, 3rd edition, Clayton R. Paul, Keith W. Whites, Syed. A. Nasar, WCB McGraw-Hill.
20. *Electromagnetism Theory & Applications*, Ashutosh Pramanik, Prentice Hall of India.

Index

T

U

V

W